Sandstone Depositional Environments

Illustrations in this book
were partly supported by
The American Association of Petroleum Geologists Foundation
with funds contributed by Allan P. Bennison

Sandstone Depositional Environments

Edited by
Peter A. Scholle and Darwin Spearing

Published by
The American Association of Petroleum Geologists
Tulsa, Oklahoma 74101, U.S.A.

Editor:
M. K. Horn
Science Directors:
G. D. Howell and E. A. Beaumont
Project Editors:
R. L. Hart and D. A. White

Printed by George Banta Company, Menasha, Wisconsin

Table of Contents:

INTRODUCTION

The study of sandstone depositional environments has an extensive existing literature. Much of this literature is oriented toward the specialist in the field rather than the generalist, and contains materials largely devoted to the hydrodynamics of sedimentation or the mechanisms of sediment movement and deposition; a minority contains well-illustrated material which is applicable to recognition of depositional environments. This volume is specifically designed for use by the non-sedimentologist, the petroleum geologist, or the field geologist who needs to use sandstone depositional environments in facies reconstructions and environmental interpretations.

Toward that purpose, this book is profusely illustrated with both diagrams and pictures of sedimentary structures in color. The text focuses on the recognition of depositional environments rather than hydrodynamic mechanisms of sediment movement. It examines assemblages of sedimentary structures and facies patterns as criteria for recognizing depositional environments in the surface and subsurface. Although individual sedimentary structures in and of themselves are not generally diagnostic criteria for recognizing environments, frequently the entire suite structures found in a rock sequence can be used to recognize environments, particularly when used in conjunction with other criteria, such as grain size distribution, lithology, associated sediments, vertical sequence of structures, and sand-body geometry. Within the concept of the use of sedimentary structures for environmental interpretation, we are including such features as primary sedimentary structures, trace fossils, and other characteristics of the early depositional and early diagenetic environment, three-dimensional facies patterns, and the geometry of large-scale sand bodies. All of these criteria are stressed in the various chapters of this book.

We would like to emphasize again that the recognition of depositional environments is not a straightforward procedure; that in very few cases are sedimentary structures alone sufficiently diagnostic to absolutely characterize a particular depositional environment. Nevertheless, in most cases one can conduct a detailed examination of such structures and come up with a reasonable environmental interpretation or at least narrow the options in terms of potential or possible depositional environments for that particular rock sequence. For example, fan facies, either submarine fans or alluvial fans, have in many ways very similar sedimentary structures largely as a consequence of the great similarity of processes acting in those two environments (despite the fact that one is dominantly sub-

aqueous and the other dominantly subaerial). Thus, the distinction of those two depositional environments is not necessarily easy. Nevertheless, an examination of sedimentary structures of those two facies will clearly allow one to distinguish those two environments from most other depositional regimes so that the distinction of a submarine fan from, for example, a barrier beach, becomes very clear on the basis of sedimentary structures. If one still is faced with the problem of having two or more possible environmental interpretations for a sedimentary unit, even based on an assemblage of primary sedimentary structures, one needs to use other criteria such as overall sedimentary-tectonic framework, trace fossils, sand size analysis, fossil assemblages, and other features to try and refine the interpretations.

Because this book is written for a non-specialist audience, it seems advisable to cite some of the extensive literature which may be of use to people who wish to delve more deeply into the details of individual depositional settings. We have listed some of the general references which pertain to the entire spectrum of depositional environments as well as a few of the more outstanding publications which deal with specific individual depositional settings, and each of the chapters in this book contains a separate bibliography of the most pertinent papers related to that particular depositional setting. Our generalized bibliography can be found at the end of this introductory section.

We would like to add a final cautionary note to this introduction. The geologic record of any depositional environment involves not only the hydrodynamic processes which are involved in the transportation and deposition of sediments and which lead to the formation of sedimentary structures, but is also influenced by tectonic patterns, fluctuations of sea level, variations of climate, variations in source rock terrain or provenance, evolutionary patterns of organisms, variations in preservation within the geologic record, and diagenetic effects which also contribute to variations in preservation of structures. All of these factors contribute greatly to the production of a large number of variations on a small number of basic themes. Although there is only a limited number of depositional environments, the factors mentioned above lead to a tremendous diversity and variety of possible sedimentary patterns. For these reasons, this book can really not be used as a simple checklist — a one-for-one master plan of how to interpret depositional environments. Every geologic study that is done, whether it involves study of depositional environments or other types of analysis, requires a great deal of interpretation on the part of the geologist. This book hopefully provides a simplified outline of the major criteria for depositional environment recognition, yet it is the responsibility of the geologist to sit down and reinterpret that material in the context of the particular example on which he is working. Each depositional environment at a specific location and deposited in a particular geologic time interval may have its own unique characteristics. Though one may recognize it or interpret it as a fluvial deposit or tidal deposit, for example, because of its setting in space and time, it may have its own unique physical characteristics not found elsewhere in deposits laid down in the same general environment.

Finally, we would like to emphasize the importance of environmental analysis to exploration and production geologists. Prediction of sandstone trends depends very strongly on the accurate interpretation of environments. Likewise, predictions of diagenetic alteration and the isolation of the most favorable reservoir bodies within an environment generally depend critically on detailed study of microenvironments. This is true in oil and gas exploration, in the search for other energy reserves (coal, uranium, etc.), and in studies of strata-bound metallic ore deposits.

Increasing concerns about improving secondary and tertiary recovery from producing oil and gas fields also has led to a growing interest in microenvironmental interpretation. Where closely spaced cores (and logs) are available, as in many major oil fields, sophisticated studies of environmentally controlled porosity and permeability patterns can be undertaken. These studies, in turn, should serve to enhance predictability of directional rock properties and lead to improved drilling patterns for injection and recovery wells.

The increased emphasis on stratigraphic traps and on the use of geophysics for identification of stratigraphic plays also brings up a major need for more sophisticated environmental interpretations. Stratigraphic analysis based on seismics now involves the recognition of sedimentary micro-environments at considerable depth and presumably requires more sophisticated analysis of existing information from wells to tie into seismic lines. In this way, more productive information can be gained from the seismic data in undrilled areas. Currently, seismic stratigraphy is involved with the actual recognition of characteristic seismic signatures of different depositional environments. Again, this cannot be done in a vacuum but requires sophisticated interpretation of core and log information from areas in which the seismic stratigraphic signatures are being interpreted. Again, this book should provide a valuable tool for people involved in these types of studies who may not be "mainline" sedimentologists.

Book Organization

To facilitate the use of this book, we have divided it into individual chapters organized by environment of deposition. Although there is some variability between chapters due to the nature of the material for each of the environments, most of the chapters are organized around a consistent outline. They start with an introduction which defines and describes the facies involved and provides a summary of the diagnostic criteria for the recognition of the envi-

ronment. The second section provides information on the relationships to lateral facies, depositional setting, and three-dimensional geometry of facies and microfacies (generally illustrated with block diagrams), the tectonic relationships and composition, texture, and other patterns characteristic of the environment. A third section, which outlines the major sedimentary structures characteristic of the environment, is heavily illustrated, and shows both physical and biological structures of primary or secondary origin. In some chapters, information is also provided on typical sediment textures and log responses. A fourth section provides information on economic considerations such as porosity potential of the overall environment and microenvironments, common diagenetic effects, overall petroleum trap potential, and the relationship of the facies to uranium, copper, lead, zinc, or other mineral deposits as well as oil and gas. In some of the chapters where extensive information on case histories of petroleum production has not previously been published and where information was available to the authors, a short section showing core photographs, seismic lines, logs, cross-sections, etc., of specific fields is included.

The book is organized with chapters in a specific order from non-marine to deep-marine settings (in essence, from the mountains to the coast and finally into deeper water environments). We have tried to match chapter length to the significance of individual environments to general exploration for oil, gas and minerals. We have included environments in this book, however, which are normally not considered to be particularly prospective for oil and gas. They have been included for a number of reasons. First, the recognition of depositional environments is really not based strictly on the potential of those environments for producing oil and gas. We would like to provide as complete a framework for recognition of environments as possible, and even units which are not necessarily producers will still be involved in facies interpretations. Second, some of the environments which are not currently considered to be economically prospective may yet contain, at some time in the future, unconventional reservoirs. We need only look back a few years to our views of lacustrine, glacial, or submarine fan sediments to see that environments which were previously completely written off as potential reservoirs have turned out, in some cases, to be very prolific ones. Lacustrine sediments, in particular, are now major oil producers in the Uinta basin, in China, and in North Africa; glacial and fluvio-glacial sediments someday may be involved in production in areas such as the Cook Inlet. Likewise, submarine fans have turned out to be productive in several areas within the North Sea Basin. This book covers only depositional environments of clastic terrigenous sediments. Equivalent sediments of a biological or chemical nature (carbonates, silicious deposits) will be covered in a separate book with parallel format.

P. A. SCHOLLE
D. R. SPEARING

BIBLIOGRAPHY

Allen, J. R. L., 1970, Physical processes of sedimentation: New York, Am. Elsevier, 248 p.

Bagnold, R. A., 1941, The physics of blown sand and desert dunes: London, Methuen and Co., 265 p.

Bouma, A. H., and A. Brouwer, eds., 1964, Turbidites: Series in Devel. in Sedimentology: Amsterdam, Elsevier.

Coneybeare, C. E. B., and K. A. W. Crook, 1968, Manual of sedimentary structures: Bur. Min. Resources, Geology and Geophysics, Bull. No. 102, Commonwealth Australia, 327 p.

Crimes, T. P., and J. C. Harper, eds., 1970, Trace fossils: Liverpool, Seel House Press, 547 p.

Duff, P. M., A. Hallam, and E. K. Walton, 1967, Cyclic sedimentation: Series in Devel. in Sedimentology, Amsterdam, Elsevier.

Dzulynski, S., and E. K. Walton, 1965, Sedimentary features of flysch and greywackes: Series in Devel. in Sedimentology, Amsterdam, Elsevier, 274 p.

Frey, R. W., ed., 1975, The study of trace fossils: New York, Springer-Verlag, 562 p.

Ginsburg, R. A., 1975, Tidal deposits: New York, Springer-Verlag, 428 p.

Glennie, K. W., 1970, Desert sedimentary environments: Series in Devel. in Sedimentology, Amsterdam, Elsevier.

Harms, J. C., J. B. Southard, D. R. Spearing, and R. G. Walker, 1975, Depositional environments as interpreted from primary sedimentary structures and stratification sequences: SEPM Short Course Notes No. 2, 161 p.

Heezen, B. C., 1977, Influence of abyssal circulation of sedimentary accumulations in space and time: Series in Devel. in Sedimentology, Amsterdam, Elsevier.

——— and C. D. Hollister, 1971, The face of the deep: New York, Oxford Univ. Press, 659 p.

Klein, G. de V., 1975, Sandstone depositional models for exploration for fossil fuels (2nd ed.): Minneapolis, Cepco Div., Burgess Pub. Co., 149 p.

McKee, E. D., ed., 1979, A study of global sand seas: U. S. Geol. Survey Prof. Paper 1052, 429 p.

Miall, A. D., ed., 1978, Fluvial sedimentology: Calgary, Canadian Soc. Petroleum Geols. Mem. 5, 859 p.

Middleton, G. V., ed., 1965, Primary sedimentary structure and their hydrodynamic interpretation: SEPM Spec. Pub. No. 12, 265 p.

Morgan, J. P., 1970, Deltaic sedimentation: SEPM Spec. Pub. No. 15, 312 p.

Pettijohn, F. J., and P. E. Potter, 1964, Atlas and glossary of primary sedimentary structures: New York, Springer-Verlag, 370 p.

Picard, M. D., and L. R. High, 1973, Sedimentary structures of ephemeral streams: Series in Devel. in Sedimentology, Amsterdam, Elsevier.

Reading, H. G., ed., 1978, Sedimentary environments and facies: New York, Elsevier North-Holland, 557, p.

Reineck, H. E., and I. B. Singh, 1975, Depositional sedimentary environments: New York, Springer-Verlag, 439 p.

Ricci Lucchi, R., 1970, Sedimentografia: Bologna, Zanichelli, 288 p.

Rigbey, J. K., and W. K. Hamblin, eds., 1972, Recognition of ancient sedimentary environments: SEPM Spec. Pub. No. 16, 340 p.

Schwarzacher, W., 1975, Sedimentation models and quantitative stratigraphy: Series in Devel. in Sedimentology, Amsterdam, Elsevier.

Selley, R. C., 1978, Ancient sedimentary environments (2nd ed.): Ithaca, Cornell Univ. Press, 287 p.

——— 1978, Concepts and methods of subsurface facies analysis: AAPG Contin. Educ. Course Notes Series No. 9, 82 p.

Shawa, M. S., ed., 1974, Use of sedimentary structure for recognition of clastic environments: Calgary, Canadian Soc. Petroleum Geols., 66 p.

Shelton, J. W., 1973, Models of sand and sandstone deposits: a methodology for determining sand genesis and trend: Norman, Okla. Geol. Survey Bull. 118, 122 p.

Spearing, D. R., 1974, Summary sheets of sedimentary deposits: Geol. Soc. America, Chart Series MC-8, 7 sheets.

Van Straaten, L. M. J. U., ed., 1964, Deltaic and shallow marine deposits: Series in Devel. in Sedimentology, Amsterdam, Elsevier.

Walker, R. G., ed., 1979, Facies models: Geoscience Canada Reprint Series 1, Toronto, Geol. Assoc. Canada, 211 p.

Characteristic Features of Glacial Sediments

Don J. Easterbrook
Department of Geology
Western Washington University
Bellingham, Washington

INTRODUCTION

Few depositional environments are as varied as those associated with glaciers, resulting in deposits of widely differing physical characteristics. Sediments deposited directly from glacial ice generally are poorly sorted and unstratified diamictons whereas those indirectly associated with ice via meltwater streams or lakes are sorted and stratified. Recognition of glacial clastic sediments is important because of climatic implications and rapid facies changes associated with glacial deposition.

DIAMICTONS

The term diamicton is used for poorly sorted, unstratified deposits of unspecific origin. The most common glacial diamictons are till and glaciomarine drift, both deposited more or less directly from ice without the winnowing effects of water. They are characterized by a heterogeneous mixture of sediment sizes, ranging from boulders to clay, and a lack of stratification. Particle size distribution is often bimodal with concentrates in the pebble-cobble and silt-clay fractions (Fig. 1). Both types of diamictons are usually massive with only minor stratified intercalations.

Till

Glacial till is deposited in direct contact with glacial ice. Although it does not make up substantial sediment thicknesses in the geologic record, till makes a discontinuous cover for as much as 30% of the earth's continental landmasses and forms significant deposits in Precambrian and Permo-Carboniferous rocks of South America, Africa, and North America (Fig 2).

Till consists of unsorted, unstratified

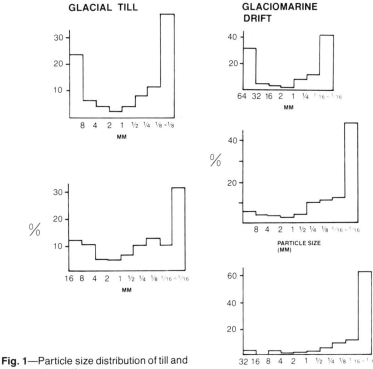

Fig. 1—Particle size distribution of till and glaciomarine drift.

pebbles, cobbles, and boulders in a matrix of sand, silt, and clay. The coarser fraction is mostly pebble size with cobbles and boulders scattered throughout.

Many pebbles are rounded to subrounded, suggesting that they were incorporated by ice riding over stream gravel; others have been faceted, striated, and polished by glacial abrasion.

Sand and silt in the matrix is usually angular to subangular quartz, with much of the fine silt consisting of quartz rock flour.

Physical properties which permit recognition of a diamicton as a till include the following: (1) poor sorting with bimodal particle size distribution; (2) lack of stratification (Figs. 2, 3); (3)

faceted, polished, and striated stones (Fig. 4); (4) fabric consisting of a preferred orientation of long axes of elongated particles; (5) compactness in close packing of constituent particles as a result of overriding ice, high bulk densities, and low void ratios; (6) erratic lithologies of stones and heavy minerals; (7) striated and polished bedrock surface beneath the till; and (8) strongly sheared, folded, structures produced by glacial movement (Fig. 5).

At least three different types of tills, lodgement, ablation, and flow have been recognized on the bases of variation in physical properties and differing depositional processes from glacial ice.

Fig. 2—Precambrian Gowganda tillite scoured by Pleistocene ice sheets, near Lake Ontario, Canada.   **Fig. 3**—Till, north of Uppsala, Sweden.

Fig. 4—Striated stones in till, Solheimajokull, Iceland.

Fig. 5—Deformation of glacial gravel by ice shove, near Bern, Switzerland.

Lodgement till — This type is deposited subglacially from basal, debris-laden ice under the influence of shear stress, as the till is plastered over the ground beneath the ice (Fig. 6). Thus, lodgement till is characterized by preferred fabric and a high degree of compaction.

The long axes of rod-shaped stones are preferentially oriented with a primary maximum parallel with the direction of ice movement and a secondary smaller maximum at right angles. Shearing and weight of ice produce compaction of the deposit evidenced by high bulk densities and low void ratios of uncemented deposits.

Ablation till — This till is deposited from englacial and superglacial debris dumped on the land surface as the ice melts away. In contrast to lodgement till, ablation till usually is only loosely consolidated and possesses a random fabric in the absence of strong shearing stresses.

Flow till — This deposit is formed by water-saturated debris flowing off glacial ice as mud flows. It hasn't the degree of compaction of lodgement till but elongate stones may have a preferred orientation as a result of flowage (Fig. 7).

Glaciomarine Drift

Deposits of glacial debris melted out of ice floating in marine water are col-

lectively termed glaciomarine drift in recognition of both the glacial and marine influence on resulting sediment. Because of their poor sorting and lack of stratification, glaciomarine drifts often resemble glacial tills in general appearance and have sometimes been referred to as "marine tills," but the term glaciomarine drift is more appropriate because sediment is not deposited in contact with glacial ice and often includes facies which are not till-like in appearance. The floating ice may consist of shelf ice or berg ice calving from a glacier margin (Figs. 8, 9).

Melting of floating ice releases clay,

Fig. 6—Lodgement till being deposited from the base of moving ice, Breidamerkurjokull, Iceland.

Fig. 7—Flow till with parallel orientation of stones, Kristineberg, Sweden.

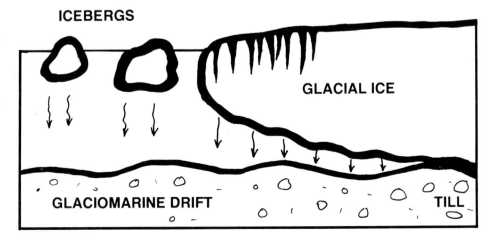

Fig. 8—Depositional environments of glaciomarine drift and basal till.

silt, sand, pebbles, cobbles, and boulders which settle to the underlying seafloor, often burying marine mollusks living on the bottom (Figs. 8, 10).

The coarser fraction of glaciomarine deposits consists of pebbles, cobbles, and a few boulders, many of which are faceted, polished, and striated, randomly scattered throughout a matrix of clay, silt and sand. The pebble to silt-clay ratio varies considerably, grading from till-like diamictons to silty clay with only a few pebbles. Typical particle size distributions are shown in Figure 1.

Glaciomarine drift is generally characterized by a lesser degree of compaction than glacial till, presumably because deposits were never under appreciable glacial loading (Fig. 11).

The glacial origin of these fossiliferous deposits is indicated by: (1) lack of sorting; (2) faceted, striated and polished pebbles showing little or no secondary rounding (Fig. 12); and (3) faceted and polished erratics of distance provenance.

Till-like deposits (Figs. 13, 14) probably represent times when debris-laden floating ice was extensive, whereas pebbly silt and clay probably represent times when floating ice was less abundant or lacked large quantities of debris.

The relatively uniform distribution of the coarser fraction throughout the clay-silt matrix indicates that a nearly continuous rain of unsorted material took place during the greater part of sediment accumulation. If berg ice was the principle transport agent, calving and melting of the bergs must have occurred rapidly to produce the relatively uniform distribution of coarse material.

Criteria for identification of glaciomarine drift includes: (1) whole, unbroken shells; (2) articulated pelecypod valves (Fig. 15); (3) preservation of mollusks in growth position (Fig. 15); (4) barnacles and other marine mollusks attached to glacially faceted surfaces of pebbles; (5) distribution of shells throughout a deposit; (6) absence of underlying fossiliferous deposits from which shells could be reworked; (7) preservation of delicate ornamentation on shells; (8) foraminifera and diatoms in matrix material; (9) sediment inside articulated shells and worm tubes indicating that organisms were living in the environment of deposition of the diamicton; (10) regional distribution of deposits; and (11) high content of Na and total exchange ca-

Fig. 9—Debris-laden icebergs from Breidamerkurjokull, Iceland.

Fig. 10—Fossiliferous Pleistocene glaciomarine drift, Penn Cove, Washington.

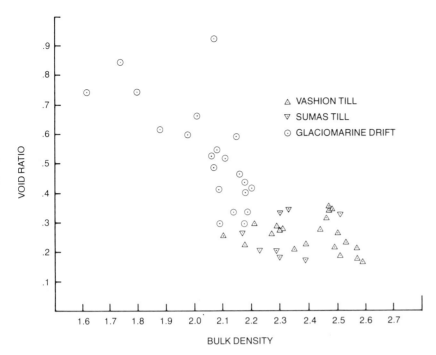

Fig. 11—Void ratios and bulk densities of Pleistocene glaciomarine drift and till, Puget Lowland, Washington.

Fig. 12—Faceted, polished and striated cobble from glaciomarine drift, Puget Lowland, Washington.

Fig. 13—Till-like texture of Pleistocene glaciomarine drift, Puget Lowland, Washington.

Fig. 15—Articulated valves of pelecypods preserved in growth positions in Pleistocene glaciomarine drift, Bellingham, Washington.

Fig. 14—Shell-bearing, till-like glaciomarine drift, Puget Lowland, Washington.

tions in clay size fractions (Fig. 16).

Measurement of exchangeable Na in clay-size fractions may be used to distinguish some glaciomarine sediments from similar appearing diamictons (Pevear and Thorsen, 1978). Sodium content in expandable clay is higher in glaciomarine drift apparently because of enrichment from sea water (Fig. 16).

ICE-CONTACT DEPOSITS

Not all debris transported and deposited by glaciers consists of till. Meltwater on, under, within, or marginal to glaciers produces detritus which, when deposited on, against, or beneath ice, forms deposits known collectively as ice-contact sediments. Because most such deposits involve

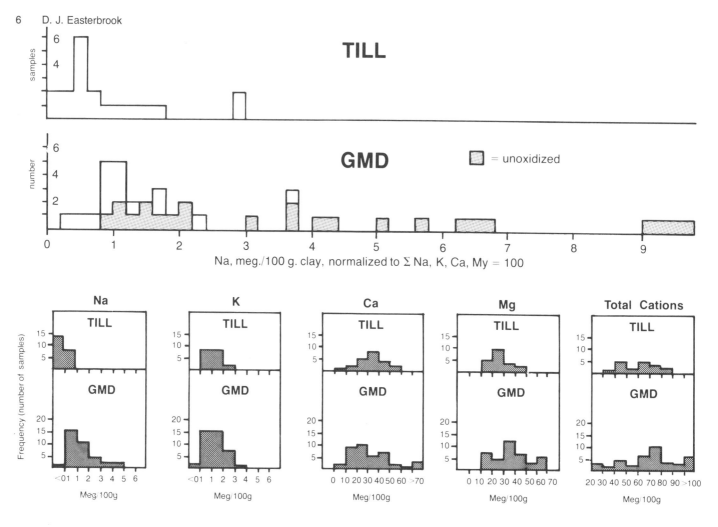

TILL

GMD   ▨ = unoxidized

Na, meg./100 g. clay, normalized to Σ Na, K, Ca, My = 100

| Na | K | Ca | Mg | Total Cations |

Fig. 16—Exchangeable sodium and other cations in glaciomarine drift and till, Puget Lowland, Washington (after D. R. Pevear, 1978).

Fig. 17—Sediment accumulating in a moulin, Breidamerkurjokull, Iceland.

Fig. 18—Deformation of ice-contact glaciofluvial deposit by collapse of supporting ice, Ostersund, Sweden.

transportation by meltwater streams, they exhibit better sorting and stratification than sediments laid down directly from ice. The finer fraction is winnowed from the coarser detritus so particle size distribution curves lose most or all traces of bimodal distribution typical of glacial till. Pebbles and cobbles are quickly rounded after only short distances of transport.

Accumulation of glaciofluvial sand and gravel in englacial or superglacial cavities of a glacier (Fig. 17) leads to deposition of irregular bodies of sediment exhibiting ice-slump deformation when supporting ice melts away (Fig. 18, 19). Post-depositional melting of ice can produce extensive internal deformation of sediments. Deformation in deposits may include a variety of collapse features, such as tilting, faulting, and folding. Often, stratified sediments and flow tills or ablation tills are interbedded (Fig. 19).

GLACIOFLUVIAL DEPOSITS

The character of sediments deposited from glacial meltwater often bears the imprint of glacial environments. Although mechanisms of transportation and deposition show similarities to other fluvial environments, large fluctuations in discharge on a daily, seasonal, or long-term basis produce abrupt particle size changes and sedimentary structures reflecting the fluctuating discharges and proximity to glaciers.

Although measurements of many glacial sediment characteristics have been made, recognition of a specific environment is often difficult because variation in rock types and recycling of detritus through several environments produce changes in pebble sizes and shapes making the influence of the final depositional environment obscure. Price (1973) found no significant differences in roundness values between moraines and eskers at Breidamerkurjokull, Iceland, presumably because many of the clasts in the moraines were formed by reworking of glaciofluvial sediments.

Glaciofluvial deposits grade into normal fluvial downstream as the glacial influence diminishes. Near the glacial margin, clasts may exhibit a lower degree of rounding than nonglacial fluvial sediments. Rounding increases abruptly, however, in transport downstream, with recycling of previously overridden material diminishing the value of rounding in distinguishing glacial and nonglacial deposits.

Pebble and cobble clasts in outwash deposits are sometimes well imbricated. Upstream dips are considered good indicators of current direction. Long axes of larger elongate clasts are often oriented transverse to main current flow.

Gravel bars on proximal portions of outwash fans consist of poorly sorted, imbricated, stratified gravel. Downstream migration of megaripples produces large-scale festoon cross-bedding, with migration of longitudinal and linguoid bars resulting in cross-bedding.

Near the glacier margin, outwash may be deposited on stagnant ice (Fig. 22). Melting of the buried ice produces significant disruption of both morphology and internal structure of the deposit (Fig. 23). Downstream, such sediments grade into kettled but relatively undisturbed glaciofluvial sand and gravel, eventually forming fluvial deposits lacking evidence of glacial origin.

Fig. 19—Interstratified ice-contact stratified sediments and diamictons, Sveg, Sweden.

Fig. 20—Glacial outwash channel and terrace, Breidamerkurjokull, Iceland.

Fig. 21—Outwash gravel in meltwater channel, Breidamerkurjokull, Iceland.

Fig. 22—Outwash sediment being deposited on stagnant ice, Breidamerkurjokull, Iceland.

Fig. 23—Collapsed outwash deposit deformed by melting of buried ice, Breidamerkurjokull, Iceland.

DELTAS AND GLACIOLACUSTRINE DEPOSITS

Where meltwater streams discharge into lakes or the sea, deltas are formed. If the glacial margin is close by, an ice-contact delta is produced (Fig. 24) and the resulting deposit will show various slump deformation structures made by melting of supporting ice in addition to normal topset, foreset, and bottomset stratification. Deltas made by glacial outwash distant from the ice terminus (Fig. 25) do not exhibit ice collapse structures, but are still subject to widely fluctuating discharges typical of meltwater streams. These result in considerable variation in sediment discharge and particle size distribution in sediments. Foreset bedding often consists of relatively parallel beds of sand and gravel inclined up to 30° (Fig. 26).

Glacially-related deltas exhibit unusually rapid sedimentation. A delta in Malaspina Lake studied by Gustavson, Ashley and Boothroyd (1975) contains approximately 9 million cu m of sedi-

ment deposited in less than 10 years. The Chicopee delta of glacial lake Hitchcock in Massachusetts was deposited at a rate of 2.2 million cu m per year. Meltwater streams carry large amounts of glacially ground rock flour which is discharged as suspended load into lakes or the seas along with bedload material. The sand-silt-clay rock flour deposits from suspension, often as parallel laminae which mimic ripples and other bedforms. Such structures are termed draped laminations by Gustavson, Ashley and Boothroyd (1975). Other common sedimentary features include graded bedding from turbidity current deposition, flow rolls, varves, ripples, load casts, "flame" structures, and involutions (Fig. 27).

REFERENCES CITED

Aario, R., 1972, Associations of bed forms and paleocurrent patterns in an esker delta, Haapajarvi, Finland: Ann: Acad. Sci. Fenn., Ser. A, pt. III, 55 p.

Allen, J. R., 1970, Physical processes of sedimentation: Allen and Unwill, London, 248 p.

Allen, P., 1975, Ordovician glacials of the central Sahara, in Ice Ages: Ancient and Modern: Spec. Issue Geol. Jour., no. 6, p. 275-286.

Andrews, J. T., 1971, Methods in the analysis of till fabrics, in Till—A Symposium: Ohio State Univ. Press, Columbus, Ohio, p. 321-327.

——— and B. B. Smithson, 1966, Till fabrics of the cross-valley moraines of north central Baffin Island: Geol. Soc. America Bull., v. 77, p. 271-290.

Ashley, G. M., 1975, Sedimentation in glacial lake Hitchcock, Massachusetts-Connecticut: in Glaciofluvial and Glaciolacustrine Sedimentation: SEPM Spec. Pub. 23, p. 304-320.

Banerjee, I., and B. C. McDonald, 1975, Nature of esker sedimentation, in Glaciofluvial and Glaciolacustrine Sedimentation: SEPM Spec. Pub. 23, p. 132-154.

Banham, P. H., 1975, Glaciotectonic structures: a general discussion with particular reference to the contorted drift of Norfolk: in Ice Ages: Ancient and Modern: Spec. Issue, Geol. Jour. no. 6, p. 69-94.

Boothroyd, J. C., and G. M. Ashley, 1975, Processes, bar morphology, and sedimentary structures on braided outwash fans, northeastern Gulf of Alaska: in Glaciofluvial and Glaciolacustrine Sedimentation: SEPM Spec. Pub. 23, p. 193-222.

Boulton, G. S., 1968, Flow tills and related deposits on some Vestspitsbergen glaciers: Jour. Glaciology, v. 7, p. 391-412.

——— 1970, On the deposition of subglacial and melt-out tills at the margins of certain Svalbard glaciers: Jour. Glaciology, v. 9, p. 231-246.

——— 1971, Till, genesis and fabric in Svalbard, Spitsbergen, in Till—A Symposium: Ohio State Univ. Press, Columbus, Ohio, p. 41-72.

———— 1972, Modern arctic glaciers as depositional models for former ice sheets: Geol. Soc. London Quart, Jour. v. 128, p. 361-393.

———— 1975, Processes and patterns of subglacial sedimentation: a theoretical approach: *in* Ice Ages: Ancient and Modern: Spec. Issue, Geol. Jour. no. 6, p. 7-42.

———— D. L. Dent, and J. Morris, 1974, Subglacial shearing and crushing and the role of water pressures in tills from SE Iceland: Geog. Annaler, 56A, no. 3, 4, p. 135-145.

Carey, S. W., and N. Ahmad, 1961, Glacial and marine sedimentation: Proc. 1st Int. Symp. Arctic Geol., v. 2, p. 865-894.

Church, M., 1972, Baffin Island Sandurs: a study of arctic fluvial processes: Canada Geol. Survey Bull., v. 216, 208 p.

———— and R. Gilbert, 1975, Proglacial fluvial and lacustrine environments: *in* Glaciofluvial and Glaciolacustrine Sedimentation: SEPM Spec. Pub. 23, p. 22-100.

———— and J. M. Ryder, 1972, Paraglacial sedimentation: a consideration of fluvial processes conditioned by glaciation: Geol. Soc. America Bull., v. 83, p. 3059-3072.

Clayton, L., 1964, Karst topography on stagnant glaciers: Jour. Glaciology, v. 5, p. 107-112.

Cook, J. H., 1946, Kame complexes and perforation deposits: Amer. Jour. Sci., v. 244, p. 573-583.

Easterbrook, D. J., 1963, Late Pleistocene glacial events and relative sea-level changes in the northern Puget Lowland, Washington: Geol. Soc. America Bull., v. 74, p. 1465-1483.

———— 1964, Void ratios and bulk densities as means of identifying Pleistocene tills: Geol. Soc. America Bull., v. 65, p. 745-750.

———— 1966, Glaciomarine environments and the Fraser glaciation in Northwest Washington: Guidebook for Annual Field Conf., Pacific Coast Friends of the Pleistocene, Western Washington State College Press, Bellingham, 52 p.

———— 1976, Quaternary geology of the Pacific Northwest: *in* Quarternary Stratigraphy of North America: Dowden, Hutchinson, and Ross, Stroudsburg, Penn., p. 441-462.

Embleton, C., and C. A. M. King, 1975, Glacial Geomorphology: Wiley and Sons, New York, 573 p.

Engeln, O. D. von, 1912, Phenomena associated with glacier drainage and wastage, with special reference to observations in the Yakutat Bay region, Alaska: Zeitschr. Gletscherkunde u. Glazialgeologic v. 6, p. 104-150.

Evenson, E. B., 1977, Subaquatic flow tills: a new interpretation for the genesis of some laminated till deposits: Boreas, v. 6, p. 116-133.

Flint, R. F., 1928, Eskers and crevasse fillings: Amer. Jour. Sci., v. 15, p. 410-416.

———— 1975, Features other than diamicts as evidence of ancient glaciations: *in* Ice Ages: Ancient and Modern: Spec. Issue, Geol. Jour. no. 6, p. 121-136.

Francis, E. A., 1975, Glacial sediments: a selective review, *in* Ice Ages: Ancient and Modern: Spec. Issue, Geol. Jour. no. 6, p. 43-68.

Fig. 24—Ice contact delta, Crillon glacier, Alaska (photo by Coastal Research Center, Univ. Mass.).

Fig. 25—Outwash delta, Baffin Island, Canada.

Fig. 26—Foreset bedding in glacial delta, Lake Cavanaugh, Washington.

10   D. J. Easterbrook

Goldthwait, R. P., 1971, Introduction to till, today: *in* Till—A Symposium: Ohio State Univ. Press, Columbus, Ohio, p. 3-26.

Gravenor, C. P., and W. O. Kupsch, 1959, Ice disintegration features in western Canada: Jour. Geol., v. 67, p. 48-64.

Gustavson, T. C., G. M. Ashley, and J. C. Boothroyd, 1975, Depositional sequences in glaciolacustrine deltas: *in* Glaciofluvial and Glaciolacustrine Sedimentation: SEPM Spec. Pub. 23, p. 264-280.

Hamilton, W., and D. H. Krinsley, 1967, Upper Paleozoic glacial deposits of South Africa and southern Australia: Geol. Soc. America Bull., v. 78. p. 783-799.

Harland, W. B., K. N. Herod, and D. H. Krinsley, 1966, The definition and identification of tills and tillites: Earth-Sci. Rev., v. 2, p. 225-256.

Harrison, W., 1957, A clay-fill fabric: its character and origin: Jour. Geology, v. 65, p. 275-308.

Hartshorn, J. H., 1958, Flowtill in southeast Massachusetts: Geol. Soc. America Bull., v. 69, p. 477-482.

Holmes, C. D., 1941, Till fabric: Geol. Soc. America Bull., v. 52, p. 1299-1354.

——— 1960, Evolution of tillstone shapes, central New York: Geol. Soc. America Bull., v. 71, p. 1645-1660.

Howarth, P. J., 1971, Investigation of two eskers at eastern Breidamerkurjokull, Iceland: Arctic and Alpine Research, v. 3, p. 305-318.

King, C. A. M., and J. T. Buckley, 1968, The analysis of stone size and shape in arctic environments: Jour. Sed. Petrology, v. 38, p. 200-214.

Krumbein, W. C., 1939, Preferred orientation of pebbles in sedimentary deposits: Jour. Geology, v. 47, p. 673-706.

Ostrem, G., 1975, Sediment transport in glacial meltwater streams: *in* Glaciofluvial and Glaciolacustrine Sedimentation: SEPM Spc. Pub. 23, p. 101-122.

Price, R. J., 1969, Moraines, sandar, kames and eskers near Breidamerkujokull, Iceland: Inst. British Geographers Trans. v. 46, p. 17-43.

——— 1970, Moraines at Fjallsjokull, Iceland: Arctic and Alpine Research, v. 2, p. 27-42.

——— 1971, The development and destruction of a sandur, Breidamerkurjokull, Iceland: Arctic and Alpine Research, v. 3, p. 225-237.

——— 1973, Glacial and fluvioglacial landforms: Oliver and Boyd, Edinburgh, 242 p.

——— P. J. Howarth, 1970, The evolution of the drainage system (1904-1965), in front of Breidamerkurjokull, Iceland: Jokull, v. 20, p. 27-37.

Rust, B. R., 1975, Fabric and structure in glaciofluvial gravels: *in* Glaciofluvial and Glaciolacustrine Sedimentation: SEPM Spec. Pub. 23, p. 238-248.

Saunderson, H. C., 1975, Sedimentology of the Brampton esker and its associated deposits: an empirical test of theory: *in* Glaciofluvial and Glaciolacustrine Sedimentation: SEPM Spec. Pub. 23, p. 155-176.

Fig. 27—Deformed sediments of a glacial lake, Lake Cavanaugh, Washington.

Introduction to Eolian Deposits

Thomas S. Ahlbrandt
Steven G. Fryberger
*McAdams, Roux, O'Connor
and Associates*
Denver, Colorado

INTRODUCTION

Deserts (arid and semi-arid regions) cover about 30% of the earth's land surface and of this area perhaps 20% is covered by sandy deserts (Cooke and Warren, 1973). Although eolian deposits appear homogeneous, they are, in fact, texturally and depositionally complex deposits. Following burial and diagenesis, they have proved to be complex (heterogenous) hydrocarbon reservoirs. This study shows the diversity of eolian sub-environments related to dune type, interdune type, and the genesis of sand seas (large dune fields), stressing the overall eolian depositional system. An improved understanding of the complexities of modern eolian systems should improve exploitation of hydrocarbon producing eolian reservoirs.

The following types of eolian and related deposits are discussed:

(1) Dune—a hill, mound or ridge of wind-blown sand;

(2) Interdune—those sediments occuring in the relatively flat areas between dunes of a dune complex;

(3) Sand Sheet—those eolian deposits occuring marginal to a dune complex that generally do not have definable dune forms (also called low-angle eolian deposits); and

(4) Extradune—includes those sediments marginal to a dune field that are not eolian, but are related to dune sediments in time and by source (Lupe and Ahlbrandt, 1979, Fig. 1A, 1B).

Neither wind-blown silt (loess) and clay on land, as summarized in Schultz and Frye (1968) and Windom (1975), nor the contribution of eolian sand to offshore marine environments, e.g. "eolian-sand-turbidites" as discussed by Sarnthein and Diester-Haass (1977), is considered. Field and wind tunnel dynamic studies supporting some seemingly speculative discussions are not treated in detail due to space limitations. The origin of some dune types, such as linear dunes, is still debated.

This paper focuses on the more common sedimentary features of warm climate eolian deposits; however, numerous small and some large dune fields occur in mid- to high-latitude regions and are subjected to cold climate processes. Distinctive tensional, compressional, and dissipation[1] sedimentary structures related to freezing, thawing, and snow melting occur in cold climate dune fields and contrast with sedimentary features of warm climate dune fields (Ahlbrandt and Andrews, 1978). Some examples of dune fields subjected, at least seasonally, to cold climate processes include: the Gobi Desert; Takla Makan Desert in the Peoples Republic of China; Peski Karakumy, Peski Kyzylkum, USSR; Arctic and Antarctic Deserts; Nebraska Sand Hills and numerous dune fields in the Rocky Mountains and Great Plains regions; and dune fields in northern, western and central Europe (the latter cover 8,500 to 13,000 sq km, Koster, 1978). The migration rate of cold climate dunes is much slower than for comparable warm climate dunes. Thus, dynamic studies of dune fields should consider not only wind regimes, but moisture and temperature factors as well.

Although studies of several dune fields have taken place, data from the Nebraska Sand Hills have been particularly instructive. The Nebraska Sand Hills are the largest dune field in the Western Hemisphere covering 22,000 sq mi (56,890 sq km). They are currently inactive, though probably representing a Holocene sand sea (<8,000 years ago), that is being preserved. The combination of preserved dune forms at the surface (some up to 400 ft [122 m] high), borehole data throughout the Sand Hills, and excellent exposures of internal structures of dunes along rivers, both originating in and dissecting the Sand Hills, is most fortuitous for eolian studies (Ahlbrandt and Fryberger, 1976, 1980).

Continuing questions in modern eolian research concern the dynamics and growth of sand seas. For example, how long does it take to deposit a compound barchan dune several hundred feet high? Age estimates range from tens of thousands of years (and much more) to much shorter times (<8,000

---

[1]Dissipation structures appear as discontinuous dark bands having a higher content of fine-grained material, predominately clay, than surrounding layers. These structures are superimposed on and obscure original eolian stratification. They are known to result from: (1) infiltration of clay-bearing waters into dunes in warm coastal regions (Bigarella, 1975); (2) soil forming processes ("B" horizon; Ruhe, 1975); and (3) infiltration of meltwater from snow and ice in cold climate dunes, locally concentrating indigenous clay coatings on the eolian sand (Ahlbrandt and Andrews, 1978).

The writers wish to acknowledge the assistance of colleagues at the U.S. Geological Survey who on many occasions provided discussions of ideas presented here. In particular we wish to thank Edwin D. McKee of the U.S. Geological Survey. In addition, the writers gratefully acknowledge the assistance of Richard F. Mast, Chief of Branch of Oil and Gas Resources, U.S. Geological Survey, Denver, who supported our research in eolian reservoirs. Sarah Andrews performed textural analyses on an Automatic Image Analyzer for which we are grateful. We wish to thank the following oil and service companies for their generous assistance: Amoco, Champlin, Chevron USA, Hilliard Oil and Gas, Mobil, Mountain Fuel, Pasco, and Schlumberger.

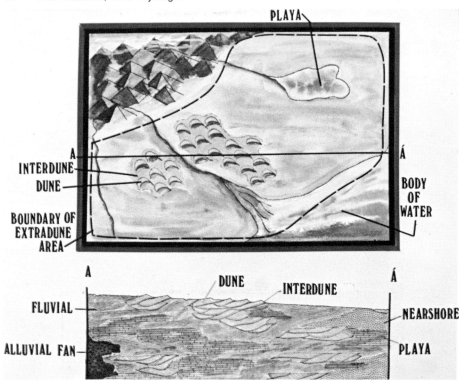

years in the Nebraska Sand Hills) for the formation of such a dune. The perplexing problem of the growth of a sand sea is compounded by the paucity of datable material in deserts, climatic fluctuation, and the inconsistent development rate of different dune types.

Recognition and interpretation of ancient eolian deposits is also controversial. Traditional eolian indicators, such as nonmarine fossils and trace fossils, raindrop imprints, mudcracks, thick, steeply dipping cross-bed sets, and well sorted texture, are increasingly debated or reinterpreted. Modern eolian analogs are important to interpreting the sedimentary features of ancient eolian deposits. Although morphologic and textural studies of modern dune fields are relatively common in geologic literature, the sedimentary structures of modern eolian deposits have been the subject of far fewer studies. Reinterpretations of the

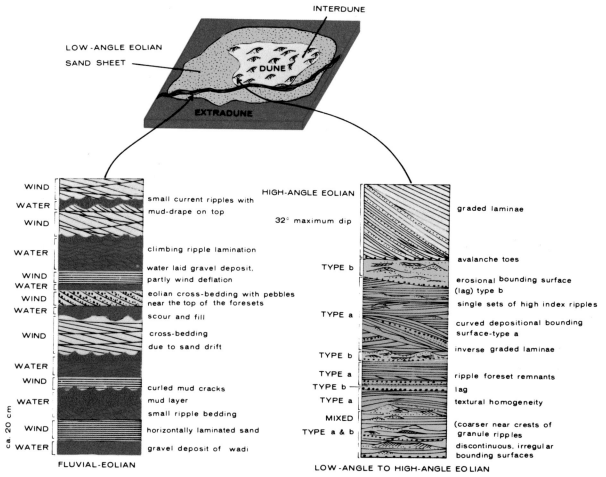

Fig. 1—A. Eolian depositional systems showing areal and stratigraphic relationships of dune, interdune, sand sheet, and extradune deposits (from Lupe and Ahlbrandt, 1979). Interdunes are the areas between dunes of the dune mass,   the sand sheet deposits are peripheral to the main dune mass; they form a transitional lithofacies between high angle eolian and "extradune" (noneolian) deposits. **B.** Sedimentary features in detailed stratigraphic   columns in fluvial-eolian (dune-extradune) and low-angle to high-angle eolian deposits (from Fryberger, Ahlbrandt, and Andrews, 1979).

depositional environment of rock units that have been classically considered eolianites have recently been proposed and subsequently rebutted by a number of authors. An explanation for such controversy is due in part to the mixing of eolian and noneolian deposits (the dune-extra dune depositional system of Lupe and Ahlbrandt, 1979), whereby interpretations of different depositional environments will not agree. Reinterpretation of ancient eolian deposits as possible marine tidal sand ridge deposits, particularly the Navajo and Nugget Sandstones (Triassic-Jurassic) in the western US by Freeman and Visher (1975), is another source of debate. Walker and Middleton (1977) summarize rebuttals to this paper and raise additional questions regarding the interpretation of vertically exaggerated sparker profiles over marine tidal sand ridges used as an alternative to the eolian model by Freeman and Visher (1975). The sedimentary features of marine tidal sand ridges are discussed elsewhere in this volume. Sedimentary features recognized in modern eolian deposits, and discussed in this paper, are easily recognized in the Navajo and Nugget Sandstone outcrops that we have studied.

There appear to be two productive approaches to recognition of ancient eolian rocks. First, the entire depositional framework of the rock units in question must be carefully evaluated, with special attention given to depositional environments gradationally overlying or underlying problematic rock units. Secondly, it is useful to combine a number of lines of evidence based on easily available data which point to an eolian origin, rather than place undue reliance on finding fossils or other unequivocal evidence which may be rare in some eolian rocks. This approach involves detailed study of individual outcrops, and in particular the detailed analysis and documentation of cross-bedding types and stratification styles.

Our work has come to rely increasingly on recognition of deposits produced by grainfall, ripple (saltation), and avalanche processes making up the bulk of eolian, particularly dune, sandstones. Eolian strata commonly contain distinctive features permitting an interpretation of the moisture content of the sand (dry, moist, wet) and the process by which strata were deposited and/or contorted. Intercalation of relatively horizontally bedded, lenticular or extensive interdune deposits (mudstones, evaporites, etc.), that commonly contain subaqueous features, should be considered compatible with eolian depositional systems. Sophisticated textural analysis such as discriminant analysis or scanning electron microscope techniques, may provide complementary evidence of eolian origin. The use of water settling velocity curves for light and heavy minerals is also known to distinguish eolian suspension deposits (Steidtmann, 1974).

Detailed analysis of eolianites has both economic and academic justification. Performance studies of eolian hydrocarbon reservoirs have documented higher oil recovery with infill drilling and selective perforation, as discussed in the economic section of this paper.

Many areas of eolian research are deserving of more investigation. However, the possiblity of hydrocarbon source beds within, or adjacent to, eolian deposits should be carefully considered. Paleointerdune deposits in the Nebraska Sand Hills are known to be potential source rocks (Ahlbrandt and Fryberger, 1976; 1980), and Malek-Aslani (1979) proposed sabkha deposits as viable hydrocarbon source rocks. An intriguing association of evaporites exists adjacent to stratigraphically controlled hydrocarbon traps in some eolianites. Could fluids discharged from such evaporites explain both the complex cementation (silica-anhydrite-carbonate) and presence of hydrocarbons in some eolianites?

EOLIAN DEPOSITIONAL SYSTEMS

Eolian sand deposits consist of three basic components—dunes, interdunes, and sand sheets (Fig. 1A, B). These

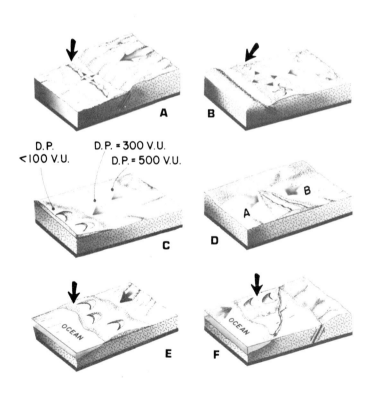

Fig. 2—Mechanisms for the accumulation of large dune fields (sand seas). Topographic (A, B), climatic (C, D), and aqueous mechanisms (E, F) are shown. The loss of wind energy related to topography either in A, the lee of barriers (or in negative relief features such as basins), or B, upwind of barriers with loss of energy due to blocking or increased wind variability. Climatic mechanisms can cause sand to accumulate by C, loss of wind energy along a gradient (in this example, sand is accumulating where Drift Potential values are less than 100 vector units), or D, by opposing wind components that cause sand to accumulate (in this example, wind regime A is predominant during one season, wind regime B during another season). Aqueous mechnisms involve water either as an obstacle to eolian sand movement, E, or a sediment source, F.

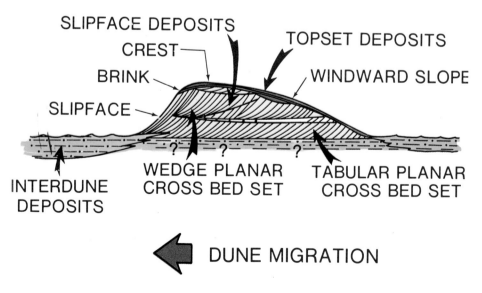

NUMBER OF SLIPFACES

0 — SAND SHEET (large scale feature) / SAND STRINGER (large scale feature) / DOME DUNE

1 — HORN / WIND / BARCHAN DUNE / BARCHANOID RIDGE DUNE / TRANSVERSE RIDGE DUNE

1 or more — VEGETATION / BLOWOUT DUNE / ARM / PARABOLIC DUNE

2 — TIME 1 WIND REVERSAL / REVERSING DUNE / TIME 2 DURING YEAR / REMNANT OF FORMER SLIPFACE / WINDS FROM 2 DIRECTIONS DURING YEAR / SYMMETRIC RIDGE / LINEAR (SEIF, LONGITUDIONAL) DUNE

3 or more — WINDS FROM SEVERAL DIRECTIONS DURING THE YEAR / STAR DUNE

BASIC EOLIAN FORMS

Compound dunes are those in which similar dunes are superimposed--e.g. small barchan on large barchan dune.

Complex dunes are those in which dissimilar dunes are superimposed--e.g. star on top of linear dunes.

SLIPFACE DEPOSITS
CREST
BRINK
SLIPFACE
TOPSET DEPOSITS
WINDWARD SLOPE
INTERDUNE DEPOSITS
WEDGE PLANAR CROSS BED SET
TABULAR PLANAR CROSS BED SET
DUNE MIGRATION

Fig. 3—A. Basic eolian bedforms as related to number of slipfaces; this terminology follows that used by McKee (1979) for ground studies of eolian deposits (modified from McKee, 1979); **B.** A cross-sectional view, parallel with wind direction, of a barchanoid ridge dune illustrating terminology used in this paper (from Ahlbrandt and Fryberger, 1980).

components or subenvironments may intercalate or be adjacent to a number of "extradune" sediments that are related to the eolian system in time, or source (Lupe and Ahlbrandt, 1979; Fryberger, Ahlbrandt, Andrews, 1979). Extradunal sediments are commonly the sources of detritus removed by eolian processes and at times deposited in dune fields or sand seas great distances (perhaps hundred of kilometers) downwind.

Three general mechanisms instrumental in formation of eolian sand seas (large dune fields) are topographic, climatic, and aqueous in nature (Fryberger and Ahlbrandt, 1979, Fig. 2). These mechanisms were deduced from studies of wind data and eolian deposits as observed primarily on LANDSAT imagery of many desert regions of the world. The wind regimes must be understood and correlated to specific dune forms to begin to understand eolian depositional systems and the mechanisms by which they accumulate

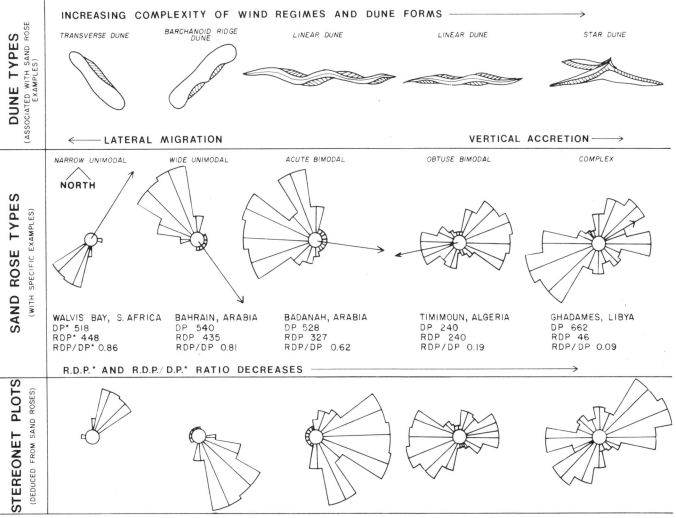

INCREASING COMPLEXITY OF WIND REGIMES AND DUNE FORMS ⟶

DUNE TYPES (ASSOCIATED WITH SAND ROSE EXAMPLES)

TRANSVERSE DUNE   BARCHANOID RIDGE DUNE   LINEAR DUNE   LINEAR DUNE   STAR DUNE

⟵ LATERAL MIGRATION   VERTICAL ACCRETION ⟶

SAND ROSE TYPES (WITH SPECIFIC EXAMPLES)

NARROW UNIMODAL   WIDE UNIMODAL   ACUTE BIMODAL   OBTUSE BIMODAL   COMPLEX

NORTH

WALVIS BAY, S. AFRICA	BAHRAIN, ARABIA	BADANAH, ARABIA	TIMIMOUN, ALGERIA	GHADAMES, LIBYA
DP* 518	DP 540	DP 528	DP 240	DP 662
RDP* 448	RDP 435	RDP 327	RDP 240	RDP 46
RDP/DP* 0.86	RDP/DP 0.81	RDP/DP 0.62	RDP/DP 0.19	RDP/DP 0.09

R.D.P.* AND R.D.P./D.P.* RATIO DECREASES ⟶

STEREONET PLOTS (DEDUCED FROM SAND ROSES)

* DP = Drift Potential; RDP = Resultant Drift Potential; RDP ÷ DP is a measure of lateral migration where 1.0 is maximum, 0.0 is minimum; Fryberger, 1978

Fig. 4—The relationship of dune forms and wind regimes. As wind regime becomes increasingly complex (multidirectional), dune forms evolve to more complex forms with greater internal complexity as expressed in the sand rose and stereonet plots. Note the change from lateral migration to vertical accretion as Resultant Drift Potential (R.D.P.) and R.D.P. ÷ D.P. (Drift Potential) values decrease.

sand (Figs. 2, 3). Wind data were evaluated using the sand rose technique of Fryberger (1979a). Sand roses (Fig. 4) are circular histograms in which arm lengths of the rose are proportional to the amount of sand that can be moved by the wind from a given direction toward the center of the rose. The Lettau equation (written commun., 1977) for the rate of sand drift was used:

$$q \propto V^2(V - V_t)$$

where q = rate of sand drift; V = wind velocity at 10 m height; and, V_t = impact threshold wind velocity at 10 m height (minimum velocity at 10 m height to keep sand in saltation).

For convenience, Fryberger (1979a) uses units of drift potential called vector units (V.U.) with three terms used to classify energy and directional properties: Drift Potential (D.P.), Resultant Drift Potential (R.D.P.) and Resultant Drift Direction (R.D.D.).

Drift Potential is a measure of the total sand-moving capability of wind expressed in vector units without regard to wind direction at the locality. The D.P. and R.D.P. values represent annual values unless otherwise noted. Resultant Drift Potential is a measure of the resultant or net sand-moving capability of wind at a locality in vector units. Strength (vector unit totals) of winds are resolved trigonometrically to a resultant, the magnitude of which is the R.D.P., the direction of which is the R.D.D. Fryberger (1979a) places wind energy levels of desert regions throughout the world in these broad categories: 0-199 vector units = low energy wind regimes; 200-399 vector units = intermediate energy; and, 400 or more vector units = high energy.

The topographic mechanism for formation of sand seas relates to decreased wind energy either in the lee of a major topographic barrier (Fig. 2A) or upwind of a topographic barrier (Fig. 2B). The climatic mechanism may cause eolian sediment to accumulate due to a gradual decrease in wind energy, with deposition then beginning (Fig. 2C), or due to seasonal variations of major effective winds, causing sand to accumulate (Fig. 2D). Aqueous mechanisms may provide either a source and/or a barrier to sand migration as schematically shown in Figures 2E, F.

The preceding general discussion will provide sufficient background for increased focus on more detailed aspects of eolian deposits.

EOLIAN DEPOSITS AND THEIR RECOGNITION

Three eolian subenvironments—dune, interdune and sand sheet—are considered in this study. Dune deposits are the most commonly discussed in the literature, and important summaries of their sedimentary features are given in McKee (1945, 1966, 1977, 1979), McKee and Tibbitts (1964), Glennie (1970), McKee, Douglass and Rittenhouse (1971), Bigarella (1972) and

Fig. 5A—Eolian depositional systems, dune-extradune deposits, the eolian Schoolhouse Tongue (ST) of Weber Sandstone (Pennsylvanian-Permian), white unit in middleground, enclosed in red alluvial sediments of the Maroon Formation near Glenwood Springs, Colorado.

Fig. 5B—Eolian depositional systems, dune-interdune deposits in the Weber Sandstone, Hog Canyon, Dinosaur National Monument, Utah, note interdune deposits (I), mostly carbonates or carbonate mudstones along diastems (D). Some contorted diastems (DC) are adjacent to contorted strata (C). The Weber is about 800 ft (244 m) thick at this locality (Fryberger, 1979b).

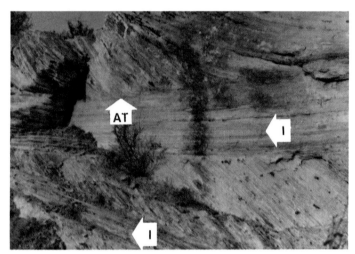

Fig. 5C—Eolian depositional systems, dune-interdune deposits. Lyons Sandstone (Permian), Owl Canyon, Colorado, note avalanche toes (AT) and interdune deposits (I) up to three ft (1 m) thick between wedge-planar crossbed sets among ancient barchan dunes.

Fig. 5D—Eolian depositional systems, dune (transverse)-wet interdune deposits, Killpecker Dunes, Wyoming.

Fig. 5E—Eolian depositional systems, complex dune forms. Transverse dunes with embryonic star dune (high dune) near Rumah, Ad Dahna Sand Sea, Saudi Arabia.

Fig. 5F—Eolian depositional systems, small scale dune with slipface, Algodones Dunes, California.

Walker and Harms (1972). Subsequently, more information has been gathered about dune deposits, some of which is summarized briefly here (Figs. 3, 4, 5).

Dunes

The basic dune types and terminology used in this paper follow that of McKee (1979) and are summarized in Figures 3 and 4. Dunes form in response to specific wind regimes of which five basic types are recognized (Fryberger, 1979a, Fig. 4). Dune types can be divided into broad groups based on form and number of slipfaces. Dunes and eolian deposits unlikely to develop a slipface include dome dunes, sand sheets, and sand stringers (an elongate sand sheet). Dunes with one slipface include barchan, barchanoid ridge, and transverse ridge dunes (Fig. 3A). Blowout and parabolic dunes which may have one or more slipfaces are related to this group, but are controlled by vegetation cover (Fig. 3A). Dunes with two slipfaces include linear (seif) dunes (symmetric ridges), and reversing dunes (asymmetric ridges; Fig. 3A). Star dunes have three or more slipfaces (Fig. 3A). Compound dunes are those in which smaller dune forms are superimposed onto larger, similar forms; e.g. small barchans on a larger barchan. Complex dunes feature different dune forms that are superimposed; e.g. star dunes and transverse ridge dunes (Figs. 3A, 4, 5E).

Several basic dune forms are illustrated in Figure 4 with corresponding sand roses and inferred stereonet plots. The evolution of dune forms is illustrated in Figures 3 and 5E and is supported by a number of studies. For example, in unidirectional wind regimes, the evolution of dome to transverse to barchan to parabolic dunes has been documented by McKee (1966) and Ahlbrandt (1974, 1975). Linear dunes, typically formed in bimodal or complex wind regimes (Bagnold, 1941; McKee and Tibbitts, 1964; Tsoar, 1978; Fryberger, 1979), evolve from barchan dunes in some deserts (Bagnold, 1941; Tsoar, 1978). As wind regimes become more complex, dune forms become more complex (Fig. 4), ultimately resulting in star dunes (McKee and Tibbitts, 1964; Fryberger, 1979a). This most complex dune type, the star dune, forms in response to winds from three or more directions as reflected in both wind data and strike and dip data (McKee, 1966; Figs. 4, 5E).

Dune types can be recognized in eolianites by measuring cross-bedding

dip directions and plotting them on stereonet (or dipmeter) plots, and applying statistical measures of strike and dip data. Mean dip and mean angular deviation of dip direction, respectively, are as follows for some modern dunes: transverse ridge dunes, 24° and 32°; barchan dunes, 22° and 54°; blowout dunes, 16° and 54° (Ahlbrandt and Fryberger, 1976; Fig. 6); and 62° mean deviation of dip direction is reported in parabolic dunes (Ahlbrandt and Andrews, 1978). The above statistical measures are derived from methods described by Till (1974).

Linear dunes (also called seif or longitudinal dunes) have bimodal clusters of dips on either side of the axis of the sand ridge (Fig. 4). In contrast to transverse ridge dunes that are oriented nor-

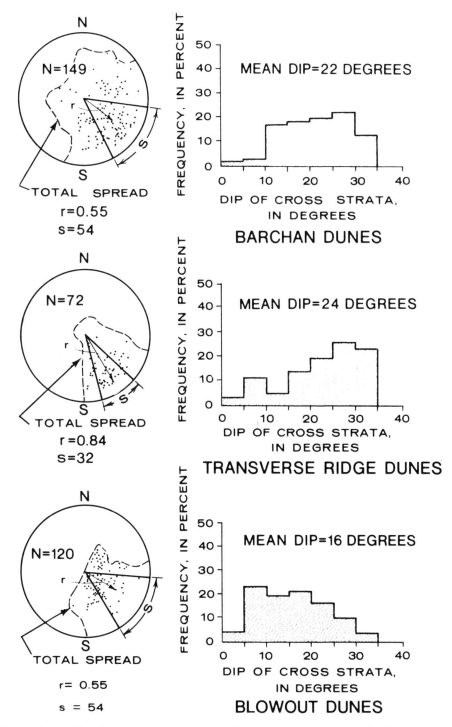

Fig. 6—Recognition of dune types using strike and dip measurements from the Nebraska Sand Hills. Mean dip angles of leeside strata in barchan dunes (22°) and transverse ridge dunes (24°) are considerably higher than blowout dunes (16°) due to the general lack of slipface development on the latter dune form (from Ahlbrandt and Fryberger, 1980).

Fig. 7A—Eolian indicators, large scale, tabular, planar cross-strata (TP) overlain by smaller scale trough cross-strata (TR), Weber Sandstone (Pennsylvanian-Permian), Sand Canyon, Colorado.

Fig. 7B—Eolian indicators, sand flow toes, forming tangential contact of cross-beds with lower bounding surface, Browns Park Formation (Miocene), 10 mi (16 km) west of Maybell, Colorado.

Fig. 7C—Eolian indicators, raindrop imprints (raised rims, depressed centers), Weber Sandstone (Pennsylvanian-Permian), Deerlodge Park, Colorado.

Fig. 7D—Eolian indicators, evenly distributed bimodal sand lags along erosional surfaces, Algodones Dunes, California.

Fig. 7E—Eolian indicators, high index ripples oriented up and down leeside eolian deposits, Lyons Sandstone (Permian), Sterling Quarry, Colorado.

Fig. 7F—Eolian indicators, desiccation cracks, basal Navajo Sandstone (Triassic-Jurassic?) Dinosaur National Monument, Colorado.

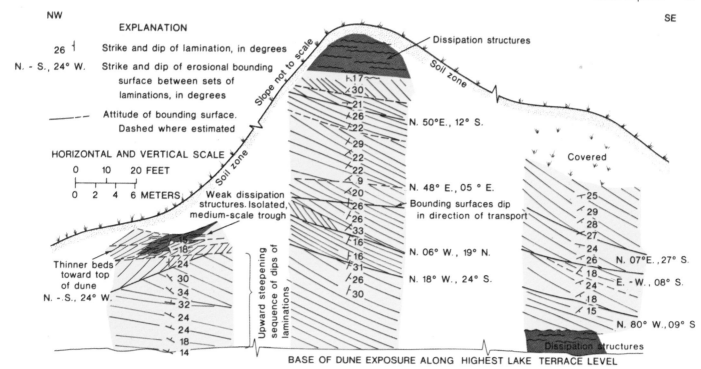

Fig. 8—Measured sections in en-echelon barchan dunes at Merritt Reservoir (Loc. 52, 54 on Fig. 11), Nebraska Sand Hills. Sections shown in close proximity on diagram were measured in different dunes along a several kilometer-long exposure of the lake front. Bounding surfaces between sets are curved and dip in direction of transport. Bed thickness and dip angle decrease higher in each dune; however, dissipation structures become more common. Dip and strike measurements of cross-strata are more variable in the upper parts of the dunes (from Ahlbrandt and Fryberger, 1980).

mal to the prevailing wind, linear dunes develop parallel with the resultant of wind components from two directions (McKee and Tibbitts, 1964; Tsoar, 1978; Fryberger, 1979a). The bimodal wind regime is reflected in laminae which dip generally perpendicular to the dune axis as seen on stereonet plots (Glennie, 1970, 1972; Fig. 4).

Sets of cross-strata in most dunes are tabular or wedge planar. Individual strata are generally concave upward and also steepen upward (Figs. 3B, 7, 8, 9, 10). Convex upward cross-strata particularly develop in vegetationally controlled dunes such as parabolic and blowout dunes. Trough cross-strata are not common in modern dunes, however the meaning of this statement should be clarified (Figs. 3B, 7, 9). Wedge planar cross-stratification as defined by McKee and Wier (1953) is very common in dunes and is generally confused with trough cross-stratification (Figs. 3B, 7, 8. 9). Figure 8 is a series of measured sections in a compound barchan dune showing internal structures that have been well exposed along Merritt Reservoir, Nebraska. Figure 8 also illustrates the vertical sequence of deposits that may be preserved within a dune having one slipface and migrating principally in one direction. A complete vertical suc-

cession would contain slipface deposits overlain successively by downwind-dipping topset and finally upwind dipping topset deposits (Figs. 3B, 8).

This Nebraska example demonstrates wedge planar cross-stratification in that bounding surfaces of cross-strata are curved and laminae within cosets meet the lower bounding surface of the set. In the example shown, bounding surfaces generally dip in the direction of sand transport and are concave upward, with the exception of blowout and parabolic dunes as previously discussed (note the moderate degree of true dips along each section in Fig. 8). Therefore, the more sinuously crested and internally complicated a dune is, the greater the occurrence within it of curved bounding surfaces. Straighter crested dunes, such as the transverse ridge dune, have tabular planar cross-bedding and thus nearly flat bounding surfaces (McKee, 1966; Ahlbrandt, 1975).

Dune types previously discussed are all common forms, and each type varies considerably in size. For example, barchan dunes may range from a few feet to several hundred feet in height. Among dunes found in wind regimes with one prevailing direction, the dome dune is probably the least common and the smallest (<50 ft [15.2 m] high).

Some deserts are dominantly comprised of unidirectional dune types, for example, the dune fields in the People's Republic of China, USSR, Thar Desert, India (where parabolic and blowout dunes are dominant), and many dune fields in the United States. Trailing arms of parabolic dunes are sometimes long (several tens of miles), and based on morphology alone can easily be confused with linear dunes. However, the internal structures of the two types of dunes are quite different. For example, parabolic dunes commonly have convex-upward, moderate angle crossbeds that may be highly bioturbated by plants whereas linear dunes may have biomodal, moderate to high angle convex-downward cross-beds.

The linear (seif, longitudinal) dune is probably the most common type in modern dune fields. It is the dominant type in Australian deserts, many deserts in Africa (e.g. Western Mauritania, Sahelian zone, Kalahari Desert, Libyan Desert, parts of the Namib Desert), and significant portions of many deserts in the Middle East. A simple linear dune may extend for a great distance, tens of miles or much more. Interdune corridors between linear dunes can be broadly catagorized into wide (many times wider than adjacent dune widths) and narrow (equal width

Fig. 9A—Types of cross-bedding in eolian deposits, tabular planar—note horizontal diastems, Navajo Sandstone (Triassic-Jurassic?) near Dinosaur National Monument, Colorado.

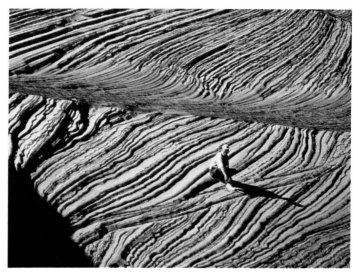

Fig. 9B—Types of cross-bedding in eolian deposits, wedge-planar, laminae are tangential to lower bounding surface, De Chelley Sandstone (Permian), Canyon De Chelley, Arizona (Photo courtesy of E. D. McKee).

Fig. 9C—Types of cross-bedding in eolian deposits, wedge-planar, Lyons Sandstone (Permian), Sterling Quarrry, Colorado.

Fig. 9D—Types of cross-bedding in eolian deposits, wedge-planar, note tangential contact along lower bounding surfaces, Browns Park Formation (Miocene), 10 mi (16 km) west of Maybell, Colorado.

Fig. 9E—Types of cross-bedding in eolian deposits, trough–laminae do not meet lower bounding surface, Browns Park Formation, just west of area shown in Fig. 9D

Fig. 9F—Types of cross-bedding in eolian deposits, internal bounding surfaces (B), in this case climbing with respect to a major diastem (D) in foreground, view to northwest, Weber Sandstone (Pennsylvanian-Permian), Yampa River Gorge, Deerlodge Park, Dinosaur National Monument, Colorado.

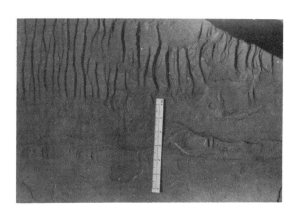

Fig. 10A—Dune stratification types, structures related to avalanche processes (sand flow), cross-section showing avalanche (sand flow) toes (AT), and obscure ("fuzzy") internal structures within sand flows, Weber Sandstone (Pennsylvanian-Permian), Hog Canyon, Utah.

Fig. 10B—Dune stratification types, structures related to avalanche processes (slumps), bedding plane view looking down on leeside deposits, Coconino Sandstone (Permian), Grand Canyon, Arizona, (photo courtesy E. D. McKee).

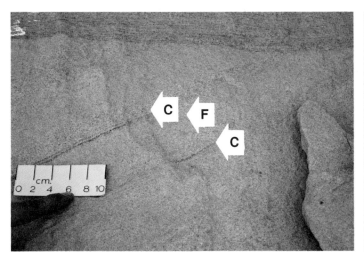

Fig. 10C—Dune stratification types, normally graded deposits, probably related to grainfall processes; Browns Park Formation (Miocene), near Maybell, Colorado. These strata are thickest near the upper bounding surface and thin abruptly downslope disappearing in two meters distance; coarse sand (C), and fine sand (F), are indicated.

Fig. 10D—Dune stratification types, leeside grainfall deposits, normally graded, Nebraska Sand Hills (Loc. 116, Fig. 11).

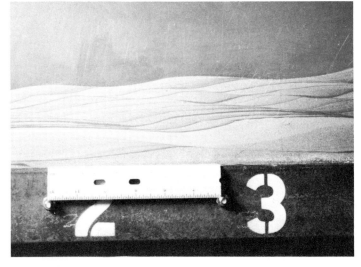

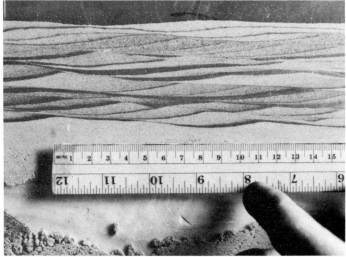

Fig. 10E, F—Dune stratification types, topset deposits produced by migrating ripples; most strata are inversely graded, darker layers are magnetite used as markers, Wind Tunnel Laboratory, U.S. Geological Survey, Denver, Colorado.

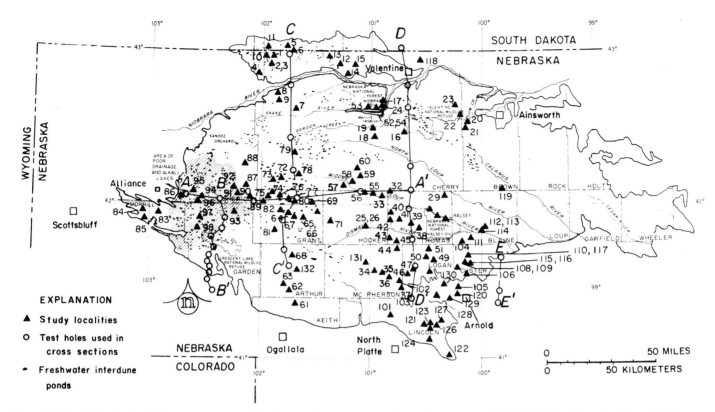

Fig. 11A—The Nebraska Sand Hills, Holocene sand sea, currently inactive. Hatches define areas of alkali lakes (evaporite interdunes in western Sand Hills). Freshwater interdune ponds are common in the western half of the Sand Hills. Dry interdune deposits are common in the eastern part due to streams with sources in the Sand Hills that drain to the North Platte River (e.g. North Loup, Middle Loup, South Loup and Dismal Rivers).

to a few times wider than adjacent dune widths). Linear dunes grow by lengthening downwind in a resultant direction of the two major wind directions affecting it (often seasonal variations). Tsoar (1978) proposes that linear dunes will evolve naturally from barchan dunes given sufficient time, sediment source, and even minor variability in wind direction. Simple linear dunes are generally less than 100 ft (30.5 m) high; however, compound linear dunes may reach heights of several hundred feet or more.

The star dune is the tallest dune type, some over 1,000 ft (305 m) high. This dune migrates very slowly, if at all; rather it grows vertically. Star dunes are significant forms, for example, in the Rub Al Khali, Saudi Arabia, the Grand Erg Oriental and Grand Erg Occidental, Northern Algeria and the Gran Desierto, Mexico.

The above discussion is an oversimplified summary of the transitions of dune types and sizes. An example of dune morphology, geometry, and preservation in a large dune field, the Nebraska Sand Hills, will provide a more detailed analysis. For a more complete morphologic mapping of dunes across the world, the reader is referred to McKee (1979). Although the thickness of eolian deposits in the Sand Hills, beneath the interdune surface, is generally less than 120 ft (36.6 m), which is relatively thin, much thicker eolian deposits (ranging to 3,860 ft, 1,200 m) are reported by Glennie (1970) in Libya.

Nebraska Sand Hills: Example of Dune Distribution and Thickness

The Sand Hills dunes can be classified primarily as simple and compound "transverse" types including dome (the small variety described by McKee, 1966; Ahlbrandt, 1975), barchan, barchanoid ridge (McKee, 1979) transverse ridge, parabolic, and blowout dunes.

Barchan, barchanoid ridge, and transverse ridge dunes up to 300 ft (91 m) high are widespread throughout the Sand Hills region (Fig. 11). A group of the largest of these dunes extends roughly north-northwest—south-southeast across the western part of the dune field. Dune heights are greatest in the central part of the dune field, ranging up to 300 ft (91 m), and decrease in size to the north and south (Fig. 11, Section D-D'). The large dunes are compound in form, for example small transverse ridge dunes occur on large transverse ridge dunes. Large bar-chans are common around the margins of the dune field, as well as in the interior. Some large dunes, straight or gently curving in plan view, superficially resemble linear or seif dunes, but their internal structures indicate that they are barchanoid ridge and transverse dunes. Some of these dunes are aligned oblique to the regional direction of sand transport. Many may have developed from en-echelon barchans and may have been preserved in a transitional stage to a linear dune type.

Many low mounds up to 16.4 ft (5 m) high, with gently dipping, curved, parallel internal stratification, occur throughout the Sand Hills. These mounds generally occur atop larger dunes and may represent remnants of windward slope or crestal deposits of older dunes, blowout deposits, or small dome dunes, although dome dunes have not previously been described in compound association with other dune types.

The Sand Hills of Nebraska form a thin "blanket" deposit unconformably overlying either a southward-thickening wedge of unconsolidated Pleistocene alluvial deposits or consolidated sediments primarily of the Tertiary Ogallala Formation (Fig. 11). In the southeastern part of the dune field

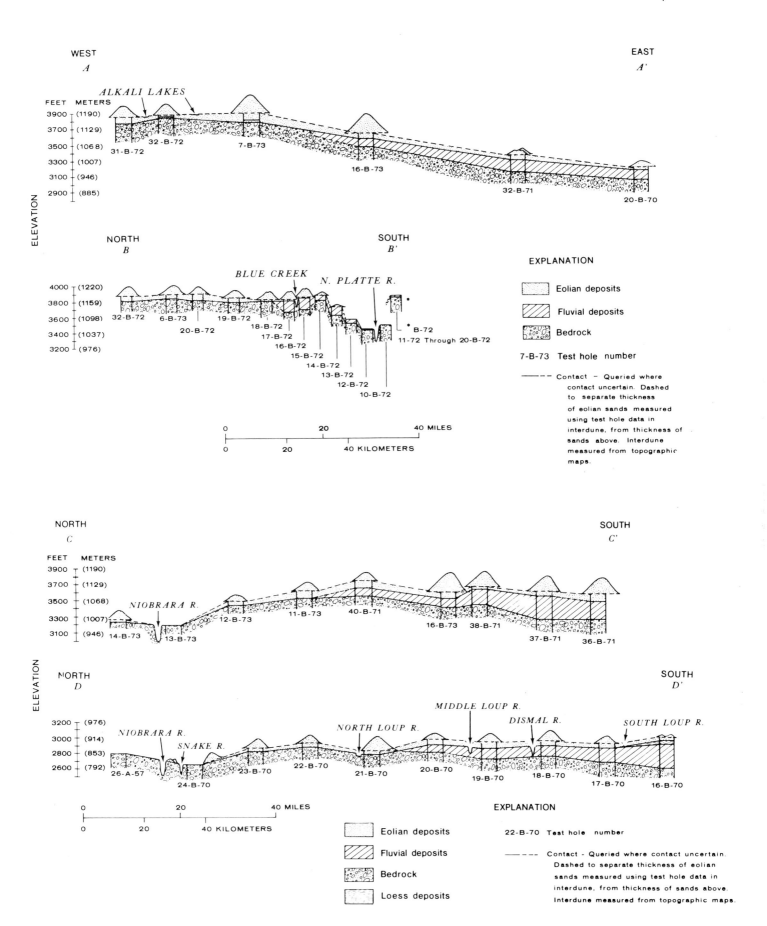

Fig. 11B—Cross-sections A-A', B-B', C-C', D-D' through the Nebraska Sand Hills using data from 35 boreholes. The height of dunes at each test-hole site is shown on each section. Vertical exaggeration = 86x (from Ahlbrandt and Fryberger, 1980).

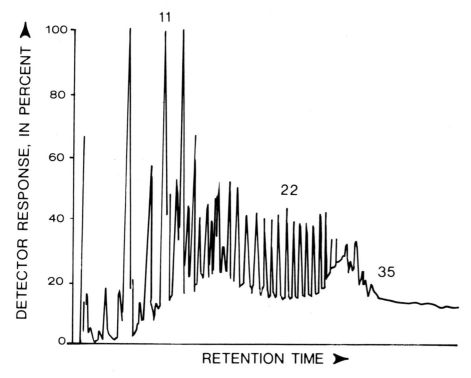

Fig. 12—Gas chromatogram of Nebraska paleointerdune deposit (C^{14} dated to 500 ± 200 years B.P.) The sample (Loc. 71, Fig. 11) was heated from 671°F (355°C) to 1472°F (800°C). Numbers along selected peaks refer to number of carbon atoms per molecule of normal paraffin hydrocarbons. The whole range of normal paraffin hydrocarbons, both oil and gas, was generated (from Ahlbrandt and Fryberger, 1980).

(Custer and Logan Counties), dune sand overlies loess in both the outcrop and subsurface (Fig. 11, Section D-D', test-hole 16-B-70). In general, the height of the dunes above the inter-dune surface is greater than the depth of eolian sand beneath the interdune surface. The thickness of eolian sand below interdune surfaces generally ranges from 30 to 80 ft (9 to 24 m), with an observed maximum thickness of 120 ft (37 m; Fig. 11, 7-B-73), based on analyses of lithologic descriptions and electric logs of 35 test-holes along four cross sections.

Ancient Dunes

Stereonet plots of strike and dip data have long been used to interpret dune types and paleowind directions in eolianites. For example, stereonet plots were used in studies of the Co-conino Sandstone by Reiche (1938) and McKee (1945). Examples of dune types that have been described in eolianites include: (1) dome (Frod-sham Member of Keuper Sandstone, Triassic, Thompson, 1969); (2) barchan (Rotliegendes, Permian, Glennie, 1970, 1972; Lyons Sandstone, Permian, Maughan and Ahlbrandt, 1978); (3) transverse (Barun Goyot Formation, Upper Cretaceous, Gradzinski and Jer-zykiewicz, 1974; Lyons Sandstone, Per-

mian, Maughan and Ahlbrandt, 1978; Weber Sandstone, Pennsylvanian-Permian, Fryberger, 1979b); and (4) lin-ear (Rotliegendes, Permian, Glennie, 1970, 1972; Lyons Sandstone, Permian, Walker and Harms, 1972). McKee (1979) discusses other dune types rec-ognized in eolianites including blowout and parabolic dune deposits.

Deformation features are commonly associated with modern and ancient eolian deposits. Many eolianites have contorted stratification, examples in-cluding: the Navajo Sandstone (Trias-sic-Jurassic?, Kiersch, 1950); the Weber Sandstone (Pennsylvanian-Permian, Fryberger, 1979); the Casper Forma-tion (Pennsylvanian-Permian, Steidt-mann, 1974); the Baron Goyot Forma-tion (Upper Cretaceous, Gradzinski and Jerzyciewicz, 1974) among others. The mechanisms and interpretation of deformational structures in modern eolianites are discussed by Glennie (1970), McKee, Douglass, and Rit-tenhouse (1971), McKee and Bigarella (1972) and Ahlbrandt and Andrews (1978), among others.

Cross-bed sets ranging up to 115 ft (35 m) in thickness have been recorded in eolianites (Walker and Middleton, 1978); however, most eolian cross-bed sets are truncated either during dune migration or subsequent erosion. The maximum thickness observed by the

writers for a cross-bed set preserved within a modern dune is 86 ft (26.2 m) in the Nebraska Sand Hills (Loc. 54, Fig. 11).

Interdunes

Interdune deposits are the second major group of eolian deposits, featur-ing three broad groupings for study. These are dry interdune deposits, wet interdune deposits, and evaporite in-terdune deposits. The three types of in-terdunes can be further classed as de-flationary or depositional (Ahlbrandt and Fryberger, 1981). Gradations from one type of interdune to another are common, even within a single inter-dune; but each type is sufficiently dif-ferent from the others to warrant sepa-rate discussion. A feature common to all three types is that their stratifica-tions are all relatively horizontal (<10° dips), in marked contrast to adjacent higher angle dune deposits. However, many interdune deposits are virtually structureless due to secondary pro-cesses.

It is difficult to make generalizations about interdune sediments. For example, the notion that interdunes are relatively thin (less than several meters) and lenticular is questionable. Our preliminary work indicates that interdunes among dunes formed in uni-modal wind regimes are fundamentally different than interdunes among dunes formed in bimodal or complex wind regimes. The mechanism of alternate dune and interdune development producing lenticular, diachronous and relatively thin, <6 ft (2 m), interdune deposits among unimodal dunes proposed by McKee and Moiola (1975) is strongly supported in our experience. However, interdune deposits among dunes formed in bimodal or complex wind regimes, i.e. linear and star dunes respectively, seem to be thicker and areally more extensive than among unimodal types (Ahlbrandt and Fryberger, 1981). The following discussion draws heavily from the writers' work in the Nebraska Sand Hills, where dry, wet, and evaporite interdune deposits occur in various parts of the field (Fig. 11).

Dry interdune deposits have re-ceived much more attention than other interdune types. McKee and Moiola (1975) demonstrated that the thickness of interdune deposits in White Sands National Monument, New Mexico, is dependent on the rate of dune migra-tion. The longer an interdune area is exposed before burial, the greater the thickness of the interdune accumulation. Thus, areas upwind in a unidirectional dunefield, where for example barchan dunes are small and migrate rapidly, may collect relatively little interdune sediment before being buried; whereas downwind, thicker interdune deposits may be preserved among parabolic dunes that migrate slowly. In interdune areas or serirs that are being deflated, poorly sorted, bimodal sand is commonly found (McKee and Tibbitts, 1964; Folk, 1971; Cooke and Warren, 1973; Ahlbrandt, 1979). The dominant depositional process in dry interdunes is ripple related, including granule ripples (very coarse crested ripples; Sharp, 1963). Sedimentary features found in sand sheet or low angle eolian deposits on the right column of Fig. 1B summarize many features found in dry interdune deposits.

Wet interdune areas, commonly containing ponds, cover 65,800 acres (266.6 sq km) in the Nebraska Sand Hills (Steen, 1961; Fig. 11). Organic silt and sand interdune deposits were 5 m thick in two lakes in Cherry County (Watts and Wright, 1966) and 3.5 m thick in Krause and Jesse Lakes in Sheridan County in the Nebraska Sand Hills. Contorted stratification, gastropods, pelecypods, diatoms, ostracodes, and pollen are common in these interdune deposits (Ahlbrandt and Fryberger, 1976, 1980; Hanley, 1980; Bradbury, 1980). An example of an interdune pond in the Killpecker Dunes, Wyoming is shown in Figure 5D.

Interdune sediments may provide an indigenous hydrocarbon source. Organic interdune deposits in the Nebraska Sand Hills generated both oil and gas when heated (Fig. 12). Interdune samples were heated from 86 to 671°F (30 to 355°C) and produced 8 to 10 major hydrocarbon compounds distilled from organic matter. A second analysis was performed on material evolved from samples at temperatures of 671 to 1,472°F (355 to 800°C). The decomposition of solid organic matter during the latter analysis produced the whole range of normal paraffin hydrocarbons and could produce petroleum upon burial and heating (G. E. Claypool, written commun., 1975).

Pyrolitic oil yield for the Nebraska interdune sample was 14.1 gal/ton (60 l/metric ton) and 0.9% organic carbon for the whole rock. This gives a whole rock pyrolitic oil yield for the original sample of .25 gal/ton (1.07 l/metric ton). Assuming a density of 0.08 lb/cu in. (2.2 g/cu cm), 436 unit volumes of source rock would be required to produce one unit volume of oil. If expulsion of oil is inefficient then two to four times more rock volume (872 to 1,744 unit volumes) would be required to produce one unit volume of oil (G. E. Claypool, written commun., 1975). Such deposits cover 65,800 acres (266.3 sq km) and evaporite ponds occur over an additional 1,000 sq mi (2,590 sq km) area in the Nebraska Sand Hills to depths of several meters. A conservative estimate of oil that could be generated from the Nebraska organic interdune ponds is approximately 200,000 barrels of oil; however, a much greater volume of hydrocarbons may be generated in sabkha evaporites (Malek-Aslani, 1979).

Pyrolysis products of interdune sediments in the Nebraska Sand Hills are comparable to hydrocarbons produced from lacustrine deposits of the Eocene Green River Formation in the western United States and resemble pyrolysis products in crude oils from the Pineview Field in Wyoming which produces from the Nugget Sandstone, considered to be an eolianite. Picard (1975) suggests an indigenous hydrocarbon source for the Nugget Sandstone in certain parts of western Wyoming. The concept for an indigenous hydrocarbon source within certain eolianites is an interesting, but little studied problem.

Evaporite interdune deposits are probably the least studied group of eolian deposits. Two types of sabkha deposits—coastal sabkhas and inland sabkhas—were recognized by Glennie (1970), and by Johnson et al (1978). Both sabkha types may be associated with eolian deposits. It is open to debate whether coastal sabkhas are more properly associated with arid nearshore marine processes or in some cases with desert processes, they may belong to both. Coastal sabkhas in Saudi Arabia studied by Johnson et al (1978), and examined by the writers, proved highly variable, complex environments of alternating, discontinuous sand, silt, carbonate mud (both calcite and dolomite), and nodular, white anhydrite layers. Some areas contained reducing black muds, abundant adhesion ripples, desiccation features, and contorted bedding (caused by growth of gypsum crystals).

Inland sabkhas briefly examined by the writers in the north central part of the Rub Al Khali (near Ramlah) were similar to coastal sabkhas in composition, texture and diversity. Both of these sabkhas (coastal and inland) are areally extensive features (not confined to one interdune corridor). We refer the reader to Johnson et al (1978) for an introduction to the sabkha problems.

Evaporite interdunes (inland sabkhas) occur in the Nebraska Sand Hills. A 1,000 sq mi (2,590 sq km) area of enclosed drainage and alkaline interdune ponds was drilled extensively in search of potash during World War I (hatched area, Fig. 11). Two or three buried hardpan layers were found in test holes, separated by sand containing alkaline brines, but no continuous hardpan horizon in the western part of the Sand Hills was found (Condra, 1918; Bradley and Rainwater, 1956). The closed basin area where the alkali ponds and brines occur produced over 100,000 tons of potash (Hicks, 1921) with brines generally alkaline (pH 10 to 11) and total dissolved solids ranging up to 148,000 ppm (Keech and Bentall, 1971). Hicks (1921) stated "commercial potash brines vary much in salinity and in composition of dissolved salts. The salts are composed of carbonates, bicarbonates, sulphates, and chlorides of sodium and potassium in varying proportions." Freshwater ponds at the surface were not in communication with underlying brines.

Gradational types of interdune deposits were found in the Killpecker Dunes of Wyoming. Water analysis from 29 interdune ponds showed a pH range of 7.5 to 9.0 with dissolved solids ≤11,000 ppm. Authigenic minerals included Na, Ca, and K carbonates, sulfates, and nitrates (for example mirabilite, thenardite and calcite; Ahlbrandt, 1974). These marls commonly contain mollusk, diatom, and plant fossils and are intercalated with sandy interdune deposits.

Ancient examples of dry and wet interdune deposits are given in fascinating detail by Gradzinski and Jerzykiewicz (1974). They describe sediments (claystones and sandstones) that intertongue with tabular planar, cross-bedded dune sands in the Upper Cretaceous Barun Goyot Formation. The writers state that "the features of the alternating claystones and siltstones suggest that these sediments were deposited in intermittent lakes, existing in interdune areas and covering the peripheral slopes of dunes." Fossils found in such "structureless" deposits of the Barun Goyot Formation include "specimens of dinosaurs, very numerous dinosaur eggs, a few crocodiles, several tortoises, a pterosaur, about eighty specimens of lizards, about fifty specimens of mammals, and specimens of diplopod myriapods and isolated sauropod teeth. ... The complete dinosaur eggs were found, as a rule, in 'structureless' sandstones, sometimes in the immediate neighborhood of mega-cross-stratified units, and often also in sandstone beds alternating with sandy claystones." (Gradzinski and Jerzykiewicz, 1974, p. 272). They similarly found all mammals and dinosaurs in 'stuctureless' sand; some of which "passed laterally into typical deposits of the mega cross-stratified unit."

Examples of dry interdune deposits (siltstones, mudstones, etc.) of the De-Chelley Sandstone (Permian) are given by McKee (1979); in the Lyons Sandstone (Permian) by Walker and Harms (1972), and Maughan and Ahlbrandt (1978); and the Weber Sandstone (Pennsylvanian-Permian) by Fryberger (1979b). Dry interdune deposits as recognized in eolianites commonly consist of nearly horizontally stratified detritus. Such deposits are commonly sandstones that are poorly sorted, finely laminated, with occasional bioturbation traces and recognizable high index ripple deposits (the ripple form can be seen).

Chemically precipitated (evaporite) interdune deposits are described in the Casper Formation (Pennsylvanian-Permian) by Hanley and Steidtmann (1973) and in the Weber Sandstone (Pennsylvanian-Permian) by Fryberger (1979b). Features in these interdune deposits include stylolization, distrupted, contorted fabric, relatively thin lenticular units, burrows, and occasional to common breccia and/or nodules. Fryberger (1979b) recognized two basic types of chemically precipitated interdune deposits in the Weber: (1) sandy brown, brecciated dolomites which were high in Mn content (>800 ppm) and had moderate Fe content (300 to 500 ppm); and (2) lime mudstones low in Mn (<200 ppm) and low in Fe (<200 ppm) using cations considered environmentally sensitive for carbonates by Friedman (1969). Analysis of Casper Formation interdune carbonates placed them in the lime mudstone category, although they had high Mn content (>800 ppm) but low Fe content (<200 ppm).

Sand Sheets

The last major group of eolian sediments are sand sheet deposits typically having low to moderate angle (0 to 20°) cross-stratification. These deposits form a transitional facies between dune and interdune deposits and noneolian deposits and may be areally extensive (Fryberger, Ahlbrandt and Andrews, 1979). Modern sand sheet deposits have a number of sedimentary features that may assist in identification of similar deposits in ancient rocks including: (1) coarse-grained, high index (width to height ratio >15) ripples (commonly horizontal); (2) gently dipping, curved or irregular surfaces of erosion (several meters in length) within the deposits; (3) abundant bioturbation traces, formed mainly by insects and plant roots; (4) patches or zones (extending up to 1 m below the ground surface) of bioturbated sand resulting from destruction of laminae by grass plant growth; (5) convex upward laminations; (6) small scale cut and fill structures, possibly due to scour around plant roots and subsequent infill; (7) gently dipping, poorly laminated layers resulting from adjacent grainfall deposition; (8) sets of normal and inverse graded laminae 1- to 4-mm thick produced by wind ripples; (9) discontinous thin 1- to 4-mm thick layers of coarse sand intercalated with fine sand; (10) intercalation of noneolian material; and (11) occasional intercalation of high angle eolian deposits (Fig. 1).

Fossils and Trace Fossils in Eolian Deposits

Numerous fossils were previously described in the Upper Cretaceous Barun Goyot Formation, and fossils and trace fossils are common in some eolianites. Specific examples are summarized for the following formations: the Navajo Sandstone (Triassic-Jurassic, Picard, 1977); the Coconino Sandstone (Permian, McKee, 1944, 1945, 1947); the DeChelley Sandstone (Permian, McKee, 1934); the Casper Formation (Pennsylvanian-Permian, Hanley, Steidtmann and Toots, 1971); the Lyons Sandstone (Permian, Walker and Harms, 1972), and the Cave Sandstone (Triassic, Haughton, 1969).

Fossils noted in eolianites include (with pertinent references) dinosaurs (Gradzinski and Jerzykiewicz, 1974; Picard, 1977) and other reptiles (Brady, 1939; Haughton, 1969); mollusks (Picard, 1977); crustaceans (Haughton, 1969); fish (Haughton, 1969); and plants (Glennie and Evamy, 1968; Haughton, 1969; Picard, 1977). Trace fossils of many types, but particularly those of arthropods, are commonly in eolian deposits (Ahlbrandt, Andrews, and Gwynne, 1978).

SEDIMENTARY STRUCTURES IN EOLIAN DEPOSITS

Examples of some typical eolian sedimentary features are shown in Figure 7. These include:

1. Large scale, moderate to high angle cross-strata, commonly wedge planar or tabular planar sets (Figs. 7A, 9) and laminae within sets generally are tangential to the lower bounding surface (Figs. 3B, 7B, 9).

2. High index ripples (crest to crest spacing of ripple divided by the height of the ripple >15) extending up and down moderately to steeply dipping cross-bed sets (the lee sides or slipfaces of dunes; Fig. 7F). A consistent ripple index from the base to the top of large bedforms is supportive of eolian, but not subaqueous deposits, because the latter commonly shows rapid ripple transitions with water depth changes. Many subaqueous ripples have ripple indices of 10 or less (Reineck and Singh, 1975). Sedimentation by high index ripples is also common in low and moderate angle eolian sand deposits.

3. Minor sedimentary features such as reptile tracks, trails (that commonly go uphill), and raindrop imprints (Fig. 7C). The article by McKee, Douglass, and Rittenhouse (1971) on deformation of lee-side laminae is critical to an understanding of small scale eolian structures.

4. Sedimentary structures related to the process of avalanching of sand down dune slipfaces, producing avalanche toes (or sand flow toes) that intercalate with lower angle crosswind deposits. The abrupt tongues of the sand flows are easily distinguished in sections parallel with sand movement (Figs. 7B, 10A), but in views perpendicular to sand movement, such deposits appear as a series of irregularly stacked small lobes with few internal laminae. The distruption of bedding, graded bedding, and abrupt downslope termination of laminae are indicators of avalanche processes. Deposits produced by sand flow may appear to be thick, homogeneous deposits (Figs. 10, 13); generally they are not. Individual avalanches are generally one to several cm thick and their traces are

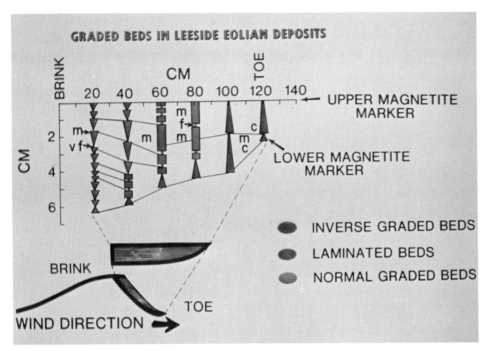

Fig. 13—Leeside graded bedding in eolian deposits. Inverse grading in upslope regions, laminated bedding in midslope and normal grading in downslope part of avalanche deposits were produced repeatedly during experiments (E. D. McKee, 1977, oral commun.) at the Sedimentation Laboratory, U.S. Geological Survey, Denver, Colorado.

seen on detailed investigation of seemingly homogeneous beds. Sand slumps (another type of avalanche) can involve thicker (up to several meters) deposits of sand such as those due to cohesion of a surficial layer by moisture or cementation (Fig. 10B).

5. Intercalated interdune deposits and/or poorly sorted, bimodal lag deposits along erosional bounding surfaces (Fig. 7E).

Less diagnostic, but common eolian sedimentary features or indicators include:

1. Large and small scale deformational structures such as break-apart structures within a fold, indicating damp, cohesive sand rather than loose saturated, sand (including dissipation structures discussed in Ahlbrandt and Andrews, 1978).

2. Normally and inversely graded, laminated and ungraded strata produced by processes of avalanching, ripple migration, and grainfall.

3. Textures, i.e. well sorted, fine sand, are only compatible, not a reliable eolian indicator. Positive skewness definitely is not a reliable eolian indicator as discussed in the texture section.

4. Burrows and root molds (dikaka) of terrestrial animals and plants, respectively (Ahlbrandt,

Andrews and Gwynne, 1978).

5. Frosting of sand grains, particularly when analyzed with a Scanning Electron Microscope.

6. Light and heavy mineral separation ratios (Steidtmann, 1974).

Some common misconceptions about eolianites include:

1. Cross-bedding is all high angle (>30°) and dip is consistent in one direction (see dips of leeside dune deposits shown in Figs. 6, 8).

2. Subaqueous deposits do not occur within eolian deposits.

3. Bioturbation does not occur in eolian deposits.

4. Eolianites are well sorted, mature (quartzose), texturally homogeneous (blanket) sandstones and thereby simple hydrocarbon reservoirs to produce.

Sedimentary Structures Within Dunes

Primary eolian sedimentary structures recognized in dune cross-strata include normally graded beds, inversely graded beds, evenly laminated beds, discontinuously laminated beds, nongraded beds, ripples, and lag deposits along bounding surfaces of sets. Slipface and topset deposits are recognized in eolian deposits as illustrated in Figure 5A. Topset deposits do not reach the angle of repose of sand (34°) and do not contain features related to sand

avalanching. It is important to note that although certain sedimentary features and lamination styles are much more commonly produced by one process than another, most laminae can be produced by more than one process. Some processes produce several different lamination styles. Although it is possible to recognize the process responsible for production of a given bed in some places, it is inappropriate to name the bed using a process.

To better interpret lamination styles, a brief discussion of processes and their resultant lamination styles is given. Three groups of eolian processes are discussed—avalanche, saltation (which for convenience will include sand creep), and grainfall. It is beyond the scope of this report to discuss details of these processes here; however, they have been observed in wind tunnel experiments, and the resulting strata described (Fryberger and Schenk, 1981). These experimental results support the discussion which follows.

The grainfall term was introduced by Hunter (1977) and is used here to refer to sand size detritus that accumulates in the lee of an obstacle or in the lee of a dune by falling from suspension, not through movement by saltation or avalanching processes. Laminae resulting from avalanche and saltation processes are more common and more easily recognized in dune cross-strata than are deposits produced by the grainfall process.

Avalanche Processes (Sand Flows and Slumps)—Avalanching of sand down lee slopes of dunes at the angle of repose (34°) produces normally graded, inversely graded, and laminated beds within a single avalanche deposit. Laboratory study results of strata produced in sand flows are shown in Figure 13. During sand flow, coarser grains are moved by shear sorting to the surface in the upper slipface, producing inverse grading there. In the mid slipface region, segregation of laminae of specific sizes occurs (no grading). The coarser grains, brought to the surface by shear sorting upslope, move to downslope margins of the sand flow and are later buried by finer grained material creating normal grading in the lower part of the deposit. Thus, an abrupt textural contrast may occur between coarse grains in the sand flow toe (concentrated there by gravity) and finer grained deposits near the base of the slipface produced by crosswind, saltation processes (Figs. 7B, 10A).

Fig. 14A—Deformational features in eolian deposits, large scale intraformational deformation, Weber Sandstone (Pennsylvanian-Permian), Split Mountain, Utah.

Fig. 14B—Deformation features in eolian deposits, bedding plane view of recumbent fold, Casper Formation (Pennsylvanian-Permian), Sand Creek, Wyoming—breccia blocks (B) indicate damp, not water saturated, condition of sand at time of deformation.

Fig. 14C—Deformation features in eolian deposits, overthrusting and flame structures (O), yellow laminae—ripple and grainfall produced strata; red layers—avalanche produced strata, Navajo Sandstone (Triassic-Jurassic), Dinosaur National Monument, Colorado.

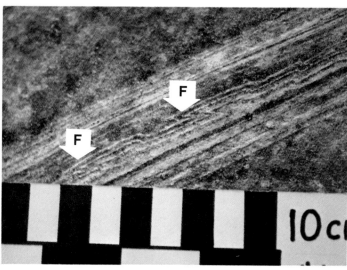

Fig. 14D—Deformation features in eolian deposits, warps and drag folds (nearly overthrust; F), Navajo Sandstone as in (C). McKee, Douglass, and Rittenhouse (1971) have described the same structures in modern dune deposits.

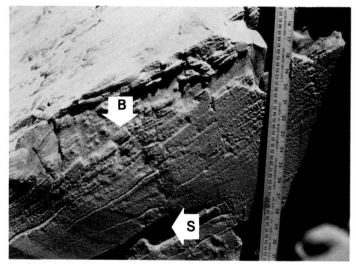

Fig. 14E—Deformation features in eolian deposits, tensional deformation due to snow melting (S), slipface parabolic dune, North Park, Colorado —deformaton has produced cohesive breccia blocks (B) and faults.

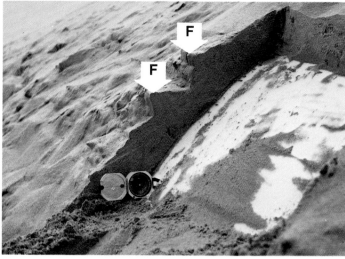

Fig. 14F—Deformation features in eolian deposits, compressional deformation downslope from 14E—note lack of cohesion and obscure bedding in folds (F) due to complete water saturation.

Fig. 14G—Deformation features in eolian deposits, trench in 10 m high barchan dune (meter stick for scale) showing convex upward snow lens, view upslope toward brink, North Park, Colorado.

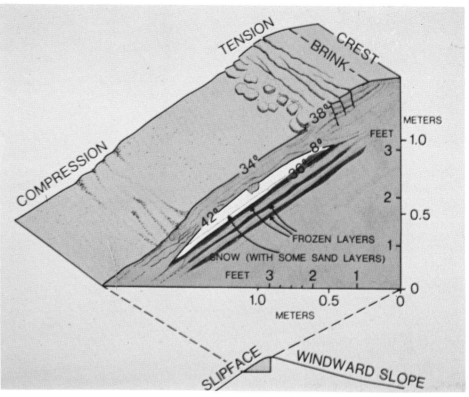

Fig. 14H—Schematic block diagram of deformational structures produced by snow melt and freezing of moisture in sand (from Ahlbrandt and Andrews, 1978).

In views normal to sand transport, sand flows appear as lenses, and in all views laminae are indistinct due to sand flowage during deposition. Individual sand flows mostly a few centimeters thick, and contain diagnostic deformation structures indicating position on the slipface (tension upslope or compression downslope), and moisture content (McKee, Douglass, and Rittenhouse, 1971). Deformational features in leeside deposits include: tensional faults (Figs. 10B, 14E); stretched laminae; fadeout laminae; overturned folds, (Figs. 14A, B); overthrusts (Fig. 14C); rotated blocks (Fig. 14B, E); pull aparts (Fig. 14C); drag folds (Fig. 14D); flames (Fig. 14C); warps (Fig. 14D); high angle asymmetric folds (FIg. 14D); breccias (Fig. 14B, E). Please refer to McKee, Douglass and Rittenhouse (1971) for a detailed discussion of these features.

A second broad group of avalanche processes are slumps. Slumps imply cohesion of sand in which blocks move downslope without internal flowage. These are easily seen in both modern and ancient dune deposits (Fig. 10B), and occur in response to wetting by dew, rain, snow (Fig. 14E, H) or cementation.

Saltation Processes — Migrating wind ripples produce a much different type of eolian bedding than do avalanche processes. Whereas most deposits produced by sandflow have indistinct laminae, deposits produced by migrating ripples are distinctly lami-

nated or bedded strata (Fryberger and Schenk, 1981). Migrating ripple lamination may be the major process responsible for what is commonly called "pinstriping" or "varving" in many eolianites. Several types of dunes are composed almost totally of deposits formed by both granule and wind ripples—for example dome, parabolic, blowout and perhaps linear dunes (the latter refers to Tsoar, 1978). Deposits on the windward slopes (topset deposits, Fig. 3B) of all dunes are produced predominately by migrating ripples, though it is uncertain how much windward slope deposition is preserved in eolianites. Migrating wind ripples leave two common deposit types—inversely graded beds (ranging to 1 cm thickness; Fig. 10E, F) and both discontinuously and evenly laminated strata (coarse layers separated by fine layers, etc.; Fig. 14C).

Although normal graded beds are not generally produced by migrating wind ripples; they have been found in dunes. Inverse grading is common because wind ripples commonly have coarser grains at their crest, or down the leeside of the ripple. Thus, coarser material high in the ripple buries finer material, producing inverse grading. If the rate of sand supply is substantially less than the rate of ripple drift, traces of ripple drift are recorded as individual laminae. Not all wind ripples are

coarse crested, and thus normally graded beds can result from deposition of wind ripples having coarser grains accumulated in ripple troughs, although this is a rare occurrence.

Whether well laminated or inversely graded, moderately dipping laminae (either texturally homogeneous or heterogeneous) may be indicative of ripple sedimentation as deduced from wind tunnel studies. Fryberger and Schenk (1981) conclude on the basis of wind tunnel studies, that climbing ripples may produce a type of strata common in topset and low angle eolian deposits. An adequate supply of sand, and wind conditions suitable for saltation transport appear to be the controlling factors in eolian climbing ripple development.

Grainfall Processes—The deceleration of wind in the lee of topographical obstacles or vegetation causes sand to fall from saltation or suspension and accumulate. As sand accumulation continues, an unstable slope eventually develops and avalanche processes begin. Thus, avalanche and grainfall processes and resultant deposits are closely related. Deposits produced by grainfall as it has been called by Hunter (1977), are the most difficult type of dune deposits to distinguish. In general, deposits produced by grainfall processes are even, parallel laminated or graded beds that thin rapidly down

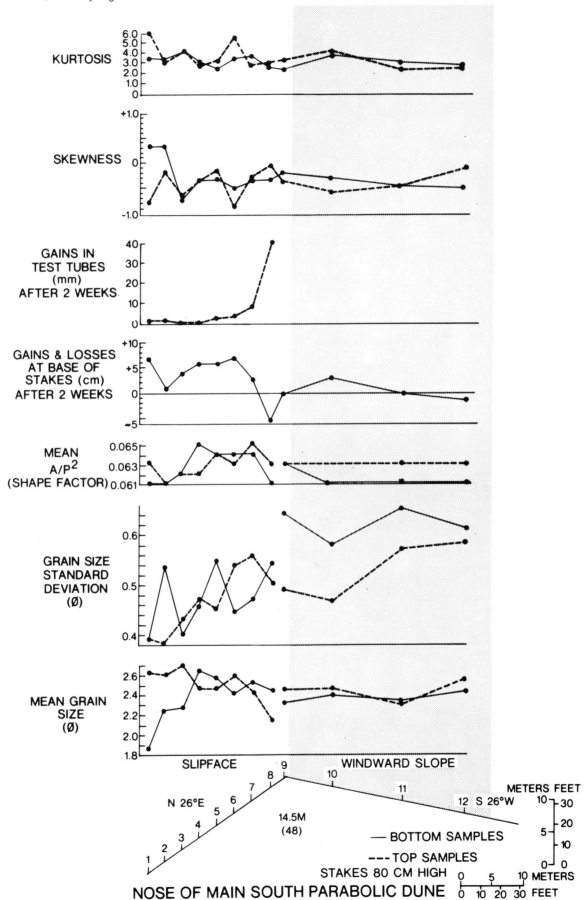

Fig. 15—Textural studies related to sediment populations, parabolic dune, North Park, Colorado. Deposits related to the grainfall processes, caught in test tubes 80 cm above the sand surface, become finer and better sorted down the lee slope; maximum accumulation is just beneath brink (gains in test tubes graph). Deposits related to avalanche processes coarsen downslope and textures reflect sediments of different lobes; they redistribute the grainfall sediment as shown by gains and losses at base of stakes graph. All data (300 grains per sample) are from Automatic Image Analysis (A.I.A.), shape factor values can be interpreted by referring to Figure 16.

the lee slope of a dune. The writers have studied grainfall deposition on dunes in North Park, Colorado, and Great Sand Dunes, Colorado and in the U.S. Geological Survey Wind Sedimentation Laboratory.

In the North Park dunes, we placed stakes at 5-ft (1.3-m) vertical increments on both the windward and leeward slope of a parabolic dune slipface (lee slope was 47 ft or 15 m; Fig. 15). Test tubes were places 80 cm above the ground to catch grainfall detritus. Textural analyses, using an Automatic Image Analyzer (A.I.A.) were made of sediment caught in test tubes and then compared to textural analyses of sand collected at the base of each stake (assuming the latter to be a mixture of sand creep, saltation, and suspension deposits). The A.I.A. technique, a numerical frequency method, was chosen due to the small amount of sediment caught in tubes on the lower slipface (Fig. 15). Several important points are illustrated in Figure 15: (1) grainfall deposition lessens rapidly beyond the brink—shown by grains in test tube plot graph; (2) deposits produced by grainfall are graded parallel with the depositional surface—coarser near the brink and finer downslope—this is opposite to the avalanched (bottom) samples that coarsen downslope as shown on the mean grain size plot (larger values reflect smaller grain sizes); (3) deposits resulting from grainfall become increasingly negatively skewed down the lee slope, compared to the bottom (avalanched) deposits; (4) grainfall deposits are increasingly better sorted downslope; bottom (avalanched) deposits do not show a consistent sorting pattern.

Grain shape, discussed in a following section, was also determined by 300 grains in each sample as plotted in Figure 15. A comparison between the value we measured (Shape Factor) versus more conventional visual sphericity and roundness values is shown in Figure 16. Sands studied (Fig. 15) are in the moderately well rounded category.

The type of grading or lamination within grainfall deposits is a function of wind velocity. For example, a rapid increase of wind velocity moving large sand grains followed by a longer decrease in velocity produces a laminated, normally graded deposit (Figs. 10C, D). If conditions are reversed, inversely graded strata may occur, or if velocity remains fairly constant, evenly laminated strata may result.

Grainfall deposits are commonly reworked by avalanching due to the in-

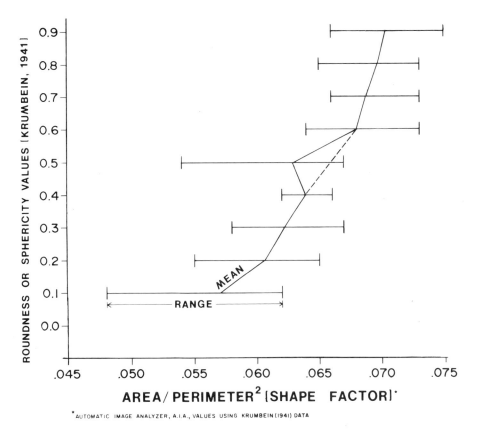

Fig. 16—Shape factor values and corresponding roundness or sphericity values, refer to Figure 15 shape factor graph.

herent instability of large amounts of sediment deposited in upslope portions of slipfaces which reach an equilibrium profile and angle of repose by avalanching. Deposits produced by the grainfall process are most easily confused with deposits produced by saltation. The result is a lack of ripple foresets or ripple forms; a presence of even, parallel, laterally continuous laminae; and a rapid downslope thinning of such deposits (seen parallel with sand transport), which are good indicators of the grainfall process.

Sedimentary Structures of Interdune Deposits

Dry interdune, wet interdune, and evaporite interdune deposits, and combinations of these are illustrated.

Dry Interdune Deposits—In interdunes that are being deflated or where little deposition is occuring, a bimodal, poorly sorted coarse-grained sand is common. If more sediment is deposited in the interdune, discontinuous, often disrupted, relatively flat-bedded laminae result. Oxidized (red) and calcified root tubules are common in dry interdunes, having the general appearance of soda straws, with hollow centers and variously cemented sandy tubes that take the form of the plant root. Adhesion ripples occur in dry interdunes that occasionally become dampened. Structures produced by adhesion ripples (discontinuous, irregular) are dissimiliar to wind ripple deposits. The adhesion ripples form by adhesion of sand to a moist surface and build into the wind, whereas wind ripples form in response to saltation and migrate downwind (see Glennie, 1970, for adhesion ripple structures). Examples of dry interdune deposits are shown in Figure 17C.

Wet Interdune Deposits—In wet interdune deposits, certain processes occur that are much less common or are absent in dry interdunes: (1) silts and clays are trapped by standing bodies of water thus preserving sediment that in a dry interdune would be removed by deflation; (2) organic activity in both plants and animals is greatly stimulated, resulting in bioturbation and organic debris in the sediment; (3) deposits are easily contorted by loading because of water saturation, which renders them much less competent; and (4) other secondary structures, such as dissipation structures, result after burial due to the finer grained component and incompetency of the sediment. Examples of wet interdune deposits are shown in Figure 17A, B, D.

Fig. 17A—Interdune sedimentary features, mud-lined interdune corridors due to periodic flooding from alluvial fans, Algodones Dunes, California.

Fig. 17B—Interdune sedimentary features, draping of mud over erosional surfaces, Algodones Dunes.

Fig. 17C—Interdune sedimentary features, dry interdune deposits with root molds (oxidized dikaka) in vertical growth position overlain by mollusk-bearing evaporite interdune deposits, Killpecker Dunes, Wyoming.

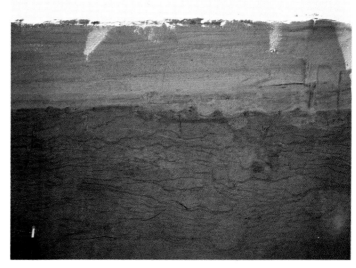

Fig. 17D—Interdune sedimentary features, wet interdune deposit with dissipation structures, overlain by dune sands, Nebraska Sand Hills (Loc. 78, Fig. 11)—note contorted bedding due to loading at top of interdune.

Fig. 17E—Interdune sedimentary features, sabkha deposits, Sabkah Ar Riyas about 10 km from the Arabian Sea in eastern Saudi Arabia—gypsum crystals underlie bounds (G), white anhydrite layer (A) near bottom of trench; and discontinuous, contorted siltstones, and marlstones and sandy siltstones near the top of trench.

Fig. 17F—Interdune sedimentary features, lenticular (about 100 m long) carbonate (interdune) in Weber Sandstone (Pennsylvanian-Permian) Sand Canyon, Colorado—nodular, white calcite within dolomite matrix.

Fig. 18A—Sand sheet (low angle eolian) sedimentary features, inversely graded strata with occasional ripple foreset laminae, Navajo Sandstone (Triassic-Jurassic), Dinosaur National Monument, Colorado.

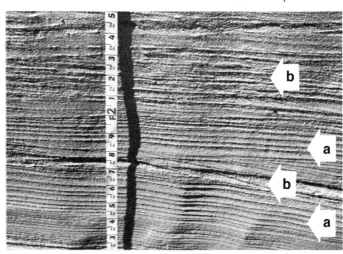

Fig. 18B—Sand sheet sedimentary features, inversely graded strata with occasional ripple foreset laminae, dome dune, Killpecker Dunes, Wyoming.

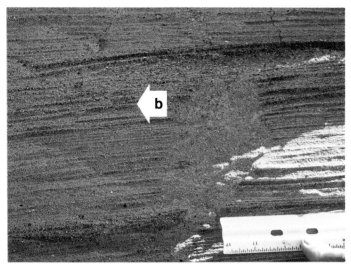

Fig. 18C—Sand sheet sedimentary features, note textural differences, coarser ripples, discontinuous laminae, and bioturbation traces in Pleistocene sand sheet deposits, Llano de Albuquerque, Albuquerque, New Mexico. Type "a" is predominant in 18A, B, and E. Type "b" deposits are predominant in 18C and D.

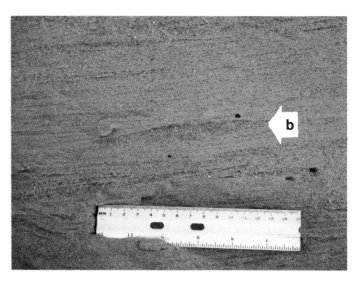

Fig. 18D—Sand sheet sedimentary features, note textural differences, coarser ripples, discontinuous laminae, and bioturbation traces in Pleistocene sand sheet deposits Llano de Albuquerque, Albuquerque, New Mexico.

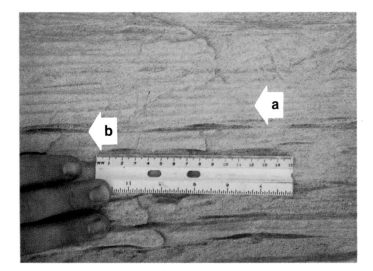

Fig. 18E—Sand sheet sedimentary features, high index, texturally bimodal (contains coarser grains than surrounding strata) ripples (type "b" deposits) preserved within inversely graded, ripple produced strata (type "a" deposits), Browns Park Formation (Miocene) near Maybell, Colorado.

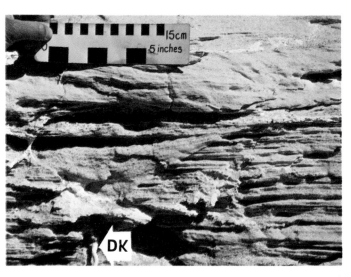

Fig. 18F—Sand sheet sedimentary features, scour and fill, possibly related to former plant, and calcified root molds (dikaka, DK) in Browns Park Formation.

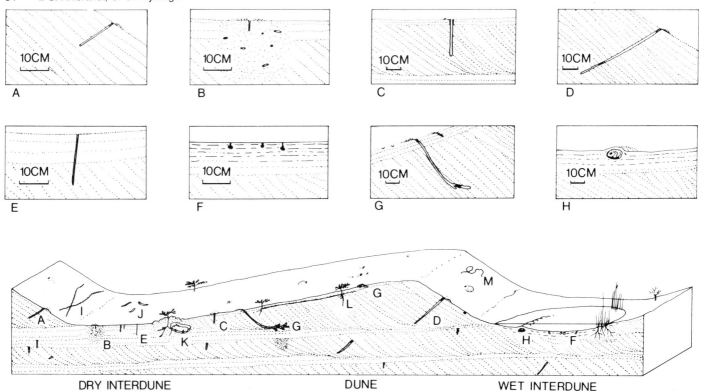

DRY INTERDUNE   DUNE   WET INTERDUNE

Fig. 19—Diagram of commonly observed bioturbation traces in eolian deposits (block diagram is not to scale): (A) beetle burrow; (B) burrowing and disruption of sediment by ants; (C) wolf spider burrow with web collar and reinforced burrow walls; (D) sand-treader camel cricket burrow, in slipface deposit, with backfilled entrance; (E) tiger beetle larva burrow; (F) aestiviating gastropods; (G) from left to right, two trial burrows, a resting burrow, and a sleeping burrow of a sand wasp; (H) toad burrow; (I) crane fly larva burrow; (J) root molds (dikaka); (K) gopher burrow with disruption of sediment by plant roots; (L) plant root traces; (M) a second type of crane fly larva burrow, (from Ahlbrandt, Andrews, and Gwynne, 1978).

Evaporite Interdune Deposits—The complex intercalation of sand, silt, clay, soluble salts, sulfates, and carbonates in modern evaporite interdune areas (Fig. 17) is analogous to complex evaporite interdune sequences observed in eolianites. Brecciated contorted dolomites and lime mudstones that likely have such an origin are illustrated in Figure 17F. Organic material found in some evaporite environments such as inland and coastal sabkhas may be a potential hydrocarbon source (Figs. 12, 20F).

Sedimentary Structures of Sand Sheet Deposits

Eolian deposits marginal to dune fields (mostly low angle and products of granule ripple, wind ripple, or grainfall deposition) are an important type of eolian deposit (Fig. 1). The eleven criteria by which low angle eolian deposits can be recognized were summarized in a previous section. Generally, two broad types of sand sheet or low angle eolian deposits can be recognized—Type a and Type b. Type (a) deposits are texturally homogeneous and relatively evenly laminated where compared to Type (b) deposits. Type (a) deposits are produced by wind ripples (Fig. 18A, B, E) or grainfall processes.

Type (b) deposits are texturally heterogenous and relative to Type (a) deposits, are more discontinuously laminated and largely the product of granule ripple sedimentation (Fig. 18C, D). Bioturbation, and scour and fill structures are common in both Type (a) and (b) deposits (Ahlbrandt, Andrews, and Gwynne, 1978; Fryberger, Ahlbrandt, and Andrews, 1979; Figs. 18F, 19, 20).

EOLIAN TEXTURES

The usefulness of textural studies in distinguishing eolian environments is debatable. Textural data seem most useful as complementary evidence for eolian interpretation. Many textural analysis techniques and interpretations of textural data are presented in this paper. The writers have studied textural data of about 1,500 eolian sand samples using different techniques including sieving on $1/4$ ϕ intervals (Ahlbrandt and Fryberger, 1976; Ahlbrandt, 1979), sieving on $1/2$ ϕ intervals (Ahlbrandt and Andrews, 1978), settling tube (Ahlbrandt, 1974, 1975), Automatic Image Analyzer, a numerical frequency device (Fig. 14), and studies for distinguishing traction, saltation, and suspension populations, graded bedding, etc. Some conclusions

reached include:

(1) Significant textural variations exist in eolian deposits. Three textural groups were recognized by Ahlbrandt (1979) and were differentiable statistically using discriminant analysis by Moiola and Spencer (1979) in 506 eolian sand samples ($1/4$ ϕ sieve data) from around the world. Moderately sorted to well sorted, fine- to medium-grained inland dune sand contrasts with well sorted fine-grained coastal dune sand which differs from moderately to very poorly sorted interdune or serir sands.

(2) There are rapid textural changes and permeability contrasts on a bed to bed basis within cross-bed sets (note adjacent dune sand samples and dune versus interdune samples, Fig. 21). Many eolian texture studies use grab samples which include several beds and thus greatly smooth textural variations.

(3) The fine-grained fraction of eolian deposits is seldom studied. Many eolian sand grains have clay coatings (Fig. 21A, D) which are not commonly considered in standard grain size analyses. If a modern eolian sand sample is ultrasonically treated prior to analysis, the clay coatings can be separated and its contribution to cumula-

Fig. 20A—Bioturbation in eolian and related deposits, burrows along bedding surfaces, Weber Sandstone (Pennsylvanian-Permian), Hog Canyon, Utah. Burrows are along former slipface bedding surface.

Fig. 20B—Bioturbation in eolian and related deposits, burrows along bedding surfaces, Weber Sandstone (Pennsylvanian-Permian), Hog Canyon, Utah. Burrows are in interdune deposits.

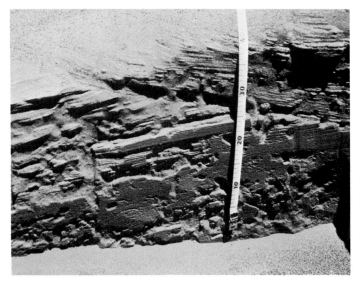

Fig. 20C—Bioturbation in eolian and related deposits, sand treader camel cricket burrows, Great Sand Dunes, Colorado.

Fig. 20D—Bioturbation in eolian and related deposits, crane fly burrows, tracks, and high index ripples on frozen sand surface, North Park, Colorado.

Fig. 20E—Bioturbation in eolian and related deposits, burrows and plant traces in low angle eolian deposits, Great Sand Dunes, Colorado.

Fig. 20F—Bioturbation in eolian and related deposits, burrows (slightly oxidized) in sabkha deposits; note organic mud and anhydrite (A), Sabkhah Ar Riyas, Saudi Arabia.

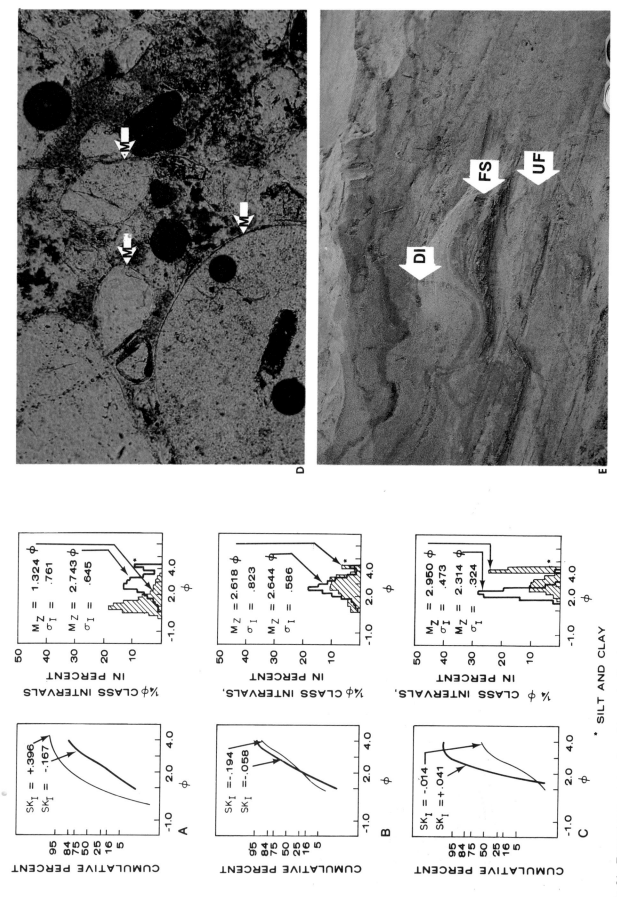

Fig. 21—Textural contrasts in adjacent types of eolian deposits in Nebraska Sand Hills. **A.** leeside deposits, light colored layer (thin line, patterned in histogram) adjacent to dark colored layer (thick line) that contains dissipation structures 5.1 cm apart, light colored sample has 2.6% silt and clay, darker layer has 16.1% silt and clay (Loc. 75, Fig. 11). **B.** leeside grainfall deposits (thin line, patterned in histogram) compared with slipface (avalanche) deposits (Loc. 40, Fig. 11). **C.** poorly sorted, sandy mud from interdune deposit (thin line, patterned to histogram) compared to slipface sand (thick line; Loc. 25, Fig. 11). Cumulative frequency curves are recalculated on the sand fraction if gravel or pan fraction is greater than 5%. **D.** photomicrograph of coarser leeside sand shown in **A;** note yellow montmorillonite (M) clay coatings (2.6%) on grains; the sand is a subfeldsarenite, $Q_{71}F_{15}L_9$, and is bimodal (coarse mode .45–1.2 mm, fine mode .1–.2 mm). Crossed nichols, blue epoxy dye. 2 mm field of view. **E.** dissipation structures (DI) forming above frozen sand layer (FS) in Nebraska Sand Hills (Loc. 54, Fig. 11) in Dec. 1975. Original dune stratification dips to the right but is obscured during formation of dissipation structures. Dissipation structure layer (DI) contains 3.1% silt and clay whereas frozen sand layer (FS) contains 0.4% sand and clay, and unfrozen sand layer (UF) contains 0.75% silt and clay (from Ahlbrandt and Fryberger, 1980).

MEASURED STRATIGRAPHIC SECTION WEBER FM.
DEERLODGE PARK, DINOSAUR NAT. MONUMENT
SEC. 20 T.6N. R.99W. COLO.

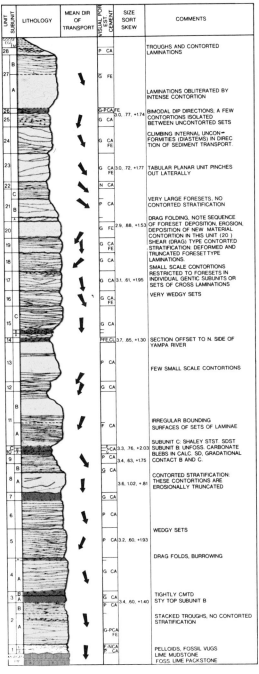

DESCRIPTION OF CORE FROM CHEVRON WELL
LARSON B-15X IN WEBER FORMATION, RANGELY OIL FIELD,
RANGELY, COLO.CW 1/2 35T. 2N. R. 102W.

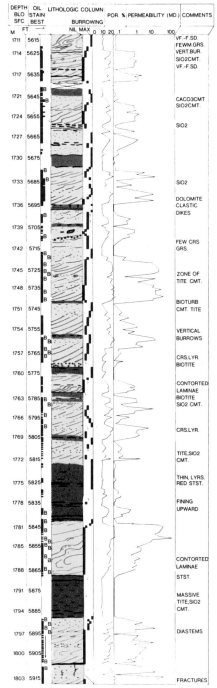

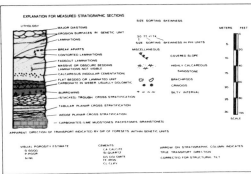

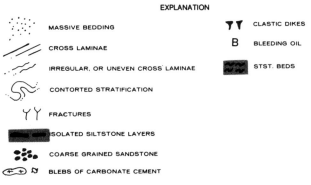

Fig. 22—Outcrop section (Deerlodge Park) and subsurface core description of Weber Sandstone (Pennsylvanian-Permian), Colorado. Flat-bedded siltstone intervals in the Larson B-15X core reflect alluvial sediments of Maroon Formation. Note rapid porosity and permeability variations in cross-bedded eolian units (from Fryberger, 1979b).

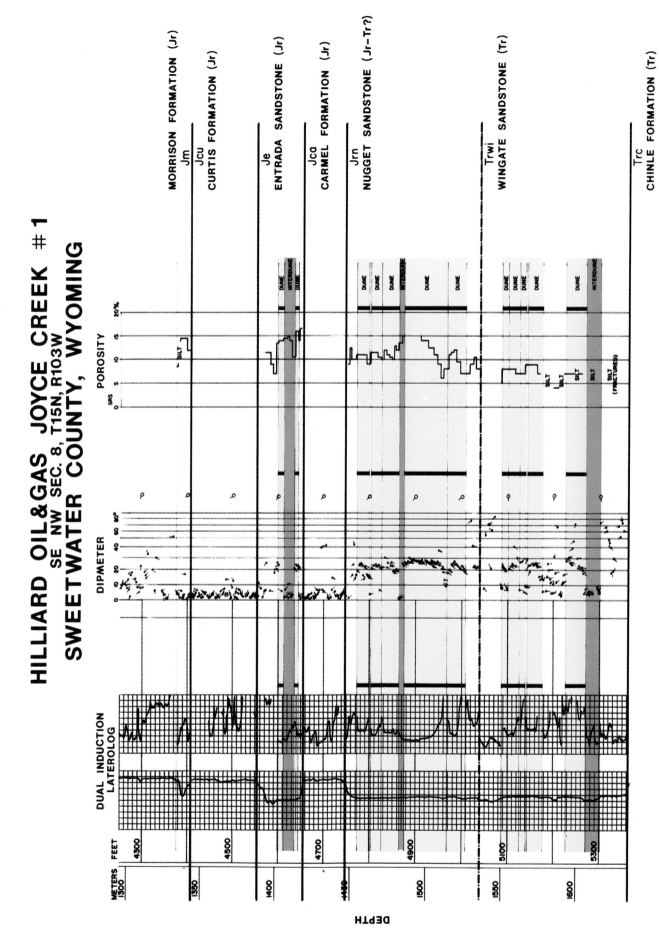

Fig. 23—Recognition of sedimentary structures and depositional environments from electric logs (Hilliard Oil and Gas, Joyce Creek #1, Sweetwater County, Wyoming). Both the dipmeter and resistivity curve of the Dual Induction Laterolog may be used to distinguish between dune, interdune, and extradune sediments. Cross-bed dips in dune intervals commonly steepen upward, and such data can be used to interpret dune types as discussed in text (modified from Lupe and Ahlbrandt, 1979).

tive frequency curves realized (Fig. 21).

(4) Simple textural analysis techniques are not adequate to distinguish eolian environments (Friedman, 1973; Ahlbrandt, 1979). The most successful eolian textural discriminator appears to be the discriminant analysis technique (Moiola and Spencer, 1968; Moiola and Spencer, 1979). The settling velocity of light and heavy minerals in water has been demonstrated to distinguish eolian from subaqueous suspension deposits (Steidtmann, 1974).

(5) The saltation population is by far the most dominant of traction, saltation, and suspension components in eolian sand deposits. The saltation population shifts to coarser grains as wind velocity increases and thus this population is not easily defined by a specific grain size range.

Positive skewness is not a reliable eolian indicator. For example, using $1/2\,\phi$ sieve data, not a single sand sample (47 samples) was positively skewed in dunes in North Park, Colorado (Ahlbrandt and Andrews, 1978). Of 113 sand samples in the Killpecker dune field analyzed using a settling tube, 68 samples were negatively skewed, 45 were positively skewed (Ahlbrandt, 1974). In the Nebraska Sand Hills, both positively and negatively skewed dune sands occur (Fig. 21A, B).

Permeability contrasts occur on a bed by bed basis subject to textural changes between them. For example, the two samples shown in Figure 18 were spaced 5.1 cm apart in slipface deposits. The fine sand bed had 2.7 darcies permeability while the medium sand bed had 6.6 darcies permeability. These textural differences cause directional fluid flow along beds rather than across them, a factor affecting cementation history following diagenesis and repeated fluid migration through the cross-bed set.

Clay coatings on sand grains (Fig. 21D) are probably responsible for the reddening of dune sand (Walker, 1967, 1979; Folk, 1976) and probably provide a source of clay for dissipation structures that obscure eolian stratification (Bigarella, 1975; Ahlbrandt and Andrews, 1978). Dissipation structures appear as discontinuous, dark bands (Fig. 17D) having a higher content of fine-grained material than surrounding sand layers (Fig. 21A, E).

Discrimination of eolian deposits using superficial textures of sand grains as viewed with the Scanning Electron Microscope has been studied. Upturned plates on sand grains are believed to represent eolian processes, and are of use when diagenesis has not obscured or altered such features (Krinsley, Friend and Klimentidis, 1978).

ECONOMIC CONSIDERATIONS OF EOLIANITES

Eolian hydrocarbon reservoirs are discussed here; however, concentrations of copper, zinc, uranium, thorium, and potash have been associated with eolianites and also merit serious consideration.

Eolianites have proven to be complex, heterogeneous hydrocarbon reservoirs. Complex porosity and permeability variations are well documented in the Weber (Tensleep) Sandstone (Pennsylvanian-Permian) in the following Wyoming and Colorado oil fields; Little Buffalo Basin (Emmett, Beaver and McCaleb, 1972), Rangely (Bissell, 1964; Larson, 1975; Fryberger, 1979b), Brady (Brock and Nicolaysen, 1975; Lupe and Ahlbrandt, 1979), Lost Soldier and Wertz (Lupe and Ahlbrandt, 1979), and Oregon Basin (Morgan, Cordiner, and Livingston, 1978). Cementation problems noted in other hydrocarbon-bearing eolianites include the Lyons Sandstone in Colorado (Levandowski et al, 1973) and the Rotliegendes (Permian) in the North Sea (Glennie, Mudd, and Nagtegaal, 1978).

Problems common in these eolian reservoirs include: (1) lateral discontinuity of reservoir zones; (2) impermeable or less permeable carbonate or flat-bedded units interspersed with more permeable cross-bedded units; (3) anisotropic permeabilities and related textural changes and cementation along individual laminae causing low transmissivity across laminae— anhydrite cement is the most common, but calcite, dolomite, and silica cement are problems in well log interpretations; (4) secondary and tertiary recovery problems related to isolated reservoirs causing reduced well spacing, commonly to 10 or 20 acres (Emmett, Beaver and McCaleb, 1972; Larson, 1975; Lupe and Ahlbrandt, 1979; Smolen and Litsey, 1977).

Sedimentary features, and strike and dip patterns within eolianites are useful in both outcrop and subsurface core and well log studies. Outcrop studies of the Weber Sandstone in Dinosaur National Monument, Colorado, can be compared to core information of the Weber within the Rangely oil field as shown in Figure 22. Intercalation of less permeable extradune sediments (alluvial deposits of the Maroon For-mation), with more permeable dune sediments forms a continental stratigraphic trap across a structural nose at Rangely field (Fryberger, 1979b). The Larson B-15X well penetrates both the alluvial and eolian sediments near the updip edge of this trap (Fig. 22).

Well logs, when calibrated with core information, can be used to recognize and interpret depositional features. For example, the resistivity curve of the electric log (here a dual induction laterolog), dipmeter data, and porosity can be correlated and utilized to distinguished eolian environments as noted in Mesozoic eolianites in Wyoming (Fig. 23). Note the decreased porosity and permeability in low angle eolian and/or interdune sequences in Figures 22 and 23.

The dipmeter is an extremely useful device for recognizing and interpreting eolianites in the subsurface. Most eolian cross-bed sets steepen upward and the combining of the spread of dips and their mean values is useful in interpreting dune bedforms and thus sandstone geometry. The dipmeter plot through the Nugget Sandstone interval in Figure 23 suggests a relatively straight crested dune (perhaps transverse ridge) migrating to the southeast. Elongation of the permeable cross-bedded reservoir sandstone perpendicular to the transport direction would, therefore, be expected. If two clusters of dips were noted (as seen in Figure 4) suggesting a linear dune, then elongation of the sand body would be expected along the axis between the clusters (see illustration by Glennie, 1970, p. 155). Other dune forms can and have been recognized in eolianites as discussed previously.

In addition to subsurface devices previously discussed, the Repeat Formation Tester (R.F.T.) has proved valuable in secondary recovery studies in the Weber Sandstone at Rangely field (Smolen and Litsey, 1977). Original formation pressures were documented with the R.F.T., and bypassed by the waterflood, indicating isolation and discontinuity of producing zones.

The Wertz 46 ABC and E well (second well from right, Fig. 24A) penetrates a dune-extradune sequence of nearshore marine-eolian deposits of the Tensleep Sandstone (Weber Sandstone equivalent). Diagnostic sedimentary features are discernible within eolian intervals as noted in Figures 24A and 25. Eolian deposits shown in Figures 24 and 25 occur in thinning upward transgressive and regressive depositional cycles. Eolian cross-bed sets

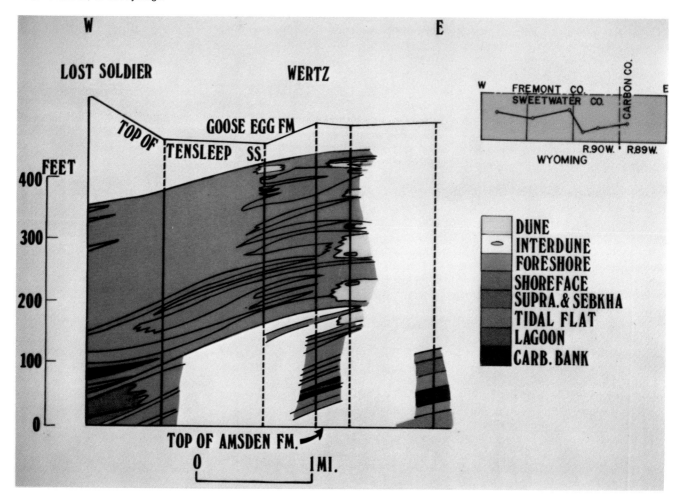

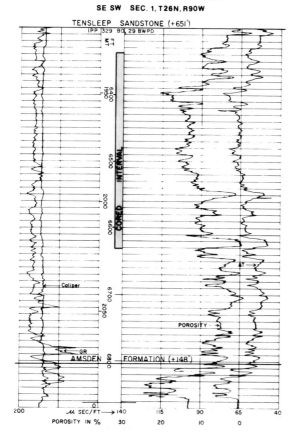

Fig. 24—A. Cross-section of depositional environments interpreted from cores of the Tensleep Sandstone (Pennsylvanian-Permian, also a Weber Sandstone equivalent) in the Lost Soldier and Wertz oil fields, Sweetwater County, Wyoming. The top of the Amsden Formation is used as a horizontal datum. Note the second well from the right is Wertz 46 ABC and E; **B.** Sonic log and cored interval of Tensleep Sandstone in Wertz 46 ABC and E.

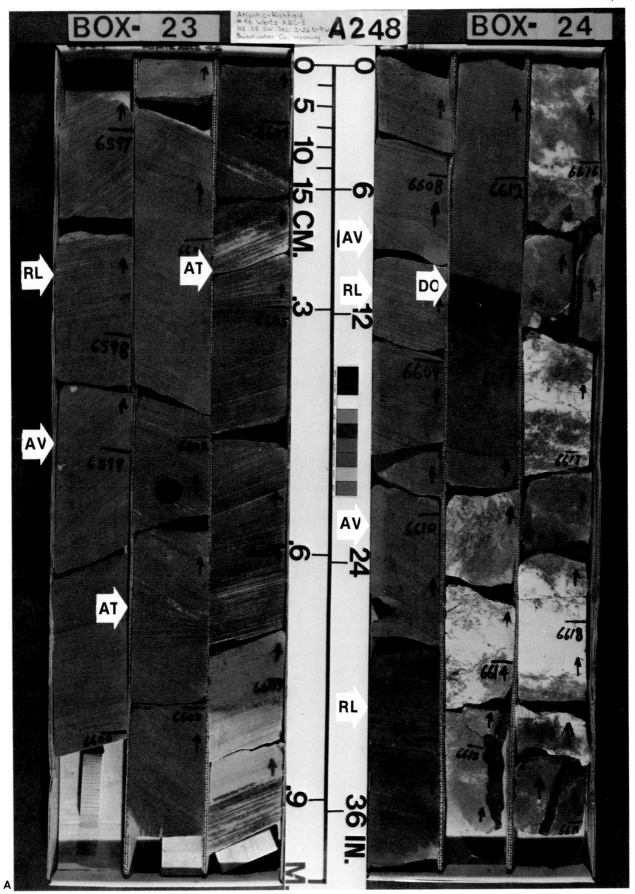

Fig. 25—Figures 25 A-D are slabbed cores in Tensleep Sandstone, Wertz 46 ABC and E well, NE SE SW, Sec. 1, T26 N, R9OW, Sweetwater County, Wyoming. Depositional environments are shown in Figure 24A. A sequence of core photographs starting near the base of the cored interval and selected examples uphole are given; all depths will refer to footages marked on cores. Detailed core description of this and other wells in the Wertz oil field are given in Reynolds, Ahlbrandt, Fox and Lambert (1975). **A.** Note: depth intervals are not converted to metric values because they are used as reference points. Dead oil stain (DO) (6612.5 to 6,613.4 ft and 6,610 to 6,611.5 ft) and live oil stain in fine grained dune sandstone. Distinct

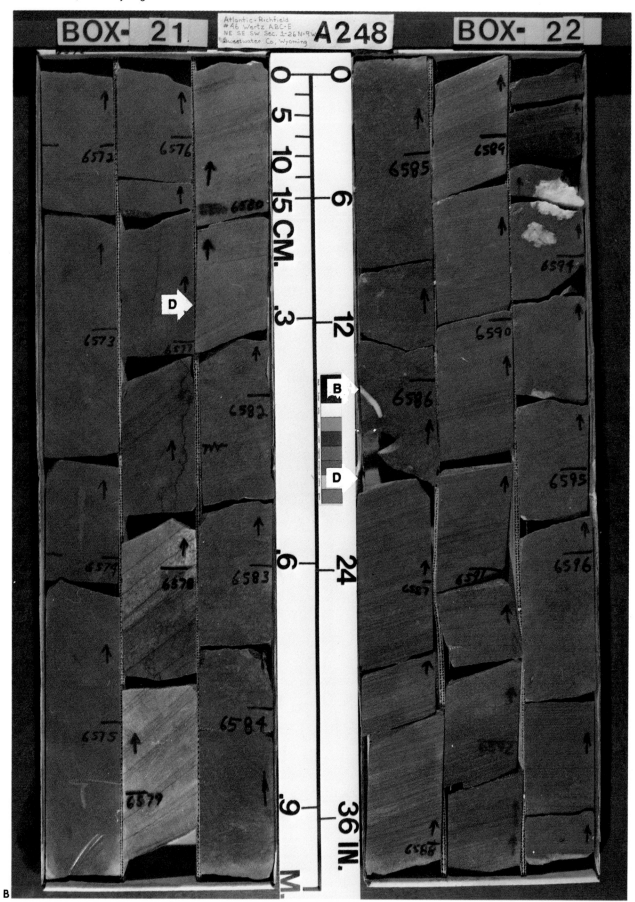

avalanche (sand flow) toes (AT) are seen at 6,602.7 and 6,604.8 ft. Bounding surfaces between cross-bed sets are predominant at 6,608.2; 6,605; 6,602.8; and 6,598.1 ft. Ripple produced laminae (RL) are well shown at

6,610.5 to 6,611.5 ft and 6,605 to 6,608 ft; contrast with avalanche deposited strata (AV) at 6,608.2 to 6,608.4 ft; 6,609.4 to 6,610.5 ft; and 6,598 to 6,599.6 ft. Note alternate anhydrite cemented laminae and dead oil stained laminae

from 6,604.5 to 6,607.5 ft. Dolomite and some calcite cement occur in live oil stained (light brown) intervals. **B.** Top of dune cross-bed (D) is 6,586.4 ft. Overlying marine shoreface deposits are silty, very fine grained sandstone and

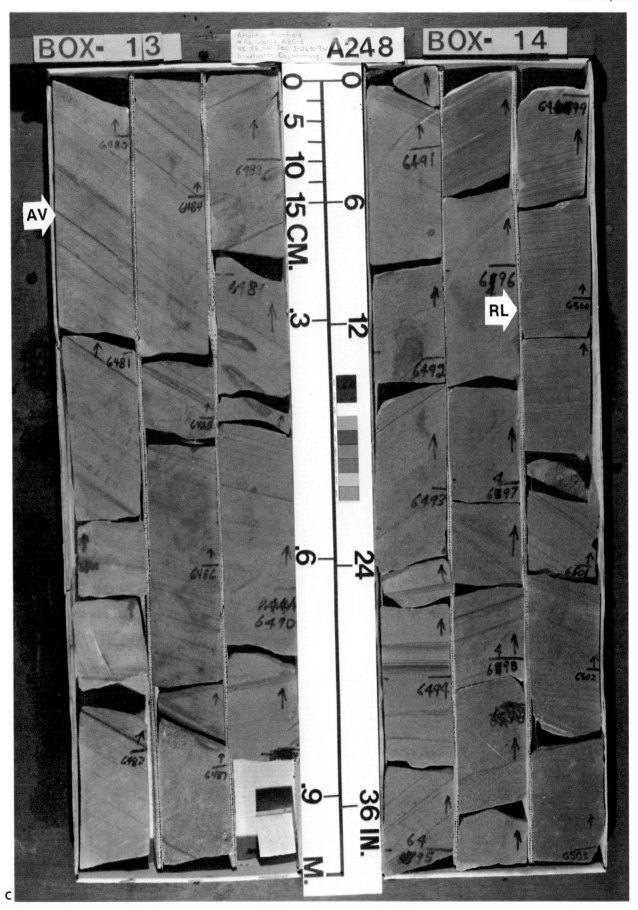

C

intensely bioturbated and distorted (6,580.7 to 6,586.4 ft). Note anhydrite filled burrow at 6,586 ft (B). Minor burrows in dune interval noted at 6,596.8 ft. Another steepening upward dune cross-bed set (D) beings at 6,580.2 ft. Gypsum nodules at 6,593.5 ft. **C.** A part of a 42-ft (12.8 m) thick dune cross-bed set, steepening upward from 6,500.6 ft. Note laminated strata probably produced by ripple processes (some ripple forms are preserved; RL) at 6,499 to 6,500.6 ft. Contrast with laminae and bedding style of avalanche (sand flow) deposited strata (AV) at 6,480 to 6,481.5 and 6,483 to 6,490 ft. The slabbed sections shown in some views are not always at the maximum dip angle; e.g. interval

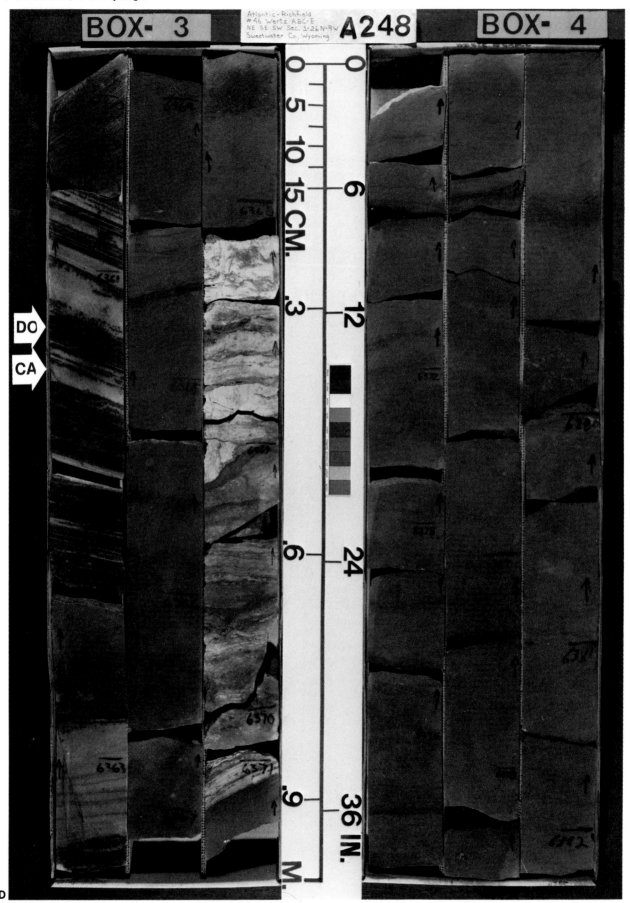

from 6,481 to 6,490.5 ft has true dips of 28°, but these appear shallower on slabbed surface. **D.** Uppermost dune cross-bed set steepens upward from 6,362.8 to 6,360 ft. Supratidal (and/or evaporite interdune) deposits with algal features, brecciated and discontinuous, and contorted bedding shown at 6,360 to 6,371 ft. The dune interval 6,362.8 to 6,360 ft has alternate dead oil (DO) and anhydrite cemented (CA) laminae. Initial textural differences and several cementation and hydrocarbon-bearing fluids have passed through the various dune sandstones, creating a very complex reservoir as evidenced by differing cementation, live and dead oil staining.

steepen upward. Alternating laminae cemented by anhydrite or stained with dead oil are common (Fig. 25), as is calcite and dolomite cement. Each eolian cross-bed set has a slightly different diagenetic history—some have several types of cement, live and dead oil, and others are completely cemented. Subtle textural variations in modern dune sands have important effects in eolianites as their different permabilities are accentuated by subsequent fluid migration and cementation (Fig. 25A, B, D). The truncation and burial of eolian sandstone (fine sand size) by a contorted, burrowed and finer-grained (coarse silt to very fine sand size) nearshore marine deposit (note prominent anhydrite-filled burrow just above dune sandstone) is demonstrated in Figure 25C.

The combined efforts of production engineers and production geologists in exploiting heterogeneous eolian reservoirs is well summarized in the case of Tensleep Sandstone Reservoir in Little Buffalo Basin field, Wyoming.

McCaleb (1979) suggested the Tensleep Sandstone there represents deposits of "braided streams cutting across a broad sand dune covered beach, very near and periodically encroached by marine environments, with wind and waves continually resorting the grains into channels, dunes, beaches and bars." He also described impermeable dolomitic and anhydritic sandstone layers and anhydritic lenses. (The latter may reflect interdune deposits, in the writers' opinion). Poor oil recovery and rapidly increasing water production prompted a detailed reservoir analysis of the field by Amoco (Emmet, Beaver, and McCaleb, 1972).

One well in particular, the Little Buffalo Basin No. 61, (Sec 1, T47N, R100W) was carefully evaluated. Ten percent of the pay in this well occurs in homogeneous, high permeability zones (average 200 millidarcys) in the upper part of Tensleep. The remaining 90% of the pay in the No. 61, occurs in cross-bedded, medium permeability sandstones (average 50 millidarcys). Amoco production personnel had also recognized directional permeabilities related to cross-bed sets within the reservoir. They postulated that higher permeability zones allowed rapid water breakthrough bypassing less permeable, but oil-bearing sections. This concept was tested by selectively perforating the medium permeability, cross-bedded zones to avoid high volume water production in the more permeable zones. McCaleb (1979) stated

that the No. 61 well was the first infill well in the field, drilled in early 1968, and was drilled among four watered out wells. It was initially completed for 1,050 b/d (oil) and 150 barrels of water, and has since produced 900,000 barrels of oil. As of May 1, 1976, daily production remained at 417 b/d (oil) and 1,892 b/d (water—82%; McCaleb, 1979).

The success of the No. 61 well caused well spacing to be reduced from 40 to 20 to 10 acres (McCaleb, 1979) and now to 5 acres (McCaleb, oral commun., 1979) in some cases without interwell interference. Field production increased from 3,000 b/d with no infill wells, to a rate of over 9,000 b/d with infill drilling and selective perforation.

The economic significance of this is best summarized by McCaleb (1979, p. 26):

> "Accelerated production paid out the total program in the first year. Current production is still holding up at 9,000 b/d. Lateral application of these ideas have given us (Amoco) production from the small satellite field of the Northwest dome. It is now estimated the recoverable reserves were increased by about 20 million barrels of oil, with an increase in the recovery factor of 10%."

Although low gravity of the oil at Little Buffalo Basin (20°API) certainly accentuates the heterogeneity of eolian reservoirs, similar heterogeneity has been reported elsewhere. The detailed analysis of eolian reservoirs by both engineers and geologists has proved to be beneficial in developing eolian reservoirs.

SUMMARY

Eolian depositional systems in modern deserts are complex and highly variable, and this complexity of modern systems is reflected in ancient rock of eolian origin. Stratigraphic complexity in eolian rocks has two basic origins. First variability is introduced in the differing spatial relationships of dunes, interdunes, and sand sheet deposits to each other as well as to extradune (noneolian) sediments. Secondly, different dune types such as barchan, linear, and star have different cross-bedding patterns, and different degrees of mobility. Differences in cross-bedding result in different fluid flow properties when the dunes are lithified. Interdune deposits, which are com-

monly impermeable in ancient rocks further complicate fluid movement in eolian reservoirs. Excessive reliance on the concept of eolian rocks as thickly cross-bedded and homogeneous deposits has hampered recognition of more lithologically complex eolian rocks that are commonly interclated with marine or noneolian continental deposits. This in turn has hampered petroleum exploration and production in rocks of eolian origin. Particular neglect has occurred in exploitation of stratigraphic traps in eolian rocks.

Recognition of eolian rocks must rely on thorough stratigraphic reconstruction and detailed study of outcrop and core, as well as borehole log data. As new information becomes available on modern eolian depositional systems, reconstruction of ancient eolian depositional systems can be quantitatively evaluated. In the future, recognition of ancient eolian deposits may increasingly rely on features which are bulk properties of the rock, such as genetic units, cross-bedding styles, and primary stratification.

REFERENCES CITED

Ahlbrandt, T. S., 1974a, Dune stratigraphy, archaeology and the chronology of the Killpecker dune field: Wyoming Geol. Survey, Rept. of Inv., N. 10, p. 51-60.

——— 1974b, The source of sand for the Killpecker sand dune field, southwestern Wyoming: Sed. Geology, v. 11, p. 39-57.

——— 1975, Comparison of textures and structures to distinguish eolian environments, Killpecker dune field, Wyoming: Mountain Geologist, v. 12, p. 61-73.

——— 1979, Textural parameters in eolian deposits, in E. D. McKee, ed., A study of global sand seas: U. S. Geol. Survey Prof. Paper 1052, p. 21-52.

——— and S. Andrews, 1978, Distinctive sedimentary features of cold-climate eolian deposits, North Park, Colorado: Palaeogeography, Palaeoclimatology, Palaeoecology, v. 25, p. 327-351.

——— ——— and D. T. Gwynne, 1978, Bioturbation in eolian deposits: Jour. Sed. Petrology, v. 48, p. 839-848.

——— and S. G. Fryberger, 1976, Structures and textures of eolian deposits in the Nebraska Sand Hills, U.S.A.: 25th International Geol. Congress, Sydney, Abstracts, v. 3, p. 829.

——— ——— 1980 Eolian deposits in the Nebraska Sand Hills: U.S. Geol. Survey Prof. Paper 1120A, p. 1-24.

——— ——— in press, Sedimentary features and significance of interdune deposits, in F. G. Ethridge and R.M. Flores, eds., Nonmarine depositional environments: models for exploration: SEPM Spec. Pub. no. 31.

Bagnold, R. A., 1941, The physics of blown sand and desert dunes: Methuen, London, 265 p.

Bigarella, J. J., 1972, Eolian environments: their characteristics, recognition and importance, in J. K. Rigby and W. K. Hamblin, eds., Recognition of ancient sedimentary environments: SEPM Spec. Pub. no. 16, p. 12-62.

——— 1975, Lagoa dune field (State of Santa Catarina, Brazil), a model of eolian and pluvial activity: Int. Symp. Quaternary, Bol. Paren, Geoci., v. 33, p. 133-167.

Bissell, H. J., 1964, Lithology and petrography of the Weber Formation in Utah and Colorado, in Guidebook of the geology and mineral resources in the Uinta Basin: Intermountain Assoc. Petrol. Geol., 13th Ann. Field Conf., p. 67-91.

Bradbury, J. P., 1980 Late Quaternary vegetation history of the central Great Plains and its relationship to eolian processes in the Nebraska Sand Hills: U.S. Geol. Survey Prof. Paper, 1120 C, p. 29-36.

Bradley, Edward, and F. H. Rainwater, 1956, Geology and ground water resources of the upper Niobrara River basin, Nebraska and Wyoming: U.S. Geol. Survey Water Supply Paper 1368, 70 p.

Brady, L. F., 1939, Tracks in the Coconino Sandstone compared with those of small living arthropods: Plateau, v. 12, p. 32-34.

Brock, W. G., and J. Nicolaysen, 1975, Geology of the Brady unit, Sweetwater County, Wyoming, in D. W. Bolyard, ed., Deep drilling frontiers of the central Rocky Mountains: Rocky Mountain Assoc. Geologists, p. 225-237.

Condra, G. E., 1918, The potash industry of Nebraska: Nebraska Bur. Publicity, Lincoln, Nebraska, 39 p.

Cooke, R. U., and A. Warren, 1973, Geomorphology in deserts: Univ. California Press, Los Angeles, 393 p.

Emmett, W. R., K. W. Beaver, and J. A. McCaleb, 1972, Pennsylvanian Tensleep reservoir, Little Buffalo Basin oil field, Big Horn Basin, Wyoming: Mountain Geologist, v. 9, p. 21-31.

Freeman, W. E., and G. S. Visher, 1975, Stratigraphic analysis of the Navajo Sandstone: Jour. Sed. Petrology, v. 45, p. 651-668.

Folk, R. L., 1971, Longitudinal dunes of the northwestern edge of the Simpson Desert, Northwest Territory, Australia, geomorphology and grain size relationships: Sedimentology, v. 16, p. 5-54.

——— 1976, Reddening of desert sands: Simpson Desert: Jour. Sed. Petrology, v. 46, p. 604-615.

Friedman, G. M., 1969, Trace elements as possible environmental indicators in carbonate sediments, in G. M. Friedman, ed., Depositional environments in carbonate rocks: SEPM Pub. No. 14, p. 193-198.

——— 1973, Textural parameters of sands: useful or useless?: Geol. Soc. America, Abstracts with Programs, v. 5, no. 7, p. 626-627.

Fryberger, S. G., 1979a, Dune forms and wind regimes; in E. D. McKee, ed., A

Study of Global Sand Seas: U.S. Geol. Survey Prof. Paper 1052, in press.

——— 1979b, Eolian-fluviatile (continental) origin of ancient stratigraphic trap for petroleum in Weber Sandstone, Rangely oil field, Colorado: Mountain Geologist, v. 16, p. 1-36.

Fryberger, S. G., and T. S. Ahlbrandt, 1979, Mechanisms for the formation of eolian sand seas: Zeitschrift fur Geomorphologie, v. 23, p. 440-460.

——— ——— and S. Andrews, 1979, Origin, sedimentary features and significance of low-angle eolian "sand sheet" deposits, Great Sand Dunes National Monument and vicinity, Colorado: Jour. Sed. Petrology, v. 49, p. 733-746.

——— and C. Schenk, 1981, Wind sedimentation tunnel studies on the origins of eolian strata: Sedimentology (in press).

Glennie, K. W., 1970, Desert sedimentary environments: Developments in sedimentology 14: Elsevier, Amsterdam, 222 p.

——— 1972, Permian Rotliegendes of northwest Europe interpreted in light of modern desert sedimentation studies: AAPG Bull., v. 56, p. 1048-1071.

——— and B. D. Evamy, 1968, Dikaka: Plants and plant-root structures associated with eolian sand: Palaeogeography, Palaeoclimatology, Palaeoecology, v. 23, p. 77-87.

——— G. C. Mudd, and P. J. C. Nagtegaal, 1978, Depositional environment and diagenesis of Permian Rotliegendes sandstones in Leman Bank and Sole Pit areas of the UK southern North Sea: Jour. Geol. Soc. London, v. 135, p. 25-34.

Gradzinski, R., and T. Jerzykiewicz, 1974, Dinosaur- and mammal-bearing aeolian and associated deposits of the Upper Cretaceous in the Gobi Desert (Mongolia): Sed. Geology, v. 12, p. 249-278.

Hanley, J. H., 1980 Paleoecology of nonmarine Mollusca from some paleointerdune deposits in the Nebraska Sand Hills: U.S. Geol. Survey Prof. Paper 1120 B, p. 25-28.

——— and J. R. Steidtmann, 1973, Petrology of limestone lenses in the Casper Formation, southernmost Laramie Basin, Wyoming and Colorado: Jour. Sed. Petrology, v. 43, p. 428-434.

——— ——— and H. Toots, 1971, Trace fossils from the Casper Sandstone (Permian) southern Laramie Basin, Wyoming and Colorado: Jour. Sed. Petrology, v. 41, p. 1065-1068.

Haughton, S. H., 1969, Geological history of South Africa: Geol. Soc. South Africa, Capetown, 535 p.

Hicks, W. B., 1921, The potash resources of Nebraska: U.S. Geol. Survey Bull. 715, p. 125-139.

Hunter, R. E., 1977, Basic types of stratification in small eolian dunes: Sedimentology, v. 24, p. 361-387.

Johnson, D. H., et al, 1978, Sabkhahs of eastern Saudi Arabia, in S. S. Al-Sayari and J. G. Zotl, eds., Quaternary Period in Saudi Arabia: Springer-Verlag, New York, 335 p.

Keech, C. F., and R. Bentall, 1971, Dunes on

the plains—The Sand Hills region of Nebraska: Univ. Nebraska, Conserv. and Survey Div., Resource Rept. No. 4, 18 p.

Kiersch, G. A., 1950, Small-scale structures and other features of the Navajo Sandstone, northern part of San Rafael Swell, Utah: AAPG Bull., v. 43, p. 923-942.

Koster, E. A., 1978, The eolian drift sands of the Velune (Central Netherlands); a physical geographical study: Univ. of Amsterdam, Ph.D. Diss., 195 p.

Krinsley, D., P. Friend, and R. Klimentidis, 1976, Eolian transport textures on the surfaces of sand grains of Early Triassic age: Geol. Soc. America Bull., v. 87, p. 130-132.

Krumbien, W. C., 1941, Measurement and geological significance of shape and roundness of sedimentary particles: Jour. Sed. Petrology, v. 11, p. 64-72.

Larson, T. C., 1975, Geological considerations of the Weber Sandstone Reservoir, Rangely field, Colorado, in D. W. Bolyard, ed., Deep drilling frontiers of the central Rocky Mountains: Rocky Mountains Assoc. Geologists, p. 275-279.

Levandowski, D. W., et al, 1973, Cementation in Lyons Sandstone and its role in oil accumulation: AAPG Bull., v. 57, p. 2217-2244.

Lupe, Robert, and T. S. Ahlbrandt, 1979, Sediments of the ancient eolian environment—reservoir inhomogeneity, in E. D. McKee, ed., A study of global sand seas: U.S. Geol. Survey Prof. Paper 1052, p. 241-252.

Malek-Aslani, M., 1979, Environmental and diagenetic controls of carbonate and evaporite source rocks: AAPG Bull., v. 63, p. 489-490.

Maughan, E. K., and T. S. Ahlbrandt, 1978, Field trip to Pennsylvanian and Permian rocks near Lyons, Colorado, in AAPG Cont. Ed. School Notes, "Clastic diagenesis—its role in reservoir quality and hydrocarbon entrapment, Field Trip," T. R. Walker, E. K. Maughan, Leaders, 18 p.

McCaleb, J. A., 1979, The role of the geologist in exploitation: AAPG Reservoir Fundamentals School, Vail Colo., July 9-13, 1979, 42 p.

McKee, E. D., 1934, An investigation of the light-colored, cross-bedded sandstones of Canyon De Chelley, Arizona: Am. Jour. Sci., v. 28, p. 219-233.

——— 1944, Tracks that go uphill: Plateau, v. 16, p. 61-72.

——— 1945, Small scale structures in the Coconino Sandsone of northern Arizona: Jour. Geology, v. 53, p. 313-325.

——— 1947, Experiments on the development of tracks in fine cross-bedded sand: Jour. Sed. Petrology, v. 17, p. 23-28.

——— 1966, Structures of dunes at White Sands National Monument (and a comparison with structures of dunes from other selected areas): Sedimentology, v. 7, p. 1-69.

——— 1979, A study of global sand seas: U.S. Geol. Survey Prof. Paper 1052.

——— and J. J. Bigarella, 1972, Deformational structures in Brazilian coastal dunes: Jour. Sed. Petrology, v. 42, p. 670-681.

———, J. R. Douglass, and S. Rittenhouse, 1971, Deformation of lee-side laminae in eolian dunes: Geol. Soc. America Bull., v. 82, p. 359-378.

——— and R. J. Moiola, 1975, Geometry and growth of the White Sands, New Mexico, dune field: U.S. Geol. Survey, Jour. Research, v. 3, p. 59-66.

——— and G. C. Tibbitts, Jr., 1964, Primary structures of a seif dune and associated deposits in Libya: Jour. Sed. Petrology, v. 34, p., 5-17.

——— and G. W. Wier, 1953, Terminology for stratification and cross-stratification in sedimentary rocks: Geol. Soc. America Bull., v. 64, p. 381-390.

Moiola, R. I., and A. B. Spencer, 1979, Differentiation of eolian deposits by discriminant analysis, *in* E. D. McKee, ed., A study of global sand seas: U.S. Geol. Survey Prof. Paper 1052, p. 53-60.

——— and D. Weiser, 1968, Textural parameters: an evaluation: Jour. Sed. Petrology, v. 38, p. 45-53.

Morgan, J. T., F. S. Cordiner, and A. R. Livingston, 1978, Tensleep reservoir, Oregon Basin field, Wyoming: AAPG Bull., v. 62, p. 609-632.

Picard, M. D., 1975, Facies, petrography and petroleum potential of Nugget Sandstone (Jurassic), southwestern Wyoming and northeastern Utah, *in* D. W. Bolyard, ed., Deep drilling frontiers of the central Rocky Mountains: Rocky Mountain Assoc. Geologists, p. 109-127.

——— 1977, Stratigraphic analysis of the Navajo Sandstone: a discussion: Jour. Sed. Petrology, v. 47, p. 475-483.

Reiche, Parry, 1938, An analysis of cross-lamination: the Coconino Sandstone: Jour. Geology, v. 46, p. 905-932.

Reineck, H. E., and I. B. Singh, 1975, Depositional sedimentary environments: Springer-Verlag, New York, 439 p.

Reynolds, M. W., et al, 1976, Description of selected drill cores from Paleozoic rocks, Wertz oil field, south-central Wyoming, Part 1: Wells 27A, B, C; 46 A, B, C and E, and W. Wertz 2: U.S. Geol. Survey Open-File Rep. 76-377, 56 p.

Ruhe, R.V., 1975, Geomorphology: Houghton Mifflin Co., Boston, Mass., 246 p.

Sarnthein, M., and L. Diester-Haass, 1977, Eolian-sand turbidites: Jour. Sed. Petrology, v. 47, p. 868-890.

Schultz, C. B., and J. C. Frye, 1968, Loess and related eolian deposits of the world: Univ. Nebraska Press, Lincoln, Nebraska, 345 p.

Sharp, R. P., 1963, Wind ripples: Jour. Geology, v. 71, p. 617-636.

Smolen, J. J., and L. R. Litsey, 1977, Formation evaluation using wireline formation tester pressure data: Soc. Petrol. Engineers Paper SPE 6822, 10 p.

Steen, M. O., 1961, Sand Hill lake survey: Nebraska Game, Forestation and Parks Commission Rept. for period July 1954, to Dec. 31, 1960.

Steidtmann, J. R., 1974, Evidence for eolian origin of cross-stratification in sandstone of the Casper Formation, southernmost Laramie Basin, Wyoming: Geol. Soc. America Bull., v. 85, p. 1835-1842.

Thompson, D. B., 1969, Dome-shaped aeolian dunes in the Frodsham Member of the so called Keuper Sandstone Formation (Scythian?-Anisian: Triassic) at Frodsham, Cheshire (England): Sed. Geology, v. 3, p. 263-289.

Till, R., 1974, Statistical methods for the earth scientist, an introduction: Wiley, New York, 154 p.

Tsoar, Haim, 1978, The dynamics of longitudinal dunes: U.S. Army, European Research Office, London, Final Technical Report, Grant No. DA-ERO 76-G-072, 171 p.

Walker, T. R., 1967, Formation of red beds in modern and ancient deserts: Geol. Soc. America Bull., v. 78, p. 353-363.

——— 1979, Red color in dune sand, *in* E. D. McKee, ed., A study of global sand seas: U.S. Geol. Survey Prof. Paper 1052.

——— and J. C. Harms, 1972, Eolian origin of flagstone beds, Lyons Sandstone (Permian), type area, Boulder County, Colorado: Mountain Geologist, v. 9, p. 279-288.

——— and G. V. Middleton, 1977, Facies models 9, eolian sands: Geoscience Canada, v. 4, p. 182-189.

Watts, W. A., and H. E. Wright, Jr., 1966, Late-Wisconsin Pollen and seed analysis from the Nebraska Sand Hills: Ecology, v. 47, p. 202-210.

Windom, H. L., 1975, Eolian contributions to marine sediments: Jour. Sed. Petrology, v. 45, p. 520-529.

Alluvial Fan Deposits

Tor H. Nilsen
U.S. Geological Survey
Menlo Park, CA 94025

INTRODUCTION

Alluvial fans accumulate at the base of a mountain front or other upland area where an emerging mountain stream deposits a sediment body whose sloping surface forms a segment of a cone (Blissenbach, 1954; Bull, 1972a, 1977; Fig. 1). The slope of alluvial fans averages about 5°, ranging from less than a degree to as much as 25°, although rarely exceeding 10° (Denny, 1965; Bull, 1977). Where the mountain front is relatively straight, a series of adjacent alluvial fans will coalesce to form a bajada or piedmont slope (Fig. 2). The term clastic wedge has been applied to these deposits, especially where the mountain front is uplifted along faults. Ancient fan deposits may form thick complexes, are indicative of a mountainous paleogeography, and are most commonly redbeds, especially where deposited under arid conditions (Fig. 3). As channels shift laterally through time, the sedimentary body develops a characteristic fan shape in plan view, a convex-upward cross-fan profile, and concave-upward radial profile (Fig. 4). Radial profiles permit subdivision of fans into upper, middle, and lower segments (Fig. 5).

Small bodies more steeply sloping that were deposited by small streams may be referred to as alluvial cones rather than fans; these may grade into scree or talus cones deposited by falling debris rather than stream transport. Fan-deltas are alluvial fans deposited into standing bodies of water (Fig. 1F); their distal facies, which may include sediments retransported by lake or ocean currents, are different from those of fans. Pediments, sloping surfaces cut into bedrock by periodic flooding that may consist of either bedrock or a thin veneer of loose sediment over bedrock, in some places flank alluvial fans in arid areas and may morphologically resemble fans.

Fig. 1—Morphology and setting of modern alluvial fans. **A.** Holocene fan, Canning River, northeastern Brooks Range, Alaska (photo by T. H. Nilsen). **B.** Holocene fan, Iteriak Creek, western Brooks Range, (photo by T. H. Nilsen). **C.** Holocene fan, Upper Sheenjek River, northeastern Brooks Range, Alaska (photo by T. H. Nilsen). **D.** Holocene fan, Steele Creek, St. Elias Range, Yukon Territory (photo by R. P. Sharp). **E.** Holocene fan, Richter Pass near Osoyoos, British Columbia (photo by J. M. Ryder). **F.** Holocene fan-delta, Lake Peters, northeastern Brooks Range, Alaska (photo by T. H. Nilsen).

I thank John A Bartow, William B. Bull, Alan P. Heward, Joseph H. McGowen, Andrew D. Miall, Brian R. Rust, Ronald J. Steel, and John R. Tinsley for helpful reviews of this manuscript and providing many useful ideas regarding both modern and ancient alluvial fan sedimentation. I thank many individuals who have supplied photographs or figures for this paper, especially J. C. Boothroyd, W. B. Bull, R. H. Campbell, S. I. Carryer, K. A. Crawford, T. D. Fouch, T. G. Gloppen, A. P. Heward, J. D. Howard, D. R. Kerr, V. Larsen, M. H. Link, J. H. McGowen, D. M. Morton, J. M. Ryder, P. R. Schluger, R. P. Sharp, K. E. Sieh, D. R. Spearing, C. J. Stuart, R. J. Wasson, and C. G. Winder.

Fig. 2—Morphology of coalesced alluvial fans.
A. Quaternary fans, west flank of Lost River Range, Idaho (photo by T. H. Nilsen). **B.** Quaternary fans, west flank of Lost River Range, Idaho (photo by D. R. Kerr). **C.** Quaternary fans, south flank of eastern San Gabriel Mountains, Los Angeles and San Bernardino Counties, California (photo by T. H. Nilsen). **D.** Quaternary fans, south-central Nevada (photo by T. H. Nilsen). **E.** Quaternary fans, Tiefort Mountains, Mojave Desert, California (photo by R. P. Sharp). **F.** Quaternary fans along Kongakut River, northeastern Brooks Range, Alaska (photo by T. H. Nilsen).

Neither fan-deltas nor glacial outwash plains are considered in detail in this paper. Many of the former are simply fan-shaped deltas, built by rivers debouching into standing bodies of water without the presence of either an adjacent highland or steep slope. Outwash plains, sometimes referred to as outwash fans or sandur plains, are generally flat and characterized by radiating distributaries emerging from different parts of a melting ice field. The reader is referred to McGowen (1971) and Wescott and Ethridge (1980) for more information about fan-deltas and to Boothroyd and Nummedal (1978) for information about outwash fans.

Size of fans ranges from less than a hundred meters (Figs. 6, 7) to more than 150 km in radius, although they average less than 10 km. The Kosi River fans of India, which derive sediment from the Himalayas (Gole and Chitale, 1966; Inglis, 1967), may be the largest modern fans (Fig. 8), although some workers consider these to be chiefly flood plain rather than alluvial fan deposits (Rust, 1979). The thickness of fan deposits can range from a few meters to perhaps as much as 25,000 m, the latter being the thickness of the Devonian Old Red Sandstone in the Hornelen basin of western Norway (Steel, 1976). However, axial migration of the basin depocenter during deposition here as a result of probable strike-slip faulting has resulted in a true maximum vertical thickness of less than 8,000 m (R. J. Steel, written commun., 1979).

Modern alluvial fans have been extensively studied from the standpoint of physiographic and surface morphology, but have received careful attention from sedimentologists only during the

Fig. 3—Colors of ancient alluvial fan deposits. **A.** Miocene and Pliocene alluvial fan and lacustrine deposits, Grant Range, Nevada (photo by T. D. Fouch). **B.** Miocene and Pliocene alluvial fan red beds, Grant Range, Nevada (photo by T. D. Fouch). **C.** Miocene and Pliocene red-weathering fanglomerates, Grant Range, Nevada (photo by T. D. Fouch).

last 10 years. Conversely, although ancient alluvial fan deposits have been extensively studied from the standpoint of clast composition and direction of sedimentary transport, they have generally not been the object of detailed facies analyses and integrated sedimentological studies. Except for some Precambrian placer gold deposits, fans have not been the object of intensive mineral or hydrocarbon exploration, and thus have not been studied in detail by industry. As a result, understanding of alluvial fan sequences is not well developed.

Several other important factors have contributed to the limited understanding of alluvial fans and fan deposits. For many years, the only modern fans studied in detail were from arid and semiarid regions. As prominent and conspicuous as these fans are, they may be characteristic only of their particular climates, whereas a large number of ancient alluvial fans may have

been deposited in temperate and humid climatic conditions. Because of vegetation cover, soil formation and human development, surprisingly little significant sedimentologic work has been done on fans in nonarid regions, particularly humid ones. Starting in the 1960s, studies of fans deposited in cold climates, including polar regions, have contributed greatly to overall understanding; these studies have been significant because many modern fans in more temperate climates may have developed primarily during much colder intervals of the Pleistocene.

Until relatively recently, the great majority of significant studies of ancient alluvial fan deposits were restricted areally to several regions and temporally to only a few time intervals, namely the Old Red Sandstone (Devonian) and New Red Sandstone (Permian and Triassic) in the Caledonian-Appalachian chain, and scattered Upper Cretaceous and Cenozoic de-

posits in the western United States. These deposits, although generally well exposed and studied in some detail, may not be typical or characteristic of most ancient fan deposits, and facies models based on them may not be comprehensive.

Recognition of ancient alluvial fan deposits has been greatly influenced by clast-size considerations. Most workers have considered fan deposits synonymous with conglomerate, and alluvial plain deposits synonymous with a composition of sandstone, siltstone and mudstone. In most paleogeographic studies, this subdivision has been convenient and has generally been accepted by the sedimentologic community. However, this procedure has inhibited study and prevented penetrating analyses of ancient deposits because it is quite clear that alluvial fans can be constructed of material of any grain size, depending simply on what type of material is brought to the

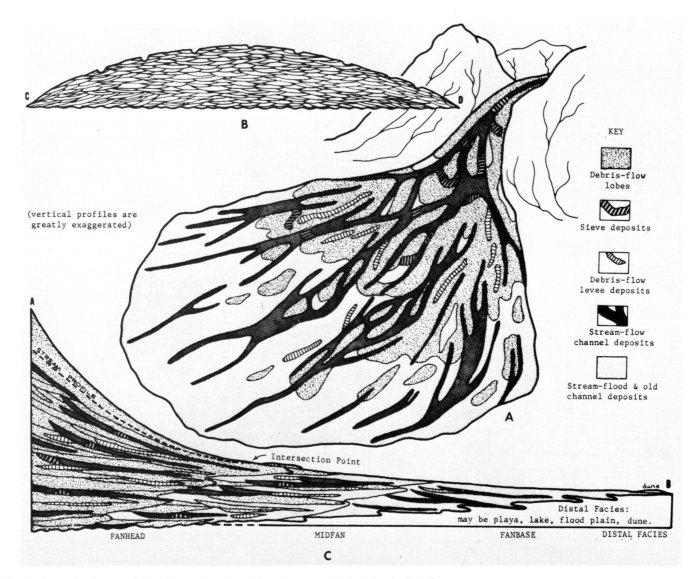

Fig. 4—Generalized model of alluvial fan sedimentation (from Spearing, 1974). **A.** Fan surface, **B.** cross-fan profile, and **C.** radial profile.

depositional basin by feeding streams. Although generally high relief is required to yield deposition of sloping alluvial fans, this condition does not require fans or all parts of fans to necessarily consist of gravel-bearing sediment. In addition, because distributary channels on alluvial fans can be braided, straight or meandering, fan channels in ancient deposits cannot be easily distinguished from alluvial plain channels, and should not be distinguished on the basis of grain size alone. Thus, literature on ancient fan deposits has few references to nonconglomeratic fan sequences, and descriptions of many fan deposits may be mistakenly attributed to alluvial plain deposits.

Another contributing factor to the limited understanding of alluvial fan deposits is the general lack of application until recently of vertical sequence analysis for facies interpretation. Interpretation of vertical sequences, developed so strongly by the petroleum industry for analysis of well logs, has greatly contributed to the understanding of fluvial, deltaic, nearshore, and deep-marine deposits. However, because (1) little vertical section is seen in modern fans, (2) coring of fan sequences has generally been difficult because of abundant conglomeratic intervals, (3) fans have not been exploration targets, and (4) many ancient conglomeratic sequences recognized as fan deposits have not appeared to contain systematic vertical changes, vertical sequence analysis has not been extensively or successfully applied to facies analyses of fan deposits.

Few published geophysical studies of alluvial fans have been made. As a result, determinations of the three-dimensional geometry, continuity of re-flecting surfaces, subsurface extent of channels, and thickness variations of modern fans is poorly known. In contrast, deep-sea fans, the marine equivalents of alluvial fans, have been the subject of numerous geophysical studies and have been thoroughly examined using vertical sequence analysis (see Howell and Normark, this vol.). Thus, although many similarities exist between alluvial and deep-sea fans, only recently have concepts developed from more detailed studies of deep-sea fans been applied to alluvial forms (see especially Heward, 1978a, b).

Late Cenozoic climatic and sea level changes have drastically altered the depositional setting of numerous modern alluvial fans. Many modern fans probably developed chiefly during the Pleistocene under different climatic conditions and during periods when lowered sea level also lowered the base level.

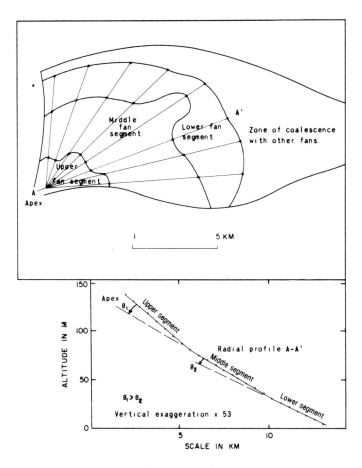

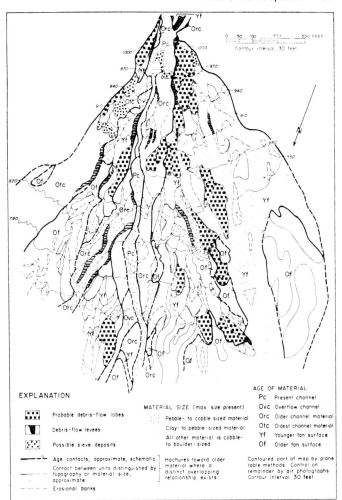

Fig. 5—Segmentation of the Tumey Gulch fan, western Fresno County, California (from Bull, 1964b).

Fig. 6—Morphology of the Trollheim fan, a small modern alluvial fan, California (from Hooke, 1967).

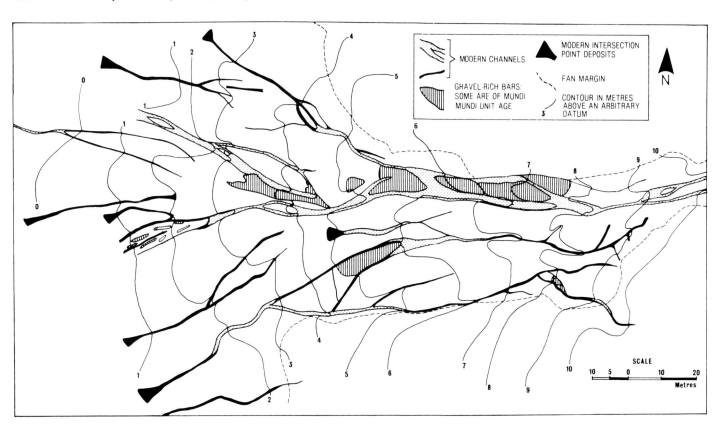

Fig. 7—Morphology of the apex region of Goat fan, a small modern alluvial fan, western New South Wales, Australia (from Wasson, 1979).

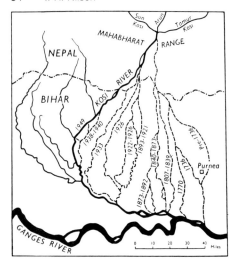

Fig. 8—Morphology of a large modern alluvial fan, Kosi River, India, showing temporal shifts of main channel (from Gole and Chitale, 1966).

Thus, modern processes may not be indicative of events leading to construction of most of the fan. Conversely, growth of fans has accelerated in some areas by Pleistocene to Holocene climatic changes. A few areas largely unaffected by either climatic, sea-level or lake-level changes may have experienced little change from Pleistocene to Holocene. Therefore, use of the present as a key to the past must be evaluated cautiously.

Nevertheless, for all the apparent shortcomings, knowledge of alluvial fan deposits is considerable, especially concerning some depositional processes which have been observed locally in great detail. In addition, the number of significant studies published during the past 3 to 4 years shows our understanding of alluvial fans to be increasing rapidly.

PREVIOUS STUDIES

Several previously published summaries of alluvial fan deposits provide useful background on the subject, and the reader is referred to them for additional and, in general, more detailed information. Allen (1965) discussed the origin and characteristics of alluvial sediments, including a brief summary of fans. Bull (1972a, 1977) provided the most integrated summaries of modern fans and compared them to some ancient fan deposits. Yazawa et al (1971) discussed most of the processes and morphologic aspects of alluvial fans. Fisher and Brown (1972) presented a general model of alluvial fan sedimentation. Spearing (1974) provided a useful well-illustrated summary sheet of the chief sedimentary characteristics of alluvial fan deposits. Schumm

(1977), Miall (1977), Collinson (1978), and Rust (1978, 1979) discussed aspects of fluvial sedimentation, including that of alluvial fans and fan-deltas. Ethridge and Thompson (1978) and McGowen (1979) summarized a great deal of information about fluvial sediments, including alluvial fans and fan deltas, with extensive discussion of mineral deposits, including placer gold and uranium, associated with fans. A bibliography of alluvial fan deposits has been prepared by Nilsen and Moore (1980).

Deposition on modern arid-region fans has been studied by Blissenbach (1954), Beaty (1963, 1970), Lustig (1965), Denny (1967), Hooke (1967, 1968), and Beaumont (1972), among others, and modern Arctic-, temperate-, and humid-region fans by Anderson and Hussey (1962), Winder (1965), Anstey (1965), Leggett et al (1966), Ryder (1971), McPherson and Hirst (1972), and Wasson (1974, 1977a, b, 1979). Experimental studies and modeling of alluvial fans have been conducted by Hooke (1968), Price (1974), Schumm et al (1977, p. 255-264), and Hooke and Rohrer (1979).

Ancient alluvial fan deposits have been studied in many parts of the world. Because they develop at the base of slopes adjacent to mountainous terrain or uplifted blocks, alluvial fan deposits have been most commonly reported from mobile belts and are most typically synorogenic or postorogenic. Within major mountain chains, they may be intramontane, intermontane or extramontane. Intramontane fans are perhaps most common, but preservation in stratigraphic records is minimal because of continued uplift of the mountains and eventual erosion of fan deposits. Intermontane and extramontane fans have a greater chance of preservation but may be interpreted as alluvial plain rather than fan deposits. Small fans, of course, may develop at bases of slopes adjacent to very small streams. Many thin fan deposits may result from minor climatic changes and represent brief depositional intervals during long periods of erosion. Some of these fans may include and not be readily distinguishable in the rock record from talus, scree, slope wash, and landslide deposits.

Ancient fan deposits have been most thoroughly described from the Devonian Old Red Sandstone (Nilsen, 1968, 1969, 1973; Miall, 1970; Schluger, 1973; Steel, 1976; Steel et al, 1977; Bluck, 1978; Steel and Aasheim, 1978) and Car-

boniferous to Jurassic red beds that include the New Red Sandstone (Krynine, 1950; Klein, 1962; Bluck, 1965; Laming, 1966; Meckel, 1967; Wessel, 1969; Deegan, 1973; Steel, 1974, 1977; Steel and Wilson, 1975; Hubert et al, 1978) in the Caledonian, Apalachian, and Hercynian geosynclines of North America and Europe. Other well-described ancient fan accumulations include Cretaceous and Tertiary deposits of the western Cordillera of North America (Sharp, 1948; Carter and Gualtieri, 1965; Soister, 1968; Wilson, 1970; Steidtman, 1971; Love, 1973; Ryder et al, 1976), and late Cenozoic deposits associated with strike-slip pull-apart basins along the San Andreas fault (Van de Kamp, 1973; Crowell, 1974; Link and Osborne, 1978; Kerr et al, 1979). Fan deposits of Precambrian age have been studied in some detail because many are associated with ore deposits (Williams, 1969: Pretorius, 1974a, b, 1975, 1976; Reid, 1974; Minter, 1976, 1978; Robertson, 1976).

ALLUVIAL FAN FACIES

Alluvial fans can most easily be divided into proximal and distal facies, and most studies of ancient fan sequences have been limited to that division. Proximal facies are deposited in the upper or inner parts of the fan, near the area of stream emergence from the upland area, and contain the coarsest sediments. Distal facies are deposited in the lower or outer parts of the fan, and contain finer grained sediments (Figs. 9, 10). Proximal and distal facies can be recognized in ancient fan deposits through use of paleocurrent data, maps showing distribution of the coarsest clasts and thickest beds, and perhaps changes in clast roundness. Some workers distinguish a proximal facies characterized by highly channeled and lenticular beds from a distal facies characterized mainly by less channelized sheetlike beds (e.g., Kerr et al, 1979).

More detailed facies analysis requires more information about the morphology and sediment distribution of modern fans. The innermost part of the fan, the apex or fanhead area, most characteristically contains an entrenched straight valley, extending outward onto the fan from the feeding mountain stream (Fig. 11). Reasons for the entrenching vary, and include much-discussed climatic and tectonic effects (Bull, 1964b; Weaver and Schumm, 1974). It is probably most directly re-

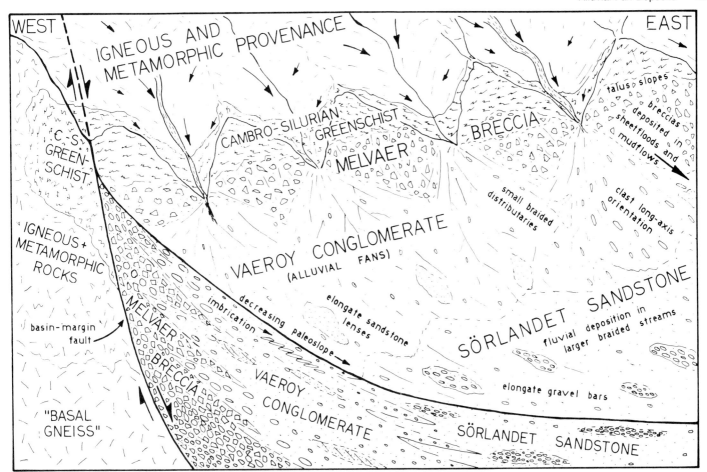

Fig. 9—Paleogeographic diagram of the Devonian Buelandet-Vaerlandet district, western Norway (from Nilsen, 1969).

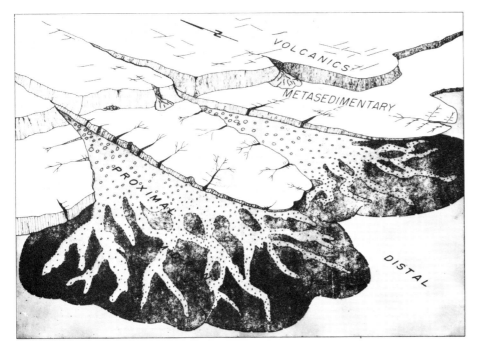

Fig. 10—Schematic plan of the alluvial fan facies of the Van Horn Sandstone, Precambrian(?), Texas (from McGowen and Groat, 1971). The coarsest material is confined to the canyons, and finer material is spread beyond the canyon mouths.

lated, however, to continued natural downcutting in the source terrain of the stream through time, and associated downcutting of the fanhead valley to maintain a suitable base level for downfan deposition. In the inner fan region facies are characterized by (1) very coarse-grained deposits of a single major channel (in some fans there may be two or three) that is generally broad and deep, and through which passes almost all sediment deposited on the fan, and (2) finer grained channel-margin, levee and interchannel deposits that may include coarse-grained landslides and debris flows derived from slopes between major drainages. If no longer an active, depositional surface, interchannel areas may form gravel-topped remnant fan-terraces with soil profiles.

A second major facies, comprising the largest area of most fans, consists of smaller distributary channels radiating outward and downfan from the inner-fan valley. Literally hundreds of channels may be present on the fan surface, but most are inactive for extended periods, as channels rapidly shift laterally with time. These channels are not as deep or wide as the inner-fan valley channel, and are sites of rapid sedimentation as channels fill during the course of gradual lateral shifting or abrupt abandonment. The channels are most commonly braided and may join and rejoin downfan. However, depending on fan gradient, sediment type, sediment supply, climatic effects and other factors, straight, braided, meandering, and anastomosing channel systems may be present. These channels and less well developed interchannel deposits make up a middle-fan facies.

A third major facies, characteristically not well preserved in arid-region alluvial fans, consists of finer grained sheetlike deposits of nonchannelized or less channelized outer fan sediment. In this part of the fan, flow emerges from channels and spreads out to deposit laterally extensive sheets or bodies of sediment. The longitudinal gradient of the fan is extremely low in this area, and deposits may be difficult to distinguish morphologically from basin-floor alluvial plain, lacustrine, or other facies. These sheetlike bodies, in some ways geometrically analogous to delta-mouth bars and outer-fan lobes of deep-sea fans, tend to be poorly developed in modern arid fans, and commonly may be truncated by basin-axis alluvial plain systems. In most fans, some shallow channels extend to the fan edge and cut through the outer-fan bodies. Holocene climatic and sea level changes have strongly affected depositional patterns of modern fans (most may at present be dominated by channel systems that extend the entire length of the fan) as a result of decreased rainfall and changes in vegetation on fan surfaces and in some source areas.

A fourth major fan facies is the fan fringe, consisting of outer-fan deposits that intertongue with other depositional systems, including fluvial, lacustrine, eolian, and marginal-marine facies. The fan-fringe facies is both the most variable and finest grained of the grouping.

DEPOSITIONAL PROCESSES
General

A limited number of depositional processes act on alluvial fans and yield a smaller variety of deposits than in most depositional settings. Fan deposits can be subdivided into two types, those resulting from streamflow and those resulting from debris flow and related processes. Streamflow processes are those acting in any fluvial system and result from deposition of sediment carried in suspension, saltation and traction by flowing water, whether of channelized or nonchannelized flow. Debris flow, mudflow and

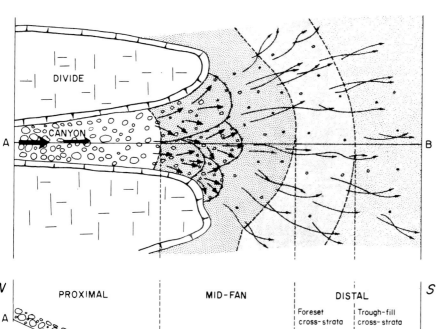

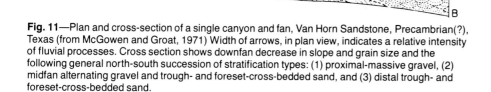

Fig. 11—Plan and cross-section of a single canyon and fan, Van Horn Sandstone, Precambrian(?), Texas (from McGowen and Groat, 1971) Width of arrows, in plan view, indicates a relative intensity of fluvial processes. Cross section shows downfan decrease in slope and grain size and the following general north-south succession of stratification types: (1) proximal-massive gravel, (2) midfan alternating gravel and trough- and foreset-cross-bedded sand, and (3) distal trough- and foreset-cross-bedded sand.

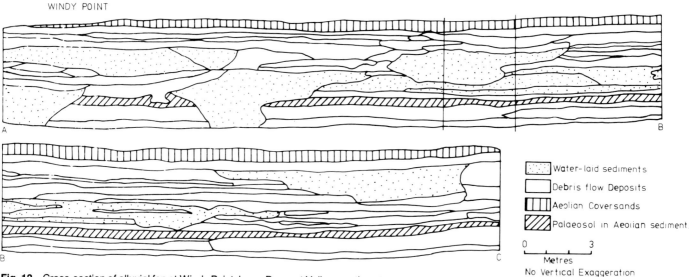

Fig. 12—Cross-section of alluvial fan at Windy Point, lower Derwent Valley, southeastern Tasmania, showing distribution of streamflow and debris-flow deposits (from Wasson, 1977b). Section is approximately perpendicular to the sediment transport direction.

Fig. 13—Characteristics of streamflow deposits in alluvial fans. **A.** Braided streams on Furnace Creek Wash, Death Valley, California, showing small bars and channels within major channel (photo by W. B. Bull). **B.** Braided streamflow deposits of Furnace Creek Wash, Death Valley, California, shown in cross-section perpendicular to flow direction (photo by W. B. Bull). **C.** Channeled and cross-stratified Miocene streamflow deposits, Fish Creek Wash, west flank of Salton trough, southern California (photo by D. R. Kerr). **D.** Nonchanneled parallel-stratified beds of line conglomerate and conglomeratic sandstone deposited by streamflow, Evanston Formation, Late Cretaceous and Paleocene, upper Echo Canyon, north-central Utah (photo by K. A. Crawford); width of photo about 3 m. **E.** Parallel-stratified streamflow conglomerate and sandstone, Old Red Sandstone, Devonian, Hornelen, western Norway (photo by V. Larsen). **F.** Medium-scale planar cross-strata and parallel strata of fan-toe streamflood deposits, New Red Sandstone, Triassic, Mull, Scotland (photo by R. J. Steel).

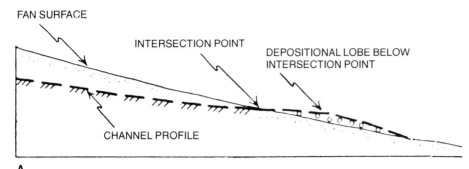

FAN SURFACE

INTERSECTION POINT

DEPOSITIONAL LOBE BELOW
INTERSECTION POINT

CHANNEL PROFILE

A

A

B

Fig. 14—Intersection-point sedimentation on alluvial fans. **A.** Idealized sketch of intersection-point relations (from Hooke, 1967). **B.** Example of intersection-point deposition of gravel from Club Lake fan, Australia (photo by R. J. Wasson).

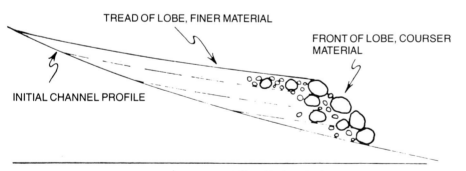

TREAD OF LOBE, FINER MATERIAL

FRONT OF LOBE, COURSER
MATERIAL

INITIAL CHANNEL PROFILE

Fig. 15—Schematic sketch of a sieve lobe deposit (from Hooke, 1967).

B

Fig. 16—Sieve lobe deposits on modern alluvial fan, Eureka Valley, California (from Bull, 1977). **A.** Hummocky topography formed by sieve lobes; slope is 16° along radial profile through fan valley and sieve lobes, 20° along older and smoother parts of the fan to the right. **B.** Unvarnished sieve-lobe deposit from same area.

related processes are sediment gravity flows in which the matrix is typically muddy or clayey and its strength supports clasts and fragments suspended in it (see Johnson, 1970, for a discussion of the mechanics of this process). Streamflow deposits behave in transport as essentially Newtonian fluids, whereas debris-flow and related deposits behave in transport as more viscous fluids. Consequently, streamflow deposits are generally well stratified, contain a variety of sedimentary struc-

tures indicative of different flow regimes, have small amounts of clay-sized matrix, are clast-supported, and contain clasts that are imbricated and oriented relative to flow direction. Debris-flow and related deposits, in contrast, are generally poorly stratified, contain few sedimentary structures, have large amounts of clay-sized matrix, are matrix-supported, and contain clasts with little or no imbrication or orientation relative to flow direction.

The abundance of debris-flow and

related deposits on alluvial fans is one of the most important features permitting fan deposit recognition in the ancient record. Although both streamflow and debris-flow deposits can be found in any part of a fan and typically intermix (Fig. 12), streamflow deposits are generally more characteristic of distal fan facies and humid-climate fans while debris-flow deposits are more characteristic of proximal fan facies and arid-climate fans.

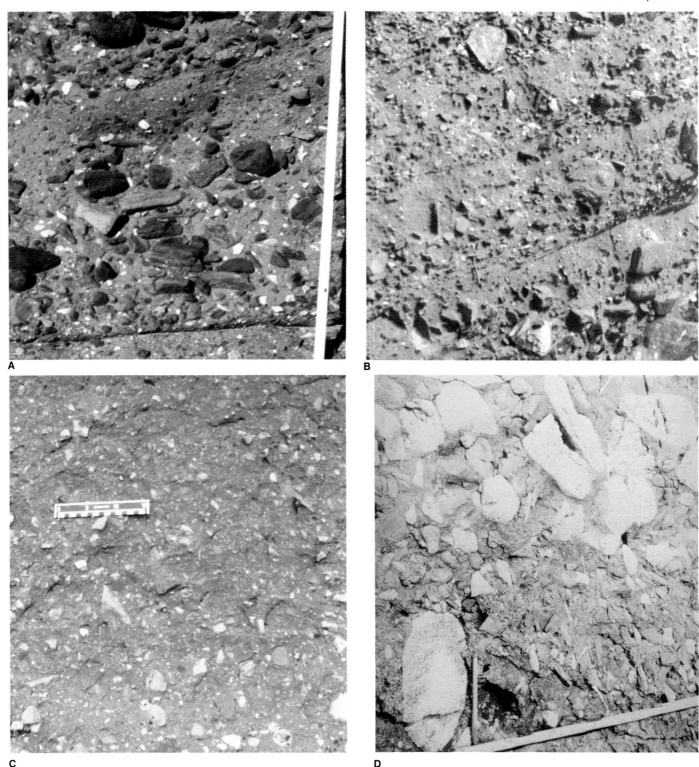

Fig. 17—Characteristics of ancient debris flows in alluvial fan deposits. **A.** Thick debris-flow deposit characterized by sharp base, inverse grading, poorly developed clast orientation, and mud-rich matrix, San Onofre Breccia, Miocene, Dana Point, southern California (photo by C. J. Stuart); stick is 1.5 m long. **B.** Stacked sequence of thick inverse- to normal-graded debris-flow deposits, San Onofre Breccia, Miocene, Dana Point, southern California (photo by C. J. Stuart). **C.** Debris-flow deposit characterized by normal grading at top, random clast orientation, and mud-rich matrix, Echo Canyon Conglomerate, Late Cretaceous, north-central Utah (photo by K. A. Crawford). **D.** Debris-flow deposit at toe of fan characterized by vertical orientation of clasts, Old Red Sandstone, Devonian, Hornelen basin, western Norway (photo by R. J. Steel).

Streamflow Deposits

Streamflow deposits are characteristic of channel fills, channel-margin and interchannel areas, and nonchannelized or less channelized outer-fan areas where sheets of sediment are de-posited. Channel fills are geometrically long and narrow and are the most coarse grained and poorly sorted streamflow deposits (Fig. 13 A-D). Channels become shallower and channel deposits finer grained downfan; channel-fill beds may be thicker than 2 m and are commonly conglomeratic. The basal surface of channel deposits is characteristically concave-upward in transverse cross-section and has erosional contacts with flanking and underlying deposits. Typical inner-fan channel deposits fill straight and entrenched channels, whereas midfan and outer-fan channel deposits most commonly fill braided channels. In ancient fan deposits, large channels may be difficult to recognize where they are shallow and present wholly within conglomeratic strata.

Channel-margin, levee, and interchannel streamflow deposits develop on most fans, although they may be poorly preserved in ancient fan sequences on which channels gradually shifted across the fans, resulting in erosion of these deposits. Where new channels develop more abruptly by avulsion, interchannel facies may be better preserved. Beds generally thin and fine away from channels toward interchannel areas, which contain thin, sheetlike beds deposited during overbank floods. Because of the high porosity and permeability of the interchannel areas of many fans, however, interchannel and levee deposits may not be laterally extensive or abundant.

Sheetflood deposits result from the spreading out of sediment-laden floodwater where it emerges from channels, most typically in lower parts of fans (Fig. 13E-F) at a point referred to (Hooke, 1967; Wasson, 1974) as the intersection point (Figs. 4C, 7, and 14A). Decrease in flow velocity associated with emergence from channels, combined with the lower slopes present in lower fan areas, results in deposition of sheets of sediment, typically as a series of low bars of gravel or sand that may be dissected by small and shallow channels cut during waning flood stages (Fig. 14B). Sheetflood deposits are typically sandy with little clay, fairly well sorted and well stratified, containing parallel stratification and cross-stratification. Sheetflood deposits can also develop in wide channels or interchannel areas.

Sieve deposits are a type of streamflow deposit consisting of permeable lobes of gravel that fine upslope and through which flood discharges completely infiltrate before reaching the fan fringe (Hooke, 1967; Fig. 15). They form on some arid-region fans where the source area supplies mostly gravel-sized detritus rather than sand, silt or clay. Although not common, sieve deposits of modern fans consist of well-sorted gravel containing relatively angular and monomict clasts (Fig. 16) that form massive, laterally extensive beds with well-developed imbrication. Sieve deposits have been mapped on modern fans, where they are generally present in proximal areas (Fig. 6).

A

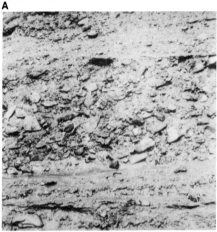

B

Fig. 18—Vertical sequences of debris-flow deposits. **A.** Stacked sequence of thin debris-flow deposits capped by beds of sandstone, New Red Sandstone, Triassic, Gruinard Bay, Scotland (photo by R. J. Steel). **B.** Sequence of thin and thick debris-flow deposits showing that maximum clast size is approximately correlative with bed thickness; clasts in this debris-flow deposit are imbricated, indicating sediment transport toward the right, New Red Sandstone, Triassic, Gruinard Bay, Scotland (photo by R. J. Steel).

Few sieve deposits have been reported from ancient fan deposits, probably because subsequent deposition and cementation eventually fill pore spaces with matrix. Reid (1974) reported their presence in a Precambrian fan deposit in Texas, and Pretorius (1975) suggested that Precambrian gold and uranium placer deposits in South Africa infiltrated previously deposited open-framework gravels which could have been sieve lobes.

Debris-Flow and Related Deposits

The dense and viscous debris-flow deposits are characteristically poorly sorted, reverse graded in their basal parts, form disorganized beds with isotropic fabrics, and may contain enough muddy matrix to render deposits relatively impermeable and nonporous (Fig. 17). Debris flows generate where

sediment sources provide abundant muddy material, where slopes are steep and vegetation is scarce, and where rainfall is either seasonal or irregular (Fig. 18). These flows can move rapidly and can transport large blocks of material, including rock, soil, trees, ice, and buildings. The abundance of debris-flow deposits depends largely on the source-area geology and climatic conditions, and debris-flow deposits are typically intermixed with steamflow deposits. Debris-flow deposits can be confined to channels or spread out laterally as sheets or lobes in interchannel or lower fan areas; they typically have well-defined margins and may transport boulders weighing many tons. Individual debris flows may have well-defined channelized axial portions and well-developed levees produced by deposition on channel flanks at times when flows were thickest (Fig. 19). Ramparts of coarse material may form along the margins or edges of debris flows.

Mudflows are similar to debris flows except that they consist almost wholly of fine-grained material, sand size and finer (Fig. 20). Mudflow viscosities can vary greatly and resulting deposits can thus range from thin widespread sheets with very thin subhorizontal margins to thick lobate bodies with thick, nearly vertical margins. Mudflows may be deposited in either channels or nonchannelized areas of fans. When mudflows solidify, the clay-rich sediment may contract, forming extensive mud cracks, especially on the surfaces of flows.

Landslide Deposits

In addition to transport of material down and along the feeding stream channel to the fan, sediment can also be transported by landsliding to the fan from inter-stream slope areas of the mountain front (Fig. 21). The landslides may include falls, slumps, flows, slides, snow avalanches, and material transported by other processes. Materials transported by landsliding include mud, sand, boulders, large blocks of bedrock, soil, vegetation, and manmade objects. Some rock avalanches (sturzstroms) travel on air cushions and move across fan surfaces by streaming (Eisbacher, 1979), resulting in deposition of debris across both proximal and distal fan areas and in channel and interchannel areas. Some of these dry avalanches can flow down channels of fans and be difficult to distinguish, especially in the ancient record, from

Fig. 19—Characteristics of recent debris flows.
A. Recent debris flow at Wrightwood, California showing ramp of debris at margins of deposit and thin mudflow around fringe of deposit (photo by D. M. Morton). **B.** Recent debris flow at Wrightwood, California showing flow margins, oriented clasts, and boulder train (photo by D. M. Morton). **C.** Holocene alluvial fan surface with large boulders transported by debris flows, Los Lobos Creek, north flank of San Emigdio Mountains, southern California (photo by T. H. Nilsen). **D.** Small-scale debris flows showing development of levees and depositional lobes, sand quarry in South Limburg, The Netherlands (photo by A. P. Heward). **E.** Debris flows with prominent levees at Wrightwood, California (photo by D. M. Morton). **F.** Debris flow on Sparkplug Canyon fan, west flank of White Mountains, eastern California (photo by W. B. Bull).

Fig. 20—Mudflow deposit on Santiago Creek fan, north flank of San Emigdio Mountains (photo by W. B. Bull).

Fig. 21—Thick and coarse deposit at top of Miocene alluvial fan deposits, Split Mountain Gorge, southern California interpreted by Kerr and others (1979) to be a subaerial landslide deposit (photo by D. R. Kerr).

muddy debris flows. Along seismically active mountain fronts, landslides may be common, especially where slopes are steep and weathering deep; where thick and repetitive, landslide deposits may form a mappable facies, and locally may form marker beds for stratigraphic analysis. Landslide deposits may be difficult to distinguish in ancient fan sequences from debris-flow and mudflow deposits transported along the fan channel system. On proximal fan margins, talus, scree, rock glacier, and other slope materials may extend for short distances out onto the fan surface.

RELATION TO OTHER FACIES

Alluvial fans generally grade laterally toward the source area into channeled fluvial deposits of the river or stream feeding the fan, although stream deposits are commonly very thin and not preserved in ancient systems because of continued uplift of or erosion into the source area (Fig. 22). They may lie on older pediment surfaces or parts of older fans, inheriting previously established fan morphology. At the base or on the lower parts of the range-front slope, the fans may grade laterally into talus deposits and associated slope wash, colluvial deposits, glacial deposits of various types, terrace accumulations, and various types of landslide deposits, including flows,

falls, slumps and slides (e.g., Blissenbach, 1954; Drewes, 1963; Nilsen, 1969). Talus accumulations can generally be recognized by their underlying contact with bedrock or soils developed on bedrock, increase in clast size downslope, and skewness toward coarser clasts (Bull, 1972a).

Alluvial fans grade laterally into other alluvial fan deposits derived from adjacent feeding streams. Because of the many possible ages, stages of activity, and sizes of adjacent fans, facies relations between coalescing fans can be quite complex (Fig. 23). Along most range fronts, there is a complex interfingering of adjacent fans to form a laterally extensive piedmont. Bull (1972a, p. 78) examined electric logs of many fans in the western Great Valley of California and concluded that boundaries of adjacent coalesced fans should change little through time, because rates of sediment yield from adjacent drainages would be equally affected by tectonic and climatic changes. Longer term tectonic effects and changes in sizes of drainage systems, however, can yield major changes in the size of adjacent fans.

Away from the source area, alluvial fans most commonly grade laterally into other alluvial deposits, typically an alluvial plain deposited by a river flowing longitudinally down a valley oriented parallel with the adjacent range and therefore perpendicular or

at high angle to the outbuilding fans. The alluvial plain facies interfingering with the fans may consist of braided, meandering, straight, or anastomosing channels, flood plains, or levees. The longitudinal river system may erode distal parts of the fans, yielding complex depositional and erosional interfingering of the fan and alluvial plain facies (Larsen and Steel, 1978). The alluvial plain deposits may be distinguishable by their better sorting and rounding, more extensive floodplain and levee deposits, larger channel systems, and general lack of mudflow, debris flow and landslide units. In some intramontane areas, the longitudinal fluvial system may consist of a larger alluvial fan building out longitudinally down the axis of the valley, interfingering or coalescing with different fans derived from different source terrains and fed by different upland drainage systems.

Other facies that fans grade into away from the source area include lacustrine (especially playa deposits in arid regions), marine (where fans are constructed on marine shelves, forming fan-deltas), estuarine, deltaic, and eolian. Lacustrine facies are normally recognized by finer grain size, offshore transitions from shoreline sand or mud to deeper lacustrine facies, fossils, and evaporitic deposits (Link and Osborne, 1978). Fans build out into lakes to form sublacustrine fan-deltas, which can have complexly intermixed subaerial and sublacustrine facies because of fluctuating lake levels. Marine and estuarine facies are distinguishable by marine fossils and trace fossils as well as many of the criteria applied to lacustrine sequences; in many areas, alluvial fans build out directly into marine coastal deposits.

In some delta systems bordered by mountainous terrain, such as the Colorado River delta, smaller alluvial fans build out into both subaerial and subaqueous parts of the delta, forming a complex pattern of interfingering facies (Meckel, 1975). In other systems, alluvial fans grade laterally offshore into large deltas that in turn can grade laterally into sublacustrine, subestuarine, and submarine turbidites deposited at different water depths. The deltaic sequence is distinguishable by its finer grain size, different petrography and paleocurrents, and associated subfacies, including fluvial, subaerial and subaqueous swamps, marginal marine, and marine facies.

Eolian deposits develop on the outer parts of fans or at fan margins, and alternating growth of fan and dune systems can yield complex interfingering relations. The eolian facies is distinguished by its textural maturity, grain-surface features, and large-scale bedforms.

GEOMETRY OF ALLUVIAL FAN DEPOSITS

Alluvial fan geometry depends greatly on the tectonic framework of the basin margin. Bull (1972a) recognized three main types of fans based on radial or longitudinal cross-section: (1) wedge-shaped bodies that are thick close to the mountain front and thin or wedge out away from the mountain front, reflecting major mountainous uplift before initiation of fan sedimentation (Fig. 24A); (2) lens-shaped bodies that thin both toward and away from the mountain front, reflecting continued uplift of the mountains during fan sedimentation (Fig. 24B); and (3) wedge-shaped bodies that are thin adjacent to the mountain front and thicker away from it, reflecting a lengthy interval of erosion and redistribution of proximal fan deposits farther downfan, commonly associated with pediment formation.

The geometry can be very complex internally, however, as most fans prograde rapidly following uplift of source areas and establishment of drainage systems. Thus, in older stratigraphic units of most fans, progradation is re-

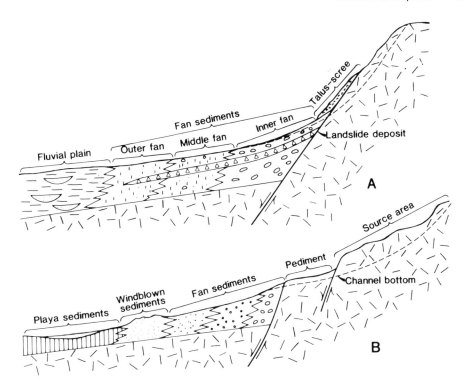

Fig. 22—Cross-sections showing relations of alluvial fan deposits to adjacent facies. **A.** Interfingering with fluvial plain deposits. **B.** Interfingering with playa and windblown deposits.

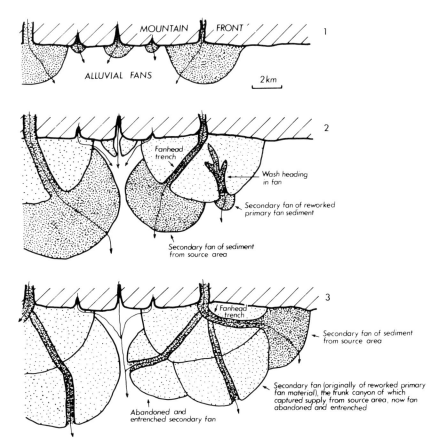

Fig. 23—Diagram showing development styles of alluvial fans (from Denny, 1967). Map 1—Small fan at base of recently elevated mountain. Map 2—Wash from mountain has dissected original fan segment and is building new fan segment. Another wash, heading in abandoned segment is building new fan segment at its mouth. Map 3—Two drainage diversions have taken place, causing wash from mountain to abandon segments and to build new segment at its mouth.

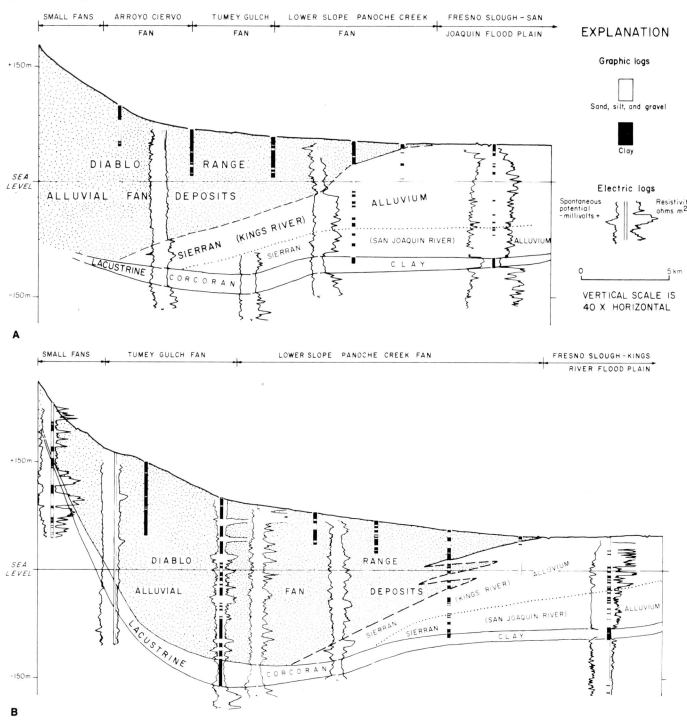

Fig. 24—Radial cross-sections of alluvial fan deposits showing different fan geometries (from Magleby and Klein, 1965, as developed by Bull, 1972a). **A.** Alluvial fan deposits that are thickest adjacent to the mountain front. **B.** Alluvial fan deposits that are lenticular.

corded, with more proximal facies overlying more distal facies, i.e., thinner and finer grained sheetlike strata overlain by thicker and coarser grained channelized strata. As the fan builds outward toward the basin, lateral shifting of channels gradually constructs a depositional body convex upward and lens-shaped in transverse cross-section (Fig. 4). Coalescing of adjacent fans produces a series of merging lens-shaped bodies, although larger and more rapidly prograding fans will en-

croach on and overlap smaller fans, especially their distal parts (Fig. 23). Eventually, new fans may begin to develop in topographically low interfan areas or zones of coalescence and interfingering of adjacent fans (Fig. 23). These new fans can be fed by newly developing drainage systems or, more commonly, by entrenchment and lateral shift of major channels from laterally adjacent older fans.

By the process of entrenchment and abandonment, a complex overlapping

sequence of lens-shaped bodies gradually builds. Unlike deltaic and deep-sea fan constructional lens-shaped bodies, which typically receive fine-grained sediments when inactive to produce a series of sandy lenses in shaly envelopes, the inactive alluvial fan surfaces undergo both weathering, which produces soil horizons, and erosion by shifting channels. This, combined with poor fossil control, great thickness of coarse-grained deposits, and lack of laterally extensive marker beds, leaves the geometry of both individual fans as well as ancient coalesced fan deposits often difficult to delineate, particularly if adjacent fans have similar composition.

The size and shape of alluvial fans are controlled by several factors, including size of drainage area, amount of sediment carried by the feeding stream, and relief and rate of tectonic uplift at the basin margin (continued uplift of the source terrane permits accumulation of greater thickness of sediment). Several studies have indicated that (1) fans with larger drainage areas generally have lower slopes than those built of the same material from smaller source areas (Hooke, 1968); (2) fans are steeper as debris size increases and sediment concentration in flows increases (Hooke, 1968; Hooke and Rohrer, 1979); (3) fans slope more gently in areas of higher precipitation and more steeply in more arid regions, probably because abundant debris-flow deposits form steeper slopes than streamflow deposits; and (4) fans that drain areas underlain by fine-grained sedimentary rocks such as shale or argillite are steeper and larger, whereas those that drain areas underlain by coarser grained sedimentary rocks such as sandstone or crystalline rocks are more gently sloping and smaller.

TECTONIC CONTROLS ON ALLUVIAL FAN SEDIMENTATION

The thickness of fan deposits, shifting of sedimentation sites, and development of characteristic progradational cycles and megasequences are strongly influenced by tectonism. Fans may undergo several major cycles of growth and nongrowth as a result of tectonic uplift of the upland source terrain. Because they are deposited close to the source area, fans are particularly sensitive to tectonic activity. It is difficult in ancient sequences to distinguish pulses of coarse-grained deposition that result from tectonism from those resulting from climatic events, although deposition usually begins as a result of tectonic uplift.

The geometry and facies distribution of fan sequences are controlled primarily by the rate of differential vertical movement between the mountains and adjacent basin, rate of erosion, and rate of sedimentation. Fan growth continues as long as uplift adjacent to the mountain front exceeds or equals the sum of channel downcutting in the mountains and deposition on the fan

apex (Bull, 1977, p. 250). New sedimentation can take place in either proximal or distal areas (Fig. 25). Rapid uplift and generation of debris flows yields new deposition in upper- and middle-fan areas, whereas slower uplift and downcutting yields channel entrenchment and new streamflow deposition in outer-fan and fan fringe areas, as well as developing new fan segments beyond the former fan limits (Figs. 23, 25A). Where strike-slip movement takes place along basin-margin faults, the geometry of fan deposits is even more complex because of lateral shifting of feeding streams and fan apexes (Crowell, 1974). Miall (1978) summarized the effects of syndepositional deformation on alluvial fan sedimentation, which include intraformational angular unconformities, broad structural warps, local thinning, and stacked sedimentary megacycles produced within the basin fill.

Various continental tectonic settings yield abundant fan deposition. Fans are generated along intracontinental rifts such as the Basin and Range province of the western United States and the east African rift system. Basement uplifts in continental interiors yield fan deposition, as seen during the late Paleozoic in Colorado (Howard, 1966) and Cretaceous and early Tertiary in the Rocky Mountains. Fan deposits are preserved in fault-bounded postorogenic basins of the Devonian Old Red Sandstone, the Carboniferous, and the Permian and Triassic New Red Sandstone of eastern North America, Greenland, and western Europe. Alluvial fans are also associated with collisional orogenies such as the modern fans in India that record suturing of India and Asia.

Alluvial fans are also characteristic of sedimentation in strike-slip regions. They have been reported from basins bounded by lateral faults in the Devonian Hornelen basin of western Norway, Carboniferous basins of northern Spain, and various basins in California adjacent to the San Andreas fault.

CRITERIA FOR RECOGNITION OF ALLUVIAL FAN DEPOSITS

Alluvial fan deposits can be recognized primarily by their physical rather than biological or chemical characteristics. Fossils are generally rare, and chemical sediments such as evaporites and carbonates are not typically present. Important physical characteristics include the types of facies

present, textural and compositional immaturity, and coarseness.

The major physical criteria for recognition of alluvial fan deposits are listed below and although each criterion is not necessarily unique to fans, collectively they can be used for the recognition of fan deposits.

1. Sediments deposited relatively close to their source area.
2. Sediments deposited by unidirectional fluid flows.
3. Sediments deposited by high-energy flows.
4. Sediments typically very poorly sorted, containing a great range in grain sizes.
5. Clasts poorly rounded, reflecting

A

B

Fig. 25—Depositional patterns on modern alluvial fans. **A.** New fan developing in the Rakaia riverbed in front of the Terrible Gully fan, Mount Hutt Range, Canterbury, New Zealand (photo by W. Sevon, from S. J. Carryer). **B.** Modern alluvial fans along Thompson Valley, British Columbia, showing proximal and distal fan development (photo by J. M. Ryder).

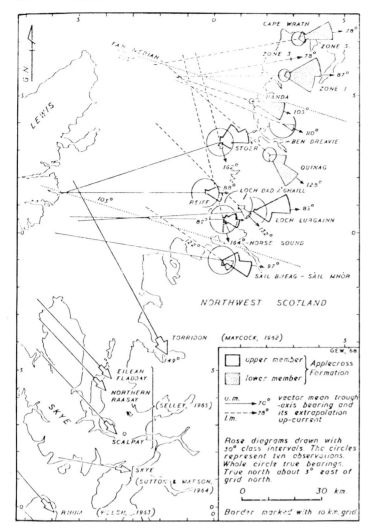

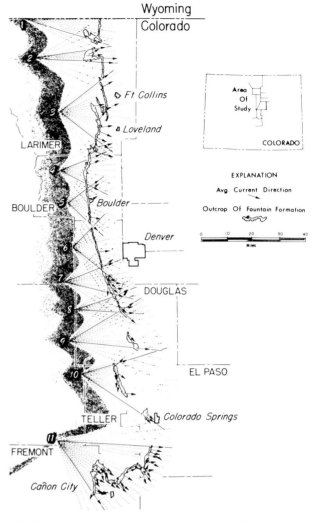

Fig. 26—Regional paleocurrent pattern of the upper and lower members of the Applecross Formation of the Torridonian Sandstone, Precambrian, Scotland, showing radial intersecting patterns related to alluvial fans shed southeastward off highlands located to the northwest (from Williams, 1969).

Fig. 27—Regional paleocurrent pattern of the Fountain Formation, Pennsylvanian and Permian, Colorado, showing arcuate pattern of paleocurrent vectors from 11 alluvial fans deposited east of the ancestral Rockies (from Howard, 1966).

the short distance of transport.

6. Sediments compositionally immature and having a great range in composition, depending on the types of rocks present in the source area.

7. Sediments characterized by major changes in lateral and vertical facies, particularly in the downfan direction.

8. Sediments characterized by rapid downfan decreases in both average and maximum clast size.

9. Sediments generally oxidized, typically never having been placed in reducing conditions; characteristic colors are red, brown, yellow, or orange.

10. Sediments generally not containing large amounts of organic matter because of oxidizing conditions of sedimentation.

11. Sediments generally not containing many fossils except for scattered vertebrate bones and plant fragments.

12. Sediments generally containing a limited suite of sedimentary structures, most commonly medium- to large-scale cross-strata and planar stratification.

13. Depositional bodies having a lenticular or wedge-shaped geometry and typically forming clastic wedges.

14. Depositional bodies consisting of a mixture of partly sorted streamflow deposits and unsorted debris-flow and mudflow deposits.

15. Depositional bodies extensively channelized, consisting dominantly of sediment-filled channels of varied depth and width with generally little lateral continuity.

16. Depositional bodies generally poorly stratified, ranging from unstratified debris flow and mudflow deposits to better stratified streamflow deposits.

17. Depositional bodies changing later-

ally in composition because each fan of a coalesced group of fans is fed typically by an individual mountain stream that may drain an area underlain by distinctive bedrock units.

18. Depositional bodies commonly associated with fault-bounded basins such as grabens, half-grabens, and various types of strike-slip fault-bounded basins.

19. Depositional bodies in which streamflow and debris-flow deposits interfinger distally or downfan with finer grained alluvial plain, eolian, lacustrine, or marine facies.

20. Depositional bodies containing abundant soil profiles and terrace surfaces.

21. Depositional bodies containing salts such as gypsum or calcite deposited interstitially, within soils, and as caliche layers.

22. Paleocurrent patterns radiating outward downfan in individual fans

A

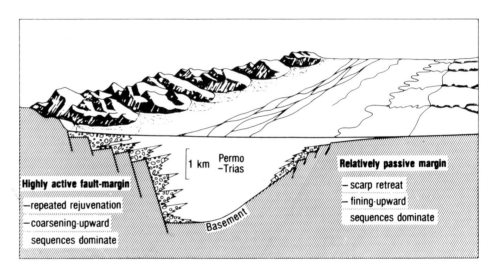

B

Fig. 28—Thickening and coarsening-upward alluvial fan sequences controlled by tectonic activity. **A.** Thickening- and coarsening-upward megasequence in Devonian alluvial fan

deposits along the northern margin of the Hornelen basin, Norway, thought to result from tectonic uplift along basin-margin strike-slip fault (photo by T. G. Gloppen). **B.** Thickening- and

coarsening-upward megasequence in Miocene alluvial fan deposits of the San Onofre Breccia, 2 km northwest of Dana Point, southern California (photo by C. J. Stuart).

and having complex radiating patterns in coalesced alluvial fans.
23. Depositional bodies that include sieve deposits, which are not characteristic of other terrestrial facies.

Additional features of alluvial fan deposits used to recognize fans in the stratigraphic record and distinguish different types of fans are discussed in subsequent sections.

PALEOCURRENT PATTERNS

Paleocurrents from an individual alluvial fan deposit should reflect the morphology of the original fan as a flow pattern radiating outward from the fan apex (Figs. 26, 27). Alluvial fan paleocurrent patterns are basically very regular and simple because currents flow directly downslope under the influence of gravity rather than being affected by other factors such as winds, waves and tides. Initial paleocurrent patterns may be more variable because of irregularities on the topographic surface over which the fan is deposited.

In ancient fan sequences, apexes are typically missing because of continued uplift along basin-margin faults, so that the radial pattern for individual fans and piedmonts may be less clear (Figs. 26, 27). Coalescing of adjacent fans and intertonguing of fans from adjacent sides of an intramontane basin often produce complex patterns. Williams (1969) suggested that paleocurrent patterns of prograding fans have wider dispersion upsequence while retro-

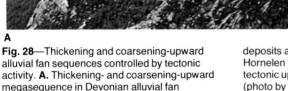

Fig. 29—Development of coarsening-upward sequences along highly active fault margins and fining-upward sequences along passive margins in Triassic basins of northwest Scotland (from Steel, 1977).

grading fans have lower dispersion upsequence.

Paleocurrents are measured from many features of alluvial fan deposits. The indicators, in probable order of reliability, include channel orientation, longitudinal and transverse bar orientation, conglomerate clast long-axis orientation and imbrication, dune orientation, large-scale cross-strata, medium-scale cross-strata, primary current lineation, small-scale cross-strata and ripple markings, and sandstone grain orientation.

Measurement of the largest clast sizes in alluvial fan sequences is also generally an excellent paleocurrent indicator, because most studies of modern fans have indicated rapid decrease

in maximum and average clast size downfan from the fan apex.

VERTICAL SEQUENCES

Alluvial fan deposits may be characterized by strongly developed overall coarsening- and thickening-upward sequences, indicative of active fan progradation and outbuilding, or by fining- and thinning-upward sequences indicative of relative inactivity or fan retreat (retrogradation). These overall sequences may be hundreds or even thousands of meters thick (Fig. 28). Triassic basins of northern Scotland have an overall thickening and coarsening-upward sequence along the tectonically active basin margins and a thinning- and fining-upward sequence

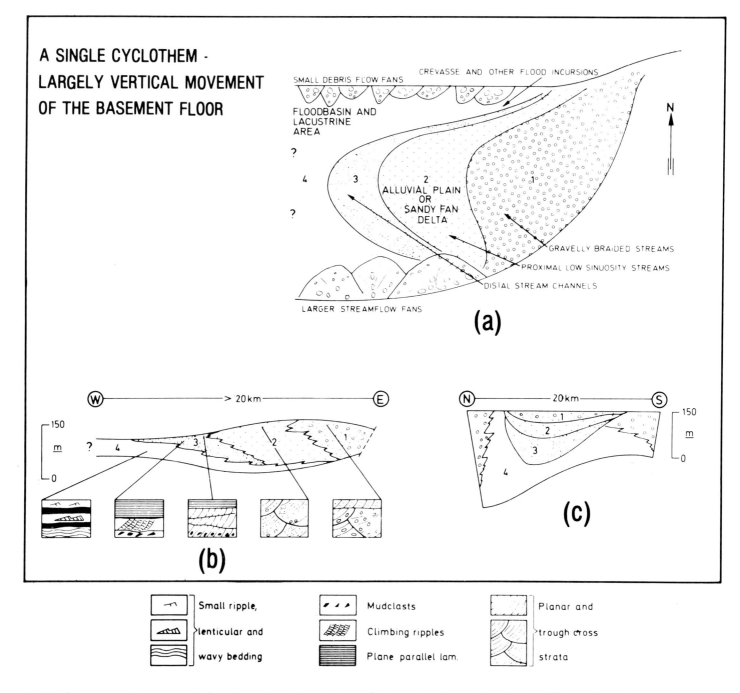

A SINGLE CYCLOTHEM -
LARGELY VERTICAL MOVEMENT
OF THE BASEMENT FLOOR

Fig. 30—Paleogeographic summary of alluvial fan sedimentation showing development of coarsening-upward sequences in the Devonian Hornelen basin, western Norway (from Steel and Gloppen, in prep.): **a.** map view; **b.** east-west cross-section; **c.** north-south cross-section.

along the less active basin margins (Steel, 1977; Fig. 29). More complex patterns of basin-margin faulting can produce more variable coarsening and fining-upward overall sequences, as well as symmetrical sequences. Major climatic changes can also produce overall cycles, primarily because of gradual changes in both the amount of precipitation and runoff, and balance between chemical and physical weathering produced by climatic change.

Tectonic effects on fan sedimentation have been well studied in Devonian basins of western Norway, where cyclic sedimentation on four scales has

been related to effects of depocenter migration, lateral faulting, vertical movements of the basin floor, and sedimentary progradational processes (Figs. 30, 31).

Smaller systematic vertical changes in coarseness or thickness of beds within alluvial fan sequences respond to briefer cycles of progradation and retrogradation that reflect either outward migration of proximal facies over distal facies or retreat of distal facies over proximal facies. These thinner cycles or megasequences generally range from 10 to 100 m in thickness. Migration of the intersection points of chan-

nels will produce megasequences.

Heward (1978a, b) demonstrated in carboniferous alluvial fan sequences of northern Spain that thinning- and fining-upward megasequences characterize the channelized middle parts of the fans and thickening- and coarsening-upward sequences the distal parts of fans (Fig. 32). Various conditions of fan sedimentation can produce megasequences (Fig. 33). This systematic pattern is similar to that of deltaic and deep-sea fan sequences, where gradual abandonment of channels yields thinning- and fining-upward megasequences with channelized bases, and progradation yields thickening- and coarsening-upward megasequences with nonchannelized bases. Inner-fan channel sequences may be complex in terms of vertical sequence because of complex patterns of entrenchment and alternate filling by streamflow, debris flows and mudflows.

SEDIMENTARY STRUCTURES AND RELATED FEATURES

General

The variety of sedimentary structures and related features in alluvial fan deposits, especially coarse-grained fans, is limited in comparison to alluvial plain, deltaic and marine deposits. This limited suite of structures is characteristic of fan deposits, where deposition is restricted to streamflow, mudflow and debris-flow processes.

Graded and inverse-graded bedding—Individual beds that fine upward, coarsen upward, and do not change upward in grain size are all

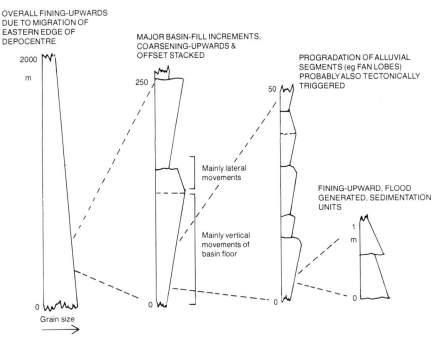

Fig. 31—Tectonic model for the Devonian Hornelen basin, western Norway, showing origin of various fining- and coarsening-upward sequences in alluvial fan deposits (form Steel and Gloppen, in prep.).

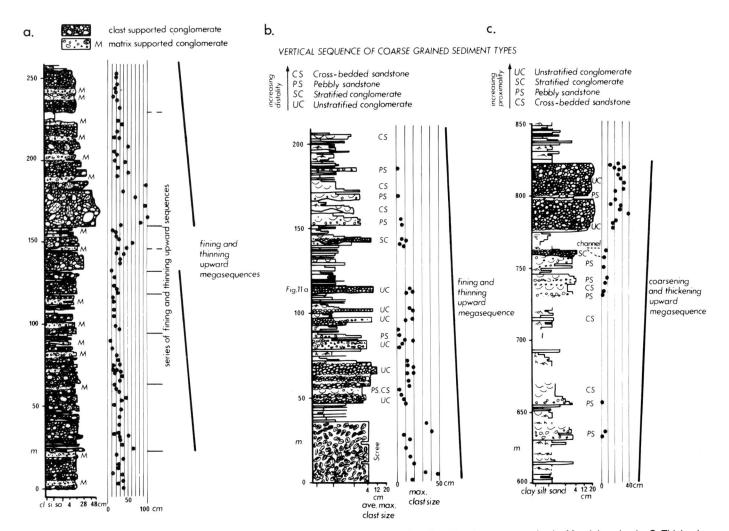

Fig. 32—Examples of alluvial fan megasequences from the Carboniferous of northern Spain (from Heward, 1978b). **A.** Thinning- and fining-upward megasequences from the Correcillas conglomeratic valley-fill at the base of the coalfield succession Matallana basin. **B.** Thinning- and fining-upward megasequences form within the coalfield succession La Magdalena basin. **C.** Thickening- and coarsening-upward megasequences form within the coal field succession, La Magdalena basin.

A *Response to Initial Topography*

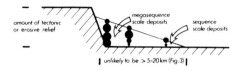

B *Short-Moderate Duration Fanhead Entrenchment : resulting from intrinsic or climatic factors (Table 4)*

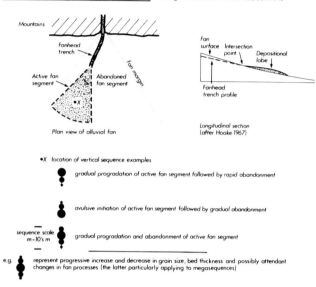

•X *location of vertical sequence examples*

gradual progradation of active fan segment followed by rapid abandonment

avulsive initiation of active fan segment followed by gradual abandonment

sequence scale m - 10's m — *gradual progradation and abandonment of active fan segment*

e.g. *represent progressive increase and decrease in grain size, bed thickness and possibly attendant changes in fan processes (the latter particularly applying to megasequences)*

C *Prolonged Fanhead Entrenchment : resulting from decreasing sediment supply, the latter perhaps due to the advanced state in the cycle of erosion; or due to lowering of base level (Table 4)*

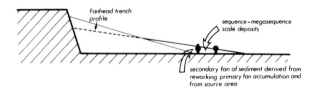

D *Scarp Retreat and Lowering of Relief :*

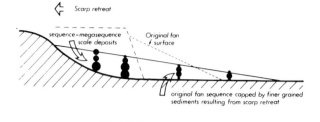

e.g. *represent progressive increase and decrease in grain size, bed thickness and possibly attendant changes in fan processes (the latter particularly applying to megasequences)*

E *Response to Tectonic Uplift : relative uplift exceeds rate of stream dissection*

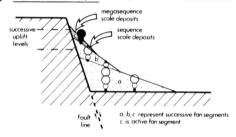

F *Response to Tectonic Uplift : prolonged entrenchment (Table 4), as stream dissection exceeds rate of relative uplift*

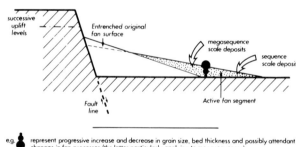

e.g. *represent progressive increase and decrease in grain size, bed thickness and possibly attendant changes in fan processes (the latter particularly applying to megasequences)*

Fig. 33—Examples of different alluvial fan settings and the types of vertical sequences that may develop (from Heward, 1978b). **A.** Response to initial topography. **B.** Fanhead entrenchment of short to moderate duration. **C.** Prolonged fanhead entrenchment. **D.** Scarp retreat and lowering of relief. **E.** Response to uplift that exceeds rate of erosion. **F.** Response to uplift that is less than rate of erosion.

present in alluvial fan sequences. Beds deposited by streamflow commonly fine upward, probably because in deposition from single flood events the fining-upward part of the bed represents deposition during the waning stage (Fig. 13). Although many beds deposited by debris flows coarsen upward in their lower part and fine upward in their upper part (Figs. 34, 35), beds that either coarsen or fine upward to the top of the bed or are ungraded may be common or more abundant in some fan deposits. These variations in character of beds deposited by debris flows result from changes in both the amount and strength of matrix relative to the

density and size distribution of clasts. Beds deposited by mudflow processes are characteristically ungraded because the matrix and framework grains are thoroughly mixed.

Channels—Channels of different size and shape are common in alluvial fan deposits, both in proximal and distal settings. The channels are commonly filled with conglomerate, but may also fill with finer grained sediments (Fig. 36). Streamflow, debris-flow and landslide material may fill the channels.

Boulder trains—Boulder and cobble trains have been noted from both modern (Bull, 1964c) and ancient allu-

vial fan deposits (Nilsen, 1968). The trains are lines of coarser clasts oriented radially outward on fan bedding surfaces (Figs. 37, 38A, B). Trains may develop from the accumulation of larger clasts in the lee of large boulders, accumulation of larger clasts in small and narrow channels, or coarse debris rolling, bouncing, and breaking up to form trains on fan surfaces from rockfalls. They can be used as paleocurrent indicators, and clasts in trains are commonly imbricated and may have long axes oriented parallel with flow (Fig. 38B).

Clast imbrication—Clasts of streamflow deposits on alluvial fans

C   **D**

Fig. 34—Fining-upward and coarsening-upward beds in alluvial fan deposits. **A.** Fining-upward pebble conglomerate and sandstone couplets, streamflow deposits, Old Red Sandstone, Devonian, near Storakersund, Solund, western Norway (photo by T. H. Nilsen).

B. Fining-upward bed of fine conglomerate and sandstone. Old Red Sandstone, Devonian, Hornelen, western Norway (photo by V. Larsen). **C.** Reverse-graded debris-flow with normal grading at top, passing upward into overlying shale-chip deposit, San Onofre Breccia,

Miocene, Laguna, southern California (photo by C. J. Stuart). **D.** Reverse-graded debris flow with normal grading at top, Echo Canyon Conglomerate, Late Cretaceous, north-central Utah (photo by K. A. Crawford).

are typically well imbricated, while those in debris-flow and mudflow deposits are typically unoriented or poorly imbricated (Figs. 17, 34). Streamflow alluvial fan deposits with long axes oriented both parallel with

(Nilsen, 1969) and perpendicular to (Rust, 1972) flow have been reported (Fig. 38C). The orientation depends on flow velocity, concentration, shape of clasts, and type of stratification, as well as other factors.

Clast features—Half-round clasts appear to be common in some fans and may be indicative of recycling of previously transported and rounded clasts (Fig. 38D). Rip-up clasts of fan deposits are common because of rapid hardening of debris-flow and mudflow deposits, mudcracking, early cementation by carbonate, and channel incision (Fig. 38E). Percussion marks are common on the surfaces of hard clasts such as quartzite because of the vigor of sediment transport.

Armored mudballs—Armored mudballs, subspherical clasts of clayey material with pebbles attached to the surface, are present in some fans, generally associated with debris-flow deposits. They range in size from 2 to 50 cm in diameter and are commonly deposited in trains (Spearing, 1974).

Synsedimentary slumps—Slumps within alluvial fan deposits are not common because (1) slopes are generally low, (2) coarse-grained deposits tend to consolidate and harden early and are thus capable of supporting relatively steep slopes, and (3) fine-grained deposits, which when wetted may become unstable, are typically absent, rare, or concentrated in lower fan areas where slopes are low and surface relief minimal. Nevertheless, slumps may occur in inner-fan channel areas, where fanhead entrenchment can produce considerable local relief and

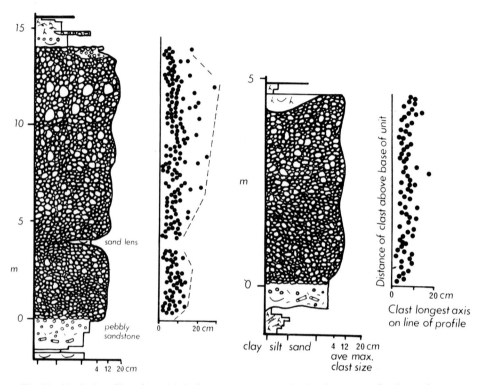

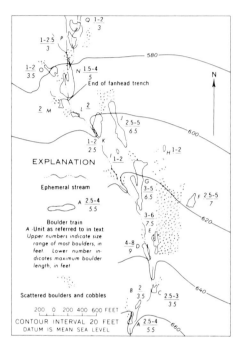

Fig. 35—Vertical profiles of two debris-flow conglomerates showing inverse grading in poorly sorted, clast-supported unstratified conglomerates deposited by debris flows. Carboniferous alluvial fan deposits of the La Magdalena coalfield, northern Spain (from Heward, 1978a).

Fig. 37—Distribution and sizes of boulder trains on a modern alluvial fan, southern Nevada (from Bull, 1964c).

A

B

Fig. 36—Channel-fill streamflow deposits in alluvial fans. **A.** Narrow conglomerate-filled Miocene channel, Fish Creek Wash, southeastern California (photo by O. R. Kerr). **B.** Broad and shallow conglomeratic sandstone-filled channel, Perry Formation, Devonian, St. Andrew's, New Brunswick (photo by P. R. Schluger).

Fig. 38—Characteristics of conglomerates deposited on alluvial fans. **A.** Pebble train on bedding surface, Old Red Sandstone, Devonian, Vetefjell, Solund, western Norway (photo by T. H. Nilsen). **B.** Boulder train on bedding surface, Old Red Sandstone, Devonian, Indre Solund, western Norway (photo by T. H. Nilsen); note orientation of boulder long axes on bedding surface roughly parallel with hammer. **C.** Clast long-axis orientation on bedding surface, Old Red Sandstone, Devonian, Drengenes, Solund, western Norway (photo by T. H Nilsen); long axes parallel with hammer. **D.** Half-round clasts that may be indicative of recycling, Old Red Sandstone, Devonian, Portercross, Ayrshire, Scotland (photo by A. P. Heward). **E.** Rounded rip-up clast of previously deposited debris-flow, Echo Canyon Conglomerate, Late Cretaceous, Echo Canyon, north-central Utah (photo by K. A. Crawford).

Fig. 39—Types of cross-stratification present in alluvial fan deposits. **A.** Large-scale cross-strata in streamflow conglomerate from fan toe area, New Red Sandstone, Triassic, Mull, Scotland (photo by R. J. Steel). **B.** Medium-scale low-angle cross-stratified sandstone lens in sequence of debris-flow conglomerate, Carboniferous coal field deposits, northern Spain (photo by A. P. Heward). **C.** Large-scale cross-strata in granular very coarse-grained sandstone of middle-fan facies, Van Horn Sandstone, Precambrian(?), west Texas (photo by J. H. McGowen); foresets about 9 feet (2.8 m) thick, overlying alternating coarse-grained sandstone and muddy sandstone that is locally injected into the foresets. **D.** Large-scale cross-strata in pebbly sandstone, Perry Formation, Devonian, St. Andrews, New Brunswick, Canada (photo by P. R. Schluger). **E.** Trough cross-strata, Old Red Sandstone, Devonian, Utvaer, Solund, western Norway (photo by T. H. Nilsen). **F.** Trough cross-strata filling channel, Old Red Sandstone, Devonian, Krakevag, Solund, western Norway (photo by T. H. Nilsen).

lateral channel erosion can undercut channel walls. The toes of fans that build out into lakes may be very unstable (Larsen and Steel, 1977).

Cross-stratification—Large-, medium-, and small-scale cross-stratification are common in gravel and sand deposited by streamflow processes (Fig. 39). The cross-strata develop in longitudinal bars, transverse bars, dunes, and ripples (Fig. 40). Different scales and types of cross-strata associated with different grain sizes characterize inner-, middle-, and outer-fan deposits, as shown by the sequence of outcrop cross-sections in Figure 41.

Longitudinal bars develop in channels and are elongate and oriented parallel with flow (Fig. 40). Although typically formed or parallel-stratified gravel, longitudinal bars may contain low-angle large-scale foreset cross-strata along the downcurrent and lateral margins of the bars.

Transverse bars are continuous-crested features extending across channels, perpendicular or at high angles to the flow direction. Transverse

bars occur both singly and in sets, migrating downstream and internally consisting of large- and medium-scale foreset cross-strata deposited as tabular cross-sets. They generally develop in sand and pebbly sand.

Dunes are bedforms that are smaller than bars and characterized by medium- and small-scale trough cross-strata. Dunes commonly develop in the distal parts of alluvial fans where sheets of sand, associated with transverse bars, are deposited by streamflow in broad shallow channels or in nonchannelized areas.

Ripple markings are not generally common in alluvial fans both because rapid flows typically transport fine-grained sand to depositional sites beyond the margins of fans and because ripples are not associated with mudflow or debris-flow depositional processes. Where present, the cross-laminae in ripples are tabular in shape where crests are straight, and trough-like where crests are undulatory (Fig. 42A). Ripples may form in distal-fan sheetflood deposits (Fig. 42B, C) and

on levee crests or interchannel areas of fans containing abundant fine-grained sediment.

Parallel stratification — Parallel stratification is characteristic both of thin sheet-like lenses of sand in gravel sequences and of some gravels deposited by streamflow in proximal parts of fans (Fig. 43). The parallel stratification results from upper flow-regime conditions that prevail during floods. Where traced across the entire length of a sandstone lens, some parallel stratification can be shown to be very low-angle cross-stratification. Primary current lineation is common on sandstone stratification surfaces. Parallel stratification in conglomerates is most common in fine conglomerates but is not restricted to them.

Antidune structures—Antidunes are produced by standing waves and generally indicate high flow velocities. They have been reported from a number of ancient alluvial fan deposits and have been noted forming on modern fans. The flow of large amounts of water confined to channels down rela-

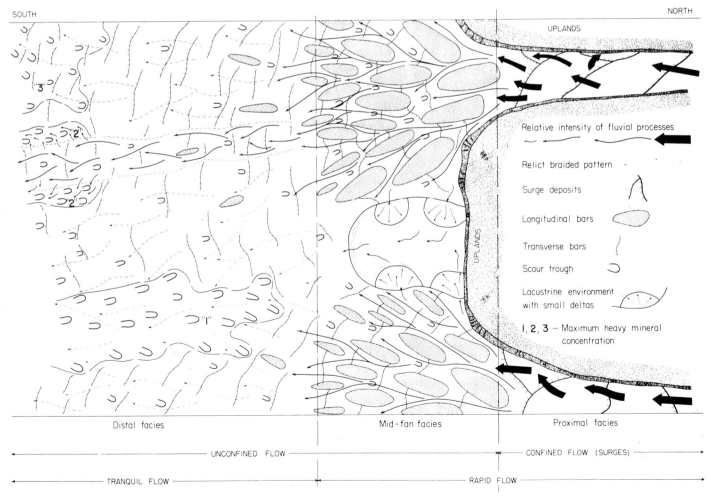

Fig. 40—Prospecting model for alluvial fan deposits of the Van Horn Sandstone, Precambrian(?), west Texas, showing distribution of various types of bars and stream patterns (from McGowen and Groat, 1971).

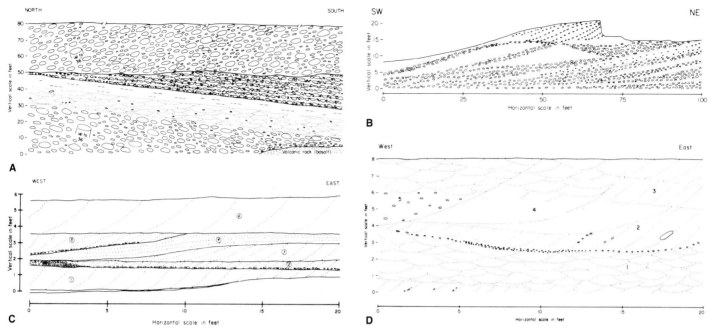

A

B

C

D

Fig. 41—Types and scales of cross-strata developed in inner-, middle-, and outer-fan deposits of the Van Horn Sandstone, Precambrian(?), west Texas. **A.** Basal part of inner-fan facies, consisting in ascending order of lower massive conglomerate, trough-cross-stratified conglomeratic coarse- to very coarse-grained sandstone, massive sandy mudstone, interbedded conglomerate and parallel-laminated sandstone, an unconformable surface, and an upper massive conglomerate. **B.** Upper part of inner-fan facies, consisting of westward-accreting foreset-bedded gravel bars and flanking shallow channels. **C.** Middle-fan facies, consisting of foreset cross-beds in sandstone developed in broad, shallow channels filled with transverse bars, thin, channel-floor gravels, and muddy fine-grained sandstone filling residual channels. **D.** Outer-fan facies, consisting of trough- and foreset-cross-bedded sandstone, with lateral discontinuity of units typical of braided channel deposits.

A

B

C

Fig. 42—Ripple markings in alluvial fan deposits. **A.** Ripple markings on bedding surface, Perry Formation, Devonian, Slipp Bay, southwestern New Brunswick (photo by P. R. Schluger). **B.** Ripple markings in sheetflood granule sandstone, Old Red Sandstone, Devonian, Hornelen, western Norway (photo by R. J. Steel). **C.** Climbing ripple markings in lower-fan sheetflood deposits, Old Red Sandstone, Devonian, Hornelen, western Norway (photo by R. J. Steel).

C

D

Fig. 43—Parallel stratification in alluvial fan deposits. **A.** Miocene streamflow deposits, Fish Creek Wash, southern California (photo by D. R. Kerr). **B.** Streamflow deposits of Old Red Sandstone, Devonian, near Polletind, Solund, western Norway (photo by T. H. Nilsen). **C.** Eocene streamflow deposits, Chivo Canyon, Simi Hills, southern California (photo by T.H. Nilsen). **D.** Thick-bedded sheetflood deposits, Old Red Sandstone, Devonian, Hornelen (photo by R. J. Steel).

tively steep gradients is ideal for development of antidunes. In ancient fan deposits, upstream-dipping cross-strata in beds of sandstone or pebbly sandstone may indicate antidunes.

Convolute bedding and elutriation structures and flame structures— Water-escape structures are uncharacteristic of alluvial fan deposits but may develop in some fans due to rapid sedimentation (Fig. 44A-C). McGowen and Groat (1971) reported upward injec-tion of muddy sandstone into trough cross-strata in distal braided distribua-try deposits of the Precambrian Van Horn Sandstone of Texas (Fig. 44A); they considered the injection to result from interruption of interal drainage by the impermeable muddy sand. Flame structures have not been commonly re-ported from fan deposits, but condi-tions for their development can occur (Fig. 44D-E).

Desiccation cracks—Desiccation cracks are common in many alluvial fans, mostly on the surface of mudflows and sometimes on the sur-face of debris flows (Fig. 45). Desicca-tion cracks are not common in streamflow deposits because these are rarely mud.

Burrows and root casts—Alluvial fans are a suitable habitat for a variety of mammals, chiefly rodents, although birds, reptiles, and insects such as bee-tles and ants may be common. Conse-

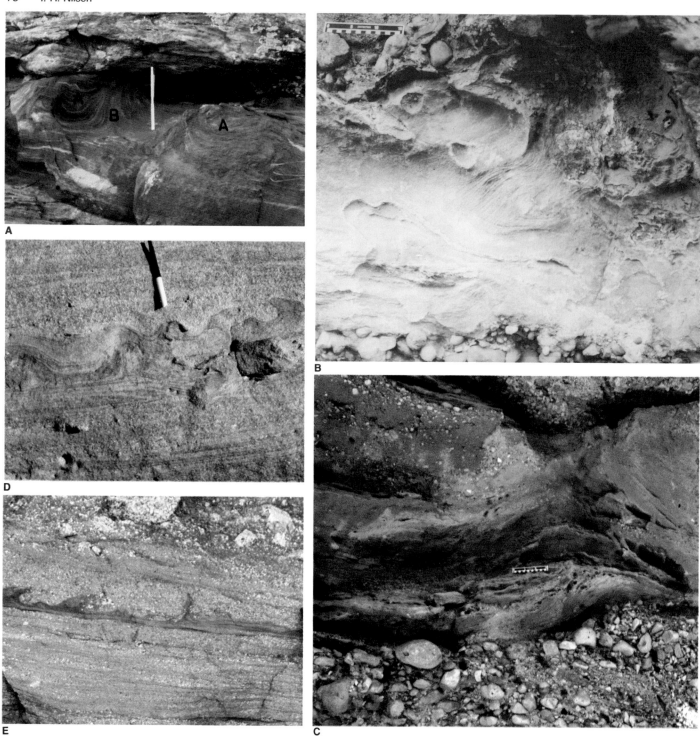

Fig. 44—Convolute laminations and fluid-escape structures in alluvial fan deposits. **A.** Convolute laminations and injection features of braided distributary deposits, Van Horn Sandstone, Precambrian(?), west Texas (photo by J. H. McGowen); trough-cross-stratified sandstone (A) injected by muddy sandstone (B).

B. Sandstone stratification contorted by fluid escape, basal conglomerate of Evanston Formation, Late Cretaceous and Paleocene, Upper Echo Canyon, north-central Utah (photo by K. A. Crawford). **C.** Sandstone deformed by fluid escape, Echo Canyon Conglomerate, Late Cretaceous, north-central Utah (photo by K. A.

Crawford). **D.** Convolute laminations, load casts, and fluid escape structures, Pliocene strata, Durmid Hills, southeastern California (photo by T. H. Nilsen). **E.** Flame structures, Miocene strata, Fish Creek Wash, southeastern California (photo by D. R. Kerry).

quently, a surprising number of burrows, large and small, are present in Mesozoic and Cenozoic alluvial fan deposits (Fig. 46). Various plants produce root casts of different types, and these may locally be difficult to distinguish from burrows. Extensive soil formation, plant growth and burrowing can destroy original sedimentary features.

COMPOSITIONAL PARAMETERS

The composition of modern fans correlates closely to that of the source area because of the relatively short transport distances and general lack of sorting within the fans and that is undoubtedly true for ancient fan deposits. Composition of fans becomes more complex distally and laterally because

of coalescence of adjacent lithologically dissimilar fans and mixing with basin-axis alluvial plains.

Conglomerates and breccias contain clasts of source-area rocks, but their relative amount will be inconsistent in fan deposits. Factors contributing to these differences include sorting that takes place on the fan and weathering

A

B

Fig. 45—Desiccation cracks in alluvial fan deposits. **A.** Desiccation cracks on bedding surface, Perry Formation, Devonian (photo by P. R. Schluger). **B.** Infilled desiccation crack in cross-section, with associated mudstone rip-up clasts, at top of mudflow deposit, Miocene and Pliocene deposits, northern Grant Range, Nevada (photo by T. D. Fouch).

A

B

Fig. 46—Burrows in alluvial fan deposits. **A.** Unusual backfilled burrow in Perry Formation, Devonian, May Island, New Brunswick (photo by P. R. Schluger). **B.** Burrows in well-stratified fine-grained Miocene and Pliocene debris-flow deposits, northern Grant Range, Nevada (photo by T. D. Fouch).

of different rock types that produces various amounts of gravel, sand, silt, and clay (depending on amount and spacing of joint and bedding surfaces, relative strength of physical and chemical weathering, and soil types).

Sandstones are both arenites and wackes, depending on whether deposition came by streamflow or mudflow processes. Compositionally, they may be quartzose, feldspathic, or lithic. The abundance of rock fragments decreases downfan as a result of abrasion during transport.

TEXTURAL PARAMETERS

Alluvial fan deposits characteristi-cally are texturally immature. Clast size and bedding thickness typically decrease downfan (Fig. 47), although changes from debris-flow to streamflow deposition must be considered. Clasts in streamflow deposits tend to be better rounded than those in debris-flow or mudflow deposits because they have been subject to more clast-to-clast collisions. Clast roundness increases downfan whereas clast sphericity shows little downfan change and seems unrelated to depositional process (Fig. 48).

GRAIN-SIZE DISTRIBUTION

Sediment in alluvial fans ranges in size from clay to boulder. Breccias and conglomerates are very common, especially in ancient deposits recognized as alluvial fans. Fan deposits contain abrupt vertical and lateral changes in sorting and maximum and mean grain size.

Sorting of debris-flow deposits is generally poor, although few grain-size analyses have been reported in the literature because of difficulties in doing analyses on such poorly sorted materials (Fig. 44). Sorting of mudflow deposits is also generally poor, typically within the clay to granule sizes (Fig. 49). Streamflow deposits are generally better sorted, but typically not as well as in alluvial plain sequences. Bull (1963) reported from (a single) modern fan in central California mean sorting values (S_o) of 1.5 for braided middle- and lower-fan stream deposits, 2.1 for incised stream channel deposits, and 9.7 for mudflow deposits. He inferred that different deposits could be readily distinguished by sorting indices from the fan, whereas Wasson (1977a) had great difficulty distinguishing various deposits for alluvial fans in Tasmania. Sorting generally increases downfan because debris-flow, mudflow, and landslide deposits tend to be most abundant in proximal parts of fans, and streamflow deposits are most common in distal parts of fans.

CM PATTERNS

CM patterns (coarsest one-percentile particle size, C. versus median particle size, M) are thought by Bull (1962, 1972a) to be distinctive for different types of alluvial fan deposits. Mudflow deposits form a pattern that slopes parallel to the C = M limit (Fig. 50A). Debris-flow deposits plot similarly to tills in the upper left part of the diagram, characterized by high C and intermediate M values (Fig. 50B). Streamflow deposits plot similar to most alluvial plain sequences. Wasson (1977b), however, found that streamflow and debris-flow deposits in Tasmania could not be distinguished by CM plots. Modern sieve deposits not infiltrated by sand or mud would plot close to the M = M limit because they contain very well sorted gravel-size material without interstitial fines.

CEMENTATION AND DIAGENESIS

Alluvial fans can have various types of cement develop at different times in the depositional and post-depositional history of the fan. Carbonate is the cementing agent in the majority of fans studied, but siliceous, iron oxide, and petroliferous cements are also noted, the first particularly in silica-rich fan sediments. Caliche zones are common in alluvial fan deposits (Fig. 51). Neither cementation nor diagenesis has been extensively studied in ancient fan deposits.

Development of fan cements is quite complex because of their subaerial exposure, extensive surface and subsurface flow of water, variable size and composition of sediments, subsidence caused by both near-surface and deeper processes, and development of various soils and desert pavements on fan surfaces. Lattman (1973), from a detailed study of near-surface calcium carbonate cementation of modern arid alluvial fans in a part of southern Nevada characterized by abundant carbonate-rich springs, concluded that six processes of pedogenic and nonpedogenic cementation produced calcic layers (see Table 1). The development of these near-surface carbonate cements depends on the composition and texture of the fan, availability of wind-blown carbonate silt and mud, and age of the fan surface.

Funk (1979), in a study of carbonate cements in Quaternary alluvial fan deposits in east-central Idaho, concluded that cementation begins soon after deposition and simultaneously occurs in the near-surface vadose zone by pedogenic and nonpedogenic processes, in the vadose zone by dissolution and incipient cementation, and in the "vadose-phreatic" zone of water-table fluctuation.

POROSITY AND PERMEABILITY

Alluvial fans typically are highly porous and permeable, forming excellent aquifers and potentially good reservoirs for oil and gas. Deposits characteristically consist of alternating zones of porous and permeable streamflow deposits and relatively nonporous and impermeable debris-flow and mudflow deposits. Modern sieve deposits in particular are highly porous and permeable. Flow of ground water is commonly aided by paleochannels, which act as conduits for flow. Aquifer characteristics vary greatly with type of deposit and location on fan (Cehrs, 1979).

Other pore space develops as intergranular voids, interlaminar voids, bubble cavities, and desiccation cracks. Distal fan deposits, although finer grained, generally consist of better sorted streamflow sediments that are porous and permeable. The greater abundance of mudflow and debris-flow deposits in proximal and medial parts of fans may result in decreased and less predictable porosity and permeability in these areas.

E-LOG CHARACTERISTICS

Little has been published regarding E-log characteristics of alluvial fan deposits because they have not generally been targets for oil exploration. The signatures of different fan facies are not well known, although it appears clear from most fan studies that proximal facies are more channelized, contain beds or groups of beds that are less extensive laterally, and contain more debris-flow deposits that may have characteristic signatures (Fig. 18) than distal facies. Heward's (1978a, b) work suggests that in fans containing a variety of sediment sizes, middle-fan deposits should be characterized by thinning and fining-upward megasequences and that outer-fan deposits should be characterized by nonchannelized thickening- and coarsening-

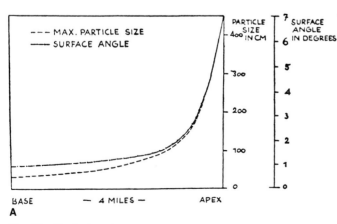

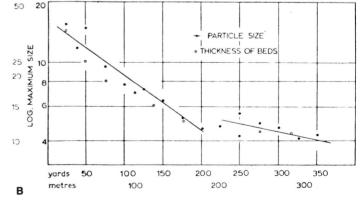

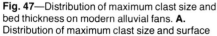

Fig. 47—Distribution of maximum clast size and bed thickness on modern alluvial fans. **A.** Distribution of maximum clast size and surface angle along a radial profile of an alluvial fan in the Santa Catalina Mountains, Arizona (from Blissenbach, 1954). **B.** Distribution of maximum clast size, bedding thickness, and inferred distance of transport on a Triassic alluvial fan deposit, Wales (from Bluck, 1965).

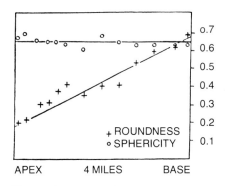

APEX   4 MILES   BASE

Fig. 48—Distribution of roundness and sphericity along a radial profile of an alluvial fan in the Santa Catalina Mountains, Arizona (from Blissenbach, 1954).

upward megasequences.

MINERAL DEPOSITS

Alluvial fan deposits are good sources of sand and gravel. They may also contain heavy mineral placers, be host rocks for roll-front uranium deposits, and contain mineable coal. They are generally not good sources of ore minerals, except perhaps where metamorphosed or subjected to faulting and emplacement of ore-bearing fluids emanating from fault zones.

Placer deposits are common in alluvial fan and associated bajada, pediment, arroyo, and fluvial deposits in many parts of the world. Ancient alluvial fan placers have not always been clearly distinguished from alluvial plain placers. Fan placers are generally developed in channels on bedrock near the heads of fans or in sandy depositional lows in outer parts of fans. Sediment reworking and sorting is required to form the placers, which have been produced experimentally by Schumm (1977) and colleagues.

Many heavy minerals can be concentrated in placer deposits. Those recovered from all types of fluvial deposits are listed in Table 2. Gold, ilmenite, magnetite, cassiterite, monazite, platinum, and diamonds have been produced from alluvial fan deposits (McGowen and Groat, 1971).

In the Precambrian(?) Van Horn Sandstone of Texas, heavy minerals are most abundant in the distal fan deposits, concentrated along slip-face laminae of transverse dunes (preserved as trough cross-strata) formed in broad, shallow channels (McGowen and Groat, 1971). In this part of the fan, heavy minerals were concentrated because of reworking of medium- to coarse-grained sands by runoff under conditions of tranquil flow (McGowen, 1979).

In the western United States, placer gold has been mined from late Cenozoic alluvial fan deposits in New Mexico, Arizona, Nevada, and adjoining areas (see summaries in Johnson, 1972a, b, 1973). Gold in these fans is generally concentrated at or near bedrock and has been eroded from mineralized bedrock and deposited after short transport distances in well-defined stream channels on alluvial fans at the edges of range fronts, or in residual hillside debris. Best production generally comes from lenses, channels and sheetlike bodies in lower gravel layers. These are characterized by better rounding, sorting and washing, and are typically found in channels at bedrock near the head of a fan or scattered in lenses in lower parts of a fan. In some districts, other placer minerals associated in economic quantities with gold have been mined, including scheelite and cinnabar in the Dutch Flat district of Nevada (Johnson, 1973, p. 30-31), and cinnabar and wolframite in the Arivaca district of Arizona (Johnson, 1972b, p. 38).

In the 2.8-billion-year-old Precambrian Witwatersrand system of South Africa, gold and uranium are concentrated as placers in sediments thought to be alluvial fan and fan-delta deposits (Pretorius, 1974a, b, 1975, 1976). The uranium occurs most commonly as uraninite distributed along both foresets and at the base of cross-strata deposited in channels of braided streams (Minter, 1976, 1978). The fans are about 45 km long and thought to have been deposited under humid conditions; they prograded into a marine or lacustrine environment and were then reworked by longshore currents. In these fans, the greatest abundance of gold is found in the upper midfan conglomeratic deposits and uraninite is located in lower midfan conglomeratic sandstone deposits. Pretorius (1975) concluded that gold and uranium were concentrated by lesser flood events that transported sand which infiltrated previously deposited open-framework gravels transported by greater flood events. Other heavy minerals present include nodular pyrite, zircon, chromite, and leucoxene.

Other Precambrian gold and uranium deposits thought to be at least in part of alluvial-fan placer origin include the Elliott Lake-Blind River deposits of Ontario (Robertson, 1976) and the Jacobina area of Brazil (Gross, 1968). In the Canadian sequence, about 2.3 to 2.5 billion years old, uraninite is associated with brannerite, pyrite, monazite, and other minerals; uraninite is concentrated in conglomerates deposited by braided streams that formed sheetlike sedimentary bodies, perhaps only in part alluvial fan deposits. In the Brazilian sequence (more than 2 billion years old), gold and uranium are concentrated in thicker conglomeratic foreset beds. These deposits have been intensely metamorphosed, and some ore mobilization apparently occurred.

Other uranium present in alluvial fan deposits occurs as roll-fronts related to diagenetic groundwater solution, transport and redeposition. The Gas Hills Uranium district of Wyoming contains uranium deposits in the arkosic sandstone facies of the lower Tertiary Wind River Formation (Soister, 1968; Seeland, 1978; Ethridge and Thompson, 1978, pt. A, p. 155-160). The arkosic sandstone facies is believed to have been deposited on a piedmont alluvial fan as middle- to outer-fan braided stream channel deposits. The most favorable sites for uranium concentrations are in permeability barriers associated with channel margins (Armstrong, 1970).

Coals are not commonly associated with alluvial fans, probably because of the sloping fan surface, but are locally thick enough in some ancient fan sequences to be profitably mined. In the Carboniferous lower La Magdalena coalfield of northern Spain, thin coals are interbedded with middle-fan deposits, and in the upper La Magdalena and two adjacent coalfields, thicker coals are interbedded with outer-fan and lacustrine deposits (Heward, 1978b). In these areas, coals are thought to have developed on abandoned and subsided parts of fans, where active sedimentation shifted for long intervals to other parts of the fans.

PETROLEUM PRODUCTION

Alluvial fan deposits are not generally reservoir rocks for petroleum because they fail to connect laterally to source rocks, are not very deeply buried, are not sufficiently extensive laterally, do not have proper seals, have low permeability and porosities following diagenesis, and generally do not contain facies that are good source rocks. Fan deltas, however, can form suitable stratigraphic traps for oil and gas, especially outer-fan facies if reworked into bars by marine currents (Fisher and Brown, 1972). Distal facies of clastic wedges, probably outer parts of fan deltas, are productive in Texas, Oklahoma, New Mexico, Colorado (Ethridge and Thompson, 1978, pt. A, p.

154-155), and California (J. A. Bartow, written commun., 1979).

ENVIRONMENTAL ASPECTS OF ALLUVIAL FANS

Alluvial fans are critically important to man as sources of ground water, especially in arid regions, and sites for farming and grazing. They also present serious environmental hazards as potential sites for flooding, landsliding, deep and near-surface land subsidence, and ground displacement and shaking from seismicity produced by range-front faults.

Alluvial fan deposits form part of the groundwater reservoir in alluviated basins throughout the world, and in many arid and semiarid regions they are the only sources of water. In the western United States, ground water pumped from late Cenozoic alluvial fan de-

posits is critically important to areas such as the Santa Clara Valley in northern California, the San Joaquin Valley in central California, the Los Angeles basin in southern California, and the basin surrounding Tuscon, Arizona. Groundwater recharge takes place primarily through alluvial fans on the margins of basins; extensive covering of fan surfaces by urban developments which increases runoff and decreases infiltration has caused groundwater recharge problems in several areas.

Alluvial fans are suitable sites for farming and grazing. Lower parts of large fans, which often consist of weathered sandy deposits, may form particularly good soils that typically overlie good aquifers (Bull, 1977, p. 2160). In this writer's opinion, many of the finest wines in California, Italy, Spain, and France are produced from

grapes grown in the well-drained soils developed on alluvial fan surfaces. Some fans, especially older ones with uplifted surfaces, may in some regions be suitable areas for development, although environmental problems can be expected.

For these beneficial aspects of fans, however, there are also attendant difficulties. Flooding is probably the major environmental hazard, often proving more serious than flooding of river systems as the rapidity and unpredictability of fan flooding presents a greater hazard to life. Active channels can shift abruptly, even during the same flood event, with sheetfloods, debris flows, and mudflows quickly covering broad areas with coarse debris. Renwick (1977) showed the effect of intense rainfall on a small fan in northern New York State as sediment was deposited

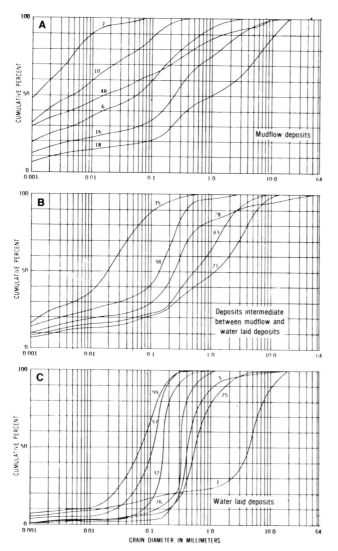

Fig. 49—Grain-size distribution plots for modern alluvial fan deposits of western Fresno County, California (from Bull, 1963). **A.** Mudflow deposits—clay content = 31%, S_o = 9.7, QD_ϕ = 3.1, and σ_ϕ =4.7. **B.** Deposits intermediate between mudflow and waterlaid deposits—clay content = 17%, S_o = 4.0, QD_ϕ = 2.0, and σ_ϕ = 3.9. **C.** Waterlaid deposits—clay content = 6%, S_o = 1.8, QD_ϕ = 0.8, and σ_ϕ = 1.4.

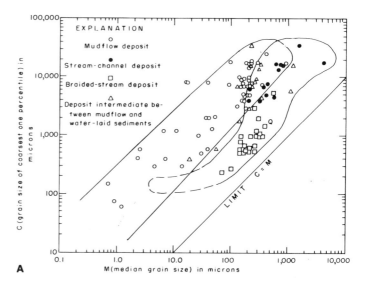

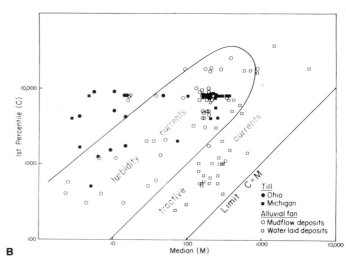

Fig. 50—CM diagrams showing characteristics of alluvial fan deposits. **A.** Mudflow, stream-channel, braided-stream, and mudflow to water-laid deposits of western Fresno County, California (from Bull, 1972a). **B.** Comparison of mudflow and waterlaid alluvial fan deposits with glacial tills from the central United States (from Landim and Frakes, 1968).

A

B

Fig. 51—Development of caliche in alluvial fan deposits. **A.** Thin caliche zone in fanglomerate of Perry Formation, Devonian (photo by P. R.

Schluger); small fault offsets zone. **B.** Caliche-cemented fanglomerate, Perry Formation, Devonian, southeast of Oven Head,

southwestern New Brunswick (photo by P. R. Schluger).

rapidly in channels, levees and inter-channel areas (Fig. 52). Flooding on fans yields both erosional and deposi-tional hazards—roads on fans can ei-ther be eroded away if constructed above the natural fan surface or cov-ered with debris if constructed below the fan surface. Aqueducts, such as those in the Great Valley of California, must be constructed so that flows pass either under or over their covered parts. Flood hazards associated with fans in various areas are summarized by Bull (1977).

Landslides develop on bedrock and colluvium-mantled slopes of the adja-cent range front and typically descend out onto alluvial fan surfaces or be-come entrained in fan channels. These locally sudden and often unpredictable events result in deposition of debris flows, mudflows, rockfalls, block glides, and other types of landslide de-bris on fans (Beaty, 1974; Eisbacher, 1979). Because of the rapid movement, some of these features can be haz-ardous to life as well as property, as demonstrated by debris flows in south-ern California (Sharp and Nobles, 1953; Scott, 1971; Morton and Campbell, 1979).

Subsidence on alluvial fans can af-fect manmade structures. Near-surface subsidence, causing settling and crack-ing of valuable farmlands, can result from wetting of clayey fan deposits, as described in the western San Joaquin Valley by Bull (1972b). Subsidence caused by compaction of deeper levels of fans because of groundwater with-drawal, especially from clay-rich or fine sandy reservoirs, has caused sur-

face fissuring, damaging aqueducts, railroads, airport runways, and roads (Schumann and Poland, 1969). More than 2 m of subsidence has been docu-mented for Arizona (Holzer, 1977) and as much as 8 m for central California (Miller et al, 1971; Bull, 1975).

Seismic hazards are often associated with alluvial fan deposits, especially those that, because of range-front fault-ing, are rapidly accumulating, coarse grained, thick and steep. Bull (1977) concluded that pedimented landscapes in southern Arizona were typical of

seismically inactive areas and alluvial fans typical of active areas.

SUMMARY AND CONCLUSIONS

Alluvial fans form at the base of up-lands where emerging streams deposit a sloping body of sediment whose sur-face forms a segment of a cone. Fans have concave-upward radial profiles and convex-upward transverse pro-files. They characteristically show rapid downfan decreases in grain size and generally contain texturally and compositionally immature sediments.

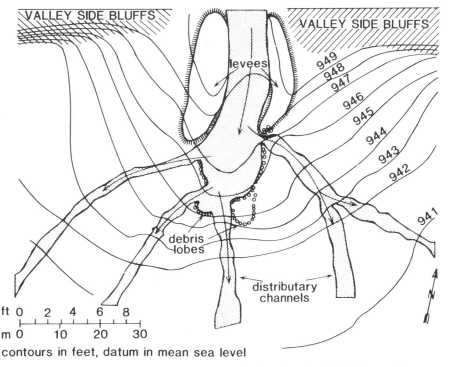

contours in feet, datum in mean sea level

Fig. 52—Map of small alluvial fan in New York State showing channels, debris lobes, and levees formed as a result of flooding in 1976 (from Renwick, 1977). Contours are in feet.

Fans can be subdivided into four major facies associations: (1) inner-fan, consisting of a major channel complex, generally incised, that is connected to the feeding upland stream; (2) middle-fan, consisting of a series of radiating distributary channels that are characteristically braided and separated from each other by interchannel and levee deposits; (3) outer-fan, a generally smooth and gently sloping area where sheets of sediment are deposited by flows emerging from channels; and (4) fan-fringe, the smooth and flat fan margin where distal fan deposits intercalate with alluvial plain, lacustrine, shoreline, eolian or other facies. Soils develop on sites of slow deposition or nondeposition.

Fan sequences consist of mixtures of streamflow deposits, debris-flow and related deposits, and landslide deposits. Streamflow deposits result from deposition of sediments transported by traction, saltation and suspension by running water; they include channel-fills, channel-margin and interchannel deposits, sheetflood deposits of lower fan regions, and, in some fans, sieve lobes, which are sheets of porous and permeable well-sorted gravel. Debris-flow and mudflow deposits result from transport of sediment as viscous muddy flows in which the muddy matrix provides support for suspended clasts; debris-flow deposits can form in the same areas as streamflow deposits, but are generally more characteristic of proximal areas and arid-region fans. Debris-flow deposits are less well stratified, contain fewer sedimentary structures, and have more isotropic fabrics than streamflow deposits. Landslide deposits can be strewn across fans where they are generated from upland slopes between streams that feed the fans.

Alluvial fan geometry is related closely to rate and style of tectonic uplift of the source area. Coarsening-upward cycles of sedimentation are most characteristic of active fans that are outbuilding, and can be found on many scales that sometimes relate to channel progradation, tectonic uplift and climatic changes. Paleocurrent patterns for individual fans are generally simple and radial in orientation, although those from piedmont-type clastic wedges composed of coalesced fan deposits can be complex.

Fans form excellent aquifers and are sources of ground water in many arid parts of the world. Although they have not been major producers of petroleum or coal, placer minerals have been mined from many fan deposits, both of Precambrian and Cenozoic age, and some roll-front uranium deposits have developed in fan deposits. Fans are excellent sources of sand and gravel. They can be good areas for development by man, but may be susceptible to flood, sedimentation, landslide, subsidence, and seismic hazards.

REFERENCES CITED

Allen, J. R. L., 1965, A review of the origin and characteristics of recent alluvial sediments: Sedimentology, v. 5, p. 89-191.

Anderson, G. S., and K. M. Hussey, 1962, Alluvial fan development at Franklin Bluffs, Alaska: Proceedings Iowa Academy of Science, v. 69, p. 310-322.

Anstey, R. L., 1965, Physical characteristics of alluvial fans: U.S. Army Natic Laboratories, Tech. Rept. ES-20, 109 p.

Armstrong, F. C., 1970, Geologic factors controlling uranium resources in the Gas Hills District, Wyoming: Wyoming Geol. Assoc. Guidebook, 22nd Ann. Field Conf., p. 31-44.

Beaty, C. B., 1963, Origin of alluvial fans, White Mountains, California and Nevada: Assoc. Am. Geographers Annals, v. 53, p. 516-535.

——— 1970, Age and estimated rate of accumulation of an alluvial fan, White Mountains, California: Am. Jour. Sci., v. 268, p. 50-77.

——— 1974, Debris flows, alluvial fans and a revitalized catastrophism: Zeitschrift fur Geomorphologie Supplement v. 21, p. 39-51.

Beaumont, P., 1972, Alluvial fans along the foothills of the Elburz Mountains, Iran: Palaeogeography, Palaeoclimatology, Palaeoecology, v. 12, p. 251-273.

Blissenbach, E., 1954, Geology of alluvial fans in semi-arid regions: Geol. Soc. America Bull., v. 65, p. 175-190.

Bluck, B. J., 1964, Sedimentation of an alluvial fan in southern Nevada: Jour. Sed. Petrology, v. 34, p. 395-400.

——— 1965, The sedimentary history of some Triassic conglomerates in the Vale of Glamoragan, South Wales: Sedimentology, v. 4, p. 225-245.

——— 1978, Sedimentation in a late orogenic basin: The Old Red Sandstone of the Midland Valley of Scotland, in D. R. Bowes, and B. E. Lake, eds., Crustal evolution in northwest Britain and adjacent regions: Geol. Jour. Special Issue no. 10, p. 249-278.

Boothroyd, J. C., and D. Nummedal, 1978, Proglacial braided outwash—a model for humid alluvial-fan deposits, in A. D. Miall, ed., Fluvial sedimentology: Canadian Soc. Petroleum Geologists Mem. no. 5, p. 641-668.

Bull, W. B., 1962, Relation of textural (CM) patterns of depositional environment of alluvial-fan deposits: Jour. Sed. Petrology, v. 32, p. 211-216.

——— 1963, Alluvial fan deposits in western Fresno County, California: Journal of Geology, v. 71, p. 243-251.

——— 1964a, Geomorphology of segmented alluvial fans in western Fresno County, California: U.S. Geol. Survey Prof. Paper 352-E, p. 89-129.

——— 1964b, History and causes of channel trenching in western Fresno County, California: Am. Jour. Sci., v. 262, p. 249-258.

——— 1964c, Alluvial fans and near-surface subsidence in western Fresno County, California: U.S. Geol. Survey Prof. Paper 437-A, p. A1-A71.

——— 1972a, Recognition of alluvial fan deposits in the stratigraphic record, in J. K. Rigby, and W. K. Hamblin, eds., Recognition of ancient sedimentary environments: SEPM Special Pub. 16, p. 63-83.

——— 1972b, Prehistoric near-surface subsidence cracks in western Fresno County, California: U.S. Geol. Survey Prof. Paper 437-C, 86 p.

——— 1975, Land subsidence in the Los Banos-Kettleman City area, California, Part 2, Subsidence and compaction of deposits: U.S. Geol. Survey Prof. Paper 437-F, 90 p.

——— 1977, The alluvial fan environment: Progress in Phys. Geography, v. 1, p. 222-270.

Carter, W. D., and J. L. Gualtieri, 1965, Geyser Creek Fanglomerate (Tertiary), La Sal Mountains, eastern Utah: U.S. Geol. Survey Bull. 1244-E, 11 p.

Cehrs, D., 1979, Depositional control of aquifer characteristics in alluvial fans, Fresno County, California: Geol. Soc. America Bull., v. 90, no. 8, pt. I, p. 709-711; pt. II, p. 1282-1309.

Collinson, J. D., 1978, Alluvial sediments, in H. G. Reading, ed., Sedimentary environments and facies: New York, Elsevier, p. 15-60.

Crowell, J. C., 1974, Sedimentation along the San Andreas fault, California, in R. H. Dott, Jr., and R. H. Shaver, eds., Modern and Ancient Geosynclinal Sedimentation: SEPM Special Pub. 19, p. 292-303.

Deegan, C. E., 1973, Tectonic control of sedimentation at the margin of a Carboniferous depositional basin in Kirkudbrightshire: Scottish Jour. of Geology, v. 9, p. 1-28.

Denny, C. S., 1965 Alluvial fans in the Death Valley region, California and Nevada: U.S. Geol. Survey Prof. Paper 466, 62 p.

——— 1967, Fans and pediments: Am. Jour. Sci., v. 265, p. 81-105.

Drewes, H., 1963, Geology of the Funeral Peak quadrangle, California, on the east flank of Death Valley: U.S. Geol. Survey Prof. Paper 413, 78 p.

Eisbacher, G. H., 1979, Cliff collapse and rock avalanches (sturzstroms) in the Mackenzie Mountains, northwestern Canada: Canadian Geotechnical Jour., v. 16, p. 309-332.

Ethridge, F. G., and T. B. Thompson, 1978, Lecture notes for the short course on the fluvial system: Fort Collins, Colorado State University, Part A, p. 211, Part B, p. 101

Fisher, W. L., and L. F. Brown, Jr., 1972, Clastic depositional systems—a genetic approach to facies analysis: Texas Bur. Econ. Geol., 211 p.

Friedman, G. M., and J. E. Sanders, 1978, Principles of sedimentology: New York, John Wiley and Sons, 792 p.

Funk, J. M., 1979, Distribution of carbonate cements in Quaternary alluvial-fan deposits, Birch Creek Valley, east-central Idaho—diagenetic model (abs.): AAPG Bull., v. 63, no. 3, p. 454.

Gole, C. V., and S. V. Chitale, 1966, Inland delta building activity of the Kosi River: Am. Soc. of Civil Engineers, Journal of the Hydraulics Division, v. 92 (HY2), p. 111-126.

Gross, W. H., 1968, Evidence for a modified placer origin for auriferous conglomerates, Canavieiras Mine, Jacobina, Brazil: Econ. Geology, v. 63, no. 3, p. 271-276.

Heward, A. P., 1978a, Alluvial fan sequence and megasequence models, with examples from Westphalian D-Stephanian B coalfields, northern Spain, in A. D. Miall, ed., Fluvial sedimentology: Canadian Soc. Petroleum Geologists Mem. no. 5, p. 669-702.

——— 1978b, Alluvial fan and lacustrine sediments from the Stephanian A and B (La Magdalena, Cinera-Matallana and Sabero) coalfields, northern Spain: Sedimentology, v. 25, p. 451-488.

Holzer, T. L., 1977, Ground failure in areas of subsidence due to groundwater decline in the United States: Proceedings, 2nd International Association Hydrological Sciences and Land Subsidence symposium, Anaheim, Calif., (Pub no. 121), p. 423-433.

Hooke, R. LeB., 1967, Processes on arid-region alluvial fans: Jour. Geology, v. 75, p. 438-460.

——— 1968, Steady-state relationships on arid-region alluvial fans in closed basins: Am. Jour. Sci., v. 266, p. 609-629.

——— and Rohrer, W. L., 1977, Relative erodibility of source area rock types, as determined from second-order variations in alluvial-fan size: Geol. Soc. America Bull., v. 88, p. 1171-1182.

——— 1979, Geometry of alluvial fans: Effect of discharge and sediment size: Earth Surface Processes, v. 4, p. 147-166.

Howard, J. D., 1966, Patterns of sediment dispersal in the Fountain Formation of Colorado: Mountain Geologist, v. 3, p. 147-153.

Hubert, J. F., et al, 1978, Guide to the Mesozoic redbeds of central Connecticut: Connecticut Geol. and National History Survey Guidebook, no. 4, 129 p.

Inglis, C. C. 1967, Inland delta building activity of Kosi River: Am. Soc. Civil Engineers Proc., Jour. of Hydraulics Div. HY1, v. 93, p. 93-100.

Johnson, A. M., 1970, Physical processes in geology: San Francisco, Freeman, Cooper and Company, 577 p.

Johnson, M. G., 1972a, Placer gold deposits of New Mexico: U.S. Geol. Survey Bull. 1348, 46 p.

——— 1972b, Placer gold deposits of Arizona: U.S. Geol. Survey Bull. 1355, 103 p.

——— 1973, Placer gold deposits of Nevada: U.S. Geol. Survey Bull. 1356, 118 p.

Kerr, D. R., S. Pappajohn, and G. L. Peterson, 1979, Neogene stratigraphic section at Split Mountain, eastern San Diego County, California, in Crowell, J. C., and A. G. Sylvester, eds., Tectonics of the juncture between the San Andreas fault system and the Salton Trough, southeastern California: Dept. Geol. Sci., Calif. Univ. at Santa Barbara, p. 111-123.

Klein, G. De V., 1962, Triassic sedimentation, Maritime Provinces, Canada: Geol. Soc. America Bull., v. 73, p. 1127-46.

Krynine, P. D., 1950, Petrology, stratigraphy, and origin of Triassic sedimentary rocks of Connecticut: Connecticut Geology and Natural History Survey Bull. no. 73, 247 p.

Laming, D. J. C., 1966, Imbrication, paleocurrents and other sedimentary features in the Lower New Red Sandstone, Devonshire, England: Jour. Sed. Petrology, v. 36, p. 940-959.

Landim, P. M. B., and L. A. Frakes, 1968, Distinction between tills and other diamictons based on textural characteristics: Jour. Sed. Petrology, v. 38, p. 1213-1223.

Larsen, V. and R. J. Steel, 1978, The sedimentary history of a debris flow-dominated, Devonian alluvial fan—a study of textural inversion: Sedimentology, v. 25, p. 37-59.

Lattman, L. H., 1973, Calcium carbonate cementation of alluvial fans in southern Nevada: Geol. Soc. America Bull., v. 84, p. 3013-3028.

Leggett, R. F., R. J. E. Brown and G. H. Johnson, 1966, Alluvial fan formation near Aklavik, Northwest Territories, Canada: Geol. Soc. America Bull., v. 77, p. 15-30.

Link, M. H., and R. H. Osborne, 1978, Lacustrine facies in the Pliocene Ridge Basin Group, Ridge Basin, California, in A. Matter, and M. E. Tucker, eds., Modern and ancient lake sediments: Internat. Assoc. Sedimentologists Special Pub. 2, p. 169-187.

Love, J. D., 1973, Harebell Formation (Upper Cretaceous) and Pinyon Conglomerate (Uppermost Cretaceous and Paleocene), NW Wyoming: U.S. Geol. Survey Prof. Paper 743-A, 54 p.

Lustig, L. K., 1965, Clastic sedimentation in Deep Springs Valley, California: U.S. Geol. Survey Prof. Paper 352-F, p. 131-192.

Magleby, D. C., and L. E. Klein, 1965, Groundwater conditions and potential pumping resources above the Corcoran Clay—an addendum to the groundwater geology and resources definite plan appendix, 1963: U.S. Bureau Reclamation Open-file Report.

McGowen, J. H., 1971, Gum Hollow fan delta, Nueces Bay, Texas: Texas Bur. Econ. Geology Rept. Inv. 69, 91 p.

——— 1979, Alluvial fan systems, in W. E. Galloway, C. W. Kreitler, and J. H. McGowen, eds., Depositional and groundwater flow systems in the exploration for uranium: Texas Bur. Econ. Geology Research Colloquium, p. 43-79.

——— and C. G. Groat, 1971, Van Horn Sandstone, West Texas, an alluvial fan model for mineral exploration: Texas Bur. Econ. Geology Rept. Inv. 72, 57 p.

McPherson, H. J., and F. Hirst, 1972, Sediment changes on two alluvial fans in the Canadian Cordillera: British Columbia Geog. Series, v. 14, p. 161-175.

Meckel, L. D., 1967, Origin of Pottsville conglomerates (Pennsylvanian) in the Central Apalachians: Geol. Soc. America Bull., v. 78, p. 223-258.

——— 1975, Holocene sand bodies in the Colorado Delta area, northern Gulf of California, in M. L. Broussard, ed., Delta Models for Exploration: Houston Geol. Soc., p. 239-265.

Miall, A. D., 1970, Devonian alluvial fans, Prince of Wales Island, Arctic Canada: Jour. Sed. Petrology, v. 40, p. 556-571.

——— 1977, Fluvial sedimentology: Canadian Soc. Pet. Geologists, Lecture notes for short course, variably paginated.

——— 1978, Tectonic setting and syndepositional deformation of molasse and other nonmarine-paralic sedimentary basins: Canadian Jour. Earth Sci., v. 15, no. 10, p. 1613-1632.

Miller, R. E., J. H. Green, and G. H. Davis, 1971, Geology of the compacting deposits in the Los Banos-Kettleman City subsidence area, California: U.S. Geol. Survey Prof. Paper 497-E, 46 p.

Minter, W. E. L., 1976, Detrital gold, uranium, and pyrite concentrations related to sedimentology in the Precambrian Vaal Reef Placer, Witwatersrand, South Africa: Econ. Geology, v. 71, no. 1, p. 157-176.

——— 1978, A sedimentological synthesis of placer gold, uranium and pyrite concentrations in Proterozoic Witwatersrand sediments, in A. D. Miall, ed., Fluvial sedimentology; Canadian Soc. Pet. Geologists Mem. 5, p. 801-829.

Morton, D. M., and R. H. Campbell, 1974, Spring mudflows at Wrightwood, southern California: Quarterly Jour. of Engineering Geology, v. 7, p. 377-384.

Nilsen, T. H., 1968, The relationship of sedimentation to tectonics in the Solund district of southwestern Norway: Norges Geologiske Undersokelse No. 359, 108 p.

——— 1969, Old Red sedimentation in the Buelandet-Vaerlandet Devonian district, western Norway: Sedimentary Geology, v. 3, p. 35-57.

——— 1973, Devonian (Old Red Sandstone) sedimentation and tectonics of Norway, in M. D. Pitcher, ed., Arctic Geology: AAPG Mem. 19, p. 471-481.

——— and T. E. Moore, 1980, Selected list of references to modern and ancient alluvial fan deposits: U.S. Geol. Survey Open-file Rept. 80-658, 53 p.

Pretorius, D. A., 1974a, The nature of the Witwatersrand gold-uranium deposits: Witwatersrand University, South Africa, Econ. Geology Research Unit Inf. Circ. 86, 50 p.

——— 1974b, Gold in the Proterozoic sediments of South Africa—systems, para-

86   T. H. Nilsen

digms, and models: Witwatersrand University, South Africa, Econ. Geology Research Unit Inf. Circ. 87, 22 p.

——— 1975, The depositional environments of the Witwatersrand goldfields—a chronological review of speculations and observations: Witwatersrand University, South Africa, Econ. Geology Research Unit Inf. Circ. 95, 47 p.

——— 1976, The stratigraphic, geochronologic, ore-type, and geologic-environment sources of mineral wealth in the Republic of South Africa: Econ. Geology, v. 17, no. 1, p. 5-15.

Price, W. E., Jr., 1974, Simulation of alluvial fan deposition by a random walk model: Water Resources Research, v. 10, p. 263-274.

——— 1976, A random-walk simulation model of alluvial-fan deposition, in D. F. Merriam, ed., Random processes in geology: New York, Springer-Verlag, p. 55-62.

Reid, J. C., 1974, Hazel Formation, Culberson and Hudspeth Counties, Texas: master's thesis, Texas Univ. Austin, 88 p.

Renwick, W. H., 1977, Erosion caused by intense rainfall in a small catchment in New York State: Geology, v. 5, p. 361-364.

Riccio, J. J., 1962, A geological and geographical appraisal of alluvial fans: Compass, v. 39, p. 87-95.

Robertson, J. A., 1976, The Blind River uranium deposits—the ores and their setting: Ontario Div. of Mines Misc. Paper 65, 45 p.

Rust, B. R., 1972, Pebble orientation in fluvial sediments: Jour. Sed. Petrology, v. 42, p. 384-388.

——— 1978, The interpretation of ancient alluvial successions in the light of modern investigations, in R. Davidson-Arnott, and W. Nickling, eds., Research in Fluvial Systems: Proc. Fifth Guelph Symposium on Geomorphology, Norwich, England, Geol. Abs., Ltd., p. 67-105.

——— 1979, Facies Models 2—Coarse alluvial deposits, in R. G. Walker, ed., Facies Models: Geoscience Canada, Reprint ser. 1, p. 9-21.

Ryder, J. M., 1971, The stratigraphy and morphology of paraglacial alluvial fans in south-central British Columbia: Canadian Jour. Earth Sci., v. 8, p. 279-298.

Ryder, R. T., T. D. Fouch, and J. H. Elison, 1976, Early Tertiary sedimentation in the western Uinta basin, Utah: Geol. Soc. America Bull., v. 87, p. 496-512.

Schluger, P. R., 1973, Stratigraphy and sedimentary environments of the Devonian Perry Formation, New Brunswick, Canada, and Maine, U.S.A.: Geol. Soc. of America Bull., v. 84, p. 2533-2548.

Schumann, H., and J. F. Poland, 1969, Land subsidence, earth fissures, and groundwater withdrawal in south-central Arizona, in L. J. Tison, ed., Land subsidence: AIHS-UNESCO Pub. No. 88, p. 295-302.

Schumm, S. A., 1977, The fluvial system: New York, John Wiley and Sons, 338 p.

Scott, K. M., 1971, Origin and sedimentology of 1969 debris flows near Glendora, Cali-fornia: U.S. Geol. Survey Prof. Paper 750-C, p. C242-C247.

Seeland, D. A., 1978, Eocene fluvial drainage patterns and their implications for uranium and hydrocarbon exploration in the Wind River Basin, Wyoming: U.S. Geol. Survey Bull. 1446, 21 p.

Sharp, R. P., 1948, Early Tertiary Fanglomerate, Big Horn Mountains, Wyoming: Jour. Geology, v. 56, p. 1-15.

——— and L. H. Nobles, 1953, Mudflow of 1941 at Wrightwood, southern California: Geol. Soc. America Bull., v. 64, p. 547-560.

Soister, P. E., 1968, Stratigraphy of the Wind River Formation in south-central Wind River Basin, Wyoming: U.S. Geol. Survey Prof. Paper 594-A, p. A1-A50.

Spearing, D. R., 1974, Alluvial fan deposits: Summary sheets of sedimentary deposits, Sheet 1: Boulder, Colorado, Geol. Soc. America.

Steel, R. J., 1974, New Red Sandstone flood-plain and piedmont sedimentation in the Hebridean province, Scotland: Jour. Sed. Petrology, v. 44, p. 336-357.

——— 1976, Devonian basins of western Norway—sedimentary response to tectonism and to varying tectonic context: Tectonophysics, v. 36, p. 207-224.

——— 1977, Triassic rift basins of N.W. Scotland—their configuration, infilling, and development, in K. G. Finstad and R. C. Selley, coordinators, Proceedings of the Mesozoic Northern North Sea Symposium: Norwegian Petroleum Soc., MNNSS-7, p. 1-18.

——— and S. M. Aasheim, 1978, Alluvial sand deposition in a rapidly subsiding basin (Devonian, Norway), in A. D. Miall, ed., Fluvial Sedimentology: Canadian Soc. Petroleum Geologists Mem. 5, p. 385-412.

——— and T. G. Gloppen, 1980, Late Caledonian basin formation, western Norway—evidence for strike-slip tectonics during infilling, in P. F. Ballance and H. G. Reading, eds., Sedimentation in oblique-slip mobile zones: Intern. Assoc. Sedimentologists Spec. Pub. No. 4, p. 79-103.

——— et al, 1977, Coarsening-upward cycles in the alluvium of Hornelen Basin (Devonian) Norway—sedimentary response to tectonic events: Geol. Soc. America Bull., v. 88, p. 1124-1134.

——— and A. C. Wilson, 1975, Sedimentation and tectonism (?Permo-Triassic) on the margin of the North Minch Basin, Lewis: Jour. Geol. Soc. London, v. 131, p. 183-202.

Steidtmann, J. R., 1971, Origin of the Pass Peak Formation and equivalent early Eocene strata, central western Wyoming: Geol. Soc. America Bull., v. 82, p. 156-176.

Van de Kamp, P. C., 1973, Holocene continental sedimentation in the Salton Basin, California, a reconnaissance: Geol. Soc. America Bull., v. 84, p. 827-848.

Wasson, R. J., 1974, Intersection point deposition on alluvial fans: An Australian example: Geografiska Annaler, v. 56, p. 83-92.

——— 1977a, Catchment processes and the evolution of alluvial fans in the lower Derwent Valley, Tasmania: Zeitschift fur Geomorphologie, Bd. 21, p. 147-168.

——— 1977b, Last-glacial alluvial fan sedimentation in the Lower Derwent Valley, Tasmania: Sedimentology, v. 24, p. 781-799.

——— 1979, Sedimentation history of the Mundi Mundi alluvial fans, western New South Wales: Sedimentary Geology, v. 22, p. 21-51.

Weaver, W., and S. A. Schumm, 1974, Fan-head trenching: An example of a geomorphic threshold: Geol. Soc. America Abs. with Programs, v. 6, p. 481.

Wescott, W. A., and F. G. Ethridge, 1980, Fan-delta sedimentology and tectonic setting—Yallahs Fan delta, southeast Jamaica: AAPG Bull., v. 64, p. 374-399.

Wessel, J. M., 1969, Sedimentary history of Upper Triassic alluvial fan complexes in north-central Massachusetts: Massachusetts Univ., Amherst, Dept. of Geology Contr., No. 2, 157 p.

Williams, G. E., 1969, Characteristics and origin of a Precambrian pediment: Jour. Geology, v. 77, p. 183-207.

Wilson, M. D., 1970, Upper Cretaceous-Paleocene synorogenic conglomerates of south-western Montana: AAPG Bull., v. 54, p. 1843-1867.

Winder, C. G., 1965, Alluvial cone construction by alpine mudflow in a humid temperate region: Canadian Jour. Earth Sci., v. 2, p. 270-277.

Yazawa, D., H. Toya, and S. Kaizuka, 1971, Alluvial fans: Kokon Shoin, Tokyo, 318 p. (in Japanese).

Lacustrine and Associated Clastic Depositional Environments

Thomas D. Fouch
Walter E. Dean
U.S. Geological Survey
Lakewood, Colorado

INTRODUCTION

Although sedimentary rocks formed in lacustrine depositional systems are common from much of the world, relatively few have been the focus of exploration for oil and/or gas. However, large accumulations of oil and gas trapped in rocks formed in ancient lake systems are known from the western part of the United States and from much of China. In addition, "shows" and oil and gas fields developed in strata of lacustrine origin are known from several other parts of the world.

Hydrocarbon source, reservoir, and trap units of Chinese and North American fields developed in continental strata were formed in large lacustrine depositional systems housing lakes or lake complexes commonly comparable in size to those of inland seas. Some of these lakes existed for several millions of years, a life span not generally believed to be characteristic of most present lake systems.

Lakes are frequently referred to as "freshwater" but they are, in fact, commonly saline. From analysis of physical, chemical, and biological data, evidence indicates that most ancient lake environments were dynamic, and their depositional facies reflect constant changes in tectonic setting, water chemistry and bathymetry. As a result, characteristics used to recognize a depositional setting existing during one period of a lake's history may differ significantly from lacustrine characteristics of a different time. Thus the distribution of sedimentary structures in lacustrine depositional facies can be expected to vary significantly from one phase of a lake's evolution to another as physical and chemical characteristics of the hydrologic basin change. For example, beds formed in quiet, relatively fresh, oxygenated water may be fossiliferous and bioturbated. Beds formed in the same depositional set-

ting but during an extreme alkaline or saline water phase of the lake's history may reflect little or no influence of indigenous biologic activity except possibly that of algae. Therefore to propose a set of physical and biological criteria that uniquely identify specific lacustrine depositional environments is difficult and potentially misleading.

This chapter illustrates some of the principal sedimentary structures in lacustrine siliciclastic rocks, and particularly those formed in large lake basins having the potential to produce oil and/or gas from sandstones. The principal oil- and gas-bearing lacustrine sandstone beds in the world are found in ancient lake basins of China. Unfortunately, little suitable information on these lacustrine rocks is available for presentation except some material from the Qaidam Pendi of western China.

The Uinta Basin of Utah and Colorado is an additional lake basin known to produce large volumes of hydrocarbons from siliciclastic lacustrine rocks. Unlike the lacustrine strata of China that are composed mainly of siliciclastic rocks, those of the Uinta Basin contain large amounts of both siliciclastic and carbonate rocks.

Many physical, chemical, and biological constituents of lacustrine rocks are not unique to lakes. As a result, a three- and preferably four-dimensional model must be developed to permit interpretation of sedimentary structures. For this reason, many illustrations in this chapter are of rocks formed in large lacustrine depositional systems, with both the lacustrine and associated peripheral facies illustrated.

Many detailed descriptions and discussions of a great variety and number of both ancient and modern lacustrine depositional systems are available in the literature. It is evident, however,

that only a few of the ancient lakes were of sufficient size, or produced and preserved enough organic matter to form petroleum in significant quantities.

Of special significance are the effects of water salinity, climate, and latitudinal position on the sedimentology, paleontology, and mineralogy of large ancient lakes. Hardie et al (1978), and Eugster and Hardie (1978) presented excellent discussions of saline lakes. Stoffers and Hecky (1978), and Muller and Wagner (1978) published new accounts or reviews of the effects of low latitudinal position and climate respectively on lacustrine sedimentation. These papers are especially significant to the formation of large ancient lake systems in which organic matter was formed and preserved in great quantities so that subsequent accumulations of hydrocarbons could form upon reaching thermochemical maturation.

Ancient lake beds found in the western part of the United States and specifically in parts of Nevada, Utah, Colorado, and Wyoming record major fluctuations in climate, tectonic setting, and water chemistry. They have been the focus of many studies illustrating and discussing details of the depositional, physical, chemical, and biological histories of the rocks. For this reason, and because of the writers' familiarity with these units and those of

The writers are indebted to Cynthia Sheehan, Patrick Anderson, Donald Orr, and Bryan Bailey of the U.S. Geological Survey EROS Data Center who provided the Landsat imagery of China. The senior author expresses his appreciation to the People's Republic of China for the opportunity to examine rocks in the Qaidam Pendi, Qinghai Province.

R. T. Ryder, J. P. Bradbury, R. M. Forester, M. A. Arthur, W. C. Butler, and J. H. Hanley, all of the U.S. Geological Survey, reviewed the manuscript and offered many excellent suggestions for its improvement.

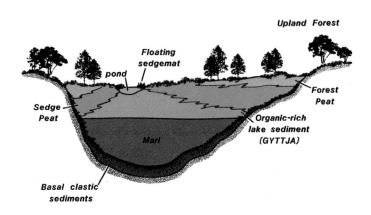

Fig. 1, 2—Typical north temperate zone lake in late stage of development. Open-lake waters are bounded by a floating mat of sedges and reeds forming a sedge-reed peat **(A)**. A bog forest **(B)** accumulating forest peat eventually replaces the sedge mat and is in turn replaced by dry land and upland forest **(C)**. As lake productivity increases, the open-lacustrine sediment becomes richer in organic matter forming an olive to black fine-grained sediment called gyttja that may or may not contain $CaCO_3$ depending upon the chemistry of the lake system. Such a lake system can produce and preserve sediments that contain both lipid-rich and herbaceous organic matter.

comparable lakes, many materials used in this chapter are taken from studies of these rocks by ourselves and colleagues.

We have illustrated features that can be expected to be typical or characteristic of many lacustrine rocks, rather than those features that are atypical but interesting. Not every sedimentary structure illustrated meets these criteria, but all are significant and critical for interpreting depositional settings in which they formed. Most illustrations are taken from previously published papers where more detailed descriptions and discussions of the processes that produced the features are available.

MODERN LACUSTRINE DEPOSITIONAL SETTINGS

Figure 1 is a diagrammatic illustration of the distribution of sedimentary facies resulting from postglacial development of a typical north temperate lake. Figure 2 is a photograph of the surface of such a lake. The physical, chemical, and biological constituents in such a lake are similar in many ways to those of some ancient lakes, and north temperate lakes commonly contain both carbonate and siliciclastic sediments. Modern north temperate lakes are in many ways poor analogues for comparison to most ancient large lacustrine systems. However, because much is known of the interrelations between chemical, biological, and physical processes operating in modern lakes, a review of this information is helpful in understanding and correctly interpreting facies relations in lacustrine rocks.

Most temperate lakes are dimictic, characterized by semi-annual overturns, one in the spring and one in the fall. Overturn logically occurs because water has a maximum density at about 4°C. Surface water therefore sinks to the bottom when cooled to 4°C in the fall and when warmed to 4°C in the spring. These overturns are the "heartbeat" of a lake and, indeed, are important in maintaining life in the lake. Overturn is the main source of oxygen to the bottom waters of a lake. Similarly, during overturns, surface waters are restocked with nutrients that accumulate in the bottom waters during summer or winter stagnation.

At fall overturn, the entire water column is isothermal at 4°C. Eventually the temperature of the surface water reaches 0°C and the lake freezes. The lake can no longer receive oxygen from the atmosphere, and oxygen production by photosynthesis is at a minimum. Oxygen consumption by respiration and decay continues and lake waters may eventually become depleted in dissolved oxygen. Oxygen depletion occurs first in the bottom waters and moves upward during the winter. If oxygen depletion occurs rapidly, or if ice remains on the lake for a long period of time, the entire water column may become deoxygenated resulting in a "winter kill" when much of the fish population suffocates.

The ice will finally melt and the surface water will heat to 4°C. At this time, the lake is isothermal and can be completely mixed by wind, carrying oxygenated surface waters to the bottom and returning nutrients stored in the bottom waters to the surface. In the summer, temperature distribution subdivides the lake into three water masses. The upper water mass or epilimnion may be isothermal as a result of wind mixing—the thickness determined largely by the depth of wind mixing. The middle water mass or metalimnion is characterized by a rapid change in the temperature-depth curve (this region of the curve is called the thermocline). The lower water mass or hypolimnion is usually characterized by a gradual decrease in temperature to the bottom. Depending on the abundance of plant life in the lake, and the extent of wind mixing, the epilimnion may be saturated or even supersaturated with oxygen from the atmosphere and photosynthesis. However, in the hypolimnion, there is only respiration and decay, and oxygen may be completely eliminated. The depth at which oxygen production by photosynthesis equals oxygen consumption by respiration and decay is called the compensation depth. The compensation depth varies daily but is usually in the metalimnion. The part of the lake bottom populated by rooted aquatic vegetation is called the littoral zone. The part of the lake bottom under the hypolimnion is called the profundal zone. A transition zone, called the sublittoral zone, is usually present.

In the fall, the surface water will be cooled to 4°C and mixed by wind causing overturn and destroying summer stratification.

Sediments of a typical, moderately deep (around 25 m), north temperate lake can be considered to be a four-component system of detrital clastic sediments, organic matter, biogenic silica, and carbonate minerals. The relative importance of each of these components changes as the lake evolves from a newly filled basin to dry land.

Basal sediments are likely to be sand, silt, and clay derived from erosion of bedrock in the drainage basin, or transported into the drainage basin as glacial drift. Very little organic matter will be present because the organic productivity of the lake at this stage is low (oligotrophic) and little vegetation is available on the surrounding land. Assuming that the drainage basin contains some limestone and/or dolomite, or calcareous glacial drift, leaching of this material will soon produce an accumulation of dissolved calcium carbonate within the lake. Once calcium carbonate has reached saturation, the lake will precipitate one or more carbonate minerals that will be mixed with clastic material forming a carbonate-rich sediment (marl) with varied proportions of carbonate and clastic material.

As these carbonate rocks are being leached from the surrounding areas, nutrients are also being leached and accumulated in the lake. Organic productivity of the lake increases and more organic debris is either incorporated in the sediments or decays, releasing more nutrients for organic growth. In addition, organic matter is also being contributed to the lake from vegetation that has become established in the drainage basin. Throughout the open-water history of a lake, the most important source of organic matter is floating microscopic algae or phytoplankton. The sediment at this highly productive (eutrophic) stage of the lake's development is likely to be a brown or black lipid-rich organic sediment called gyttja. The organic content is usually greater than 20% (about 10% organic carbon) and may be as high as 50% (about 25% organic carbon).

If the lake is highly eutrophic with a hypolimnion that is anoxic much of the year, or if the lake is permanently stratified (meromictic) with permanently anoxic bottom waters (monimolimnion) that contain hydrogen sulfide and/or methane, sediment is likely to be a black ooze called sapropel, that contains abundant lipid-rich organic matter and ferrous sulfide.

Analyses of organic matter in Lake Ontario and Lake Erie by Kemp (1971) shows bitumens accounting for 3-6% of the organic matter, humic and fulvic acids 19-27%, and kerogen 35-49%; organic carbon contents of these sediments ranged from 1.6 to 3.4%.

As the lake becomes shallower, the littoral zone of rooted aquatic vegetation becomes wider and a floating mat

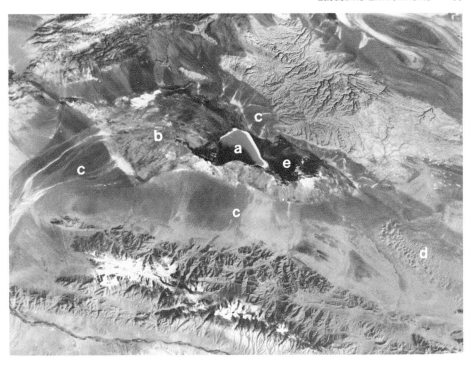

Fig. 3—Gas Lake, Qaidam Pendi, Qinghai Province, China. Digitally enhanced and merged Landsat satellite image where blue color outlines the lake **(A)** in which both terrigenous and evaporite sediments are being deposited in a permanent lake of an arid region. A delta has formed at the left (west) margin and is prograding into the deeper part of the lake (black). White band bounding the lake is composed of evaporite salts; light blue color marks shallower water. Vegetation (red) on salt marsh, alluvial plain, and interchannel bars **(B)** bound the lake. Coalescing coarser-grained alluvial fans **(C)** and eolian dunes **(D)** are peripheral to the lake. The dark colored area **(E)** is an area of low relief that is periodically submerged. Lake is about 12 km long.

of sedges begins to grow outward into the lake. The algal gyttja around the lake margins is replaced by sedge peat. The sedge peat eventually covers the gyttja as the lake continues to become shallower and smaller in area. Once the sedge peat provides a stable substrate, a bog forest can develop. Eventually the sedge peat will be covered with a layer of bog-forest peat. The resulting relations among lacustrine sedimentary facies would be similar to those illustrated in Figure 1, where basal clastic sediments grade upward into marl as sediments become increasingly rich in calcium carbonate. Marl grades upward into algal gyttja recording increased productivity of the lake. Gyttja grades into sedge peat and finally bog-forest peat as algal organic matter is replaced by herbaceous organic detritus from aquatic and forest vegetation. Such a lake system has the capacity to generate and preserve lipid-rich and herbaceous organic materials that are potential sources of both oil and natural gas.

An additional perspective for reconstructing the depositional setting of rocks formed in lacustrine environments can be gained by examination of other modern lake systems. Figure 3 is a digitally enhanced and merged Landsat image of Gas Lake, Qinghai Province, China. The lake center is underlain by evaporite mineral crusts and clastic sediments, the latter originating primarily from a well-defined delta formed at the south margin of the ponded body of water. The lake is bounded by deltaic plain, alluvial fan, and salt marsh environments developed in peripheral settings.

Gas Lake is a highly saline lake in an arid setting at an altitude of more than 3,000 m in a basin surrounded by high mountains. It is located in the Qaidam Pendi and represents present-day lacustrine sedimentation in an area underlain by lacustrine rocks ranging in age from Jurassic to Holocene. Although Gas Lake is saline and relatively small, it contains and is bounded by depositional environments similar in part to those of its much larger precurser lakes.

Pyramid Lake, Nevada (Fig. 4) is also a modern remnant of the much larger Pleistocene Lake Lahonton. The lake presently occupies more than 2,000 sq km and is up to 100 m deep. It is presently the site of terrigenous sedimentation, but carbonate deposits are being formed near spring-fed areas at the margin of the water body. Born (1972) described the Quaternary and Holocene sedimentation of this lake system. Well-defined terrigenous depositional

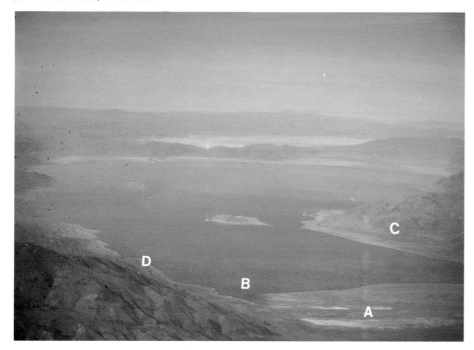

Fig. 4—Pyramid Lake, Nevada. The Truckee River crosses a delta plain **(A)** and enters the lake at **(B)**. Beach deposits and terraces mark present and ancient shorelines **(C)** some of which contain tufa mounds and pinnacles. Alluvial-fans bounded the lake during higher water stages **(D)** and subaqueous fan toes formed fan deltas. Pyramid Lake presently occupies more than 2,000 sq km and has a maximum depth of 100 m. Distance from **(B)** to island is 13 km.

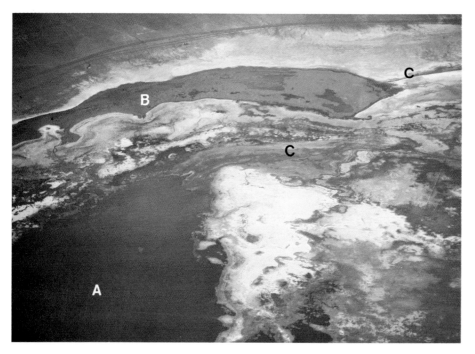

Fig. 5—Toulon and Humboldt Lakes, Nevada. An algae-dominated alkaline lake system that is flanked by beach ridge, small pond, and eolian dune deposits covered with or consisting of alkaline salts. The open water body **(A)** is part of Humboldt Lake and the arcuate water body **(B)** is part of Toulon Lake. The lakes are fed by the Humboldt River channels **(C)** and have no outlets. The lakes represent the waning stages of a larger lake system that was the site of terrigenous sedimentation. 1 cm = 1.6 km.

environments such as alluvial fans and deltas are exposed around the margin of the lake and each is composed of sediments with characteristic sedimentary structures.

Lakes Toulon and Humboldt, Nevada (Figure 5) also represent remnants of Pleistocene Lake Lahonton. Both lakes are currently fed by the Humboldt River, are highly alkaline, and contain abundant microscopic and macroscopic algae. In the past, both lakes received great volumes of terrigenous sediments and were "freshwater." Al-

though the central part of Humboldt Lake is permanent, the peripheral environments are subaerially exposed most of the time and covered by crusts of evaporite minerals. Eolian dunes are present locally on the exposed, playalike flats surrounding the lakes.

Ruby Lake, Nevada (Figure 6) also represents the waning stages of a formly much larger lake in the arid Basin and Range Province. However, in contrast to the alkaline waters of Lakes Toulon and Humboldt, alkalinity varies greatly in the marsh and lake system of Ruby Lake, a phenomenon that obviously characterized many ancient lakes. In addition, open-lake areas are commonly surrounded by extensive areas of marshland covered or nearly covered by aquatic plants, grasses, and some small shrubs. Evaporite minerals are being deposited in isolated ephemeral ponds or along shorelines away from freshwater influx areas. The pattern of present sedimentation in Ruby Lake is in many ways comparable to that of a north temperate lake discussed earlier.

UINTA BASIN DEPOSITIONAL MODEL FOR COMPARISON

Figure 7 is a diagrammatic illustration of the distribution of interpreted depositional environments of the western part of Lake Uinta, Utah, as it existed in the early Eocene (modified from Ryder et al, 1976). The model is most appropriate for an early "freshwater" permanent lake stage of Lake Uinta's existence in northeastern Utah. Kerogen-rich lake beds of the Green River Formation as tyically developed in the Green River basin of Wyoming and the Piceance Creek basin of Colorado are also present at several different horizons of the Green River Formation in the Uinta Basin in Utah, but represent a significant variation from the depositional conditions that existed throughout much of the history of Lake Uinta in northeastern Utah.

Although the grouping of rocks shown in Figure 7 is not appropriate for all lacustrine rocks, it has served as a practical model for interpreting both surface and subsurface geochemical, biological, and physical rock data. In addition, it separates the rocks into depositional facies containing probable source, reservoir, and trap units for hydrocarbons. However, for purposes of this paper, the facies relations diagrammed in Figure 7 serve as a guide for the illustration and interpretation of many of the physical and biological features found in terrigenous rocks of

Fig. 6—Ruby Lake, Nevada. A lacustrine system in the arid Basin and Range province that consists of marsh lakes. Some lakes are connected by channels with slow-moving currents. Open-lake settings are commonly surrounded by marshlands covered by emergent aquatic plants, grasses, and some small shrubs. Alkalinity of the waters varies with distance from springs or inflowing streams. Evaporite mineral crusts form in some areas in shallow water or on beaches peripheral to the permanent lakes.

other large ancient lakes and lake complexes of the world.

The continental sedimentary rocks of the early Eocene of the Uinta Basin can be divided into three major depositional facies (1) alluvial, (2) marginal lacustrine, and (3) open lacustrine. Ryder et al (1976) indicated that each of these facies contains a suite of sedimentary structures and biologic constituents that collectively indicates the depositional environments of different stages of Lake Uinta's existence. These facies can, in turn, be subdivided into subfacies such as alluvial fan, high mud flat, interdeltaic, deltaic, lower deltaic plain, carbonate mud flat, and nearshore and offshore open lacustrine. Each subfacies can be further separated into many genetic subdivisions.

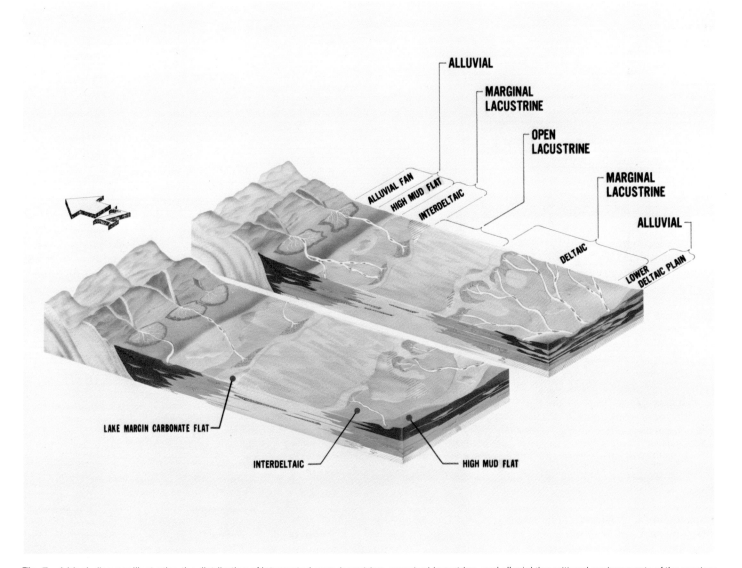

Fig. 7—A block diagram illustrating the distribution of interpreted open-lacustrine, marginal-lacustrine, and alluvial depositional environments of the western part of Lake Uinta, Utah, as it existed during the early Eocene. Diagonally striped blue pattern = grain-supported carbonate rock; solid blue pattern = mud-supported carbonate rock; yellow pattern = sandstone; claystone units are shown in their natural colors. Some thin graded siliciclastic units of probable turbidite origin are present in delta fronts and in the open lacustrine faces. Much of the open lacustrine facies is composed of kerogen-rich carbonate units. The width of Lake Uinta in the diagram is about 40 km. Vertical exaggeration is between 15 and 20. Diagram is modified from Ryder and others (1976).

Distribution of biologic constituents in lakes is greatly influenced by water temperature, chemistry, and circulation (Bradbury and Hanley, 1979). Fossils of organisms (including some species of rays, ostracodes, and dinoflagellates) generally considered to be indicators of marine water have been reported from some ancient lake sequences with no known marine connection. These lakes appear to have been characterized by a physical and chemical environment with traits temporarily similar to those of oceanic environments. However, such fossils are usually atypical of the entire lacustrine depositional complex because most of the intercalated strata contain fossil plants and animals considered more characteristic of freshwater or other nonmarine aquatic environments.

Although the distribution of organisms in lacustrine depositional systems is variable among modern and ancient lakes, and commonly varies within the same lake system with time, for purposes of this discussion we will use fossil associations taken from many lakes and compare them to the major depositional facies of ancient Lake Uinta. However, many of the plants and animals discussed have not been reported from Lake Uinta.

Many beds representing open-lacustrine environments probably accumulated under anoxic conditions. Lack of oxygen may have occurred only within the sediments or may have extended some distance into the overlying water column. Fossils preserved in this environment are likely to have been transported from aerobic environments near the edge of the lake, from subaerial sites outside the lake, or from oxygenated waters near the surface of the lake. The common fossils recovered from open-lacustrine strata are generally from one or more of the following groups: (1) aquatic vertebrates; (2) some ostracodes; (3) planktic and (or) benthic algae: (4) palynomorphs; or (5) mollusks that lived in nutrient-rich oxygenated waters of nearshore open-lacustrine environments.

Plants and animals that lived in or were transported and preserved in marginal-lacustrine deposits can be abundant and commonly include: (1) bivalves, and terrestrial and aquatic gastropods; (2) ostracodes; (3) charophytes, diatoms, and other algae; (4) aquatic and terrestrial vertebrates; (5) aquatic and terrestrial plants; (6) conchostracans, and (6) palynomorphs.

Rocks interpreted to have formed in alluvial environments may contain: (1) fossil aquatic and terrestrial gastropods; (2) bivalves (may be very abundant); (3) aquatic ostracodes; (4) conchostracans; (5) aquatic charophytes; (6) rare aquatic diatoms; (7) terrestrial and aquatic plants; (8) palynomorphs; and (9) terrestrial and aquatic vertebrates. However, these fossils are commonly not abundant.

Trace fossils can be formed and preserved in most aerobic environments, but are most commonly reported from deposits that accumulated in alluvial and marginal-lacustrine depositional environments.

OPEN LACUSTRINE FEATURES

Open lacustrine features are illustrated in Figures 8 through 23 with explanatory captions.

Fig. 8—Eocene part of the Green River Formation, Hells Hole, Uinta Basin, Utah. A view of the bedding character of open-lacustrine rocks that formed in a low energy setting of Lake Uinta. The Parachute Creek Member (Tgp) in this part of the Uinta Basin contains several oil-shale zones, the principal one being the Mahogany zone (arrows). This exposure contains over 300 m of nearly continuous strata that contain abundant lipid-rich organic matter. These open-lacustrine rocks grade laterally and downward into marginal-lacustrine rocks of the Douglas Creek Member (Tgd).

Fig. 9—Eocene Parachute Creek Member, Green River Formation, Uinta Basin, Utah. Outcrop bedding character of horizontally laminated, kerogen-rich "oil shale" that was preserved in an anoxic, low-energy, open-lacustrine environment of Lake Uinta. The rocks shown are part of the Mahogany ledge exposed near Hells Hole, Utah. 1 cm = 0.33 m.

Fig. 10—Pliocene Ridge Basin Group, California. Offshore ferroan dolomite and analcime interbedded with dark-colored mudstone rich in organic matter. Beds contain even and continuous varve-like laminae (Link and Osborne, 1978). Note ripple-laminae below the bottom of the coin. Coin is 2 cm in diameter. Photograph by M. H. Link.

Fig. 11—Pliocene Ridge Basin Group, California. Offshore ferroan dolomite and analcime with mudstone rich in organic matter. Beds are thinly laminated and laminated; some of the laminae have been deformed (Link and Osborne, 1978). Coin is 2 cm in diameter. Photograph by M. H. Link.

Fig. 12—Lower Oligocene(?) part of the Kinsey Canyon Formation, Schell Creek Range, Nevada. Continuous- and even-laminated tuffaceous mud-supported carbonate rock with convolute and contorted bedding. The beds formed in a low-energy, quiet-water setting with little or no inflora or infauna probably caused by anoxic or near-anoxic conditions in the sediments. Coin is 2 cm in diameter.

↑ **Fig. 14**—Green River Formation (Eocene), Soldier Summit, Utah. Polygonal pull-apart or dessication cracks filled with mud in unfossiliferous kerogen-rich laminated "oil shale". Some cracks are partly filled with shallow water mineral containing aquatic mollusks (*Goniobasis* sp.). These beds formed beneath shallow, fresh, oxygenated waters containing aquatic plants, and away from terrigenous clastic influx sources (Fouch et al, 1977). Lack of an infauna and preservation of kerogen-rich laminae indicate anoxic conditions in the bottom sediment. Scale is 15 cm.

← **Fig. 13**—Bird's-nest zone, Parachute Creek Member, Green River Formation, Uinta Basin, Utah. Deformed and contorted beds and laminae of "oil shale" exposed along the trace of a mined gilsonite vein. Man's finger indicates the position of a cavity that resulted from leaching of a nodule of water-soluble nahcolite ($NaHCO_3$; Fouch et al, 1976).

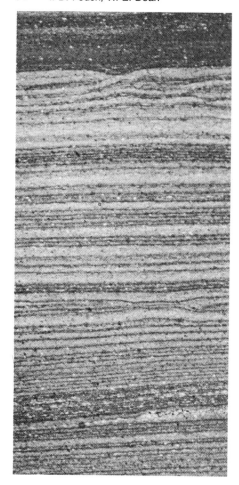

Fig. 16—Eocene saline facies of the Uinta Formation, Indian Canyon, Utah. Salt-crystal molds in nodular chert, mud-supported carbonate rock, and calcareous claystone. Beds are laminated to thin-bedded, although alteration of saline minerals has imparted a "lumpy" texture or wavy, irregular bedding to many units. These beds formed in a low-energy, alkaline and highly saline environment that was peripheral to the main depositional center of the lake. Bedded salt minerals, including halite, have been recovered from cores of temporally equivalent rocks formed at the lake's center (Fouch et al, 1976). Coin is 2 cm in diameter.

Fig. 15—Parachute Creek Member, Green River Formation, Piceance Basin, Colorado. Well-laminated, kerogen-rich, clayey mud-supported dolomite ("oil shale") typical of much of the open-lacustrine facies of the Green River Formation that is rich in organic matter. Continuous and discontinuous, even, parallel laminations are the characteristic stratification types. Disrupted thin beds and laminae are broken by low- and high-angle, small-scale faults and boudinage pull-apart structures commonly called "loop-bedding." Light bands are composed mostly of altered tuffaceous minerals and dark bands are rich in organic matter. Bar = 5 mm.

Fig. 17—Pliocene Ridge Basin Group, California. Thin-bedded sandstone and interbedded mudstone of Link and Osborne's (1978) turbidite facies. Rocks are commonly graded, contain Bouma Ta, Tab, Tbe, and Tcde intervals, sole marks, and rip-up clasts, and some are burrowed or ripple marked. Photograph by M. H. Link.

Fig. 18—Pliocene Ridge Basin Group, California. Graded conglomeratic sandstone and interbedded mudstone of Link and Osborne's (1978) turbidite facies. Note the irregular loaded base of the sandstone at the base of the hammer handle. Photograph by M. H. Link,

Fig. 19—Pliocene Ridge Basin Group, California. Intraformational sedimentary slump structure including contemporaneous folds and faults in a sandstone unit of Link and Osborne's (1978) turbidite facies. Mudstone beds that bound the deformed units are relatively undeformed and are laterally continuous. Photograph by M. H. Link.

Fig. 20—Pliocene Ridge Basin Group, California. Slump structure in sandstone and interbedded mudstone of Link and Osborne's (1978) turbidite facies. Photograph by M. H. Link.

Fig. 22—Pliocene Ridge Basin Group, California. Dish structures with pillar columns in a medium- to coarse-grained sandstone. Base of the bed has scour structures and the unit grades upward into a mudstone. Associated beds are graded, internally deformed, and are locally channeled. These beds are part of a turbidite facies (Nilsen et al, 1977; Link and Osborne, 1978). Scale is 15 cm. Photograph by M. H. Link.

Fig. 21—Pliocene Ridge Basin Group, California. Block showing the base of a sandstone bed with load features and grooves. The sample is from the turbidite facies of Link and Osborne (1978). Scale is 15 cm. Photograph by M. H. Link.

Fig. 23—Walker Lake, Nevada. Sand beach with lakeward dipping accretion beds **(A)**, beach ridges **(B)**, beach ponds **(C)**, ripple zones **(D)**, and pebble-marked former shorelines **(E)**. The area to the upper left (behind the beach) is presently filled with desiccated algal mats and crusts of evaporite minerals. This area is a bay or lagoon during prolonged higher-water stages. Photograph by M. H. Link.

MARGINAL LACUSTRINE FEATURES

Marginal lacustrine features are illustrated in Figures 24 through 62 with explanatory captions.

Fig. 24—Pyramid Lake, Nevada. Shoreline-connected sand bar with beach ridges **(A)**, spits **(B)**, and a lagoon or bay **(C)** between the bar and shoreline. Vegetated sand-flats **(D)** are part of the the delta plain of a local delta. Photograph by M. H. Link.

Fig. 25—Pyramid Lake, Nevada. Sand bar developed near the edge of the lake. A small bay or lagoon is present between the bar and the shoreline. Bars and vegetated lacustrine flats in the background are flooded during periods of shoreline transgression. Photograph by M. H. Link.

Fig. 26—Pliocene Ridge Basin Group, California. Sandstone beds intercalated with mollusk-bearing mudstone. Each sandstone bed has a sharp base and a convex-upward upper boundary, and is formed of low-angle cross-strata dipping away from the sandstone body. Link and Osborne (1978) interpret these beds to be a lacustrine bar. Photograph by M. H. Link.

Fig. 27—Pliocene Ridge Basin Group, California. Two major sandstone bars interbedded with mollusk-bearing sandstone and mudstone (Link and Osborne, 1978). Figure 26 is from the uppermost bar. Photograph by M. H. Link.

Fig. 28—Pliocene Ridge Basin Group, California. Low-angle (accretionary) cross-stratification in a sandstone bar. Link and Osborne (1978) indicate that the bar is convex upward with onshore- and offshore-dipping cross-stratification. Photograph by M. H. Link.

Fig. 29—Miocene Barstow Formation, California. Truncated ripple marks on a sandstone from a shoreline facies. Photograph by M. H. Link.

Fig. 30—Pliocene Ridge Basin Group, California. Climbing-ripple lamination in the middle part of the photograph from a beach in the shoreline facies of Link and Osborne (1978). Photograph by M. H. Link.

Fig. 31—Topaz Lake, Nevada. Sand spillover lobes probably formed by wind-driven waves modifying eolian dunes. Interlobe ponds are located behind the beach between lobes. Eolian dunes are formed on these sand spillover lobes. Width of lake is 3 km. Photograph by K. O Stanley.

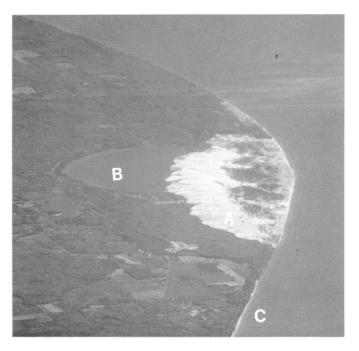

Fig. 32—Michigan and Silver Lakes, Michigan. Wind-blown sand (A) at the east edge of Lake Michigan. View is toward the southwest near Silver Lake (B). Dominant wind direction is from the west (right) and eolian dunes are transgressing lacustrine sediments in Silver Lake. Subaqueous sand benches in Lake Michigan are faintly visible near the shoreline (C). Scale is 2.54 cm = 1.6 km.

Fig. 34—Great Salt Lake, Utah. Evaporite mineral flat (A) with halite crusts. Open lake waters (B) surround the setting. An artificial barrier is now partly covered by oolites that are forming in a beach setting (C).

Fig. 33—Dead Sea, Israel. Salt ridges (A) and salt encrusted beach ridges composed primarily of sand grains (B) formed in a modern shallow-water lacustrine evaporite setting near the lake shoreline. A small protected evaporite pool exists behind the salt ridges at (C) and in a brine pool on the beach at (D). Salt "pillows" are forming in shoal waters at (E). Photography by P. A. Scholle.

Fig. 35—Great Salt Lake, Utah. Growth polygons created by the formation of salt minerals at the evaporative margin of the lake. This setting is playa-like but it is peripheral to a permanent lake. The setting is usually subaerial and is partly covered by a thin mantle of wind blown salt and mineral matter. Photograph by P. A. Scholle.

Fig. 36—Great Salt Lake desert, Utah. Crystal growth polygons developed in an artificial evaporative brine pool. The features represent a volumetric increase of mineral matter. Pressure structures are formed at the boundary between individual polygons. Cross-section views of the polygon boundaries indicate filling of the cracks by displacive salt crystals that form in the cracks. Scale is 1 cm = 0.25 m.

Fig. 37—Paleocene part of the Flagstaff Member, Green River Formation, Price River Canyon, Utah. Polygonal pull-apart cracks, asymmetric ripple marks, and vertical and horizontal burrows exposed on the base of a channel-formed sandstone. The beds formed on an exposed carbonate lacustrine flat during a period of regressive sedimentation. The sandstone is intercalated with mollusk- and ostracode-bearing lime wackestone units, some of which have a well developed soil profile on the upper surface. Scale is 15 cm.

Fig. 38—Early Eocene Douglas Creek Member of the Green River Formation, Three-Mile Canyon, Utah. Salt molds developed on a stromatolitic algal boundstone. Feature formed near a margin of a saline and alkaline water phase of Lake Uinta (Fouch et al, 1976).

Fig. 39—Walker Lake, Nevada. Coating of tufa on beach boulder. Photography by K. O. Stanley.

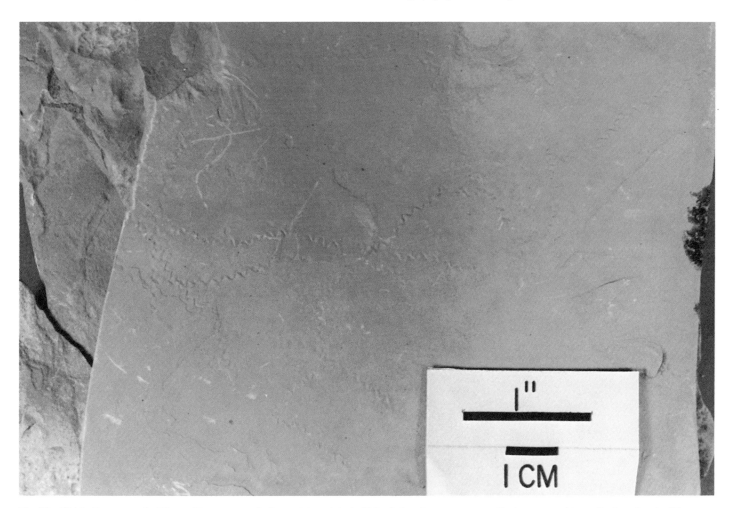

Fig. 40—Middle Eocene part of Green River Formation, Soldier Summit, Utah. Symmetrical, angular locomotion trails on a tan calcareous and silty laminated claystone. Beds containing trails are intercalated with beds bearing polygonal cracks (pull-apart) and bird tracks. The trails may be the trace fossil *Belorhaphe* Fuchs, 1895, which is known from the Cretaceous and lower Tertiary flysch of Europe (Fuchs, 1895). Moussa (1970) interpreted the trails as having been left by nematodes in very shallow water.

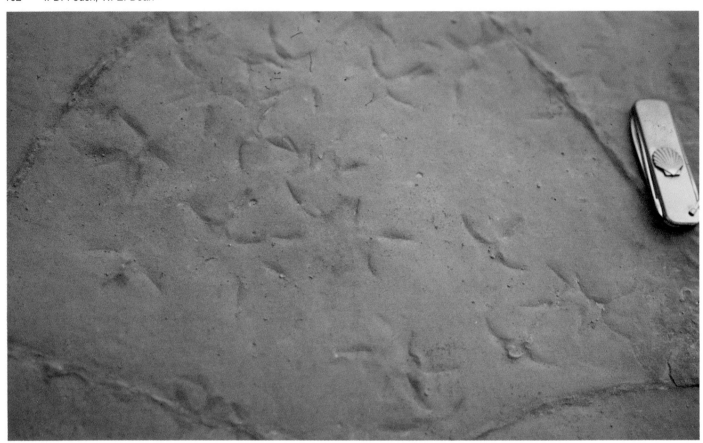

Fig. 41—Green River Formation (Eocene), Slab Canyon, Utah. Polygonal pull-apart (dessication) cracks with bird tracks on very thin-bedded calcareous siltstone. The depositional setting of this feature is interpreted to be part of a lacustrine mud flat that was periodically exposed. Photography by R. T. Ryder.

Fig. 42—Middle Eocene part of the Green River Formation, Soldier Summit, Utah. Syneresis structures in an argillaceous, laminated siltstone. These rocks are interbedded with units bearing bird tracks, the trace fossil *Belorhaphe*, polygonal pull-apart cracks, and very small-scale, asymmetrical ripple laminations. Syneresis is generally thought to represent subaqueous dehydration. Some structures termed "syneresis" in laminated, kerogen-rich rocks of the Green River Formation occur on bedding planes along the traces of "loop-beds" (Fig. 15).

Fig. 43—Middle Eocene part of the Green River Formation, Soldier Summit, Utah. Linear small-scale ripple marks developed on a buff, continuously laminated, clayey siltstone. These beds are part of a depositional complex containing bird tracks, syneresis structures, polygonal pull-apart cracks, and the trace fossil *Belorhaphe*. Physical and biological constituents collectively indicate that the rocks were formed in a shallow, mud-dominated, low-energy lacustrine setting such as a lacustrine mud flat that was periodically exposed.

Fig. 44—Green River Formation (Eocene), Slab Canyon, Utah. Polygonal pull-apart cracks developed in a calcareous siltstone that were filled with sandstone. These beds are part of a marginal-lacustrine complex characterized by laterally continuously bedded siltstone and claystone with rare channel-form sandstone units. The complex was formed in an area of a fluctuating lacustrine shoreline. During periods of exposure, stream channels distributed grains to areas that were previously sites of lacustrine sedimentation. Photograph by R. T. Ryder.

Fig. 45—Late Quaternary Truckee River delta, Pyramid Lake, Nevada. Gently-dipping topset **(A)** and steeply-dipping foreset beds **(B)** of an overall coarse-grained Gilbert-type delta. Steep foresets, parallel and small-scale cross-lamination, and convolute laminations are common sedimentary structures. Photograph by K. O. Stanley.

Fig. 46—Late Quaternary Truckee River delta, Pyramid Lake, Nevada. Parallel and small-scale cross-laminated sandstone and siltstone beds of the delta foresets. Convolute laminations are common structures in these beds. Born (1972) included these rocks in a delta-slope setting. Photograph by K. O. Stanley.

Fig. 47—Laney Member of the Green River Formation, Wyoming. Deltaic complex showing Gilbert-type delta bottomset **(A)**, foreset **(B)**, and topset **(C)** beds. The depositional sequence is conformable with the topset beds increasing in dip and grading into the foreset beds (Stanley and Surdam, 1978). Photograph by K. O. Stanley.

Fig. 48—Bonneville Lake Beds, Lehi, Utah. Gravel and coarse-grained topset **(A)**, and foreset **(B)** beds in sediments that G. K. Gilbert (1885, 1890) described as having been formed as a part of a lacustrine delta. These sediments represent the transition from subaerial to subaqueous parts of an alluvial fan at the lake margin adjacent to the Wasatch Mountain front. Coarse-grained fan deltas with steeply dipping foresets are commonly called "Gilbert-type deltas." The deposits are thought to form by flow of sediment in steep gradient streams into adjacent "deep" lake waters.

Fig. 49—Bonneville Lake Beds, Lehi, Utah. Gastropod-rich, poorly-sorted, gravelly, coarse-grained sand in low-angle, trough cross-beds of the Gilbert-type delta topset beds. The beds were deposited by braided stream flow on the subaerial part of a fan delta.

Fig. 50—Bonneville Lake Beds, Lehi, Utah. Poorly sorted sand, gravel, and silt in steeply dipping foreset beds of a Gilbert-type delta (Fig. 48) as observed along depositional strike. Low-angle, large-scale trough cross-strata occur within individual foreset units. Avalanche slip faces occurred in some foresets and initiated periodic turbid density flow into deeper water.

Fig. 51—Bonneville Lake Beds, Lehi, Utah. →
Trough cross-stratified gravel, sand, and silt foreset beds of a Gilbert-type delta (Fig. 48) as observed looking up depositional dip.

Fig. 52—Lanye Member of the Green River Formation, Wyoming. Foreset beds on a Gilbert-type delta in which individual beds decrease upward in grain-size. The vertical sequence in ascending order consists of ripple-bedded sandstone **(A),** parallel-laminated sandstone **(B),** and parallel-laminated mudstone **(C)** (Stanley and Surdam, 1978). These features constitute incomplete Bouma cycles. Photograph by K. O. Stanley.

Fig. 53—Green River Formation, Utah. Base of a lower Eocene sandstone bed with flute and tool marks. Similarly marked beds in the Flagstaff Member are commonly graded, and contain incomplete Bouma cycles. The beds are interpreted as having formed at the front margin of small deltas at the shoreline of the lake. Photograph by R. T. Ryder.

Fig. 52

Fig. 53

Fig. 54—Upper Paleocene part of the North Horn Formation (?), Price River Canyon, Utah. A bed of algal coal 5 cm thick with well-preserved, nontransported gastropods and bivalves located along discontinuous, wavy laminae. Numerous similar fossiliferous coal beds are intercalated with beds of channel-formed sandstone and lacustrine mollusk wackestone. Plant and animal fossils recovered from these rocks lived in clear, fresh, oxygenated, quiet-water, lacustrine environments. Lack of an infauna and the preservation of organic matter and laminae indicate that the sediments were anoxic.

Fig. 55—Upper Paleocene part of the North Horn Formation, Price River Canyon, Utah. Smooth-walled burrows formed by an unidentified animal. The burrowed rock is part of a horizontally bedded sandstone underlying lipid-rich coal and lime wackestone that have a high oil yield (Fouch and Hanley, 1977).

Fig. 56—Late Paleocene part of the North Horn Formation, Price River Canyon, Utah. Smooth walled burrows formed by an animal of uncertain identity. Burrowed rock is part of a flatbedded sandstone that underlies kerogenous coal and lime wackestone that have a high oil yield. Coin is 2 cm.

Fig. 57—Upper Paleocene part of the North Horn Formation, Price River Canyon, Utah. Channel-formed sandstone **(A)**, mollusk-bearing algal coal **(B)**, and mollusk-, charophyte-, and ostracode-bearing, thin-bedded to laminated lime wackestone that contains lipid-rich organic matter **(C).** The bed sequence can be traced on outcrop for more than 6.5 km and in the subsurface for more than 73 km (Fouch and Hanley, 1977). Such a depositional sequence is a significant complex of potential petroleum source and reservoir rocks formed in a continental setting. Hammer for scale.

Fig. 58—Douglas Creek Member of the Green River Formation, Bull Canyon, Utah. Flat-bedded depositional complex of discontinuously laminated, mud-supported carbonate beds (**A**) locally capped by stromatolitic algal boundstones with polygonal pull-apart cracks, interbedded with thin-bedded to structureless, green, argillaceous siltstone (**B**). These units are cut by channel-formed, fine-grained sandstone beds (**C**). The sandstone units grade upward from large-scale, low-angle tabular to trough cross-strata near base to medium-scale trough cross-strata to small-scale cross lamination at the upper margin of each channel-form bed. This sequence represents cyclic, shallow lacustrine, deltaic, and subaerial delta-plain sedimentation during periods of shoreline regression.

Fig. 59—Pliocene Ridge Basin Group, California. Link and Osborne (1978) indicate that these rocks are deltaic beds (**A**) consisting of a laterally-continuous sequence of sandstone, conglomerate, and mudstone beds. Units are locally channeled and contain fossil ostracodes, charophytes, mollusks, and vertebrates. Underlying turbidite units (**B**) are commonly graded, contain rip-up clasts, dish structures, and sole marks, and they are slump-folded. Photograph by M. H. Link.

Fig. 60—Green River Formation, Hay Canyon, Utah. Channel-formed sandstone beds with deep scour and abundant large- and medium-scale cross beds cut into near-horizontally bedded siltstone and claystone units that are locally altered to a red color. Some thin stromatolitic and mollusk- and ostracode-bearing, mud-supported carbonate beds are exposed between sandstone units. These rocks are the product of sedimentation on a lower delta plain in an area of frequent lacustrine shoreline transgressions. Oil and gas are recovered from siliciclastic rocks of this setting in the Uinta Basin (Fouch and Cashion, 1979).

Fig. 61—Douglas Creek Member of the Green River Formation, Three-Mile Canyon, Utah. Deformed sandstone, siltstone, and claystone of a fluvial and shallow lacustrine origin. Terrigenous units are overlain by stromatolitic algal boundstone. This depositional complex was formed both on a lower deltaic plain and on a carbonate-flat during a saline phase of Lake Uinta (Fouch et al, 1976).

Fig. 62—Douglas Creek Member of the Green River Formation, East Tavaputs Plateau, Utah. Stromatolitic algal boundstone **(A)** overlain by an ostracodal, curviplanar, laminated to small-scale cross-laminated, calcareous sandstone **(B)**. The sandstone is overlain by a laterally continuous, small-scale cross-laminated, ostracodal, grain-supported limestone **(C)** up to 1 m thick that is turn overlain by stromatolitic carbonate rock. This depositional complex represents cyclic carbonate shoal deposition overlain by terrigenous strata deposited in deltaic and lower delta-plain environments.

ALLUVIAL FEATURES

Alluvial features are illustrated in Figures 63 through 68 with explanatory captions.

Fig. 63—Green River Formation, Willow Creek Canyon, Utah. Channel-form sandstone with well-developed, large-scale, low-angle cross-beds separated by clayey siltstone units and with scour structures and logs at the base of the unit. The overlying flat-bedded units consists of tan to gray, ostracodal, mud-supported carbonate rocks and red, green, and gray claystone and siltstone. The channel was formed in a meandering steam on a lower interdeltaic plain adjacent to Lake Uinta that was periodically transgressed (Ryder et al, 1976). Note men at base of channel for scale.

Fig. 64—Qaidam Pendi, Qinghai Province, China. Paleogene red beds exposed along a fault scarp near Leng hu. Red-colored rocks consist of thin- to thick-bedded, argillaceous sandstone, siltstone, and claystone. Light gray rocks are pebble conglomerate that grades upward to fine-grained sandstone. These units have scour structures, burrows, and large- and medium-scale trough cross-beds at the base and asymmetrical small-scale cross-laminae with root traces in the upper part of the sandstones. The rocks were formed in clay-dominated alluvial stream and overbank environments. They are in part similar in rock type and bedding character to units in the upper part of the Colton Formation near Willow Creek in the western Uinta Basin (Fouch et al, 1976). At this locality, the Colton Formation contains some bituminous sandstone and represents alluvial plain sedimentation peripheral to Lake Uinta away from major terrigenous inflow areas. Scale is 1 cm = 6 m.

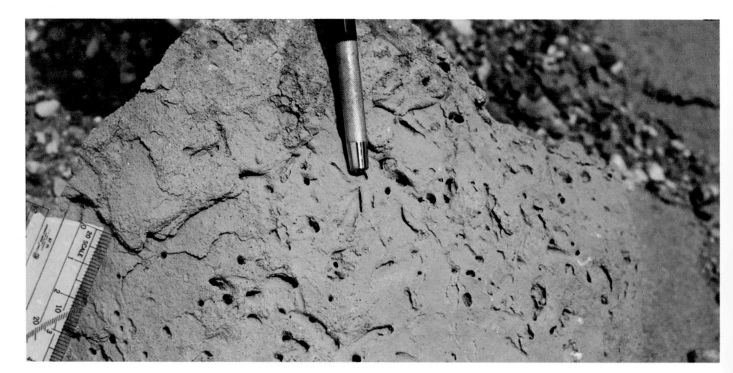

Fig. 65—Paleogene sedimentary rocks near Leng hu, Qaidam Pendi, Qinghai Province, China. Molds of oblique and horizontal smooth-walled burrows of an unidentified animal developed in the base of a 1 m thick channel-formed moderately well-sorted fine-grained sandstone. Channels are encased in red claystone, and argillaceous sandstone and siltstone units. The beds are interpreted to have been formed in an alluvial interdeltaic setting peripheral to an ancient lake.

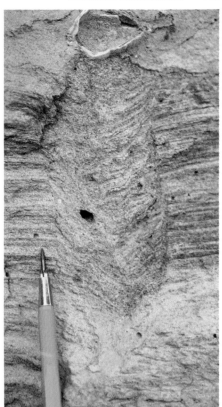

Fig. 66—Tongue River Member (Paleocene) of the Fort Union Formation, Powder River Basin, Wyoming. Sandstone bed that is laterally discontinous and grades upward into clayey siltstone. The fabric of bivalves and gastropods and sedimentary structures indicate that the beds were formed in flowing water. This bed is part of an overbank splay depositional sequence and is adjacent to a large stream channel. Photograph by J. H. Hanley. Coin is 2 cm in diameter.

Fig. 67—Tongue River Member of the Fort Union Formation, Powder River Basin, Wyoming. Escape structure produced by the burrowing of a bivalve in response to burial by rapid sedimentation. Photograph by J. H. Hanley.

ECONOMIC CONSIDERATIONS

Siliciclastic rocks of lacustrine origin are known to contain oil, natural gas, and bitumens as well as oil shale; they are commonly interbedded with beds composed of saline minerals such as halite, trona, nahcolite, and dawsonite. In addition, marginal-lacustrine and related alluvial siliciclastic strata may contain uranium and coal.

The best known petroleum-bearing lacustrine rocks are those of the Uinta Basin, Utah (Fouch, 1975) and of the giant oil fields of China (Meyerhoff and Willums, 1976). In these lacustrine depositional systems, the primary reservoirs are siliciclastic rocks. Although oil and gas in the Uinta Basin and in China is produced from lacustrine rocks, much is recovered from rocks that formed outside the ancient lakes in depositional settings at the fluctuating margin of the lake or in environments well removed from the lake. Hydrocarbons found in nonlacustrine beds are believed to have formed from lacustrine source rocks and migrated into beds of the peripheral depositional facies.

Figure 69 is a generalized structure-stratigraphic section from exposures

Fig. 68—Horse Camp Formation (Neogene), Grant Range, Nevada. Depositional sequence typical of those formed in closed internal-drainage basins of the Basin and Range structural province of the western United States. Rocks near the fault-bounded mountain front consist of **(A)** horizontal- to near-horizontal-bedded, boulder to pebble conglomerate and coarse sandstone deposited by sheet-flow and in braided streams on alluvial fans that grade into **(B)** near-horizontal-bedded to channel-formed siltstone and sandstone with plant-root and stem casts formed on an alluvial plain to **(C)** gypsum-bearing, stromatolitic and pisolitic, recrystallized lime mudstone and grainstone and ostracode-, mollusk-, and wood-bearing siltstone and sandstone of lacustrine and alluvial-plain origin deposited at the depositional center of the basin.

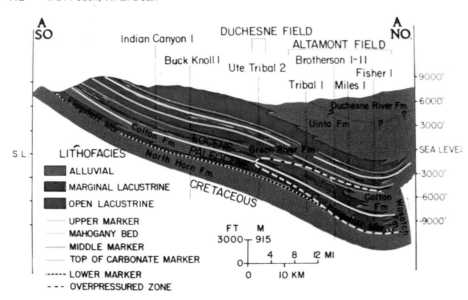

Fig. 69—Generalized structural-stratigraphic cross section from outcrops on the southwest flank of the Uinta Basin, through Duchesne and Altamont-Bluebell oil fields, to the north-central part of the basin. The Uinta Formation includes lacustrine saline facies and other lacustrine rocks assigned to the Uinta by Dane (1954) (Fouch, 1975).

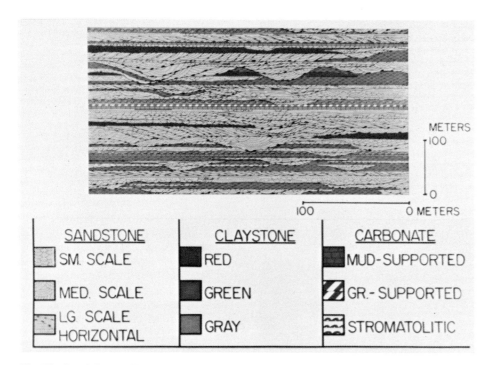

Fig. 70—Panel diagram illustrating the interpreted distribution of rock types and bedforms based on a study of cores and surface exposures of beds that produce oil and gas from "unconventional" reservoirs in the southeastern part of the Uinta Basin. Pitman and Fouch (1978) concluded that the beds were deposited in both alluvial and lacustrine environments during the early Eocene.

Bituminous sandstone beds in the Uinta Basin are interpreted to be dominantly of the alluvial and marginal lacustrine facies. Stream channels which developed on the periodically subaerially exposed lake margin distributed terrigenous grains to areas that were previously sites of lacustrine sedimentation. Siliciclastic channel sandstone units interbedded with lake beds are the principal reservoirs for bitumen on surface exposures and for much of the oil in the subsurface.

Carbonate-rich open-lacustrine rocks of the Uinta Basin are interpreted to be the sources of hydrocarbons for most of the oil and associated gas fields in the basin (Fouch, 1975; Tissot et al, 1978). In addition, open-lacustrine rocks contain the principal deposits of saline minerals and kerogen-rich dolomitic "marlstone," the so-called "oil shale" of the Green River Formation (Figs. 8, 9, 13, 14, 15).

The alluvial facies contains some accumulations of oil, gas, and bitumen, but it is not the major productive depositional facies in deeply buried Tertiary rocks in the Uinta Basin. Figure 70 is a panel diagram illustrating rock types and bedforms in some lower Tertiary rocks of the southeastern part of the Uinta Basin. The diagram was developed from a study of surface exposures (Figs. 58, 60), subsurface cores, and borehole geophysical logs in intervals that produce gas from siliciclastic low-permeability reservoir rocks in the southcentral part of the basin. Natural gas and some oil is recovered from secondary porosity developed in channel-formed units in the so-called "unconventional reservoir" province of the basin (Pitman and Fouch, 1978).

Abrupt variation in lithologies of both carbonate and siliciclastic units, and the variable distribution of secondary porosity in reservoir beds have historically limited the use of borehole geophysical logs for identifying genetic units and evaluating hydrocarbon content in rocks of the basin. For these and other reasons, petroleum exploration in the Uinta Basin and in other lacustrine depositional systems has proved difficult.

Sufficient information to permit accurate description and illustration of the nature of petroleum-bearing lacustrine rocks of China is not available. However, much of the oil is produced from large structural traps developed in clastic rocks (Meyerhoff and Willum, 1976). The ancient lakes of China were very large and resultant oil fields include reservoirs developed in turbidite,

along the south-central flank of the Uinta Basin to subsurface oil and gas fields in the north-central part of the basin. The marginal lacustrine facies contain the principal reservoir rocks of the major oil and gas fields. Secondary and some primary porosity developed within delta-front, fluvial-channel and overbank sandstone units provide most of the storage capacity for hydrocarbons (Fouch, 1975; Pitman and Fouch, 1978). Intercalated marginal-lacustrine and open-lacustrine rocks provide contrasts in ductility between brittle, carbonate-rich rocks of the open lacustrine facies and relatively ductile claystones of the marginal lacustrine facies. Interbedding of rocks of contrasting ductility has influenced the frequency of fractures that provide much of the permeability needed to drain clastic reservoir rocks (Lucas and Drexler, 1976).

lacustrine bar, and delta settings. Paleogene red beds in the Qaidam Pendi of Qinghai Province contain bitumen-bearing, channel-formed sandstone beds on surface exposures (Figs. 64, 65). Examination of these exposures indicates that the red beds are similar in rock type and bedding character to those exposed in a part of the Colton Formation near Willow Creek in the western part of the Uinta Basin (Ryder et al, 1976; Fouch et al, 1976). At the Willow Creek locality, the Colton contains bituminous sandstone and the rocks are interpreted to represent alluvial-plain sedimentation peripheral to ancient Lake Uinta in an area removed from a major terrigenous sediment inflow area.

SELECTED REFERENCES

Baer, J. L., 1969, Paleoecology of cyclic sediments of the lower Green River Formation, central Utah: Brigham Young Univ. Geol. Studies, v. 16, pt. 1, p. 3-96.

Begin, Z. B., A. Ehrlich, and Y. Nathan, 1974, Lake Lisan, the Pleistocene precurser of the Dead Sea: Geol. Survey Israel Bull. 63, p. 1-30.

Bradbury, J. P., and J. H. Hanley, 1979, Paleoecology, inland aquatic environments, in R. W. Fairbridge, and David Jablonski, eds., The encyclopedia of paleontology, Encyclopedia of earth sciences, Vol. VII, Stroudsburg, Pennsylvania, Dowden, Hutchinson and Ross, Inc., p. 541-551.

Born, S. M., 1972, Late Quaternary history, deltaic sedimentation, and mudlump formation at Pyramid Lake, Nevada: Center for Water Resources Research, Desert Research Institute, Univ. Nevada, Reno, 97 p.

Bradley, W. H., 1929, Algae reefs and oolites of the Green River Formation: U.S. Geol. Survey Prof. Paper 154-G, p. 203-223.

—— 1931, Origin and microfossils of the oil shale of the Green River Formation of Colorado and Utah: U.S. Geological Survey Prof. Paper 168, 58 p.

—— 1964, Geology of the Green River Formation and associated rocks in southwestern Wyoming and adjacent parts of Colorado and Utah: U.S. Geol. Survey Prof. Paper 496-A, 86 p.

—— 1973, Oil shale formed in desert environment: Green River Formation, Wyoming: Geol. Soc. America Bull., v. 84, p. 1121-1124.

—— and H. P. Eugster, 1969, Geochemistry and paleoclimatology of the trona deposits and associated authigenic minerals of the Green River Formation, Wyoming: U.S. Geol. Survey Prof. Paper 496-B, 71 p.

Callen, R. A., 1977, Late Cainozoic environments of part of northeastern South Australia: Geol. Soc. Australia Jour., v. 24, pt. 3, p. 151-169.

Callender, Edward, 1969, Geochemical characteristics of Lakes Michigan and Superior sediments: Proc. 12th Conf. on Great Lakes Research, p. 124-160.

Chatfield, J., 1972, Case history of Red Wash field, Uinta County, Utah, in Stratigraphic oil and gas fields—classification, exploration methods, and case histories: AAPG Mem. 16, p. 342-353.

Cole, R. D., and M. D. Picard, 1975, Primary and secondary sedimentary structures in oil shale and other fine-grained rocks, Green River Formation (Eocene), Utah and Colorado: Utah Geology, v. 2, no. 1, p. 49-67.

Dean, W. E., and Eville Gorham, 1976, Major chemical and mineral components of profundal surface sediments in Minnesota lakes: Limnol. Oceanography, v. 21, no. 2, p. 259-284.

Desborough, G. A., 1978, A biogenic-chemical stratified lake model for the origin of oil shale of the Green River Formation: an alternative to the playa-lake model: Geol. Soc. America Bull., v. 89, p. 961-971.

Eardley, A. J., 1938, Sediments of Great Salt Lake, Utah: AAPG, Bull., v. 22, p. 1305-1411.

Eugster, H. P., and L. A. Hardie, 1975, Sedimentation in an ancient playa-lake complex; the Wilkins Peak Member of the Green River Formation of Wyoming: Geol. Soc. America Bull., v. 86, p. 319-334.

—— and R. C. Surdam, 1973, Depositional environment of the Green River Formation of Wyoming: A preliminary report: Geol. Soc. America Bull., v. 84, p. 115-120.

—— and L. A. Hardie, 1978, Saline lakes, in Abraham Lerman, ed., Lakes—chemistry, geology, physics: New York, Springer-Verlag, p. 237-293.

Fouch, T. D., 1975, Lithofacies and related hydrocarbon accumulations in Tertiary strata of the western and central Uinta Basin, Utah, in D. W. Bolyard, ed., Symposium on deep drilling frontiers in the central Rocky Mountains: Rocky Mtn. Assoc. Geologists, p. 163-173.

—— 1979, Character and paleogeographic distribution of Upper Cretaceous (?) and Paleogene nonmarine sedimentary rocks in east-central Nevada, in J. M. Armentrout, M. R. Cole, and Harry Ter-Best, eds., Cenozoic paleogeography of the western United States, Pacific Coast Paleogeography Symposium 3: Pacific Section, SEPM, p. 97-111.

—— in press, Chart showing distribution of rock types, lithologic groups and depositional environments for some lower Tertiary and Upper Cretaceous rocks from outcrops at Willow Creek-Indian Canyon through the subsurface of the Duchesne and Altamong oil fields, southwest to north-central parts of the Uinta Basin, Utah; U.S. Geological Survey Oil and Gas Investigations Chart, OC-81, 2 sheets.

—— et al, 1976, Field guide to lacustrine and related nonmarine depositional environments in Tertiary rocks, Uinta Basin, Utah, in R. C. Epis, and R. J. Weimer, eds., Studies in Colorado field geology: Colo. School of Mines Prof. Contributions, no. 8, p. 358-385.

—— and J. H. Hanley, 1977, Interdisciplinary analysis of some petroleum source rocks in east-central Utah—implications of hydrocarbon exploration in nonmarine rocks of western United States (abs.): AAPG Bull., v. 61, no. 8, p. 1377-1378.

—— and W. B. Cashion, 1979, Distribution of rock types, lithologic groups and depositional environments for some lower Tertiary, Lower and Upper Cretaceous, and Upper Jurassic rocks in the subsurface between Altamont oil field and San Arryoyo gas field, north-central to southeastern Uinta Basin, Utah: U.S. Geol. Survey Open-file Rept. 79-365, 2 sheets.

Fuchs, Theodor, 1895, Studien uber Fucoiden und Hieroglyphen: Same, Den-Prschr., v. 62, p. 369-448, 9 pl.

Gilbert, G. K., 1885, The topographic features of lake shores: U.S. Geol. Survey Fifth Ann. Rept., p. 69-123.

—— 1890, Mongrr. U.S. Geological Survey 1.

Hanley, J. H., 1976, Paleosynecology of nonmarine mollusca from the Green River and Wasatch Formations (Eocene), southwestern Wyoming and northwestern Colorado, in R. W. Scott, and R. R. West, eds., Structure and classification of paleocommunities: Stroudsburg, Pa., Dowden, Hutchinson, and Ross, p. 235-262.

Hardie, L. A., J. P. Smoot, and H. P. Eugster, 1978, Saline lakes and their deposits: a sedimentological approach, in Albert Matter, and M. E. Tucker, eds., Modern and ancient lake sediments: International Assoc. Sedimentologists, Spec. Pub. no. 2, p. 7-42.

High, L. R., Jr., and M. D. Picard, 1971, Nearshore facies relations, Eocene Lake Uinta, Utah: AAPG Bull., v. 55, p. 343.

Kemp, A. L. W., 1971, Organic carbon and nitrogen in the surface sediments of Lake Ontario, Erie, and Huron: Jour. Sed. Petrology, v. 41, no. 2, p. 537-548.

Koesoemadinata, R. P., 1970, Stratigraphy and petroleum occurrences, Green River Formation, Redwash field, Utah: Colorado School of Mines Quart., v. 65, 85 p.

La Rocque, A., 1960, Molluscan faunas of the Flagstaff Formation of central Utah: Geol. Soc. America Mem. 78, 100 p.

Lerman, Abraham, ed., 1978, Lakes-chemistry, geology, physics: New York, Springer-Verlag, 363 p.

Link, M. H., and R. H. Osborne, 1978, Lacustrine facies in the Pliocene Ridge Basin Group: Ridge Basin, California in Albert Matter, and M. E. Tucker, eds., Modern and ancient lake sediments: International Assoc. Sedimentologists, Spec. Pub. no. 2, p. 169-187.

Lucas, P. T., and J. M. Drexler, 1976, Altamont-Bluebell, a major, naturally fractured stratigraphic trap, Uinta Basin, Utah, in North American Oil and Gas Fields: AAPG Mem. 24, p. 121-135.

Lundell, L. L., and R. C. Surdam, 1975, Playa-lake deposition—Green River Formation, Piceance Creek basin, Colorado: Geology, v. 3, p. 493-497.

Matter, Albert, and M. E. Tucker, eds., 1978, Modern and ancient lake sediments: International Assoc. Sedimentologists, Spec. Pub. no. 2, 290 p.

Meyerhoff, A. A., and J. O. Willums, 1976, Petroleum geology and industry of the People's Republic of China: United Nations ESCAP, CCOP Tech. Bull., v. 10, p. 103-212.

Moussa, M. T., 1969, Green River Formation (Eocene) in the Soldier Summit area, Utah: Geol. Soc. America Bull., v. 80, p. 1737-1748.

——— 1970, Nematode fossil trails from the Green River Formation (Eocene) in the Uinta Basin, Utah: Jour. Paleontology, v. 44, p. 304-307.

Muller, G., and F. Wagner, 1978, Holocene carbonate evolution in Lake Balaton (Hungary): a response to climate and impact of man, in Albert Matter, and M. E. Tucker, eds., Modern and ancient lake sediments: International Assoc. Sedimentologists, Spec. Pub. no. 2, p. 55-80.

Murray, D. K., and L. C. Bortz, 1967, Eagle Springs oil field, Railroad Valley, Nye County, Nevada: Bull., v. 51, no. 10, p. 2133-2145.

Nilsen, T. H., et al, 1977, New occurrences of dish structure in the stratigraphic record: Jour. Sed. Petrology, v. 47, no. 3, p. 1299-1304.

Normark, W. R., and F. H. Dickson, 1976, Sublacustrine fan morphology in Lake Superior: AAPG Bull., v. 60, no. 7, p. 1021-1036.

Picard, M. D., 1972, Paleoenvironmental reconstruction in an area of rapid facies change, Parachute Creek Member of Green River Formation (Eocene), Uinta Basin, Utah: Geol. Soc. America Bull., v. 83, p. 2689-2708.

——— and L. R. High, Jr., 1970, Sedimentology of oil impregnated lacustrine and fluvial sandstone, P. R. Springs area, southeast Uinta Basin, Utah: Utah Geol. and Mineralogical Survey Spec. Studies 33, 32 p.

Pitman, J. K., and T. D. Fouch, 1978, Mineralogic characteristics of lower Tertiary low-permeability reservoir rocks, south-central Uinta Basin, Utah (abs.): AAPG Bull., v. 62, no. 5, p. 891.

Roehler, H. W., 1974, Depositional environments of Eocene rocks in the Piceance Creek basin, Colorado, in Guidebook to the energy resources of the Piceance Creek basin, Colorado: Rocky Mtn. Assoc. Geologists 25th Ann. Field Conf., p. 57-64.

Russell, I. C., 1885, Geological history of Lake Lahontan: A Quaternary lake of northwestern Nevada: U.S. Geol. Survey Mon. 11, 288 p.

Ryder, R. T., T. D. Fouch, and J. H. Elison, 1976, Early Tertiary sedimentation in the western Uinta Basin, Utah: Geol. Soc. America Bull., v. 87, no. 4, p. 496-512.

Sims, J. D., 1973, Earthquake-induced structures in sediments of Van Norman Lake, San Fernando, California: Science, v. 182, p. 161-163.

Solomon, B. J., E. H. McKee, and D. W. Anderson, 1979, Stratigraphy and depositional environments of Paleogene rocks near Elko, Nevada, in J. M. Armentrout, M. R. Cole, and Harry TerBest, eds., Cenozoic paleogeography of the western United States, Pacific Coast Paleogeography Symposium 3: Pacific Section, SEPM, p. 75-88.

Stanley, K. O., and J. W. Collinson, 1979, Depositional history of the Paleocene-lower Eocene Flagstaff Limestone and coeval rocks, central Utah: AAPG Bull., v. 63, no. 3, p. 311-323.

——— and R. C. Surdam, 1978, Sedimentation on the front of Eocene Gilbert-type deltas, Washakie Basin, Wyoming: Jour. Sed. Petrology, v. 48, no. 2, p. 557-573.

Stoffers, P., and R. E. Hecky, 1978, Late Pleistocene-Holocene evolution of the Kivu-Tanganyika Basin, in Albert Matter, and M. E. Tucker, eds., Modern and ancient lake sediments: International Assoc. Sedimentologists Spec. Pub. no. 2, p. 43-54.

Surdam, R. C., and K. O. Stanley, 1979, Lacustrine sedimentation during the culminating phase of Eocene Lake Gosiute, Wyoming (Green River Formation): Geol. Soc. America Bull., Part I, v. 90, p. 93-110.

Swain, F. M., 1977, Paleoecological implications of Holocene and Late Pleistocene Ostracoda, Lake Lahonton Basin, Nevada, in Heinz Loffer, and Dan Danielopo, eds., Aspects of ecology and zoogeography of Recent and fossil Ostracoda: Sixth International Ostracod Symposium, Saalfelden, Austria, p. 309-320.

Thomas, R. L., A. L. W. Kemp, and C. F. M. Lewis, 1972, Distribution, composition and characteristics of the surficial sediments of Lake Ontario: Jour. Sed. Petrology, v. 42, no. 1, p. 66-84.

Tissot, B., G. Deroo, and A. Hood, 1978, Geochemical study of the Uinta Basin: formation of petroleum from the Green River Formation: Geochim. et Cosmochin. Acta, v. 42, p. 1469-1485.

van de Kamp, P. C., 1973, Holocene continental sedimentation in the Salton Basin, California: a reconnaissance: Geol. Soc. America Bull., v. 84, p. 827-848.

Williamson, C. R., and M. D. Picard, 1974, Petrology of carbonate rocks of the Green River Formation (Eocene): Jour. Sed. Petrology, v. 44, p. 738-759.

Wolfbauer, C. A., 1973, Criteria for recognizing paleoenvironments in a playa-lake complex—the Green River Formation of Wyoming: Wyoming Geol. Assoc. Guidebook to the geology and mineral resources of the greater Green River Basin, p. 87-91.

Fluvial Facies Models and Their Application

Douglas J. Cant
Alberta Geological Survey
Edmonton, Alberta, Canada

INTRODUCTION

Fluvial sediments are deposited by activities of rivers, and are common in the geologic record. They form a quantitatively important part of the many sedimentary basins; for example, they comprise about 65% of the sandstones in the central Appalachian geosyncline according to Pettijohn et al (1973). Fluvial sediments ranging in age from Archean to Quaternary contain concentrations of oil, gas, coal, gold and uranium in different parts of the world.

In the last 25 years considerable research has been done on development of fluvial facies models (summaries of sedimentary environments and their deposits) based on sedimentary structures and grain-size trends. These are valuable in understanding and predicting important properties of sedimentary bodies deposited in these environments. Application of these facies models to exploration problems depends on the geologist's ability to recognize assemblages of sedimentary structures, these being more diagnostic than any individual structures. Correct application of a reliable facies model allows prediction of sand body geometry, lateral facies relationships, and potential reservoir characteristics from limited data, obtainable from conventional cores or less reliably from well logs.

DEFINITIONS AND DESCRIPTIONS OF RIVER TYPES

One important conclusion drawn from previous research is that river sediments are highly variable in many aspects and cannot be characterized by any single facies model. Much of this variablity arises because different types of rivers occur in nature, and al-though a continuous spectrum of river types exists, they can be catagorized into a number of discrete types— straight, anastomosing, meandering, and sandy or pebbly braided (Fig. 1).

This chapter focuses on criteria for recognition of each type and implications following interpretation of each. In addition, entrenched valley fill systems and their importance in the geologic record is discussed.

Factors which determine river types are not well understood, but experiments by Schumm and Khan (1972) have shown that increasing valley slope causes channels to adopt straight, then meandering, then braided patterns. Anastomosing rivers, those divided by permanent alluvial islands, are favored by dense vegetation which stabilizes river banks. Braiding is also promoted by rapid, large fluctuations in river discharge, a high rate of coarse sediment supply, and easily ero-

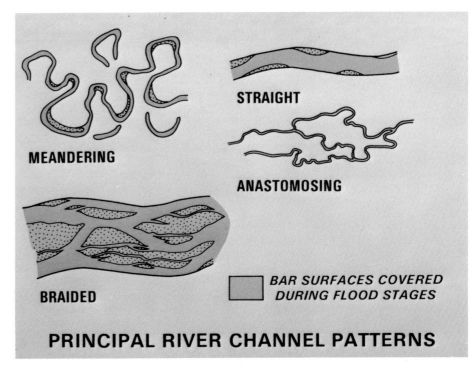

MEANDERING

STRAIGHT

ANASTOMOSING

BRAIDED

BAR SURFACES COVERED DURING FLOOD STAGES

PRINCIPAL RIVER CHANNEL PATTERNS

Fig. 1—Illustration of river types in plan view. Deposits of meandering and braided rivers have been most widely recognized in the ancient record. After Miall (1977).

The author wishes to thank J. B. Dunham, Union Oil Research; F. J. Hein, University of Southern California; P. J. McCabe, University of Nebraska at Lincoln; N. D. Smith, University of Illinois at Chicago Circle; and C. R. Williamson, Union Oil Research, for contributing photo-graphs. N. D. Smith and D. G. Smith, University of Calgary, allowed the author to use material from a paper not yet published. L. Baker and G. Masson, Shell Canada Ltd., helped locate cores from the Sunburst Formation.

dible, noncohesive banks. Braided rivers were probably more common before land plants became widespread in the Early Devonian (Schumm, 1968).

Straight and anastomosing rivers are relatively rare in nature, and their deposits have not been widely recognized in ancient records. Straight rivers have well-defined single channels with stable banks, whereas anastomosing rivers have several channels flowing around permanent vegetation-covered alluvial islands (Smith and Smith, in press).

In plan view, meandering rivers are composed of a series of wave-like loops with depositional areas termed point bars on the inside of each bend (Fig. 2). Many empirical equations relating channel width, depth, slope, meander wavelength and amplitude have been developed by geomorphologists, but these are difficult to apply to the subsurface and are not reported here. The interested reader is referred to Leopold et al (1963). Channels of meandering rivers are relatively narrow and deep. Their beds and lower parts of point bars are covered at most river stages by sinuous-crested dunes up to about 1 m in height (Jackson, 1976) depending on depth of the river. In some rivers, upper flow regime plane beds may form in places low on the point bar. Farther up this depositional slope, small-scale ripples are the most common bedform types (Bernard et al, 1970) but larger straight-crested tabular bars may also be present (Jackson,

1976). In meandering rivers with highly varied discharges, minor channels cut across the tops of the point bars, terminating at the downstream end of the bar in large foresets or chute bars. Low depositional ridges termed levees develop along the channel margins, and floodplains or backswamps with muddy sediment accumulation and vegetation growth extend away from the channel. The lateral migration of meander loops leads to point-bar deposits overlain by fine-grained floodplain deposits.

Braided rivers have unstabilized mid-channel bars of sand (Fig. 3) or gravel (Fig. 4) emergent during periods of low to moderate discharges. Channels are relatively wide and shallow, and in sandy systems beds are covered by dunes (Cant and Walker, 1978), but in some very large rivers, plane bed may be present at high stage (Coleman, 1969). The beds of these channels may aggrade rapidly during waning flood stages and form vertically accreted channel deposits. Most sandy braided rivers have tabular bars, either linguoid or straight-crested in plan view (Collinson, 1970a; Smith, 1970). In gravelly braided rivers, channel beds are commonly flat-bedded imbricate gravel, but slipface-bounded tabular bars (Smith, 1974) and small bedforms resembling gravel sheets a single clast high above the surrounding bed (Hein and Walker, 1977) have been recorded. Large braid bars in both sandy and gravelly rivers are complex deposi-

tional areas formed by coalescence and accretion of smaller bedforms, resulting from many different flood events. Bar morphologies are complex because of repeated periods of erosion and deposition. Braided rivers aggrade vertically by deposition of sediment in one part of the system with diversion of flow to other parts of the system. Muddy floodplain deposits are present in many of these rivers, but are not as thick or extensive as those of meandering streams. Because of this lack of fine-grained cohesive sediments, braided rivers commonly migrate laterally across wide areas.

Incised rivers are those flowing in valleys cut into older sediments or rocks. Incision results from steepening of the river course by tectonics or base level (usually sea level) fall. Where continued incision occurs, the river is erosional with no net deposition of sediment. A period of rapid incision followed by tectonic and base level stability can cause a river to aggrade within a valley, resulting in valley-fill sediments.

These differences in river types and their long term patterns of migration and aggradation lead to differences in vertical sequences of sedimentary structures, grain size trends, sand or gravel body geometry, and potential reservoir characteristics. These aspects are discussed in detail to aid in recognition of fluvial sediments, interpretation of river type involved with the fluvial deposit, and most importantly, prediction concerning important properties of the sediments.

SETTING OF FLUVIAL SEDIMENTS

Tectonic Relationships

Formation of fluvial deposits requires an elevated source area in close proximity to a platform or shallow basin above sea level. These conditions are fulfilled in several tectonic environments, most notably near young fold mountains, in fault-bounded basins unconnected to orogenic belts, and on tectonically stable, aggrading coastal plains.

The ongoing uplift of large volumes of rock in young fold mountains leads to shedding of synorogenic and postorogenic clastic wedges into foreland or intermontane basins. These sediments have been termed molasse by analogy to a suite of syn- and postorogenic nonmarine sediments of the same name in the Swiss Alps (Van Houten, 1974). Some well known ex-

Fig. 2—Oblique aerial photo of a curved point bar on the inside of a meander bend, Brazos River, Texas. Accretion ridges on the top of the point bar can be seen in the exposed sediments and in the lines of trees. Flow is right to left (Photo courtesy of C. R. Williamson).

Fig. 3—Oblique view of the South Saskatchewan River showing channels, exposed sand flats, and vegetated islands and floodplains. The upstream end of the sand flat has been overridden by a large bar which has caused vertical and lateral accretion. The sand flat is about 300 m in length; flow to the right.

Fig. 4—North Saskatchewan River in Alberta showing gravelly channels with a braided pattern. Large areas of the floodplain are heavily vegetated. Where islands are present in this river, it is anastomosed because of this plant cover; flow away from the camera (Photo courtesy F. J. Hein).

amples of molasse sequences are the Devonian Old Red Sandstone of the North Atlantic region, the Tertiary deposits of the North American western interior, and the presently active Indus-Ganges-Brahmaputra alluvial wedges being shed from the Himalayas. Where molasse wedges spread onto foreland areas, they may show relatively simple dispersal patterns and fining trends away from the mountains. Structural re-entrants in the mountain belt, however, may localize the input of sediment to the basin. This has been well documented for the Kootenay-Blairmore (Upper Jurassic to middle Cretaceous) and Belly River-Paskapoo (Upper Cretaceous to Oligocene) molasse wedges shed from the Columbian orogen into Alberta. Each of these is composed of a series of laterally amalgamating alluvial wedges which radiate outwards from discrete sources (Eisbacher et al, 1974).

Intermontane basins may show a great complexity of paleocurrent patterns because they are commonly fault bounded, and several different sources may contribute to the basin fill. Because of their presence within the mountain belt itself, sources of sediment input are susceptible to change due to tectonic activity. During the early Tertiary, sediment input to the Sustut basin of British Columbia was rearranged from a transverse to a longitudinal pattern, which reflected the tectonic activity in the orogen (Eisbacher et al, 1974).

Because molasse deposits and those nonmarine sediments deposited in other (nonorogenic) fault-bounded basins are so closely associated with ac-

tive tectonism, they can be markedly affected by this deformation. Rapid uplift in the source area, or conversely rapid subsidence of the basin, can lead to increasingly coarser sediment being shed farther into the basin. Long periods of tectonic quiescence cause facies to develop progressively closer to the depositional margin area because of erosion of the elevated source area. Repeated tectonic events can induce a form of cyclicity in facies patterns and grain size on the scale of formations, which can be seen on well logs. These cycles are fundamentally different in scale and origin from those cycles due to sedimentary processes.

As previously discussed, tectonic events in intermontane basins which change locations of structural re-entrants also cause major changes in dispersal patterns. In addition to source area effects, tectonism in the basin also influences the nature of molasse sediments commonly resulting in synsedimentary faults, folds, and intraformational unconformities (Miall, 1978). In many areas, the loading of sediment into the basin may have been responsible for some of the tectonism which modified the molasse sequence. Details of responses of molasse sediments to tectonism in source areas or in basins are given by Miall (1978) in his review of molasse and nonmarine basins.

Tectonism plays another role in the formation of fluvial deposits. Together with climate, base-level control (sea level in most cases), and pre-existing geology of the area, tectonism influences the type of river formed and, therefore, the types of sediments depo-

sited. Rapid source area uplift creates large volumes of coarse sediments and forms steeper regional gradients, favoring formation of braided rivers. Because of the laterally unstable nature of braided rivers and, in some cases due to tectonic tilting of the basin, they may migrate laterally over long distances leaving sheets of sand or gravel over wide areas. Coleman (1969) documented the persistently westward migration of the Brahmaputra River since the Pleistocene. Conversely, lack of tectonic activity causes reduction of relief, lowering of regional gradients, and in humid climates, an increase in chemical weathering. Because of the presence of lower slopes and abundant clay sediments, rivers tend to develop stable cohesive banks and adopt meandering or anastomosing patterns. These river types are much less likely to migrate laterally and deposit sheets of sediment, instead forming much more linear sand bodies. Details of the geometries of the deposits of different river types are discussed in a later section.

Relationships to Lateral Facies

In some areas, fluvial sediments bear only erosional relationships to the sediments around them. Rapid steepening of a river slope due to regional tectonic uplift or base level fall can cause incision of the stream into the substratum over which it flows. Rivers have cut deep valleys during many periods, most recently during the sea-level drop in the Pleistocene. The resulting valleys are infilled with sediments as base level rises, and the valley-fill sediments bear an erosional relationship to the

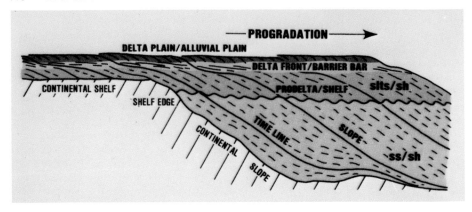

Fig. 5—Diagrammatic illustration of a prograding clastic wedge in which fluvial sediments (in red) have advanced over the top of deltaic deposits. This type of sequence can commonly be recognized in seismic sections. After Brown and Fisher (1977).

Fig. 6—Point bar on the Brazos River, Texas, and an oxbow lake which once was an active river meander. The lake is almost filled by clay sediments and has much vegetation growing within it. Clay plugs created by the infilling of these lakes restrict the lateral migration of the river, flow to the right.

rocks or sediments below them. These filled valleys have been documented in many places in the geologic record, e.g. the Pleistocene Mississippi system (Fisk, 1944) and the Permian Cook sandstone (Bloomer, 1977).

More commonly, fluvial deposits are part of larger scale facies tracts where fluvial sediments grade basinward into deltaic, shoreline, and shelf facies which in turn may be succeeded by slope and basin facies. This linking of facies into larger-scale units has been particularly well illustrated by techniques of seismic stratigraphy. Sangree and Widmier (1977) have shown that reflections from fluvial sediments of the Cretaceous Menefee Formation of the San Juan basin can be traced laterally into those from shoreline sands and marine shales. This lateral variation in depositional environment is pre-

served in lateral facies variation in the rocks. The Cretaceous Castlegate Formation of Utah shows a lateral transition from braided-river conglomerates in the west to deltaic sandstones and coal, and finally shallow-marine shales in the east (Van De Graaff, 1972). Brown and Fisher (1977) illustrate progradation of fluvial deposits over inclined deltaic beds, which themselves overlie prodelta and offshore marine deposits (Fig. 5). Many reports exist in the literature of complete sequences of facies with turbidites at the bottom, slope deposits above these, then shallow-marine sediments and fluvial sandstones at the tops (e.g. Collinson, 1970b).

On a smaller scale, fluvial deposits commonly are laterally associated or interbedded with alluvial fan, eolian, lacustrine, swamp, and deltaic sedi-

ments. Alluvial fans are commonly sediment sources for rivers. In many small interior drainage basins of the western United States, fans emerge from mountain fronts transverse to the basin, while ephemeral or braided streams drain longitudinally down the axis of the basin. Because streams are only slightly reworking and winnowing the alluvial fan sediments, the difference in paleocurrent orientation may be significant for distinguishing the two types of continental deposits. Eolian deposits may be closely associated with rivers and affect them in a number of ways. Where rivers traverse areas with abundant eolian sands, they pick up much sandy bed load, and may adopt a braided pattern. Eolian processes can also rework fluvial sediments which are exposed by falling river stages. Eolian dunes up to 2 m in amplitude have formed on floodplains, islands, and some large braid bars of the South Saskatchewan River (Cant and Walker, 1978). These dune deposits were trapped and stabilized by vegetation, and could probably be preserved in aggrading systems. Interbedding of fluvial and eolian sediments is probably more common in the geologic record than has been previously recognized.

Shallow lakes and swamps form in many areas on floodplains, some due to actions of the rivers themselves. For example, abandoned meander belts, channels, or swales (between accretionary ridges on point bars) form small lakes and ponds which eventually fill with fine-grained sediment creating swamps (Fig. 6). In other areas, lakes form because of rapid rises in base level or blockages of the river system (i.e. by glacial advance across a valley). Some shales interpreted as floodplain deposits may actually be laid down in these relatively ephemeral lakes, ponds and swamps which are not directly related to fluvial processes.

Deltas are formed at the distal ends of rivers discharging into standing bodies of water. The transition from fluvial to deltaic deposits may be gradational if the river progrades over the delta top. Transition may be abrupt if marine processes rework fluvial sediment as in a wave-dominated delta.

GEOMETRIES OF FACIES

Because different river types generate sedimentary bodies with different geometries, each river type deserves separate discussion. The most common cases for each type are described, but it should be realized that a complete gradation among the types exists.

Braided River Deposits

Braided rivers are laterally unstable because floodplain banks are thinner, less cohesive, and more erodible than those of other rivers; braided rivers also have high discharge peaks in areas of steeper slope, which increase their eroding power. Both sandy and gravelly braided rivers migrate laterally leaving sheet-like or wedge-shaped deposits of channel and bar complexes

preserving only minor amounts of floodplain material.

Lateral migration coupled with aggradation leads to sheet sandstones or conglomerates with thin, impersistent shales enclosed within coarser sediments (Fig. 7). This has been well documented by Campbell (1976) for the Westwater Canyon Member of the Jurassic Morrison Formation of New Mexico. He showed sandy braided river deposits consisting almost entirely of sandstones with laterally coalesced channels, smaller ones of 180 m width and 4 m depth enclosed in

larger ones of 11 km width and 15 m depth (Fig. 8). These probably resulted from lateral migration of channels within the braided system, and migration of the river itself. A sheet sandstone was formed, 60 km wide and elongated down the paleoslope.

Great thicknesses of braided river deposits have been noted in the geologic record (Fig. 9). The Ivishak Formation (Triassic; Fig. 10) of the Alaskan North Slope, reservoir rock of the Prudhoe Bay Field, contains 200 m of braided river sandstone with only minor shale lenses scattered throughout.

Meandering River Deposits

Meandering rivers are generally more laterally stable than braided rivers because they have thicker, more heavily vegetated, cohesive floodplain deposits which are difficult to erode; they occur in areas of lower slope; and they show more regular discharge patterns. These rivers are also confined laterally by abandoned mud-filled meander loops, known as oxbow lakes, which are common on the floodplains of most meandering rivers (Fig. 6). Presence of many of these clay plugs on a floodplain restricts any tendency toward lateral migration of the meander belt. The channel freely migrates within the dominantly sandy meander

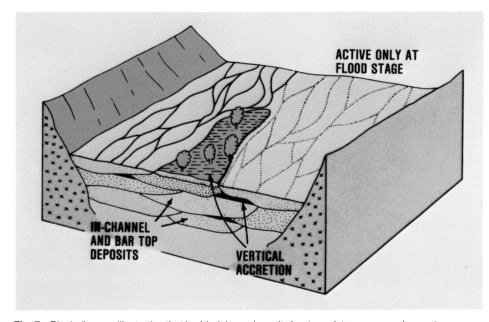

Fig. 7—Block diagram illustrating that braided rivers deposit sheet sandstones or conglomerates restricted laterally only by the width of the area undergoing alluviation. The preserved shales are thin and impersistent.

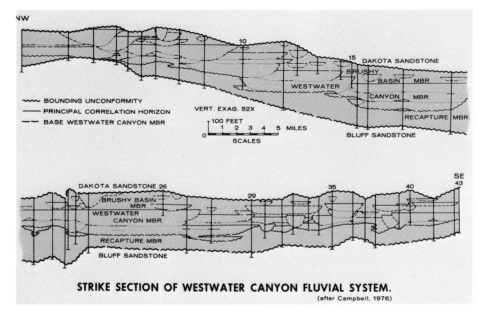

Fig. 8—Cross section of sheet sandstone of the Westwater Canyon Member of the Morrison Formation (Jurassic) of New Mexico. The sandstone body is 60 km wide across the paleoslope and consists of a series of laterally coalesced channels which are filled by trough crossbeds.

Fig. 9—Thick section of sandstone laid down by a braided river. Little fine material is present in the units except as small clay plugs or drapes, Castlegate Sandstone (Cretaceous), Utah.

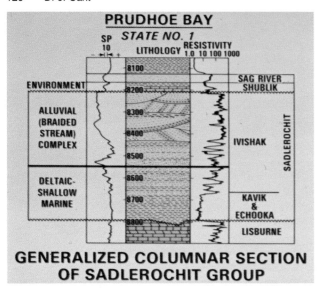

Fig. 10—SP and resistivity logs showing the irregular patterns characteristic of braided river deposits. The Ivishak Formation is the reservoir rock for the Prudhoe Bay field.

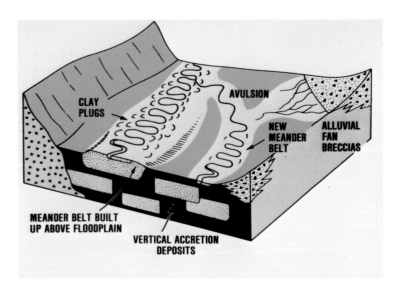

Fig. 11—Block diagram of meandering stream deposits showing linear sand bodies oriented down the depositional slope. These porous facies are laid down by the meander belts, and are enclosed within impermeable floodplain shales.

belt it established, but cannot easily widen the belt. Judging from the presence of meander scars on floodplains of many modern rivers, the meander belt is likely to be about twice the average meander amplitude. This lateral restriction of the river causes formation of a linear or shoestring sand body oriented down the depositional slope (Fig. 11) which may curve or meander somewhat on a large scale. This sand body is commonly surrounded by fine-grained floodplain sediment which may act as a covering after abandonment (Fig. 12). This impermeable covering is formed by overbank sedimentation from another active channel in the alluvial system.

As a river aggrades, it builds above the level of its own floodplain—an unstable situation. Eventually, a levee break during a flood results in abandonment of the meander belt and establishment of a new one. This process is known as avulsion and leads to the formation of a series of sand bodies oriented down the depositional slope in a matrix of impermeable floodplain deposits (Fig. 11). The degree of interconnectedness of these alluvial sand bodies depends on several factors, such as the rate of aggradation, frequency and distance of avulsion, and depth of the river involved. Allen (1978) suggested from theoretical studies that for meandering stream deposits with less than 50% sandstone in a vertical sequence, the likelihood is that sandstone bodies are not connected. In meandering stream deposits with more than 50% sandstone in a vertical section, probability of the sandstone bodies being connected rises with the in-

crease in the amount of coarse sediment. While these studies are theoretical and have yet to be confirmed by observation they may be useful in reservoir development because sandstone thicknesses can be easily measured on well logs. These studies may help predict whether a sequence of fluvial deposits of this type will act as a single production unit, or whether each sand body within the sequence will be isolated and require separate drainage.

Anastomosing River Deposits

These rivers have semipermanent islands dividing the flow and well developed floodplain and backswamp areas extending away from the river. Smith and Smith (in press) have shown that channel sediments of several anastomosing rivers in the Canadian Rockies consist of very coarse grained gravels, but that alluvial islands and floodplains are composed of fine silts and peats. Because of the cohesive nature of this sediment and high degree of root binding, islands and floodplains are very stable and only slowly change in position and size. Smith and Smith (in press) were able to show by augering to well dated ash layers that silts have been deposited on the sites of some islands and floodplains for 6,600 years or more, a period during which up to 8 m of sediment accumulated. Because of this long-term stability of channels, islands, and floodplains, individual facies accreted vertically, stacking deposits in thick piles (Fig. 13). The islands and floodplains occupy much wider areas than channels, so peaty silts are the dominant facies deposited

by this type of river. The sand or gravel bodies resulting from aggradation of the channels are very narrow proportional to their thickness and divide and unite around silt plugs deposited on the islands. These bodies of coarse sediment are more narrow in proportion to their thickness than those in meandering-stream deposits.

Incised Rivers

Any of the river types described previously may be involved in the filling of a valley. The anastomosing rivers described by Smith and Smith (in press) are all aggrading within valleys; during the post-Pleistocene rise in sea level, the Mississippi valley was partly infilled by sandy braided-river deposits while the modern river is strongly meandering. The geometry of river deposits aggrading within a valley is determined by the form of the valley itself. In most cases, valleys are cut over a long period of time and are widened and straightened by the streams migrating against their walls. Thus, most valleys are relatively straight or gently curving, but in a few areas where rapid uplift occurred (San Juan River on the Colorado Plateau) the original meandering pattern of the river is preserved in the form of the valley.

Valley fills are well known in the ancient record and most commonly are relatively narrow, linear bodies (Fig. 14) which may extend for long distances in the subsurface. Depending on the type and size of river within the valley, the fill may be dominantly coarse or fine grained. Abandoned valleys may be infilled by shale which makes them efficient hydrocarbon traps.

MICROFACIES OF RIVER SEDIMENTS

Individual sedimentary structures are far less diagnostic of differing fluvial environments than are assemblages of structures. This section details microfacies and assemblages of sedimentary structures helpful in recognizing deposits of each river type.

Braided Rivers

Very proximal braided river sediments consist of mainly horizontally bedded, clast-supported gravels (Fig. 15), deposits of channels and large complex braid bars. In most braided rivers the pebbles have their long (a) axes transverse to flow (Fig. 16) and upstream imbricated intermediate (b) axes. This pebble fabric may be useful in distinguishing these from deep-sea conglomerates which have imbricated long (a) axes aligned parallel with flow (Harms et al, 1975). This criterion may be helpful in dealing with relatively fine conglomerates in conventional cores, but measurements of a statistically valid sample of pebbles should always be made—from 100 to 300 depending on the variance of the pebble orientations.

In a few channels of some of these rivers, large gravelly crossbed sets, either of trough (Fig. 17) or tabular types, may form by migration of gravelly dunes or flat topped bars. These may be preserved in some places, but they form only a minor part of the deposit. Crossbeds form more commonly in topographically higher parts of river deposits, where wedge-shaped gravel crossbeds with wide paleocurrent dis-

Fig. 12—Large outcrop of meandering stream deposits in the Sespe Formation (Oligocene) of California which shows light colored sandstones and red shales. The large sandstone bed in the center of the photograph thins and splits laterally. The more continuous bed above this is about 3 m thick (Photo courtesy J. B. Dunham).

persion are deposited on bar margins (Fig. 18) around core facies of horizontally bedded gravels (Smith, 1974; Hein and Walker, 1977; Rust, 1978). Wedges of sand are present in similar situations in many modern rivers, but are rarely preserved.

A graphic model for this type of river deposit is presented by Miall (1977) and shown in Figure 19a as the Scott Type model. Deposits consist mainly of horizontally bedded gravels which fill stacked wide, shallow channels. Minor amounts of crossbedded gravels are also present. One distinguishing feature of these deposit types is their failure to show development of cyclicity,

largely beause the coarse gravel moves only during high river discharges and relief on bars is small compared to the depth of flow. Flow velocities are, therefore, similar in all parts of the river, and consistent vertical grain size segregations do not occur.

Braided rivers with mixed bedloads of sand and gravel develop some degree of vertical zonation, with bars standing one to three meters above channel beds. The lower parts of channels fill with imbricated, horizontally bedded or roughly crossbedded gravels, which may be supported by a sandy matrix (Fig. 20) depending on the bed load composition. The upper

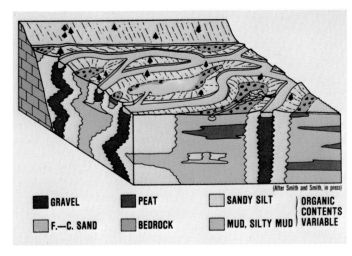

Fig. 13—A block diagram of an anastomosing river showing the vertical stacking of channel sandstones and conglomerates. These coarser facies are bounded by floodplain silts, muds, and peats.

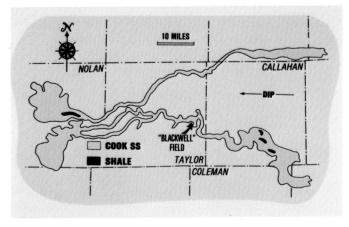

Fig. 14—Map of the valley fill deposits of the Cook Sandstone (Lower Permian) of the Midland basin. The valleys are incised about 30 m into shales and limestones and are filled by fluvial sandstones and capped by shales. They form stratigraphic traps in several places. After Bloomer (1977).

Fig. 15—Well-imbricated fluvial conglomerate with streaks of openwork gravel, here cemented by white calcite. In many deposits, these openwork streaks provide good porosity and permeability, and placer minerals may be trapped in these. The scale is extended 66 cm. Triassic of New Brunswick.

Fig. 16—The top of a gravel bar in the Kicking Horse River, British Columbia, illustrating extreme bimodality, with silt occurring between large cobbles. Current lineations in the silts show the current direction, upper left to lower right, while the clasts show a transverse long axis fabric.

Fig. 17—Trough crossbeds developed in pebbly sandstone at the base of a channel in the Triassic Moss Back Member of the Chinle Formation, Utah. This braided river deposit locally contains uranium.

parts of channels or minor channels are floored by dominantly sandy, sinuous crested dunes depositing sets of trough crossbeds. Braid bars in these rivers may show horizontally bedded gravels, but in their upper parts they are also composed of sandy planar crossbed sets and parallel laminated sand. Because of the topographic elevations of bars and floodplains, areas on the tops of these are protected from high velocity flood flows, and rippled fine sands and muds may accumulate. The deposits of these rivers may therefore show rough fining-upward sequences (Williams and Rust, 1969) which are commonly channelled out, small grain size reversals due to fluctuating river stages, and highly variable thicknesses. They begin at the base with horizontally bedded or cross-stratified gravels which fine upward into roughly cross-stratified gravelly sand. Toward the top, each sequence may be highly irregular with sandy tabular crossbed sets or parallel laminations and capped by fine grained rippled sands and muds. However, in many places, sequences will be incomplete due to nondeposition or erosion and any of the channel or lower bar facies may comprise the entire sequence. Because rivers of this type show relatively variable discharges, cyclicity is poorly developed (Fig. 21, 22) with coarser flood deposits commonly overlying finer-grained material (Fig. 23). Miall (1977) attempted to summarize these deposits in his Donjek-type model (Fig. 19a).

Sandy braided rivers also show a variety of morphologies which to a great extent affect the degree of vertical cyclicity development in their deposits. Those with the most variable discharge (including ephemeral streams) rarely show any topographic differentiation between channels and bars, and during high stages, flows are unconfined, and sweep over the entire alluvial area as sheet floods. Deposits consist of flood sequences of upper flow regime parallel-laminated sands (Fig. 24) with minor crossbeds and small amounts of sandy climbing ripples. No vertical segregation of grain size occurs, and resultant deposits consist of sands of variable thicknesses in these flood

Fig. 18—View of an exposed gravel bar or flat in the Kicking Horse River of British Columbia. Small gravelly, slipface-bounded bars are present where the channels discharge into the wider area downstream of the exposed area, flow toward the camera.

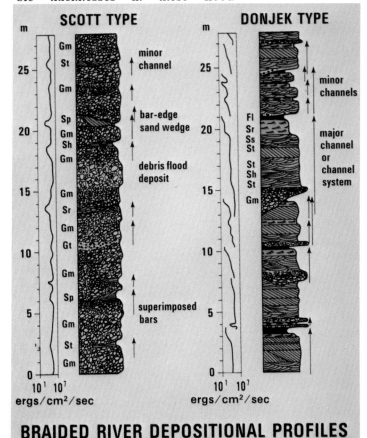

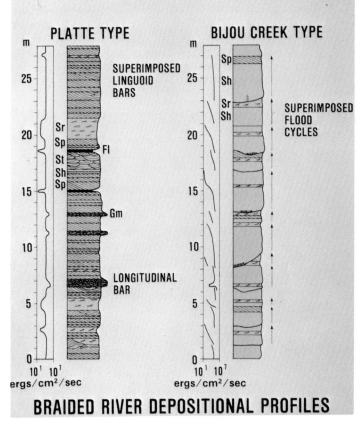

Fig. 19—Miall's (1977) models of braided river sediments. The graphs beside each diagram represent an estimate of the energy involved during deposition of the sediments. Part a: The Scott-type model shows noncyclic conglomerate, and Donjek-type model consists of conglomerates and sandstone which show a rough form of cyclicity. Part b: The Platte-type model is made up of sandstone with stacked planar-crossbed sets laid down by linguoid bars, and the Bijou Creek-type model is composed of superimposed horizontally bedded sheetflood deposits.

Fig. 20—Section of core from the Cadomin Conglomerate (Lower Cretaceous) of Alberta, a braided river deposit. Good porosity exists in this unit, and it occurs in oil and gas producing areas. However, because of its sheet-like geometry and the absence of shales, the Cadomin has few stratigraphic traps in which hydrocarbons might have accumulated.

Fig. 21—Pleistocene braided river deposits near Canmore, Alberta. They have clast-supported, well-imbricated conglomerates interbedded with horizontally stratified gravelly sands (lighter color) and lenses of sand. No cyclicity exists in these deposits; they fit Miall's Scott-type model. About 3.5 m of section shown.

Fig. 22—Poorly sorted bed of conglomerate from the Silurian Shawangunk Conglomerate of Pennsylvania. Thin beds of conglomerate and crossbedded sandstone suggest a Platte River-type of braided stream.

sequences. Miall (1977) summarized these deposits in his Bijou Creek- type model (Fig. 19b).

Sandy braided rivers with more steady discharges and more topographic differentiation develop large linguoid, sinuous or straight-crested tabular bars at high stage which are transverse to flow. These may be considerably modified by low-stage dissection, but planar crossbeds (Fig. 26) deposited by these bars are the stratification type most likely preserved (Fig. 27). Even where floodplains are developed, they are inundated at high stages, with tabular bars driven onto them (Fig. 28). Muddy floodplain sediments deposited at moderate river stages may be preserved in a few areas (Fig. 29), but more commonly are ripped up to make cohesive mud intraclasts which are scattered, forming mud-clast conglomerates at some channel bases. Deposits of these river types are summarized in Miall's Platte- type model (Fig. 19b).

Fig. 23—Interbedded conglomerate and a cohesive silt layer in an exposed bank area of the braided Kicking Horse River, British Columbia. This river lays down deposits with little cyclicity in grain-size trends (Photo courtesy N. D. Smith).

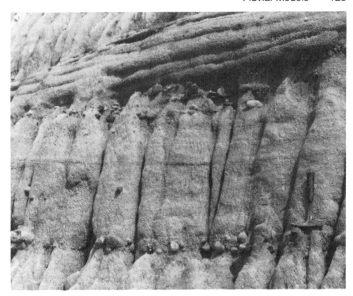

Fig. 24—Flood sequence from braided-river deposits of the Rosarito Beach Formation (Miocene) near Tijuana, Mexico. Although sandstone predominates, large cobbles are also present. They are concentrated mainly at the base of the sequence, but a few isolated individual cobbles are scattered throughout the bed. These deposits appear to resemble Miall's Bijou Creek-type model (see Fig. 19).

Fig. 25—Small sand flats in the South Saskatchewan River which have been partly overriden by a large bar. The bar has built forward into the channel between the sand flats and extends diagonally upstream and downstream. The planar crossbeds deposited by this bar are extensive and have a highly complex paleocurrent pattern, flow to the right. The front sand flat is about 100 m in length.

Fig. 26—A trench in a sand flat from the South Saskatchewan River showing a planar-crossbed set with scattered pebbles. Some vertical as well as downstream accretion has occurred. These deposits were laid down during the waning part of a large flood, flow was to the right.

In contrast to the shallow streams described, some sandy braided rivers show more steady discharges and higher degrees of topographic differentiation with large complex depositional braid bars (termed sand flats by Cant and Walker, 1978) and well developed floodplains. These rivers are described in some detail because their deposits can be easily confused with those of meandering rivers. The most intensively studied example of this river type is the South Saskatchewan River in central Canada (Fig. 30). In this river, channels average 3 to 5 m in depth below the level of nearby sand flats and are floored by sinuous-crested dunes (Fig. 30) which build up at flood stage (Fig. 31) and deposit sets of sandy trough crossbeds (Figs. 32, 33) up to 1 m in thickness. Large flat-topped oblique bars form in channels where flow expands and velocities decline, as at channels junctions. These bars deposit sets of planar crossbeds averaging about 1 m in thickness, but in extreme cases may be up to 3 m thick. The large sandy depositional areas termed sand flats (Figs. 34, 35) to distinguish them from the bars just described consist of stacked planar crossbed sets which are deposited during floods and have minor amounts of vertical-accretion parallel-laminated sands (Fig. 36). In all topographically high areas of the river (whether from channel filling or bar growth) minor sandy crossbed sets resulting from migration of sand waves, and muddy rippled sands are deposited. Cohesive vertical accretion deposits consisting of 1 meter thick silty and sandy muds with convoluted and disturbed interbeds of sand are present on floodplains and islands (Fig. 37). This river sedimentary regime is summarized in Figure 38.

Fluvial sequences may be generated by channel aggradation and diversion of flow to another part of the braided river, sand flat growth, or combination of the two processes. Because of this, a series of vertical profiles can result (Fig. 39) depending on the relative importance of each process. It should be emphasized that the braided nature of the sequences is indicated by planar tabular crossbed sets (Fig. 40) within the sand body. Ancient sequences similar to each of these have been described from the Devonian Battery Point Formation (Cant and Walker, 1976), and summarized as a single vertical sequence (Fig. 41). Sandy braided rivers of this type deposit vertical sequences which can be confused with sequences of meandering rivers. Care must be taken to log cores in detail to attempt to distinguish the types of deposits. Braided river deposits have more planar crossbed sets within the sand body, more irregularities in grain size trends, and in many cases less fine overbank material.

Fig. 30—Many small dunes with well developed scour troughs in which fine gravel is concentrated. The dunes show great variations in amplitude and have falling-stage ripples superimposed on them, from the South Saskatchewan River. Flow was toward the camera.

Fig. 27—Stacked planar crossbed sets of the Nubia Formation (Lower Cretaceous) of Egypt. Deposits of this kind are laid down by shallow rivers such as the Platte River which contain many large tabular bars at flood stage. During periods of high flow, the river is essentially straight with no mid-channel bars emergent. The sets average about 30 cm in thickness (Photo courtesy C. R. Williamson).

Fig. 28—An area of overbank sediment near the Platte River, Nebraska, where small bars have spilled out of the river depositing layers of planar to wedge shaped crossbeds (Photo courtesy F. J. Hein).

Fig. 29—Interbedded sands and organic-rich rooting muds forming floodplain deposits of the Platte River, Nebraska. These noncohesive sandy floodplain deposits do not restrict lateral migration of the braided river (Photo courtesy N. D. Smith).

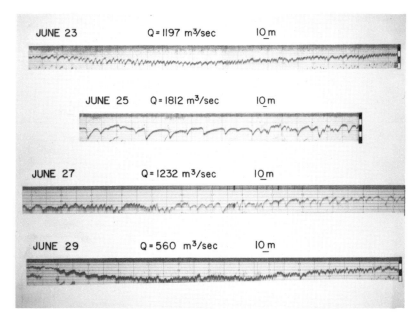

Fig. 31—Four echo sounder profiles made down a channel in the South Saskatchewan River at different times during a flood event. Each profile represents approximately the same length of channel and the scale on the right is in meters. The profiles illustrate the response of the dunes on the bed to the rising flood (June 23 and 25) and the subsequent fall in stage (Jule 27 and 29). The large bedforms formed at the peak flow probably deposit sets of trough crossbeds up to 1 m thick.

Fig. 32—Small trough crossbeds in a boxcore taken in a minor channel of the South →
Saskatchewan River. These 10 cm-thick troughs were deposited by the migration of small dunes with amplitudes of 15-20 cm. The scale is in cm and in inches.

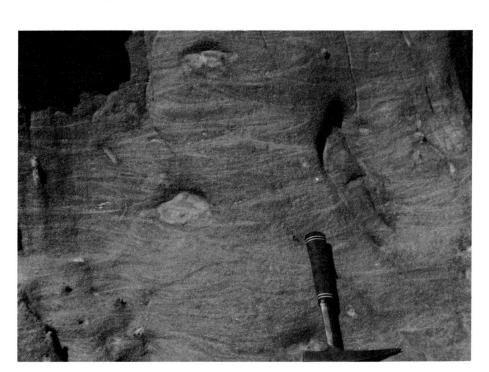

Fig. 33—Small trough crossbeds developed in coarse, reddened sandstone with light-colored concretions. These sediments are part of a braided river sheet sandstone, Simmsboro Sandstone of the Wilcox Group (Cretaceous) of Texas.

Meandering Rivers

Meandering rivers have a great deal of topographic differentiation (Fig. 42) giving their deposits highly developed cyclicity. Beds of channels and lower point bars are covered by large dunes (Fig. 43) which deposit sets of trough crossbeds. In very sinuous rivers with relatively fine sediment, the scale of the crossbeds and grain size of sands decline upward through the trough zone. Locally, on the slope of the point bar, upper flow regime plane beds may form, depositing lenses of parallel-laminated sands intercalated with the trough crossbedded sands. These lenses presumably occur because of the presence of local areas of high flow velocity during flood stages. Near the top of the point bar, rippled sands (Fig. 44) and mud drapes (Fig. 45) are deposited at moderate river stages. These sinuous rivers have thick floodplain muds, which commonly contain roots or caliches (Figs. 46, 47) depending on the climate. Lateral river migration deposits a sequence which fines upward from lag gravel and intraclast conglomerate (Fig. 48) to sand, to mud (Fig. 49), the classic fining-upward model of Bernard et al (1970) and Allen (1970; Fig. 50). This model of meandering river deposits has been in use for many years. In some fining-upward sequences, successive positions of point-bar surfaces are recorded by fine beds which drape the surface. These lateral accretion surfaces or epsilon crossbeds (Fig. 51) are useful indicators of these relatively small, sinuous meandering rivers. Some cores from meandering stream deposits are illustrated in Figure 52.

Recent work on coarse-grained, less sinuous meandering rivers, however, has shown that several variants of this model can exist. McGowan and Garner (1970) studied coarse-grained point

bars on several rivers and concluded that deposits showed little fining-upward trend. They observed large chute bar development across tops of the point bars at the ends of high-stage channels. These chute bars consist of large (up to 3m thick) planar or wedge shaped crossbed sets which occur at the top of the sand body directly under any floodplain deposit. Jackson (1976) has shown in another meandering river which does not deposit fining-upward sand bodies, that large planar crossbed sets are deposited on the tops of point bars by flat-topped transverse-type bars. Again these are developed at the top of the sand body directly under the floodplain material. Careful core logging is necessary to distinguish this type of meandering river deposit from braided river deposits described in the

Fig. 34—Area on the top of a sand flat in the South Saskatchewan River covered mainly by current ripples. Extending diagonally across the foreground is a bar front which has been modified by falling stage flow, with ripples and falling water marks on its slipface. Flow was left to right, diagonally across the bar front.

Fig. 35—A much modified bar front on a sand flat in the South Saskatchewan River. Original irregularities in the shape of the bar front have been eroded by falling stage flow, forming two scour holes. These are being filled by mud and organic debris.

Fig. 36—A trench in sand flat sediments of the South Saskatchewan River. The section shows a well developed set of planar crossbeds overlain by upper flow regime parallel laminations, flow to the left. The scale is extended 30 cm.

last section. Even in these coarse-grained meandering river deposits, however, overbank fine-grained deposits may be thicker and have a greater chance of preservation than in braided river deposits. However, little data exist on this problem and it is currently a controversial subject (Collinson, 1978).

Anastomosing Rivers

Channels in these rivers contain sediments similiar to channel sediments in other river types discussed, (flat-bedded gravels or trough crossbedded sands) depending on the type of load transported. The channel deposits show relatively little fining-upward tendency because they cannot migrate laterally, and cannot easily switch to other parts of the alluvial plain. The most distinctive feature of anastomosing river deposits is the presence of thick peat-rich silts and clays deposited on islands and floodplains and the development of the silts and clays laterally to the coarse-grained channel facies. Fine-grained deposits may be interbedded with thin graded beds or coals and are otherwise very similar to backswamp sediments of meandering stream deposits. It is the arrangement of facies which is distinctive of stream type, not the facies themselves.

SEDIMENT TEXTURES AND LOG RESPONSE

Texture

The textures developed in river sediments are as varied as the facies themselves. Fluvial conglomerates range from clast-supported, matrix-free examples, through clast-supported gravels (Fig. 53) with interstitial sandy matrix, on to sandy conglomerates with dispersed clasts. Grain size distributions of all these types are different. The matrix-free conglomerates are reasonably well sorted and unimodal, the clast-supported gravels with a sandy matrix show bimodal size distributions, and sandy conglomerates are unimodal but with poor to moderate sorting. All these types may occur within a few meters of section depending on the mechanisms of deposition.

Fluvial sands range from very coarse with scattered pebbles and little interstitial clay (Fig. 54) through fine to medium-grained well-sorted examples, to silty or muddy sands with a large proportion of fine material. These textural characteristics are again primarily controlled by the mechanism of deposition. Channel, lower point bar, and braid-bar sands commonly are moderately well sorted to well sorted, with little fine matrix because preserved deposits are mostly laid down at high river stages when fine sediment is held in suspension. Higher in point bar deposits or in those braided river deposits which show cyclicity, the sediments reflect deposition in quieter water and have a more silty or muddy matrix.

Plotting of grain size cumulative curves (in phi units) on a probability scale commonly yields a series of straight-line segments indicating the distribution is made up of overlapping populations, each with a normal distribution. These populations have been interpreted as reflecting the several transport processes which operate on sediment at the same time. For fluvial

Fig. 37—Bank in the South Saskatchewan River showing crossbedded sands at the base followed by rippled sands. A scour was cut into the surface, draped with mud, and infilled with sand (possibly blown in). The section is capped by muddy floodplain deposits, flow to the left. The scale is in cm and in inches.

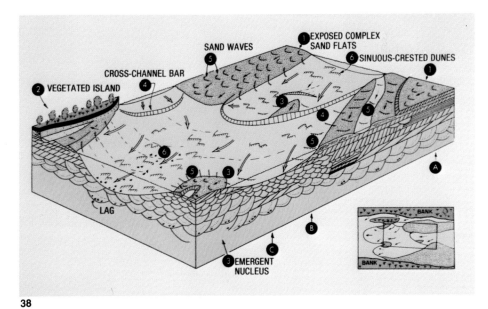

38

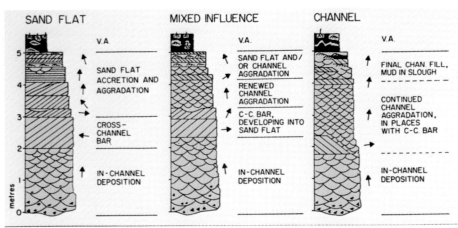

Fig. 38, 39—Block diagram illustrating the microfacies of the South Saskatchewan River with a plan view of the same hypothetical reach shown in the inset. The circled letters A, B, and C, refer to the sand flat, mixed influence, and channel sequences respectively in Figure 39. This river would generate cyclic sandy braided stream deposits. After Cant and Walker (1978).

sands, these straight-line segments have been interpreted as resulting from traction, saltation and suspension processes (Middleton, 1976). Not all of these populations are necessarily present in every sample of fluvial sand. Little use has been made of this approach for interpretation of ancient sediments, partly because characteristics of modern sands in all depositional environments have not been well documented. The same is true for attempts to use plots of statistical parameters derived from cumulative curves to interpret depositional environment. For modern sediments, graphs of skewness versus sorting discriminate between environments in many cases, but appear to be of little use in dealing with rocks. Problems occur because of recrystallization of the fine fraction, authigenic minerals and overgrowths, and measurement error in thin sections. In addition, it is unlikely that any textural plot uniquely represents an environment in all cases.

Log Response

Electric and gamma ray logs from fluvial sediments commonly show the different types of cycles or sequences previously discussed. On the smallest scale, meandering fining-upward sequences generate a bell shaped SP or gamma ray curve. This has been well illustrated from the Brazos valley (Fig.

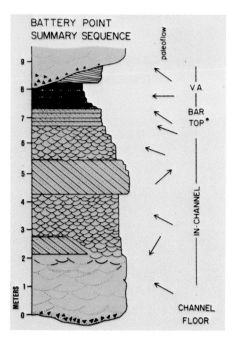

Fig. 41—Summary sequence from the Battery Point Formation (Devonian) of Quebec which shows interbedded trough and planar crossbeds, smaller crossbeds above, and minor amounts of mud and rippled- to parallel-laminated sands at the top. This sequence is believed to result from bar advance, sand flat formation, and channel aggradation in a braided river. After Cant and Walker (1976).

Fig. 40—Large planar crossbed set with a flat base and top in the Devonian Battery Point Formation of Quebec. The laminations within the set are thick (up to 4 cm) and straight in most cases, but some are slightly tangential at the base. The scale is 30 cm long.

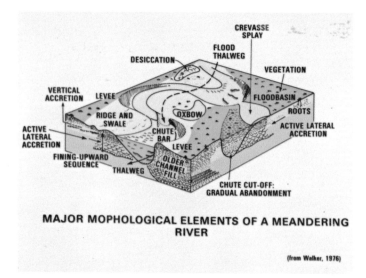

Fig. 42—Block diagram illustrating the geomorphological elements of a meandering river showing the relationships between the active environment and the preserved deposits. The flood thalweg is the path followed by the flow at high stage.

Fig. 43—Large dunes on the lower point bar of one meander bend of the Brazos River, Texas. These are typical of the channel and lower half of the point bar and deposit sets of trough crossbeds. Large logs and other organic debris are also present. Flow away from the camera.

55) by Bernard et al (1970) and from many ancient deposits. Stacking of these sequences above one another may generate repeated bell-shaped curves, but channelling of one sand body into another may cause a more complex curve (Fig. 56). Braided river deposits may show some upward-fining, as discussed previously, but often the grain-size variation is insufficient to generate a recognizable log pattern. The almost random preservation of smaller units of muddy flood-plain deposits causes the logs to mainly indicate coarse sediment with some finer interbeds, but no easily recognizable patterns emerge (Fig. 10).

In all types of fluvial sediments, dipmeters may register a dominant direction indicating crossbed dip direction and therefore give a measure of channel direction and in some cases sand body elongation. However, presence of side slopes into trough crossbeds and obliquely directed planar crossbeds (Cant and Walker, 1976) makes interpretation of dipmeter data somewhat more uncertain. In meandering sequences, dipmeters may register lateral-accretion surfaces or epsilon cross-stratification (Fig. 51) because these beds commonly are composed of fine-grained sediments. The directions indicated by these crossbeds will vary around the meander, but in no case will they indicate downcurrent direction.

Also remaining is the more general problem of interpretation of dipmeter data. Tectonic dips, fractures, and spurious instrument readings make interpretation of crossbed dip direction very difficult, if not impossible, in many cases.

On a larger scale, tectonic events may cause changes in the types of sediments deposited, and therefore the log patterns obtained. Coarsening-upward fluvial sequences several hundreds of meters thick that result from uplift of the source area can be recognized on logs in some areas. Within these sequences, which may encompass several formations, smaller sequences generated by lateral and vertical accretion may be recognizable (McLean and Jerzykiewicz, 1978).

Mineralogy

Because fluvial sediments inherit the mineralogy of the sediment source, no mineral suite is diagnostic of fluvial deposits. Alteration of iron-rich minerals to hematite occurs in fluvial as well as other oxygenated environments, therefore, the red color of sediments is not as indicative of the depositional envi-

Fig. 44—Trench in the upper part of a point bar of the Brazos River, Texas. The trench shows crossbedded and rippled sands at the base and convoluted, deformed laminations above. This deformation took place when the sediment was water saturated. This is commonly interpreted in the ancient record as a result of earthquake shock, but is clearly not so in this case.

Fig. 45—Floodplain deposits capping the point-bar sand of a Brazos River meander belt. These deposits consist of thin polygonally cracked muds with abundant plant roots.

Fig. 46—Red fine-grained floodplain deposits of the Triassic Moenkopi Formation of Arizona. The whitish beds are caliche deposits and light-red sandstones at the top are crevasse splays. Many reduced green bands can be seen. A large swale or scour fill is present at the base. About 6 m of section is present.

ronment as it is the conditions of diagenesis. Fluvial environments may be ruled out by mineralogy where the presence of abundant glauconite indicates marine conditions.

Paleontology

Fluvial sediments contain a very restricted suite of fossils compared with many marine sediments. Fish fossils are relatively common in some fluvial deposits, but it is not always possible to determine whether the fish lived in fresh or salt water. The same problem exists with the use of gastropods, pelecypods, and other invertebrates. Floodplain sediments may show vertebrate remains or other trails, but these are not restricted to fluvial sediments.

By far the most common fossils in fluvial deposits are plant fossils. Large

Fig. 49—A fining-upward sequence with gravelly crossbeds at the base, followed by cross-bedded sandstones, and red shales at the top. This represents the deposits of a shallow stream which has a meandering pattern. The book is 20 cm long. Shepody Formation (Carboniferous) of New Brunswick.

Fig. 47—A caliche zone in red floodplain sediments of the Enrage Formation (Carboniferous) of New Brunswick. Irregular caliche nodules are present above the more continuous zone. The book is 20 cm long (Photo courtesy of P. J. McCabe).

Fig. 48—Basal conglomerate showing mudstone intraclasts and quartz, feldspar, and rock fragment clasts. The mudstone intraclasts are larger, more elongated and imbricated than the other fragments. The red blocks on the notebook are 5 cm long. Boss Point Formation (Carboniferous) of New Brunswick (Photo courtesy P. J. McCabe).

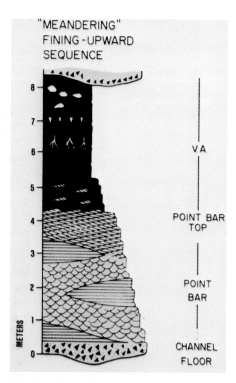

Fig. 50—Classic fining-upward sequence deposited by lateral migration of a meandering river. The sequence begins with intraclast-rich conglomerate at the base, then trough crossbedded to parallel laminated sands, then finer rippled sands, and finally levee and overbank muds. After Allen (1970).

Fig. 51—Part of a fining-upward sequence which shows large scale, low-angle, lateral accretion crossbeds (epsilon crossbedding). These form as a point bar accretes laterally, with sands and finer sediment laid down during periods of different discharges. These surfaces are commonly cited as a relatively reliable criterion of meandering stream deposits. Carboniferous of Kentucky (Photo courtesy P. J. McCabe).

logs may be deposited as channel lags, and comminuted plant debris occurs in some muddy floodplain sediments. In many Devonian or younger fluvial deposits, *in situ* plants recognizable by their vertical stems are present in upper parts of fluvial sequences. However, plant fossils may be coalified and specific identifications cannot be made. Fossil pollen is also common, but because it is wind transported, it is not restricted to fluvial sediments. However, a large amount of pollen and lack of marine microfossils may help confirm or deny some hypothesis.

ECONOMIC CONSIDERATIONS

Hydrocarbon Potential

Because of the coarse-grained nature of fluvial sediments, they may form potentially good reservoir rocks for oil or gas. Ancient fluvial sediments with up to 30% average porosity and

Fig. 52—Cores of meandering stream deposits from the oil-producing Sunburst Formation (Lower Cretaceous) of Alberta. Two major sequences are represented, and the base of each is indicated by an arrow. The lower begins with coarse sandstone with scattered cobbles (a), then crossbedded sandstone (b), a thin mudclast-rich zone (c), and is capped by fine sandstone with climbing ripples and rare coal fragments (d). The second has coarse structureless sandstone (e) at its base, then coarse crossbedded sandstone (f), well-laminated sandstone with mudclast zones (g), and mottled green siltstones (h) at the top. This core is probably from the middle of a meander belt, and represents stacked point bar and levee deposits. The red siltstone at the top is floodplain material.

permeabilities of thousands of millidarcys have been reported. At the base of channel deposits, higher values can be expected, though these can be considerably reduced by diagenesis. During diagenesis, fluids circulating in the subsurface may react with the detrital grains resulting in clay cementation. Some lesser amounts of clay cement may result from recrystallization of the fine grained fraction of the original sediment. Hematite-stained clays form the groundmass between the framework grains in many red fluvial sediments. Other types of cement common in fluvial sediments are calcite and quartz. These may be deposited by groundwater early in the burial history of the sand, or may result from migration of fluids deep in the subsurface. These diagenetic events are poorly understood and difficult to predict.

Fluvial sediments are commonly highly oxidized because of exposure to oxygenated water during early diagenesis. They are also located on the margins of basins, relatively far from marine source rocks. These considerations have tended to lessen exploration interest in fluvial sediments of many basins. In spite of this, both structurally and stratigraphically trapped hydrocarbons have been discovered in fluvial deposits. Braided-river deposits offer excellent reservoirs in many cases, but have little potential for stratigraphic traps because of their paucity of thick, continuous fine-grained sediment (Figs. 57, 58). Thus, prospecting for structural traps would seem the optimum exploration strategy. Meandering stream deposits, with their abundant impermeable floodplain shales and laterally restricted sand bodies, are most likely to form stratigraphic traps of limited size. Because fluvial sediments are commonly associated with plant material and coal, they are commonly considered more likely to contain gas than oil.

Coal

Although deltaic sediments are the most prolific sources of coal, it is also mined from some fluvial sequences. Coals may be associated with meandering or anastomosing river deposits, accumulating in backswamps lateral to the channel. The coal seams may be split by thin crevasse-splay sands deposited during floods. The channels themselves may cut out coal seams, forming washouts causing difficulties in extraction. Hacquebard and Donaldson (1969) have documented the lateral splitting and pinching out of seams in a fluvial sequence in the Carboniferous of the Sydney basin of Nova Scotia. The ability to predict sand body geometry and trend may save considerable costs during coal mining operations. This is emphasized by Horne et al (1978) who concluded that some Carboniferous coals in Appalachia were formed in backswamps of rivers. These coals are associated with linear, lenticular bodies of sandstone (15 to 25 m thick and 1.5 to 11 km wide) which fine upwards and show lateral-accretion surfaces. These sandstones are deposits of meander belts and interfinger with backswamp siltstones, thin sandstones, shales containing abundant plant fossils, and coals. The coals are commonly 1.5 to 8 km in lateral extent and split by crevasse-splay sandstones which thicken toward the meander belt. In Appalachia these backswamp coals are low in sulfur content compared to deltaic coals.

Placer Gold and Uranium

During the Proterozoic, low oxygen pressure in the atmosphere allowed detrital grains of uranium to exist. Because of the high density of uranium minerals, they accumulated along with gold as placer deposits in fluvial sediments of this age. Proterozoic braided river conglomerates at Blind River in Canada and Witwatersrand basin of South Africa contain important uranium deposits of this kind. In South Africa, detrital gold occurs with uranium and is the primary objective of mining. Minter (1978) has documented that the ore minerals concentrate at the bases of channels which have a dendritic pattern. Knowledge of the channel pattern could lend to more efficient extraction of ore minerals.

Chemical Precipitates

Because of the porosity and permeability of fluvial deposits, they act as conduits for movement of groundwater and other low temperature solutions which bear complex metal ions in solution. These solutions are affected by the presence of organic matter which acts as a strong reductant, and also by primary facies changes which act as barriers to migration. Uranium is the most important metal deposited by

Fig. 53—Well-sorted fine-grained conglomerate which in places is openwork with no matrix, Cedar Mountain Conglomerate (Cretaceous) of Utah.

Fig. 54—Almost structureless sandstone with scattered granules and pebbles from the base of a meandering fining-upward sequence. Some vague inclined laminations are present in a few places. From the Devonian Battery Point Formation of Quebec. The scale is extended 30 cm.

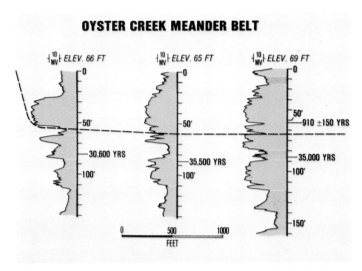

OYSTER CREEK MEANDER BELT

Fig. 55—Three SP logs made in Oyster Creek Meander Belt in the Brazos valley. The modern meander belt (outlined by the dashed line) has cut down into older deposits, forming amalgamated or multi-story sands in the two sections on the right. The log on the left shows that near the edge of the meander belt the sands are separated by a mud interval. After Bernard et al (1970).

Fig. 56—A fluvial sequence developed on a sharp channelled base consisting mainly of parallel laminated sandstone. The sandstone shows little fining-upward trend and may contain a scour surface (arrowed) indicating that it is an amalgamated sandbody. The floodplain deposits are relatively thin. This sequence is the deposit of a meandering river. Shepody Formation (Carboniferous) of New Brunswick.

this type of process. In the United States, it is estimated that 95% of the uranium deposits are hosted by continental sediments, mostly as chemical precipitates of this type. Sources of most of the uranium consist of weathered tuffs or other volcanics. These deposits occur as replacements of logs and other plant debris which reduced uranium carboxyl ions in solution and caused their deposition in a number of different ore minerals. In other areas, so called "roll-front" uranium deposits occur where uranium-bearing groundwater migrates downdip and comes into contact with fluids which are somewhat reducing. As ura-

nium is supplied to it, the reaction front moves downdip, causing the deposit to migrate slowly in this direction. Development of uranium leaching techniques has made it economically feasible to exploit many of these deposits covered by considerable overburden. Copper deposits may also form where solutions contact reducing agents (Fig. 59). However, because of the price of copper and abundance of other sources, this type of deposit is not important at present.

EXAMPLES OF OIL AND GAS PRODUCTION FROM FLUVIAL SEDIMENTS

The Cut Bank field of northern Montana and southern Alberta produces oil from meandering stream deposits. Lower Cretaceous Cut Bank Sandstone occurs on the margin of the Sweetgrass Arch at a depth of about 900 m and forms a series of meandering stream bodies (as defined by isopach maps) which extend in a linear belt some 140 km through Montana and Alberta (Conybeare, 1976). The rivers depositing this sand flow northward, and the sand bodies bifurcate in this direction, indicating that avulsion of the channel occurred periodically. The Cut Bank field occurs where hydrocarbons are trapped against the updip edge of chan-

Fig. 57—Sandy braided river deposits with two thin, impersistent shales lining the channels. These shales are not extensive enough to form barriers to hydrocarbon migration. Crossbeds can be seen in the sandstones near the top of the photograph. The photograph shows about 6 m of section. Castlegate Sandstone (Cretaceous) of Utah.

Fig. 58—A mud-filled channel (light colored) forming a fine-grained plug in sandy braided river deposits of the Simmsboro Sandstone, Wilcox Group (Cretaceous) of Texas. This is a sheet sandstone with rare shale plugs of this kind.

Fig. 59—Green reduction spots in fine-grained floodplain deposits. These spots are common in fluvial sediments and may be localized around some organic matter. These show copper enrichment at their centers. Enrage Formation (Carboniferous) of New Brunswick (Photo courtesy P. J. McCabe).

nel sandstone truncated against floodplain shales and siltstones. Each channel shows a fining-upward log pattern and cores verify that channel sediments range from fine-pebble conglomerate at base to fine sandstone at top. Unstratified sandstone and conglomerate are present at the bases of some channels, but much of the sandstone is crossbedded. In higher parts of the meandering sequences, ripples, mud cracks, burrows, and penecontemporaneous deformation structures are present (Shelton, 1967). The average porosity of the sandstone is about 15%, but ranges between 12 and 19%, and permeability averages 110 md with a range of 19 to 306 (Shelton, 1967). The field will yield about 4.8 MM cu m (30 MMb) of 38° API gravity oil, and 2,200 cu m (8×10^4 cf) of gas (Conybeare, 1976).

This relatively small field illustrates the type of sand bodies and traps common in meandering stream deposits. Linear sand bodies form excellent stratigraphic traps because they are surrounded by shales.

Valley fill deposits of incised rivers are important reservoirs in some areas. Bloomer (1977) has shown that Lower Permian Cook Sandstone of the Midland basin, Texas, consists of narrow fluvial sand bodies which have filled valleys up to 30 m deep in the underlying Waldrip shales and limestones. Mapping of these valleys by electric log correlation shows straight to meandering patterns with downstream bifurcation (Fig. 14). Stratigraphic or combination stratigraphic and structural traps are developed in various places along the fluvial systems, and one particular example given by Bloomer (1977) illustrates clearly the role of the valley in development of the reservoir. The Blackwell field is localized in a valley meander at a depth of 670 m with an 18 m thick point bar sand partly infilling the valley. When abandoned, muds filled its course, including the last active channel southeast of the sandstone. The regional dip of 0.5° west caused the pooling of oil in the point bar sandstone against the impermeable clay plug. The fourteen productive wells in the field have produced an average of 1.6×10^4 cu m (100,000 bbl) each.

The largest recently discovered oil field in North America, Prudhoe Bay, produces mainly from braided river deposits of the Late Permian—Early Triassic Ivishak Formation of the Sadlerochit Group (Jones and Speers, 1976). Ivishak sandstone ranges from 90 to 200 m in thickness, most of this variation occurring because of later erosion which created an unconformity truncating the unit. The formation is part of a prograding clastic wedge shed from a source to the north. The wedge includes marine shales and sandstones at its base overlain by deltaic deposits and succeeded finally by fluvial sediments. The braided river de-

posits make up 55% of the Ivishak (Eckelmann et al, 1976) and are mainly sandstones and conglomerates with only minor shale lenses. The lower part of the unit contains stacked channel sequences from 3 to 6 m thick, which fine upward from conglomerate to fine sandstone. Lateral migration of rivers in which these sequences were deposited has caused formation of a sheet sandstone at least 50 km wide with only 5 to 15% shale.

Sandstones and conglomerates comprising the braided-river deposits are excellent reservoirs with little matrix. Porosities are commonly about 30% and permeabilities reach 1 darcy in some zones. The shales are of limited lateral extent and do not play a major role in blocking fluid migration. As noted earlier, braided-river deposits do not commonly form stratigraphic traps because of the lack of shales. In this case, the hydrocarbons are trapped because of the southward dip of the Ivishak, and presence of impermeable Cretaceous shale lying unconformably over the unit. This shale may also have acted as source rock. The formation has an oil column of 137 m with reserves of 1.5×10^9 cu m (9.6 billion bbl) of oil and 7.2×10^{11} cu m (26 Tcf) of gas (Jones and Speers, 1976).

These examples illustrate the potential of meandering and incised stream deposits for forming small stratigraphic or combined stratigraphic-structural reservoirs, and of braided river deposits for forming structural reservoirs. While major undiscovered braided river structural reservoirs may be present only in frontier areas, many small stratigraphic reservoirs created by meandering streams likely remain undiscovered in many mature basins. Both types of reservoirs can be successfully discovered and exploited by subsurface mapping and utilization of facies models presented here.

REFERENCES CITED

Allen, J. R. L., 1970, Studies in fluviatile sedimentation: A comparison of fining-upwards cyclothems, with special reference to coarse-member composition and interpretation: Jour. Sed. Petrology, v. 40, p. 298-323.

——— 1978, Studies in fluviatile sedimentation: An exploratory quantitative model for the architecture of avulsion-controlled alluvial suites: Sed. Geology, v. 21, p. 129-147.

Bernard, H. A., et al, 1970, Recent sediments of southeast Texas: Texas Univ. Bur. Econ. Geology Guidebook Number 11.

Bloomer, R. R., 1977, Depositional environments of a reservoir sandstone in west-central Texas: AAPG Bull., v. 61, p. 344-359.

Brown, L. F. and W. L. Fisher, 1977, Seismic-stratigraphic interpretation of depositional systems: examples from Brazilian rift and pull-apart basins: in C. E. Payton, ed., Seismic stratigraphy—applications to hydrocarbon exploration: AAPG Mem. 26, p. 213-248.

Campbell, C. V., 1976, Reservoir geometry of a fluvial sheet sandstone: AAPG Bull., v. 60, p. 1009-1020.

Cant, D. J. and R. G. Walker, 1976, Development of a braided-fluvial facies model for the Devonian Battery Point Sandstone, Quebec: Canadian Jour. Earth Sci., v. 13, p. 102-119.

——— ——— 1978, Fluvial processes and facies sequences in the sandy, braided South Saskatchewan River, Canada: Sedimentology, v. 25, p. 625-648.

Coleman, J. M., 1969, Brahmaputra River: channel processes and sedimentation: Sed. Geology, v. 13, p. 129-239.

Collinson, J. D., 1970a, Bedforms of the Tana River: Geog. Annaler, v. 52A, p. 31-56.

——— 1970b, Deep channels, massive beds and turbidity current genesis in the central Pennine basin: proc. Yorkshire Geol. Soc., v. 37, p. 495-519.

——— 1978, Vertical sequence and sandbody shape in alluvial sequences: in A. D. Miall, ed., Fluvial sedimentology: Canadian Soc. Pet. Geologists Mem. 5, p. 577-586.

Conybeare, C. E. B., 1976, Geomorphology of oil and gas fields in sandstone bodies: New York, Elsevier Pub. Co., 341 p.

Eckelmann, W. R., W. L. Fisher, and R. J. DeWitt, 1976, Prediction of fluvial-deltaic reservoir geometry, Prudhoe Bay, Alaska: in T. P. Miller, ed., Recent and ancient sedimentary environments in Alaska: Alaska Geol. Soc. Symposium, p. B1-B8.

Eisbacher, G. H., M. A. Carrigy, and R. B. Campbell, 1974, Paleodrainage pattern and later orogenic basins of the Canadian cordillera: in W. R. Dickenson, ed., Tectonics and sedimentation: SEPM Spec. Pub. 22, p. 143-192.

Fisk, H. N., 1944, Geological investigation of the alluvial valley of the Lower Mississippi River: Mississippi River Commission, 78 p.

Hacquebard, P. A. and J. R. Donaldson, 1969, Carboniferous coal deposition associated with flood-plain and limnic environments in Nova Scotia: in E. C. Dapples, and M. E. Hopkins, eds., Environments of coal deposition: Geol. Soc. America Spec. Paper 114, p. 143-192.

Harms, J. C., J. B. Southard, D. R. Spearing, and R. G. Walker, 1975, Depositional environments as interpreted from primary sedimentary structures and stratification sequences: SEPM Short Course 2, Dallas, 161 p.

Hein, F. J. and R. G. Walker, 1977, Bar evolution and development of stratification in the gravelly, braided Kicking Horse River, British Columbia: Canadian Jour. Earth Sci., v. 14, p. 562-570.

Horne, J. C., J. C. Ferm, F. T. Caruccio, and G. G. Baganz, 1978, Depositional models in coal exploration and mine planning in Appalachian region: AAPG Bull., v. 62, p. 2379-2411.

Jackson, R. G., 1976, Depositional models of point bars in the lower Wabash River: Jour. Sed. Petrology, v. 46, p. 579-595.

Jones, H. P. and R. G. Speers, 1976, Permo-Triassic reservoirs of Prudhoe Bay Field, North Slope, Alaska: in J. Braunstein, ed., North American oil and gas fields: AAPG Mem. 24, p. 23-50.

Leopold, L. B., M. G. Woman, and J. P. Miller, 1964, Fluvial processes in geomorphology: W. H. Freeman and Co., 522 p.

A. D. Miall, 1977, A review of the braided-river depositional environment: Earth Sci. Rev., v. 13, p. 1-62.

——— 1978, Tectonic setting and syndepositional deformation of molasse and other non-marine-paralic sedimentary basins: Canadian Jour. Earth Sci., v. 15, p. 1613-1632.

McLean, J. R. and T. Jerzykiewicz, 1978, Cyclicity, tectonics and coal: some aspects of fluvial sedimentology in the Brazeau-Paskapoo Formations, Coal Valley area, Alberta, Canada: in A. D. Miall, ed., Fluvial sedimentology: Canadian Soc. Pet. Geologists Mem. 5, p. 441-468.

Middleton, G. V., 1976, Hydraulic interpretation of sand size distributions: Jour. Geology, v. 84, p. 405-426.

Minter, W. E. L., 1978, A sedimentological synthesis of placer gold, uranium, and pyrite concentrations in Proterozoic Witwatersrand sediments: in A. D. Miall, ed., Fluvial sedimentology: Canadian Soc. Pet. Geologists Mem. 5, p. 801-829.

Pettijohn, F. J., P. E. Potter, and R. Siever, 1973, Sand and sandstone: New York, Springer Verlag, 618 p.

Rust, B. R., 1978, Depositional models for braided alluvium: in A. D. Miall, ed., Fluvial sedimentology: Canadian Soc. Pet. Geologists Mem. 5, p. 605-625.

Sangree, J. B. and J. M. Widmier, 1977, Interpretation of clastic depositional facies: in C. E. Payton, ed., Seismic stratigraphy—applications to hydrocarbon exploration: AAPG Mem. 26, p. 168-184.

Schumm, S. A., 1968, Speculations concerning paleohydrologic controls of terrestrial sedimentation: Geol. Soc. America Bull., v. 79, p. 1573-1588.

——— and H. R. Khan, 1972, Experimental study of channel patterns: Geol. Soc. America Bull., v. 83, p. 1755-1770.

Shelton, J. W., 1967, Stratigraphic models and general criteria for recognition of alluvial, barrier-bar, and turbidity-current sand deposits: AAPG Bull., v. 51, p. 2441-2461.

Smith, N. D., 1970, The braided stream depositional environment: comparison of the Platte River with some Silurian clastic rocks: Geol. Soc. America Bull., v. 81, p. 2993-3014.

——— 1974, Sedimentology and bar formation in the upper Kicking Horse River, a braided meltwater stream: Jour. Geology, v. 82, p. 205-223.

——— and D. G. Smith, in press, Sedimentation in anastomosed river systems: examples from alluvial valleys near Banff, Alberta: Jour. Sed. Petrology.

Van De Graaf, F. R., 1972, Fluvial-deltaic facies of the Castlegate Sandstone (Cretaceous), east-central Utah: Jour. Sed. Petrology, v. 42, p. 558-571.

Van Houten, F. B., 1973, Meaning of molasse: Geol. Soc. America Bull., v. 84, p. 1973-1976.

Walker, R. G., 1976, Facies models 3. Sandy fluvial systems: Geosci. Canada, v. 3, p. 101-109.

Williams, P. F. and B. R. Rust, 1969, The sedimentology of a braided river: Jour. Sed. Petrology, v. 39, p. 649-679.

Deltaic Environments of Deposition

J. M. Coleman
D. B. Prior
Coastal Studies Institute
Louisiana State University
Baton Rouge, LA

INTRODUCTION

Deltaic depositional facies result from interacting dynamics processes (wave energy, tidal regime, currents, climate, etc.), which modify and disperse riverborne (fluvial) clastic deposits. The term delta was first applied by the Greek philosopher Herodotus (490 B.C.) to the triangular land surface formed by deposits from Nile River distributaries. In the broadest sense deltas can be defined as those depositional features, both subaerial and subaqueous, formed by fluvial sediments. In many instances the deposition of fluvial sediments is strongly modified by marine forces such as waves, currents and tides, and depositional features found in deltas therefore display a high degree of variability. Depositional features include distributary channels, river-mouth bars, interdistributary bays, tidal flats, tidal ridges, beaches, eolian dunes, swamps, marshes, and evaporite flats (Coleman, 1976).

A significant deltaic accumulation necessarily requires the existence of a river system carrying substantial quantities of clastic sediment from an inland drainage basin to the coast, where the deposits form the delta plain. Modern deltas exist under a wide range of environmental processes; some deltas form along coasts experiencing negligible tides and minimal wave energy, whereas others form in areas where tide ranges are extreme and wave energy is intense. Despite the environmental contrasts, all actively prograding deltas have at least one common feature—a river supplies clastic sediment to the coast and adjacent shelf more rapidly than it can be dispersed by marine processes, and thus a regressive sedimentary deposit forms.

A delta plain generally can be subdivided into physiographic settings. Every delta plain consists of a sub-

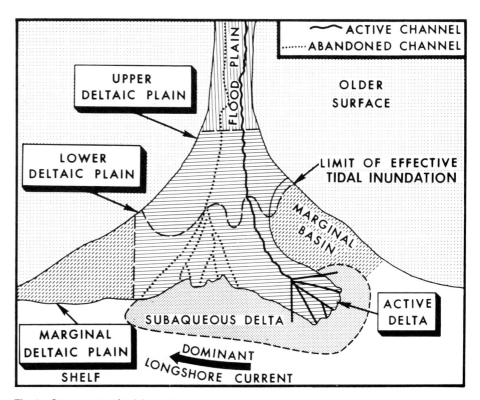

Fig. 1—Components of a delta system.

aerial and subaqueous component. The subaerial component is often divided into upper and lower delta plains (Fig. 1), the upper plain normally being the older part of the subaerial delta and existing above the area of significant tidal or marine influences. Unfortunately, in ancient rock sequences, only faunal evidence can be used to separate these major components. The upper plain is commonly the continuation of an alluvial valley and is dominated by riverine processes. The lower delta plain lies within the realm of river-marine interaction and extends landward from the low-tide mark to the limit of tidal influence; thus a lower delta plain is most extensive in areas where tidal ranges are large and seaward gradients and topographic relief are low. The sub-

aqueous delta plain is that part of the delta lying below the low-tide water level and contains relatively open marine fauna. It is the foundation across which subaerial delta deposits must prograde. The subaqueous delta is most commonly characterized by a seaward fining of sediments, sands and coarser clastics being deposited near the river mouths and finer grained sediment dispersing farther seaward. In most deltas, the coast of the delta plain receives dissimilar rates of sediment introduction, and it is not unusual for one part of the delta shoreline to be rapidly prograding seaward, while other parts are subjected to reworking by marine processes. If wave and current reworking are intense, the delta shoreline undergoes landward transgression, and coastal barriers, beach or dune complexes will often be formed. In other places, particularly when subsidence rates are high, marine waters encroach rapidly across the subsiding delta mass and shallow-water marine deposits will directly overlie the regressive delta deposits with no significant marine reworking taking place.

To systematically describe the sedimentary characteristics of deltaic deposits, it is convenient to subdivide the sedimentary facies into the following categories:

Upper Delta Plain
   (a) Migratory channel deposits— braided-channel and meandering-channel deposits;
   (b) Lacustrine delta-fill and flood-plain deposits.
Lower Delta Plain
   (a) Bay-fill deposits (interdistributary bay, crevasse splay-natural levee, marsh);
   (b) Abandoned distributary-fill deposits.
Subaqueous Delta Plain
   (a) Distributary-mouth-bar deposits (prodelta distal bar, distributary-mouth bar);
   (b) River-mouth tidal-range deposits;
   (c) Subaqueous slump deposits.

In each category, vertical stratigraphic sequences, internal characteristics, sand isopach trends, electric-log characteristics, and continuous and interval photographs are utilized to illustrate major characteristics of the sedimentary facies.

UPPER DELTA PLAIN

The upper delta plain lies above the level of effective saltwater intrusion and is unaffected by marine processes. Most of the sediments comprising this part of the delta plain originate from the migratory tendency of distributary channels, overbank flooding during annual highwater periods, and periodic breaks in the river banks, in which "crevassing" into adjacent lake basins occurs. The major environments of deposition include braided channels, meandering channels (point bars and meander-belt deposits), lacustrine delta fill, backswamps, and flood plains (swamps, marshes, and freshwater lakes).

Braided-Channel Deposits

Braided channels are marked by successive divisions and rejoinings of the flow around alluvial islands. Most braided river channels are characterized by a dominant bedload transport of sediment, high variations in water discharge, high downstream gradients, and large width-depth ratio of channels. Most braided rivers display rapid and continuous shifting of sediment and position of channels. These channels are found in all climate zones, but, because of their dependence on erratic discharge and high bedload, they are most common in arid and arctic settings. Braiding characteristics of the channel often extend all the way into the delta plain. However, one of the largest braided channels in the world, the Brahmaputra River, has formed in a humid climatic setting. Lateral migration of channels can be dramatic, as in the Brahmaputra River, where lateral migration rates of several thousand meters during a single flood are not uncommon (Coleman, 1969). In the Rosi River (a tributary of the Ganges River), lateral migration of the channel over the past two centuries has been about 170 km; in a single year the channel may shift over 30 km laterally (Holmes, 1965, in Reineck and Singh, 1973). Because of high lateral migration rates and shallow depths of scour, most braided-channel deposits display high lateral continuity but are rather thin (rarely over 30 m thick).

Individual channels split around numerous mid-channel islands or braid bars. During floods erosion occurs on the upstream ends and lateral sides of bars and eroded material is added to the downstream side of the bar. Because each channel has a different depth, lateral migration of the channel results in scour to differing depths. Multiple fining-upward cycles (which are commonly truncated) occur within the resulting sand body. Each fining-upward cycle (owing to deposition of a laterally migrating channel) is characterized by a scoured base and overlying sets of large-scale cross-bedding in which individual sets are up to 1 m thick (Fig. 2A). This lower sequence of large-scale cross-bedding can attain thicknesses of up to 7 m. Small-scale ripple bedding, scour and fill structures, organic trash, and clay layers are occasionally found.

Overlying this unit is a zone displaying finer grain size and composed of lenticular-shaped units of large-scale cross-bedding (trough type) intercalated with zones of climbing ripple laminations, horizontal laminations, and ripple cross-bedding (Fig. 2B). In some places small-scale laminations are

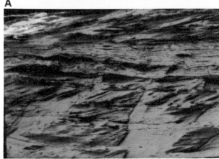

Fig. 2—Photographs of bedding in a braided channel deposit. **A.** Large-scale cross-bedding in the lower part of a fining-upward cycle on a braided channel. **B.** Trough-shaped cross-bedding in lenticular sets that form the overlying zone in a fining-upward cycle of a braided channel. **C.** Ripple drift bedding separated by parallel sand laminations.

present within certain zones. The uppermost unit of the fining-upward cycle displays horizontal laminations separating well-defined, near-horizontal sets of steeply dipping ripple-drift bedding (Fig. 2C). Small-scale convolute laminations and burrowed sand and silt layers are common in the uppermost part of the unit. However, scouring by later migration of the channel often removes this upper burrowed section.

Figure 3 summarizes the major characteristics of braided-channel deposits, including lateral relationships (block diagram in upper left); typical vertical sequence (upper right), including grain size, directional properties, dip angles, relative porosity, and sedimentary structures; sand body isopach map (lower left); and representative electric logs (lower right) at selected sites to show variation in log shape. As this diagram illustrates, the typical vertical sequence is characterized by mul-

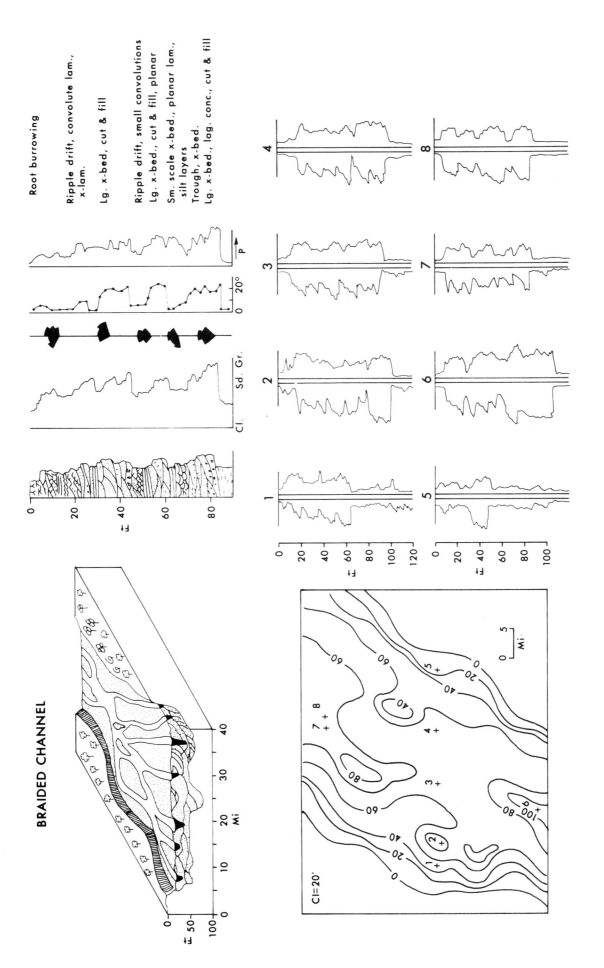

Fig. 3—Summary diagrams illustrating the major characteristics of braided channel deposits (letters on the vertical section refer to core or outcrop photographs).

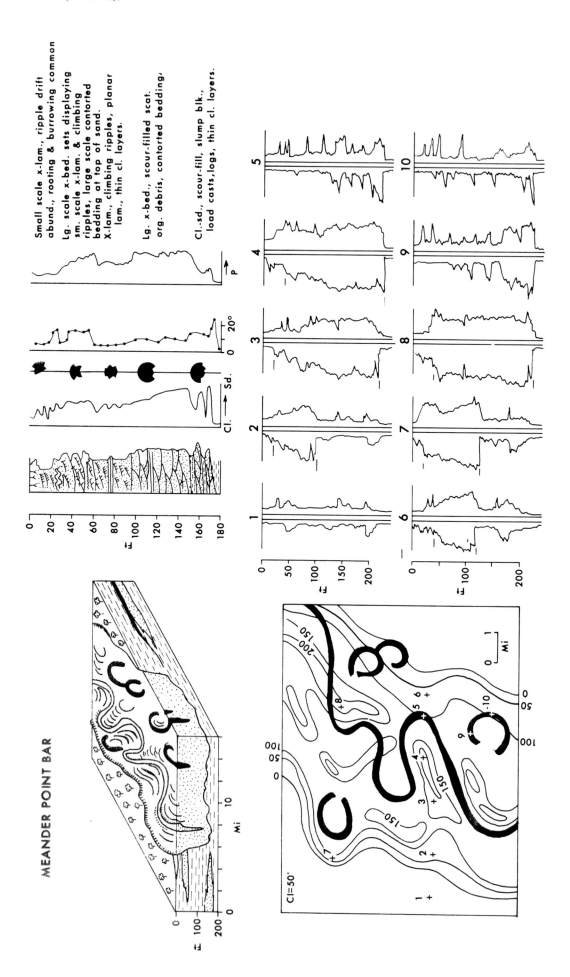

Fig. 4—Summary diagram illustrating the major characteristics of meandering point-bar deposits.

tiple stacked fining-upward cycles of deposition (each representing deposition by a migratory channel). Directional properties within each cycle often display narrow directional spread and are fairly representative of the long-axis orientation or downstream direction of the channel. High dip angles are most often associated with large-scale cross-bedding and distorted layers.

The isopach map shows a laterally continuous sand body, often extending 20-50 km laterally in a direction perpendicular to the downslope channel direction. Most braided channels display rather uniform thicknesses across the entire sand body (averaging 15 to 25 m thick) and localized deeper sand-filled scoured pods. Log response often shows an overall blocky shape, with numerous sharp "kickouts" representing local coarse-sand-filled scours. Locally numerous fining-upward cycles can be well defined on the logs. Although distinctive cycles can often be discerned from log data, it is very likely that individual units cannot be carried laterally any great distance, and presence of the numerous thin silt and clay layers discourages thinking that reservoir continuity may extend for great distances. Exposures in some of the tar sands in Canada show a lack of reservoir continuity as tars are concentrated along distinct layers within the overall sand body. Potential for porosity traps is great in the braided-channel environment.

Faunal remains are rare in braided-channel deposits, but locally freshwater pelecypod and gastropod shells are present. Ostracodes are sometimes encountered in the thin clay zones. Plant remains and root burrows are common in temperate and tropical areas and normally are associated with the topmost unit of a depositional cycle. Leaf impressions are found along bedding planes in the silts and clays, and large logs and other organic trash are present near the lower basal scour. Early diagenic inclusions are normally rare, but calcium carbonate nodules and iron oxide pellets are found in braided channels in the arid environment.

Literature on braided-channel deposits include articles by Clapp (1922), Leopold and Wolman (1957), NEDECO (1959), Wright (1959), Chien (1961), Doeglas (1962), Krigstrom (1962), Are (1963), Fahnestock (1963), Allen (1965a, b), Visher (1965, 1972), Coleman (1969), Williams and Rust (1969), Collinson (1970), Smith (1970), Rust (1972), Schumm (1972), Fahnestock and Bradley (1973), Shelton and Noble (1974), and Cant and Walker (1976).

Meandering-Channel Deposits

A meandering river is one with a channel pattern that displays high sinuosity in plan view (sinuosity greater than 1.5). Meanders are most commonly associated with rivers displaying nonerratic flooding characteristics, high suspended-sediment load (generally fine-grained bedload), and low downslope gradient. Most meandering rivers occur in tropical and temperate climates, where suspended loads and high annual flooding are common, but they will also occur in arctic and arid climates.

Meandering rivers are most common in alluvial valleys of river systems, but in several modern river deltas meandering sections of the distributaries are present, especially in rivers where distributaries are influenced by high tidal ranges or in rivers carrying extremely large coarse bedload sediment.

The channel profile normally displays an asymmetric V shape, the steep cutting bank is referred to as the cut bank; the shallow-sloping depositing side is termed the point bar. Maximum flow velocities are found near the steep concave bank, while lower velocities are present along the point-bar side of the channel. Undercutting of the cut-bank side of the channel during a flood causes oversteepening and slumping of the channel wall. Locally, where the channel scours into sandy deposits, high flood levels (excessive hydraulic head) force river water into the buried sand body. A sudden drop in the river level results in excess pore-water pressure in the buried sand, and water and sand flow back into the river, causing slumping in the overlying bank deposits. In either case, the slumping results in an increase of the channel's cross-sectional area, thereby reducing velocity and causing deposition on the point-bar side of the channel.

Although numerous other processes are active, slumping on the cut-bank side and deposition on the point bar cause channel migration, and continued repetition of this process results in formation of a point-bar or meander-belt sand deposit. In most places, meander-belt deposits fail to migrate entirely across a valley, instead forming finite-width features whose dimensions are narrow compared to the sand-body length. The width of the meander-belt deposit is dependent on the discharge and channel depth. In the lower Mississippi River the channel depth averages 50 m (hence average sand body thickness is 50 m) and the meander-belt width averages 16 km; river depth in the Niger River is 10-15 m and meander-belt width is 10 km; the Sabine River (Louisiana-Texas) is 6-8 m deep and the meander belt is 5-6 km wide.

The channel abandoned when a river switches its course to another site is most commonly filled with silty clays, organic clays, and organic trash. The clay-filled channel, therefore, results in numerous nearly isolated sand bodies in the overall meander belt providing innumerable stratigraphic trapping possibilities. This is illustrated in the upper left diagram of Figure 4. The vertical sequence in the meander point-bar sand body is illustrated in the upper right diagram of Figure 4.

Most commonly, the vertical sequence shows a fining-upward grain size relationship; a few coarser layers are found near the upper one-third of the sand body. The sand body has a scoured base, and often coarse, organic trash (logs, limbs and clay clasts) is found intercalated with the sandy units. Thin clay and silt layers often separate coarse sandy units. Above this basal unit is normally a massive, thick sand unit displaying large-scale cross-bedding with occasional contorted layers and thin laminations of organic debris. The large-scale bedforms migrate primarily during periods of high flood.

Overlying the large-scale cross-bedding is a well-sorted sand sequence composed of repeated cyclic sedimentation units of climbing ripples, convolute laminations, and parallel well-sorted sand laminations (Fig. 5A). Each of the cyclic units represents deposition during a single flood and a change from lower flow to upper flow regime. Such units range in thickness from 0.5 to over 1.5 m. In some areas the flood deposit is poorly sorted and consists primarily of silts and sands displaying small-scale ripple laminations (Fig. 5B).

During times of extremely high flooding, rapid deposition often takes place on the point bar because of river currents cutting across the meander loop, attempting to shorten its course. In those instances, thick sequences of rapidly deposited, well-developed climbing ripples capped by contorted laminations are common (Fig. 5C). Deposition of 1 to 2 m can take place over only a few hours to a day. Most commonly the distorted bedding simply caps the sand unit, but the entire

Fig. 5—Photographs of bedding in a meander point bar. **A.** Cyclic flood deposits in a point bar. **B.** Small-scale cross–stratification and organic debris. **C.** Climbing ripple sequence capped by convolute laminations. **D.** Highly contorted bedding in point-bar deposits. **E.** Soil zones alternating with ripple laminations in upper part of point-bar deposits.

bed can be rendered into a fluid state and massive contortions can be present (Fig. 5D). The zone of maximum contorted bedding often closely approximate the low-water river level and extends upward into the zone covered by flood waters. The uppermost unit of the meander point bar often consists of silt and sand beds (few centimeters thick) displaying ripple laminations alternating with silty-clay and clay beds exhibiting root burrows (Fig. 5E). In these uppermost units iron oxide and calcium carbonate nodules usually are abundant.

Sand bodies display wide variations in current directional properties, primarily because of the sinuous nature of the channel cand its migrational variations. This is especially true in units consisting of large-scale cross-bedding. The upper climbing-ripple part of the deposit shows much less scatter in directional properties. Dip angles are generally low and primarily reflect cross-bedding, except in the upper contorted bedding unit, where erratic dips of 15 to 20° are common.

An isopach map of a meander-belt deposit is illustrated in the lower left diagram of Figure 4, with characteristic variations in log response shown in the lower right. Most of the logs show a bell-shaped, fining-upward sequence, but locally blocky sands (BH8) are present. Thicker pods of sand within the overall meander-belt deposit tend to show this blocky aspect. Borings 5, 9 and 10 show log response in the abandoned channel fill and indicate that little or no major sand buildup has formed.

Faunal remains are generally absent in the sands and silts of point-bar deposits. Plant remains are sometimes common, but those in the lower part of the sand body represent transported organic debris and are not usually representative of the local flora. In the uppermost units some leaf remains are present along the bedding planes, but most of the organic materials that are deposited are quickly oxidized.

Literature on meandering rivers and their characteristics is extensive, and the following references are but a few examples of the types of studies available: Fisk (1944), Sundborg (1956), Leopold and Wolman (1960), Frazier and Osanik (1961), Harms et al (1963), Allen (1965a), Steinmetz (1967), Bernard et al (1970), McGowen and Garner (1970), Bluck (1971), Fisher and Brown (1972), Jackson (1975), and Lewin (1976).

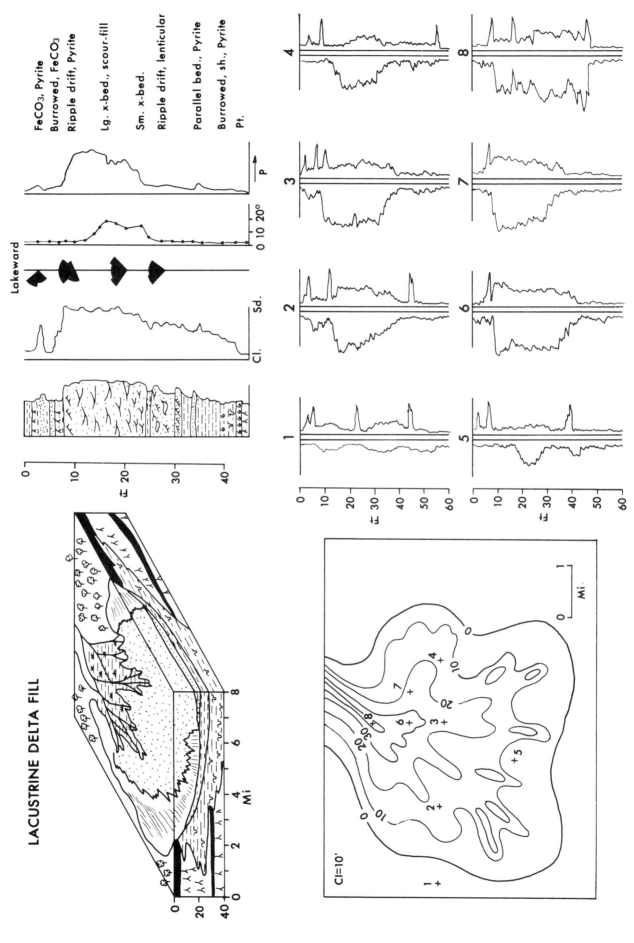

Fig. 6—Summary diagram illustrating the major characteristics of lacustrine delta-fill deposits in the upper delta plain.

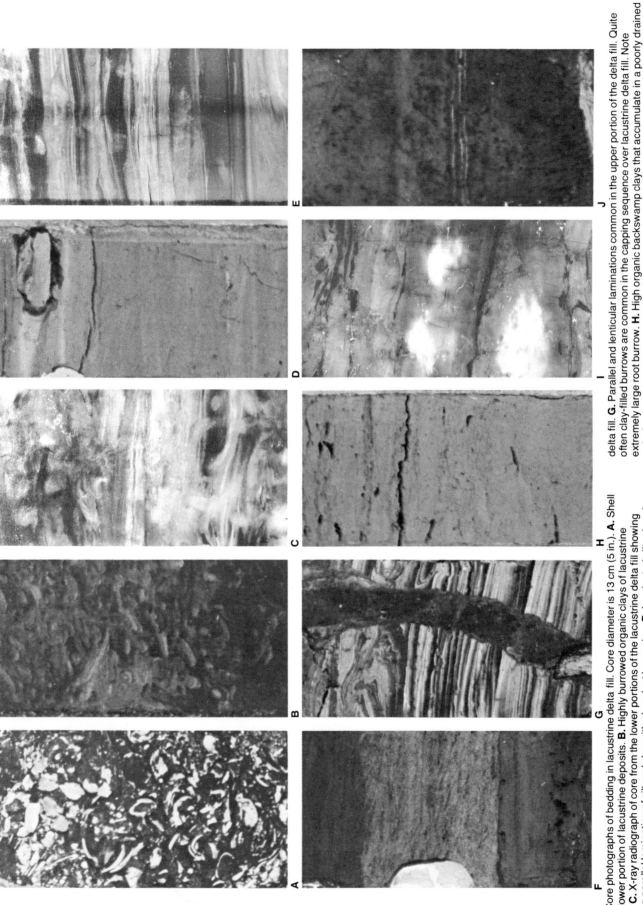

Fig. 7—Core photographs of bedding in lacustrine delta fill. Core diameter is 13 cm (5 in.). **A.** Shell debris in lower portion of lacustrine deposits. **B.** Highly burrowed organic clays of lacustrine deposits. **C.** X-ray radiograph of core from the lower portions of the lacustrine delta fill showing alternating parallel lamination of silt and clay with abundant burrowing. **D.** Laminated silty clays in lower portion of lacustrine delta fill. Note the elliptical Fe_2CO_3 nodule. **E.** X-ray radiograph of clays in lower portion of lacustrine delta fill in which high sedimentation rates preclude burrowing organisms. **F.** Well-stratified silty and sandy deposits of the coarser sediments forming the bulk of the lacustrine delta fill. **G.** Parallel and lenticular laminations common in the upper portion of the delta fill. Quite often clay-filled burrows are common in the capping sequence over lacustrine delta fill. Note extremely large root burrow. **H.** High organic backswamp clays that accumulate in a poorly drained reducing swamp environment. Organic stringers and peat deposits are common in this environment. **I.** X-ray radiograph of core taken in backswamp deposit. Note the stringers of organic debris (dark layer) and the early formation of siderite nodules (Fe_2CO_3). Pyrite is abundant in this setting. **J.** Silty clays with abundant iron oxide and calcium carbonate nodules that form in well-drained oxidizing swamp environment.

Lacustrine Delta-Fill Deposits

Lakes are standing water bodies filled primarily with fresh water. These shallow inland-restricted interdistributary lakes constitute an important environment in many upper delta plains and occur in all climatic settings. The climate strongly affects the characteristics of deposits forming in lakes, with those in arid climates often composed almost entirely of evaporative sequences. Lakes associated with delta plains, however, are subjected to overbank crevassing or channel diversion, which infill the lake with freshwater deltaic deposits. Interdistributary deltaic lakes range in size from only a few square kilometers to 80 to 100 sq km. Most lakes in such environments are extremely shoal, and even the larger ones rarely attain depths greater than 15 to 20 m. The process by which these lakes form has not been studied in any great detail, but they obviously are closely related to channel pattern and local subsidence and compaction.

Once a depression is formed between two distributaries, standing water accumulates. With time, compaction and subsidence associated with the delta plain result in enlargement of most of these lakes, as well as localized deepening. Growth of the lake will normally stop when a stream diverts into the lake and an influx of sediment begins the infilling process.

Quiet-water deposition, reducing conditions, abundance of burrowing organisms (especially soft-bodied organisms such as polychaete worms), and occasional wave and current action are characteristic of this environment. Deposits within the lake bottoms consist of dark-gray to black organic-rich clays containing scattered silt lenses. In some lakes organic debris, such as large accumulations of freshwater shell, is present where overturning of the lake waters is a common process.

Elsewhere near-anoxic conditions exist and deposits consist of extremely organic-rich, fine-grained clay. The most common types of stratification include parallel and lenticular laminations, intense bioturbation, and occasionally distorted primary structures.

Although some of the parallel laminations result from alterations in textural properties, most are the product of alternating flocculated and nonflocculated layers. Within the flocculated layers there is commonly an abundance of small cracks and fractures. Most cracks are oriented perpendicular to bedding, but differing orientations are occasionally encountered. These are probably what has been termed syneresis cracks, having developed from expulsion of fluids when internal forces of attraction between particles are greater than internal forces of repulsion between solid phase particles. Microfaunal remains are usually abundant within lacustrine facies but consist of only a small number of ostracod species. Charophytes are encountered in some samples. Early diagenetic inclusions include vivianite, which is normally associated with drifted plant remains or burrow fills; and pyrite, which is extremely common and generally occurs as small cubes or isolated drusy masses.

Diversion of a stream channel into one of the lake basins causes an appreciable increase in the sedimentation rate. Because of the infilling process, sedimentation rates increase within the lake sequence, and both grain size and thickness of laminations increase vertically upward. The process of infilling seems rapid if judged by present lakes where historic documentation indicates fill within only a few tens of years. Fisk (1952) has illustrated the phenomenal rate of growth of the Mississippi River delta into Grand Lake (Atchafalaya Basin) in recent years. In just over 20 years, more than 150 sq km of land surface formed within the lake area. The thickness of the delta fill varies, but ranges up to 7 and 8 m.

Figure 6 illustrates the major characteristics of the lacustrine delta-fill sequence. The upper left diagram is a schematic representation of a small distributary that has been diverted into a shallow freshwater lake. The schematic shows lateral relationships of the various facies, indicating that the bulk of the delta-fill forms a wedge of coarse clastics within an overall deposit consisting of fine-grained organic-rich clays from lacustrine and backswamp deposits. The upper right diagram of Figure 6 illustrates a typical vertical sequence, which normally displays a coarsening-upward trend. The lowermost units consist of lacustrine deposits. Quite often these lowermost units of the lacustrine delta consist of large accumulations or biomasses of shell (Fig. 7A). Normally the shell debris is encased in a matrix of fine-grained organic clays containing abundant pyrite inclusions. As the sedimentation rate within the delta increases, there is generally a decrease in the amount of coarse faunal remains. Organic activity, however, does not normally cease, and burrowing and abundant bioturbation of soft-bodied organisms can totally obliterate the primary structures (Fig. 7B).

In most fill deposits, the burrows are sand filled due to the physiological processes of polychaete worms, which burrow through the muds, concentrating sands within their bodies and leaving behind essentially a sand- or silt-filled burrow. As the sedimentation rate continues to increase, laminations of thicker silt and sandy silt layers alternate with thinly laminated organic clays (Fig. 7C). Burrowing persists upward within the deposits, but not to the point of masking primary stratification. Quite often the parallel-laminated clay deposits consist of extremely organic-rich clays, and leaf remains are common along the bedding planes. Color laminations and inclusions of iron carbonate nodules are often common within this part of the delta fill (Fig. 7D).

Iron carbonate or siderite nodules often display smooth, elliptical forms along bedding planes of the slightly coarser deposits. Figure 7E is an X-ray radiograph of a core taken within the lower part of a silty clay delta fill. Parallel laminations are extremely well developed. Such laminations can be reminiscent of varve-like deposition. As the delta continues to prograde and infill the lake, sediments become appreciably coarser and depositional slopes become much steeper. Within the major part of the sand body itself, small-scale cross-laminations form the bulk of the stratification (Fig. 7F). Scour and fill structures, occasional small slump structures, and load casts are common, and organic debris is generally prevalent. In the upper part of the delta-fill, large-scale cross-beds are very common, and it is not unusual in seismic sections to see essentially steep foreset beds dipping into the lake at angles of 10 to 12°.

When the delta completes its infilling process, swamp vegetation actively begins encroachment onto the filled lake surface, and root burrows extend down into the silty clays and sands often forming a distinctive horizon that caps the sand deposit (Fig. 7G). Most often the root-filled burrows consist of clay and organic-rich infills. With the cessation of high depositional rates, organic clays begin to accumulate over lacustrine delta sands. Thin stringers of organic debris, both transported and in situ, commonly form organic-rich or near-black clays, which cap the sand body (Fig. 7H). The organic-rich clays are deposited in a poorly drained, reducing backswamp environment and commonly contain abundant pyrite and iron carbonate inclusions (Fig. 7I). The iron carbonate nodules begin to form almost immediately after deposition and often follow root burrows. As a result, nodules can display extremely knobby shapes.

With continued but slow accumulation, the deposit eventually fills to a level where the surface of the swamp is exposed to oxidizing conditions for several months during the year. Thus, oxidation begins to remove the organic content from the sediment. Movement of groundwater fluids through the sediments results in formations of numerous small, rounded pellets or iron oxide and calcium carbonate, commonly encrusting the root burrows. The clays within this unit often display a slight red coloration, illustrated in Figure 7J.

The lower two diagrams in Figure 6 illustrate a typical isopach map of a small lake delta infill and variations in log response through such a deposit. The isopach tends to indicate a wedge of coarse clastic sediments sandwiched between lower freshwater organic lacustrine deposits and overlying swamp and marsh deposits. Most often the logs will document the coarsening-upward sequence of this delta fill. Within the sand body, dip angles are often quite high, reaching 10 to 15°, and result primarily from the foresets of the rapidly prograding lacustrine delta. Often, resistivity kicks are extremely common within such a setting and are responses to lignite, coal, and iron-rich seams which form within these essentially red bed deposits. The sand body itself normally displays a graded base; however, in some areas, generally near breaks in the riverbank, thick, sharp-based sands can often accumulate immediately within the region of the actual crevassing.

Modern lacustrine deposits have not been extensively researched, and most literature on lakes has been concentrated on extremely large examples such as Lake Geneva and Lake Constance. Studies of small lacustrine deposits within delta environments have received only minimal attention. In ancient rock sequences, few extensive studies exist that adequately document the characteristics of ancient lacustrine deposits. The major literature on lacustrine deposits is as follows: Twenhofel (1950); Fisk (1952); Bradley (1964); van Houton (1964); Coleman and Gagliano (1965); Coleman (1966); Axelson (1967); Coleman and Ho (1968); Reeves (1968); Muller (1971).

Fig. 8—High-altitude air photograph of a relatively large bay-fill environment in the lower delta plain of the Mississippi River delta.

LOWER DELTA PLAIN

The lower delta plain lies within the realm of river-marine interaction and extends landward from the shoreline to the limit of tidal influence. Large areal extent of the lower delta plain is common where tidal range is large and seaward gradients of the river channel and delta are low. Most commonly in this environment, channels become more numerous and often show a bifurcating or anastomosing type of plan view pattern. Environments between distributary channels comprise the largest percentage of the lower delta plain and consist of actively migrating tidal channels, overbank splays (natural levees), interdistributary bays, bay fills (crevasse splays), marshes, and swamps. Deltas having an extremely high tidal range and an arid climate are often characterized by interdistributary evaporite or barren salt flats, where intricate networks of tidal creeks occur and are separated by broad evaporative sequences. From the standpoint of sand-body formation, the major environmental sequence consists of bay-fill deposits, which of-

ten form thin clastic wedges stacked one on top of another and separated by interdistributary bay and marsh deposits. The major environmental sequences described consist of bay-fill deposits (interdistributary bay, crevasse splay-natural levee, and marsh) and abandoned distributary-fill deposits.

Bay-Fill Deposits

One of the major facies associated with many deltas is the large areal extent of bay fills or crevasses that break off of main distributaries and infill the numerous interdistributary bays within the lower delta plain. These sequences form the major land areas in the lower delta plain. Crevasse deposits build into shallow bays between or adjacent to major distributaries and extend themselves seaward through a system of radial bifurcating channels similar in plan to the veins of a leaf.

The interdistributary bays into which the crevasses prograde are normally open bodies of water, often completely surrounded by marsh or distributary channels. More often, however,

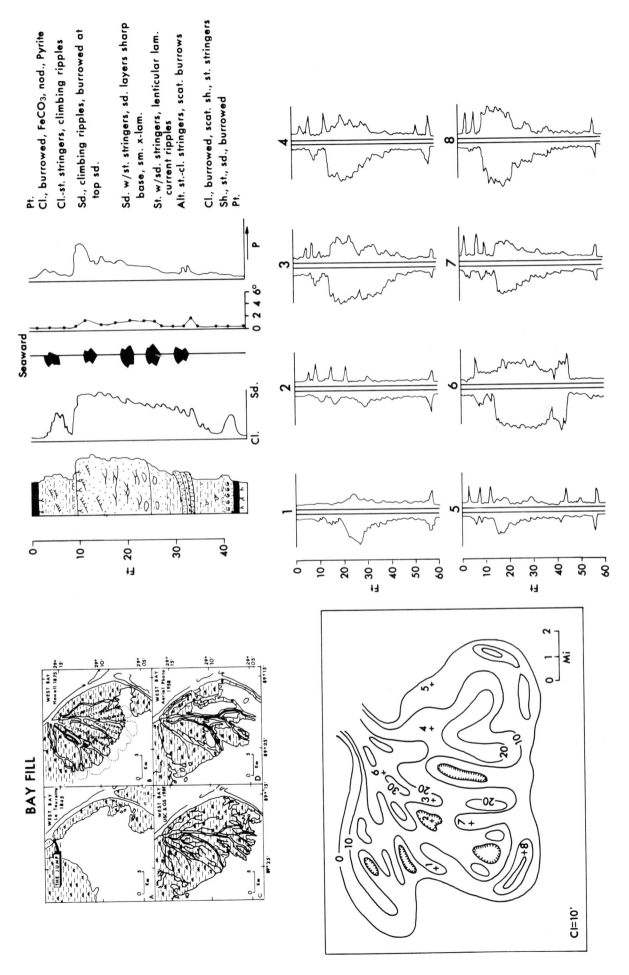

Fig. 9—Summary diagram illustrating the major characteristics of the bay-fill deposits in the lower delta plain.

they partially open to the sea or connect to it by small tidal channels. Most of the bays are shallow-water bodies, rarely exceeding 7 to 8 m in depth and averaging approximately 4 m, containing brackish to marine waters. The bays are commonly elongate, with their longest dimension ranging in size from a few hundred meters to approximately 15 to 20 km. The process of crevassing or infilling has been recorded historically in many cases (Coleman and Gagliano, 1964). Each bay fill forms initially as a break in the major distributary channel during flood stage, gradually increases in flow through successive floods, reaches a peak of maximum deposition, wanes, and becomes inactive. As a result of subsidence, the crevasse system is inundated by marine waters, reverting to a bay environment, and thus completing a sedimentary cycle.

The mass of sediment resulting from crevassing process ranges in thickness from 3 to 15 m, this sequence forming in a period of approximately 100 to 150 years. Although the individual bay fills are relatively thin, continuing subsidence and repeating of similar processes result in stacking of one bay fill on top of another, eventually building a thick sequence of lower delta-plain deposits.

Figure 8 shows a high-altitude photograph of a relatively large bay-fill environment in the lower delta plain of the Mississippi River delta. This particular crevasse, Cubits Gap Crevasse, occurred in 1862, when a break in the levee diverted some 6 to 8% of the Mississippi River water through the crevasse opening (Welder, 1959). By 1870 only a small number of low shoals were visible on the maps. By 1903 essentially the entire bay area had been infilled with crevasse-splay or bay-fill sediments. Since that time, subsidence and deterioration have resulted because of a lack of clastic sedimentation, and presently only the major distributary patterns are obvious on high-altitude photographs.

Most of the former land surface has now reverted to extremely shallow-water bays, and in approximately 30 to 40 years the entire landmass will have subsided below sea level and the environment will have reverted to an interdistributary bay. The upper left diagram of Figure 9 illustrates, by a series of historic maps, development of the West Bay fill. The break in the major Mississippi channel occurred in 1839, and maps from that time show water depths of 7 to 10 m in the adjacent bay.

After the initial levee break, coarse sediment was dumped subaqueously in the vicinity of the break, and no new subaerial land developed. However, with continued deposition and a general shoaling in the bay near the break, a bifurcating channel pattern and infilling of the bay sequence developed rapidly. This stage of development is illustrated by the map in 1875; most of the channels were still actively prograding, and nearly the entire bay had been filled with sediments or a regressive sequence of delta deposits.

The map from 1922 shows that many of the channels had been abandoned. They had prograded far enough seaward to lose their gradient advantage, and only a few of the major channels continued to deliver sediments to the bay. Much of the newly exposed land had been converted into luxuriant marsh growth, and organic-rich clays were capping the top of the regressive sequence. With time, plant growth could no longer maintain its productivity because of encroaching marine waters, and slowly the marsh began to break into numerous small lakes and bays. Wind-generated waves in the shallow bays, coupled with subsidence, began to destroy the marsh surface, and by 1958 much of the original land buildout had subsided below sea level, the area reverting to a bay.

By 1978 the entire region had been inundated by marine waters, and the West Bay complex had reverted to a shallow-marine interdistributary environment. Given time, another crevasse will eventually form on the bank of the Mississippi and another period of progradation will ensue, again filling the interdistributary bay with detrital sediments. It is this process of repeated filling, alternating with periods of marsh destruction, that forms the bulk of cyclic deposits in the lower delta plain.

The upper right-hand diagram in Figure 9 illustrates the typical vertical sequence resulting from bay infilling. As can be seen, it is a coarsening-upward sequence, with shallow brackish water clays and organic debris forming the lower part and well-sorted clastics forming the upper sand body. The upper unit is essentially distributary-mouth bar deposits associated with the prograding distributary. The lowermost part of the bay fill generally consists of alternating silts and silty clays, with the clays often showing silt- and sand-infilled burrows (Fig. 10A).

With continued and increasing sedimentation in the interdistributary bay, coarser particles and more rapid depo-

sition occur. Then silty and sandy stringers begin to intercalate with silty clays. Burrowing is generally reduced (Fig. 10B). As the distributary system advances farther into the bay, delivery of coarser grained clastics begins at the site of the vertical section, and often sands and silty sands alternate with thin silty clay laminations. Graded bedding and some small-scale climbing ripple structures are the most common types of lamination (Fig. 10C). As the distributary mouth progrades closer to the site of the vertical section, small-scale cross-bedding and occasionally organic trash within the sandy deposits become the most common types of stratification (Fig. 10D).

The lower part of the distributary-mouth bar itself is often characterized by cross-bedded sands alternating with sandy silts and silty clays. In general little or no scouring is evident in this part of the bay-fill sequence (Fig. 10E, F). The bulk of the sand deposit associated with the advancing distributary-mouth bar is composed of cross-stratified sands and sandy silts, displaying a wide variety of climbing ripples and small-scale festoon-type cross-bedding (Fig. 10G, H). A high mica content and transported organic debris along bedding planes are common in this part of the vertical sequence.

Once the distributary-mouth bar has prograded across the site of the vertical section, it is capped by small overbank crevasse splays, which often display alternating silt, sand, and clay units, the silts and sands including well-developed small-scale ripple laminations (Fig. 10I). In some instances, scour into the underlying deposits is apparent; however, the scour planes are of extremely low angle. The uppermost unit capping the bay-fill sequence is essentially an interdistributary bay or a marsh deposit. Interdistributary bays normally consist of highly burrowed silts and silty clays, whereas marsh deposits generally display a sequence of highly burrowed organic clays (Fig. 10J). In some deposits iron carbonate or siderite nodules are common, and in most cases pyrite is abundant, replacing plant debris.

The lower two diagrams (Fig. 9) illustrate an isopach map of a bay-fill sequence and variations in log response that can occur within such a sand body. The isopached sand body generally displays a fan-shaped wedge, with the thickest sands generally being found near the initial break in the distributary channel. Often, sands in this vicinity

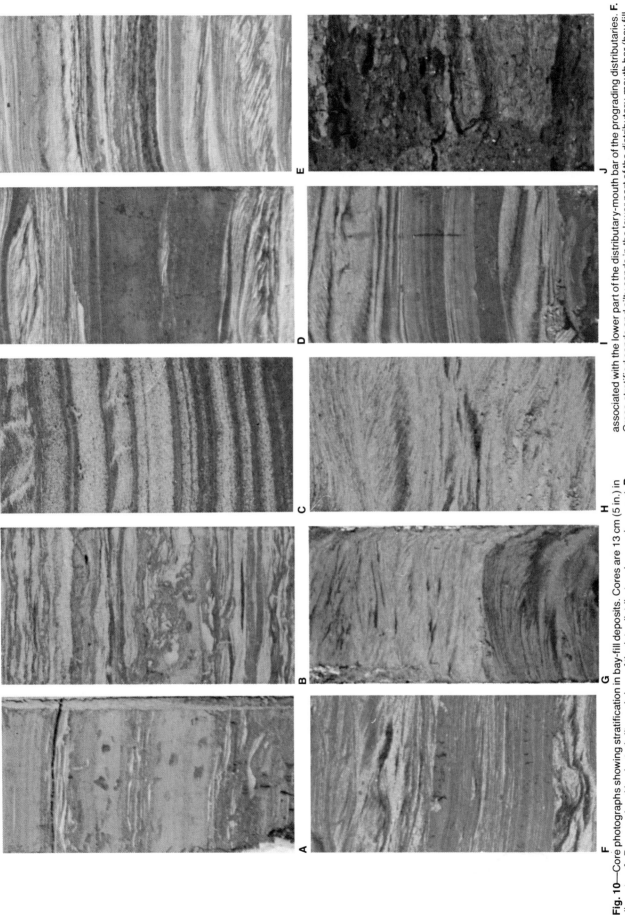

Fig. 10—Core photographs showing stratification in bay-fill deposits. Cores are 13 cm (5 in.) in diameter. **A.** Burrowed and laminated silts and clays of the interdistributary bay environment. **B.** Ripple laminations and burrowed zones in silts and silty clays in the lower part of the bay fill sequence. **C.** Well-laminated silts and silty sands of the crevasse infilling. The sandy layers often display small climbing ripple sequences. **D.** Cross-laminated silts and sands of the crevasse infilling sequence. **E.** Well-sorted and cross-laminated sand layers alternating with silts and silty clays

associated with the lower part of the distributary-mouth bar of the prograding distributaries. **F.** Cross-stratified sands and silty sands in the lower part of the distributary-mouth bar (bay fill sequence). **G.** Small-scale cross-stratified sands of the distributary-mouth bar (bay fill sequence). **H.** Well-sorted cross-stratified sands of the upper part of the distributary-mouth bar (bay fill sequence). **I.** Alternating silts and silty clays of the overbank splays that cap the bay fill sequence. **J.** High organic clays that form in the marsh environments and cap the bay fill sequence.

Fig. 11—Cored boring through the West Bay fill sequence. Boring is located near the distal end of the bay infill. Diameter of core is 8 cm (3 in.).

Fig. 12—Cored boring through the West Bay bay fill sequence. Boring is located in central part of the bay-fill sequence. Diameter of core is 8 cm (3 in.).

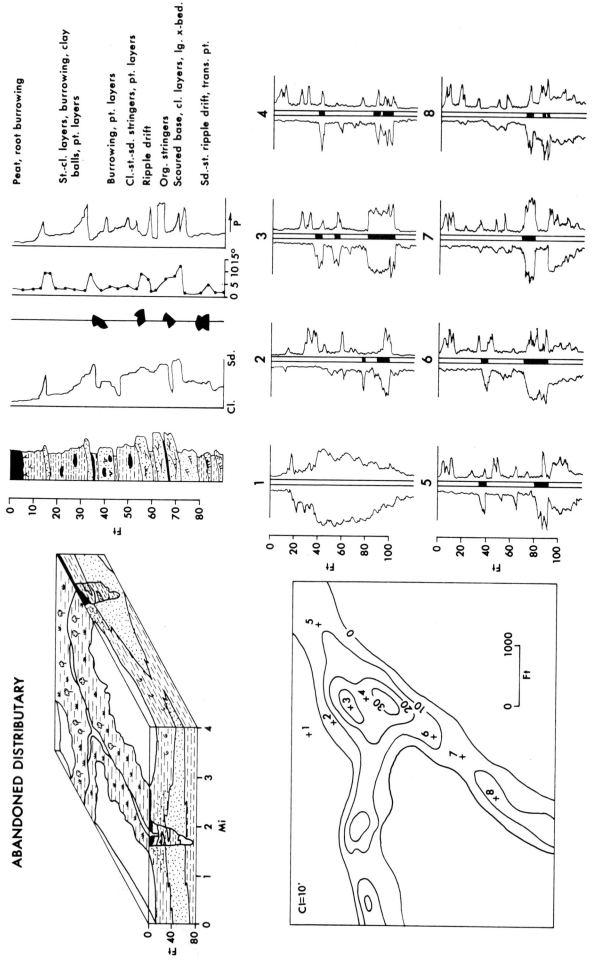

Fig. 13—Summary diagram illustrating the major characteristics of the abandoned distributary deposits in the lower delta plain.

display a sharp base scoured into the underlying interdistributary bay and marsh deposits. Away from the initial break, however, the typical coarsening-upward sequence (or inverted bell-shaped logs) becomes the most common type of log response. Within the overall sand body there are areas where sands have not accumulated to any great thickness, and therefore zones (bore hole 2) in which virtually no sand can be found and the entire sequence consists of interdistributary-bay silts and clays, grading upward to marsh deposits.

Figures 11 and 12 illustrate a continuous core through the West Bay complex. Location of the boring shown in Figure 11 is toward the outer fringes of the sand body, and the distributary-mouth-bar sands become much thicker and are more apparent in the cored borings. Comparison of these continuous cored borings with the vertical profile in Figure 9 indicates that stratification from the lower part of the bay fill to the upper part changes from parallel laminated and burrowed clays at the base to a higher and higher percentage of those structures associated with current flow near the top of the infilled sequence. Because of position within the overall bay fill, however, thickness of the sand body can vary considerably.

Literature on bay-fill sequences is extensive. The following articles are representative of some of the published papers: Russell (1936); Fisk et al (1954); Scruton (1955); Welder (1959); McEwen (1963); Coleman and Gagliano (1964, 1965); Coleman et al (1964); Donaldson (1967); Fisher and McGowen (1967); Morgan (1967); Dapples and Hopkins (1969); LeBlanc (1972); Arndorfer (1973); Reineck and Singh (1973); Bagans et al (1975); Coleman and Wright (1975); Hobday and Matthews (1975); Humphreys and Friedman (1975); Visher et al (1975); Coleman (1976); and Horne et al (1978).

Abandoned Distributary Deposits

The distributary channel is the natural flume which accommodates and directs a part of the water and sediment discharged from the parent river system to the receiving basin. In most deltas, the distributary channels are rather stable and do not display a tendency toward lateral migration, thereby preventing the formation of point-bar or meander-belt deposits. In some deltas, for example, with high bedload streams or in those environmental settings where tidal range is high,

migration of the distributary channel can take place, resulting in formation of deposits similar to channel deposits described in the upper delta plain.

Although little research has been conducted, the lack of channel migration in the lower delta plain is undoubtedly due to the fact that most river channels scour down through their distributary-mouth-bar deposits into underlying marine clays. This scouring provides an entrenchment of the distributary channel with minimal tendencies for lateral migration. Active distributary channels vary considerably in size, some only a few meters wide and 1 to 2 m deep, and others of a large major river delta system with channels reaching 1 km in width and 30 m in depth. Depth within the channel decreases rapidly as the river-mouth bar is approached, and water depths over most distributary-mouth bars rarely exceed 3 m.

Abandonment of a distributary channel is an extremely complex process, and in many areas is simply an accident. Log jams, loss of gradient advantage, infilling during a catastrophic period (such as a hurricane), or changes in the upstream part of the river will cause the channel to deteriorate and infill. Deprived of an active influx of sediment and water, the channel will then undergo an infilling process in which only local sediments derived from both upstream and downstream will infill the abandoned hole in the ground. The lower parts of the channel are commonly filled with poorly sorted sands and silts containing an abundance of transported organic debris. As a channel shoals, the water becomes more stagnant, and lower current velocities are maintained; soon, fine-grained materials begin to infill the channel proper. With time and continued subsidence the channel often fills entirely with fine-grained, poorly sorted sediments. Organic debris, logs, and clays with extremely high water content often form the upper part of the channel fill. Thus in many deltaic regions of low tides and high suspended sediment load, no process is available to infill the channels with sand or other coarse debris.

This aspect of the infilling of a distributary channel is schematically illustrated in the upper left diagram of Figure 13. This diagram illustrates the infilling of two distributary channels, showing the channel deposits themselves cut through the major sand body, which is essentially the distributary-mouth bar.

The upper right diagram, based on numerous cores through abandoned channels, illustrates a typical vertical sequence. The major characteristic is the erratic nature of the thin sand and silt layers that alternate with clays, forming the bulk of the deposit. Core hole after core hole indicates no orderly plan to the infilling process. About the only common attribute is a tendency for any substantial amount of sand (should substantial amounts be present) to concentrate near the base of the channel or at points where the channel bifurcates. Because of this lack of organized infilling it is often difficult to pick the base of the channel itself. Generally the only applicable method is use of faunal remains, because most of the distributary channels scour down into marine deposits, whereas clays infilling the channel generally contain little or no fauna and virtually no open marine fauna.

Near the base of the infilled channel, often erratic and contorted clay layers are concentrated as clast within the sand body (Fig. 14A). Within the deposited sands themselves, contorted structures are very common. Slump-type structures, distorted bedding, clay infills, and occasionally fairly persistent layers of organic trash sandwiched between sand layers are also common. Figure 14B illustrates some of the contorted bedding probably associated with localized slumps within a channel-fill deposit. Silts and silty clays that are deposited in the central part of the channel fill often display thin silty and sandy layers that intercalate with highly burrowed clays. Most commonly the sandy laminations display extremely sharp upper and lower surfaces (Fig. 14C). Convolute laminations and other types of distorted bedding such as flow rolls, ball and pillow structures, etc. (Fig. 14D), are common within the sand bodies. The uppermost part of the fill consists primarily of organic-rich clays (Fig. 14E) that generally show intense root burrowing. Occasionally, thin seams of shell debris and silt-infilled animal burrows can be detected.

Although grain size has a general tendency to show a fining-upward sequence within any one sandy unit, in some deposits there is virtually no change in grain size. Dip angles can also be extremely erratic, resulting primarily from the large number of disturbed and distorted structures found within the channel-fill deposits. Undoubtedly, the rapidity of the infilling process and extremely high porewater

Fig. 14—Core photographs showing stratification in an abandoned distributary fill deposit. Diameter of cores is 13 cm (5 in.). **A.** Clay clasts in a sand matrix found near the base of the channel. **B.** Highly contorted and disrupted bedding in sandy matrix in channel-fill deposits. **C.** Irregular bedding and burrowed structures in silty clay layers of a channel-fill deposit. **D.** Contorted bedding and flow rolls in sands and sandy silt deposits of the channel fill. **E.** High organic clays forming the uppermost part of the channel-fill deposits.

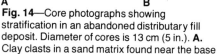
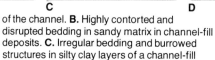

content of the clays and silts are responsible for the large amount of distorted bedding.

Figure 15 is a continuously cored boring through an old distributary channel that existed off Southwest Pass, Mississippi River delta, in the early 1800s. The channel on old maps was approximately 9 to 10 m deep and about 100 to 110 m wide. The base of the channel in the cored boring is identified by the letter A and occurs at a depth below mean sea level of 10.5 m. The clays directly below this scour plane, where burrowing is present, contain marine microfaunal remains. Just above the scour surface is a sand approximately 1 m thick. It contains fairly well sorted fine- to medium-grained sands displaying small-scale current structures. Organic trash and other debris are common along the bedding planes.

Above this unit are a series of alternating sands, silty clays, and clays displaying a large amount of contorted bedding, load casting, and small-scale slumping (Fig. 15). The central part of the fill consists of silty clays alternating with thin silt laminations. Although not especially well displayed in this cored boring, one of the more characteristic features of the laminations within abandoned channels is the extremely sharp nature of both the upper and lower bounding planes of silt and sand laminations within the overall fine-grained channel fill. Graded bedding is rarely observed. The uppermost section of the fill in this distributary chan-

nel consists of fine-grained silts and silty clays containing large amounts of organic debris along the bedding planes. Distorted laminations within the clays and silty clays are apparent (uppermost cores, Fig. 15). The uppermost meter in the abandoned channel in Figure 15 consisted primarily of organic-rich clays and peats, which, because of their extremely high water content, were unable to be preserved and adequately cored and thus are not illustrated in this boring.

The lowermost two diagrams in Figure 13 illustrate an isopach map and variations in electric-log response across an abandoned channel-fill sequence. As shown in the isopach map, if any sand buildup is apparent in the abandoned channel, it most commonly occurs where the channel tends to display a bifurcating nature. Often 5 to 8 m of sand will be concentrated in this particular part of the channel fill. Electric logs generally give extremely erratic and ragged appearances in the fill itself. The only significant buildup of sand generally occurs near the base of the channel. Sand layers above this unit often show little high lateral continuity, and logs only a few hundred meters apart cannot be accurately correlated.

SUBAQUEOUS DELTA PLAIN

The subaqueous delta is that part of the delta plain lying below low-tide level and extending seaward to that area actively receiving fluvial sediment. This area ranges in water depth from 50 to 300 m and in width from a

few kilometers to tens of kilometers. It is the foundation across which progradation of the subaerial delta must proceed. Most commonly, the subaqueous delta is characterized by a seaward fining of sediments, sands and coarser clastics being deposited near the river mouths, and finer grained sediments settling farther offshore from suspension in the water column or as a result of downslope mass-movement processes. The seawardmost section of the subaqueous delta is commonly referred to as the prodelta environment and is composed of the finest material deposited from suspension.

Prodelta silts and clays grade landward and upward vertically into the coarser silts and sands of distal bar. Directly at the mouths of the active distributaries lie the coarsest sand deposits. These deposits are commonly referred to as the distributary-mouth bar. If sediments deposited seaward of the river mouth accumulate faster than subsidence or removal of sediments by marine processes occurs, deltaic progradation will take place and the subaerial deltaic deposits will overlie the uppermost parts of the subaqueous delta, forming a complete delta sequence.

Distributary-Mouth-Bar Deposits

At river mouth, seaward-flowing water leaves the confines of channel banks and spreads and mixes with ambient waters of the receiving basin. This point is the dynamic dissemination for sediments which contribute to

Fig. 15—Continuously cored boring through abandoned distributary channel off Southwest Pass, Mississippi River delta. Diameter of cores is 8 cm (3 in.).

continued delta progradation and is responsible for forming one of the major sand bodies associated with deltaic sequences—the distributary-mouth-bar deposits.

The geometry of the river mouth and the distributary-bar topography are influenced by effluent dynamics and marine processes. The resulting geometry and distribution of river-mouth-bar sand bodies is determined by riverine flow conditions, density contrast between issuing ambient water, bottom slope seaward of the mouth, tidal range and tidal currents within the lower river channel, and the ability of waves and other forces to redistribute the fluvial clastics. Comprehensive studies, both theoretical and field, of river-mouth processes have been published by Bates (1953), Scruton (1960), Jopling (1963), Borichansky and Mikhailov (1966), and Wright and Coleman (1971, 1974).

The most important river-mouth processes are those enabling river effluent to interact with ambient marine waters, resulting in deconcentration of outflow momentum and consequent loss of transporting ability. These interactions yield specific geometry to the resulting distributary-mouth-bar deposit. The geometry of the resulting sand body depends on the relative roles of three primary forces: (a) the inertia of the issuing water and associated turbulent diffusion; (b) friction between the effluent and the bed just seaward of the mouth; and (c) buoyancy resulting from density contrasts between issuing and ambient fluids.

In the absence of high tidal range and extremely strong marine energy, the distributary channel pattern is often one of seaward bifurcation (Fig. 16). Because of this bifurcating channel pattern, the distributary-mouth bars at each of the river mouths often merge and form a near-continuous sand strip around the entire periphery of the delta. Shallow offshore slopes and low wave and tide action favor this type of distributary pattern, and turbulent diffusion within the water mass becomes restricted to the horizontal. Bottom friction plays a major role in causing effluent deceleration and expansion. Initially a broad, arcuate radial bar will form at the mouth. However, as deposition on the bar continues, natural subaqueous levees will develop beneath the lateral boundaries of the expanding effluent, where velocity gradients are generally steepest. Development of subaqueous levees tends to inhibit further increases in effluent expansion, so

Fig. 16—High-altitude photograph of bifurcating distributary channel in the Mississippi River delta.

that with continuing bar accretion continuity can no longer be maintained simply by increasing effluent width. As the central part of the bar grows upward, channelization develops along the threads of maximum turbulence, which tend to follow the subaqueous levee. This process results in formation of a bifurcating channel, which has a triangular middle ground shoal separating diverging channel arms. This type of channel pattern is well displayed in the high-altitude photograph shown as Figure 16.

The upper left diagram of Figure 17 illustrates schematically the process of a bifurcated channel. In the cross sections on this panel, the three major units illustrated include lower prodelta clays, overlain by alternating silt and sand units of the distal bar, and uppermost, distributary-mouth-bar sands. As the diagram shows, distributary-mouth-bar sands form a nearly continuous sand body extending laterally large distances.

The upper right-hand diagram shows the most common vertical sequence within the distributary-mouth-bar deposits and some of its characteristics. The unit generally displays a coarsening-upward sequence in which depositional dips are extremely low, rarely exceeding 1° except in areas where slump deposits result in high angles within the mass-moved sediment. Figure 18A and B illustrates some of the characteristic structures found within the lowermost units of prodelta clays. Parallel, colored clay laminations, thin graded silt and silty clay parallel laminations, bioturbation (generally confined to the clay laminations), and slump structures are common within the prodelta deposits. Microfaunal remains generally indicate marine depo-

sition, and diversity of species is generally quite high, indicating an open, inner to outer shelf depositional environment. These deposits display most of the characteristics of normal marine shelf deposits and are differentiated only by their rate of accumulation.

For a given length of time, prodelta deposits are thicker (because of high sedimentation) than an equivalent section of normal marine shelf deposits. Rapid deposition often results in slightly lower amounts of bioturbation and deposits displaying excess pore fluid pressures. Graded bedding zones are sometimes present in the prodelta deposits, but are rarely found in normal marine shelf deposits. X-ray radiography of the cores reveals that many of the parallel laminations are defined either by inclusions of diagenetic origin or extremely slight differences in textural characteristics. Because of the high rates of deposition normally associated with prodelta deposits, intense bioturbation is usually confined only to the lowermost parts of the deposit, where it grades downward into normal open-marine shelf environments, that often show intense bioturbation.

Progressing upward within the vertical sequence, the distal bar (also referred to as delta front or delta platform) overlies the prodelta facies. This environment is the seaward-sloping margin of the advancing delta sequence. Increase in sedimentation rates and coarseness of sediments are characteristic of this environmental sequence.

Lithologically, the sediments primarily consist of laminated silts and clays containing numerous thin-graded and cross-laminated sand-silt layers. Depositional slopes, although higher than in any other subaqueous delta environment, rarely exceed 0.5 to 0.75°. Core photographs in Figure 18D-F show some of the more common structural types associated with distal-bar deposits. Lower parts of the distal bar show small burrows and shell remains generally scattered throughout the deposits, that often result in partial destruction of parallel and lenticular laminations (Fig. 18D). Higher in the distal bar deposits, a wide variety of structures associated with both oscillatory and unidirectional currents become much more common. Sedimentary structures such as small-scale cross laminae, starved current ripples, small scour and fill, and graded sand units become much more common. In general the amount of clays and silty clays begins to diminish within the section.

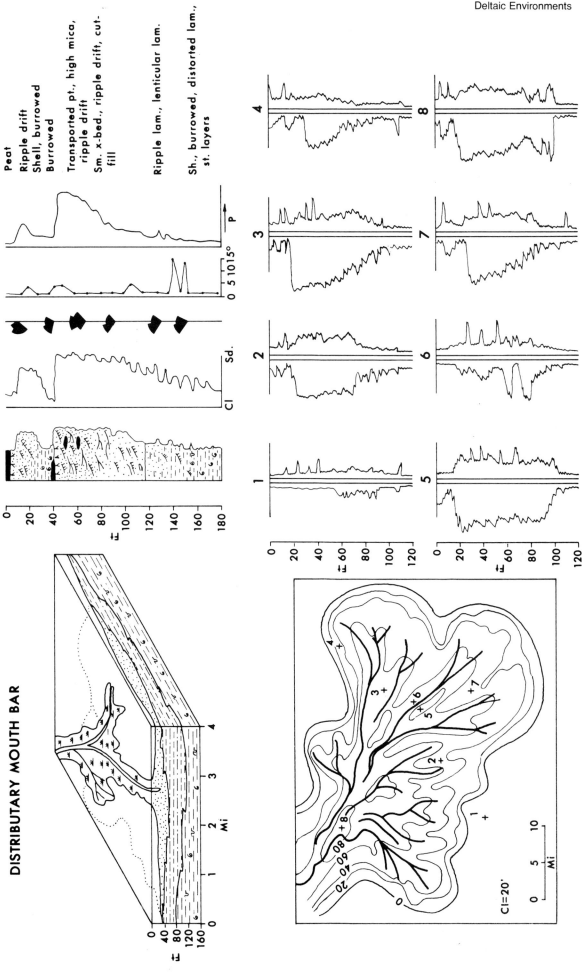

Fig. 17—Summary diagram illustrating the major characteristics of the distributary-mouth bar deposits in the subaqueous delta plain.

Figure 18F illustrates the wide variety of sedimentary structures found within only a 20-cm core from the uppermost parts of this environmental sequence. The clays, when present, are generally extremely thin, with lateral continuity of individual clay laminations usually low. Distal-bar deposits represent potential reservoirs in the subsurface, and numerous gas fields in such deposits have been located in the Pleistocene deposits of offshore Texas.

The distributary-mouth bar is the area of shoaling associated with the seaward terminus of a distributary-mouth channel. Shoaling is a direct consequence of a decrease in velocity and a reduction in carrying power of a stream as it leaves the confines of its channel. Accumulation rates are extremely high, probably higher than in any other environment associated with the delta. In some places depositional rates of coarse sediments at the mouth of the Mississippi reach 1 to 3 m per year.

The sediments are constantly subjected to reworking, not only by stream currents but by waves generated in the open-marine waters beyond the channel mouth. A general understanding of the processes and mode of formation of the distributary-mouth bar is critical to understanding the evolution and vertical relationships illustrated in Figure 17. As the low-density, turbid, fresh river water flows out of the distributary mouth over denser saline marine waters, the lighter effluent waters expand and lose velocity. Coarser sediments (the sands) settle rapidly, both from suspension and bedload migration, and almost all of the sand is deposited within the vicinity of the distributary mouth.

Because of variations in turbulence at the river mouth and different process intensities between low river stage and high river stage, silts and clays will occasionally be deposited with sands in this environment. However, reworking by marine and riverine processes results in cleaning and sorting of the sediments. As a result, the distributary-mouth bar commonly consists of clean, well-sorted sand and thus is obviously a potential reservoir rock for hydrocarbons. The remaining finer grained suspended load carried by the river is distributed widely by the expanding river effluent and forms distal bar and prodelta environments.

The most common sedimentary structure consists of a variety of small-scale cross laminae and current ripple drift types (Fig. 18G-I). Quite often

mass-movement processes such as small localized slumps result in distorted laminations. This is particularly true near lower sections of the distributary-mouth-bar sequence, where overpressured sediments are much more common. Figure 18G illustrates one type of slump structure commonly seen. Near the top of the distributary-mouth bar large accumulations of river-transported organic debris are often present. Water-saturated logs and other organic debris are transported down the rivers in times of flood and discharged into the nearshore zone, where wave action grinds down the coarser wood particles into large concentrations of organic debris. Figure 18I is a core showing some of the organic laminations present within the upper part of the sand body.

Directly capping the distributary-mouth bar are a variety of overbank splays and shallow-bay deposits. These overbank splays are commonly referred to as natural levees. During annual river floods, the river stage normally tops the distributary channel, and numerous small overbank or crevasse splays result. Each splay will be maintained for one or several years until it builds up or grades the natural levee to flood level, and then it will cease activity. Thus, capping the distributary-mouth bar is commonly a series of small coarsening-upward sequences representative of these splay deposits. Deposits are characterized by a wide variety of sedimentary structures, with climbing current ripples and small-scale cross laminae being the most common in the coarser grained parts. Figure 18J illustrates some of the ripple-type structures present in this environment.

Figures 19, 20 and 21 show parts of a continuously cored boring (approximately 130 m in depth) taken through the distributary-mouth-bar deposits in the Mississippi River delta. Figure 19 illustrates the sedimentary structures in the core representing lower sections of the distal-bar deposits. Even within this 10 m section of core the change from a predominance of clay in the lower part of the distal bar to a predominance of fine, silty and sandy laminations is apparent. Several regions show highly contorted bedding believed to be primarily due to slumping rather than coring processes.

Figure 20 represents approximately 7 m of core taken in the lower segment of the distributary-mouth-bar deposits. The predominance of sand laminations intercalating with silty clay and silt

laminations is apparent. Darker colored zones in lower parts of the core represent transported organic debris. Also shown in this sequence is a section of highly contorted core, once more believed to be the result of slumping rather than core disturbance. The predominance of small-scale ripple laminations is obvious within the sequence. Figure 21 represents 7 m of cored boring taken from the uppermost part of the distributary-mouth-bar deposit. The core consists primarily of sand deposits, with sand-sized particles making up 80 to 90% of the total unit. Dark layers are stratification representing transported organic debris. Note that most of the stratification is nearly flat-lying, and few or no steep dips exist, except for small-scale cross-bedding. One core indicates tremendous disturbance and distortion of lamination and undoubtedly represents a portion of a slump block. Ripple laminations, climbing ripples, and larger scale trough cross-laminations are the most dominant sedimentary structure types.

The lower two diagrams on Figure 17 represent a sand isopach map of a distributary-mouth-bar system in which individual distributary-mouth bars have merged, forming a delta-front sand body type. The distributary pattern is shown as solid dark lines. A boring through the distributary channel itself would show very erratic sand distribution in the distributary channel. Electric-log responses and their variations are shown in the lower right-hand diagram of Figure 17. Most borings show a coarsening-upward sequence, with the sand body varying in thickness, depending on the location of the core with reference to the distributary channels themselves. In general, the nearer the boring to the axis of the distributary, the sharper the base of the sand body, and gradational contacts become less well defined. Distally (away from the distributary-channel axis), the sequence displays a much greater tendency toward a large transition from distal bar to distributary-mouth bar.

Published literature on distributary-mouth-bar environments is extensive, and far beyond the scope of this paper to document all contributions. The major papers that deal with the distributary-mouth-bar sequence are as follows: Johnston (1922); Arnborg (1948); Fisk et al (1954); Fisk (1955, 1961); Fisk and McFarlan (1955); Kruit (1955); Scruton (1955, 1956, 1960); Shepard (1956); Shepard et al (1960); Allen (1965c, 1970); Kolb and Van Lopik

(1966); Mikhailov (1966); Shirley (1966); Fisher and McGowen (1967); Oomkens (1967, 1974); Coleman et al (1970); Ferm (1970, 1977); Bagans et al (1975); Belt (1975); Maldonado (1975); Coleman (1976); and Horne et al (1978).

River-Mouth Tidal-Ridge Deposits

The distributary-mouth-bar sequence just described represents the most common type of vertical sequence within many deltaic deposits. In deltas debouching into macro- or high-tidal ranges and into narrow, elongate basins, however, tidal processes play an important role in dispersing and redistributing fluvial clastics. Where rivers debouch into narrow, elongate basins where tidal range creates strong bidirectional transport, large linear tidal ridges within and just seaward of the river mouth are common topographic forms associated with the distributary-mouth-bar environment. These ridges are commonly composed of coarse fluvial sand and vary in size, some displaying heights exceeding 30 m above the adjacent swales. The ridges are commonly oriented parallel with the river channels forming elongate prominent sand deposits.

Although in present world deltas, river-mouth tidal ridges form one of the major sand bodies associated with rivers prograding into basins displaying macro- or high-tidal ranges, literature on these river-mouth tidal ridges is sparse. Many past river deltas undoubtedly existed in similar environmental settings, and they must therefore form prominent sand bodies in many ancient-rock deltaic sequences. Major literature associated with these types of features includes that of Off, (1963), Keller and Richards (1967), Reineck and Singh (1967), Houbolt (1968), Klein (1970), Ludwick (1970), Meckel (1975), and Wright et al (1975).

River distributaries debouching into high-tidal regions commonly display a funnel-shaped configuration, with widths attaining several kilometers. Linear, elongate tidal ridges aligned parallel with each other in the direction of tidal flow are the most prominent channel and river mouth accumulation forms. They appear to be directly related to bidirectional sediment transport patterns, high-tidal amplitudes, and tidal-current symmetry. Tidal ridges described by Coleman (1976) and Wright et al (1975) in the Ord River are typical of the study shoals found in river-dominated deltaic distributaries.

Those tidal ranges in the Ord River range in relief from 10 to 22 m and compositely account for over 5×10^6 cu m of total sand accumulation. Tidal ridges average roughly 2 km in length and 300 m in width, with crests emergent or near the surface at low tide. A few are permanently emergent and vegetated by mangrove. In some deltas, tidal ridges attain extreme lengths of 10 to 15 km. Deltas displaying similar types of shoals have been described by Off (1963) and Meckel (1975). Meckel (1975) referred to tidal ridges at the mouth of the Colorado delta, Gulf of California, as tidal bars. Ridges at the mouth of the Colorado display relief of 7 to 10 m and a crest to crest spacing of several kilometers. In cross-section they vary from roughly symmetrical to distinctly asymmetrical, with steep sides commonly facing the downstream direction of tidal propagation.

Although little coring has been done on these types of deposits, Figure 22 is an attempt to summarize data presently available. The upper left diagram illustrates the distribution of some of the tidal ridges seaward of the mouth of the Shatt-el-Arab River delta, which flows into the Persian Gulf. Lengths of the tidal ridges at the river mouth range from 5 to 15 km, with some of the larger ridges displaying widths of 2 km. General spacing of the ridges across the distributary-mouth-bar area ranges from a minimum of about 2 km to slightly over 5 km. The upper right-hand diagram shows a typical vertical sequence resulting from river-mouth progradation and lateral migration of ridges. In general, the coarsening upward sequence displayed agrees quite well with data presented from the lower Colorado delta (Meckel, 1975) and Ord River delta (Coleman, 1976).

Sand units are generally well sorted and display a variety of small-scale and large-scale cross-stratifications. One of the more common sedimentary structures within sand bodies is the small-scale bidirectional or herring-bone stratification type. Shell debris is generally common, both scattered throughout sand deposits and concentrated into thin lag-type deposits. Parallel sand layers are common throughout the entire sequence of sandy deposits and probably result from deposition during the upper flow regime, especially during low tide, when water depths across the shoals are quite low and velocities are quite high.

Thompson (1968) measured flood and ebb currents of 100 to 135 cm/sec, with maximum velocities of more than 200 cm/sec, in bars at the mouth of the Colorado. Although exposures are generally limited within the tidal ridges, shallow pits and box cores near the tops of many tidal ridges have large-scale trough cross-bedding, with the probabilities that within the uppermost sequences, large-scale cross-bedding could be preserved. Directional properties throughout the sequence generally show a net downstream direction; however, upstream-oriented cross-stratification is not uncommon, and thus current roses would probably show the bidirectional pattern.

The lower left diagram in Figure 22 depicts the probable sand isopach associated with a river-mouth tidal-ridge environment. This particular isopach is based on limited data and is patterned after the Ord River mouth. Sand thickness throughout the isopached interval would undoubtedly vary and be concentrated into the linear type of ridges seen topographically in modern deltas. Log response (lower right diagram, Fig. 22) displays extreme variation because of sand thickness; the base of the sand deposit displays a gradational contact to a rather abrupt basal scour plane associated with those ridges of prominent scour. In general, the ridges tend to display the coarsest and best sorted sand units and are illustrated by core holes 3, 5 and 7.

In an upstream direction, depicted by core hole 8, it is highly probable that the base of the sand represents a scoured surface, and thus the sand body displays an extremely sharp base. In drop cores and samples from inter-ridge-ridge areas, sediments tend to be much more poorly sorted. Clay clasts, organic trash, and shell lags are common. It is probable that electric-log response would show extremely erratic and ragged types of patterns as indicated in bore holes 4 and 6.

Although data are sparse from this type of environmental setting, the writers believe such sand bodies to be indeed common in ancient rock sequences, and until a larger number of cored borings and cores on and through these river-mouth tidal ridges are obtained, data presented in Figure 22 remains somewhat speculative. However, observations from literature cited above indicate that the general nature of the deposits is as illustrated in the vertical sequence of Figure 22.

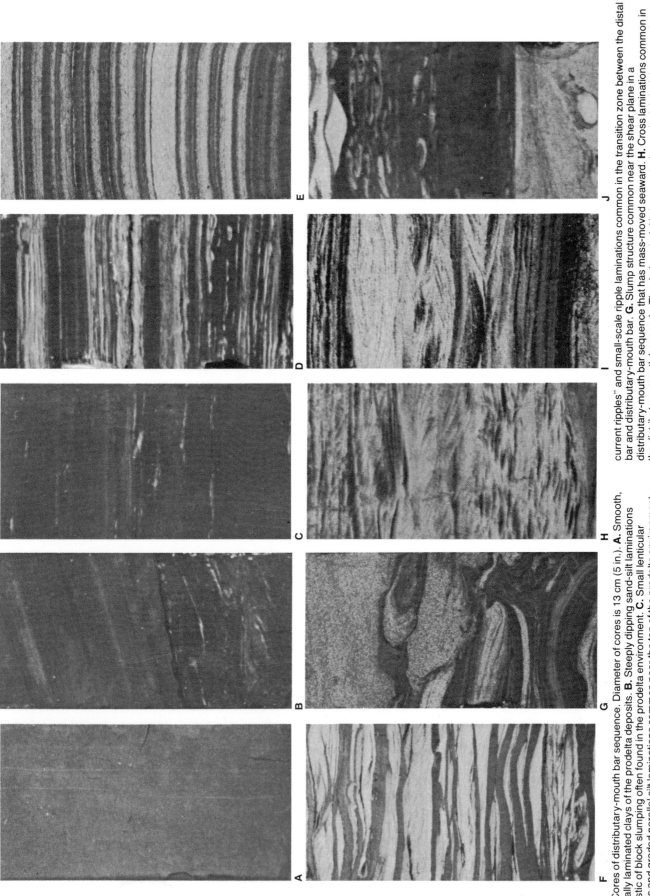

Fig. 18—Cores of distributary-mouth bar sequence. Diameter of cores is 13 cm (5 in.). **A.** Smooth, gray, partially laminated clays of the distal bar and distributary-mouth bar. **B.** Steeply dipping sand-silt laminations characteristic of block slumping often found in the prodelta deposits. **C.** Small lenticular laminations and graded parallel silt laminations common in the prodelta environment. **D.** Alternating sand, silt, and silty clay laminations in the lower part of the distal bar environment. **E.** Well-developed parallel silt and sand laminations showing graded bedding and small-scale ripple laminations common in the distal bar deposits. **F.** Lenticular sand laminations representing "starved current ripples" and small-scale ripple laminations common in the transition zone between the distal bar and distributary-mouth bar. **G.** Slump structure common near the shear plane in a distributary-mouth bar sequence that has mass-moved seaward. **H.** Cross laminations common in the distributary-mouth bar sands. The dark material is transported organic debris. **I.** Large-scale cross laminations common near the top part of the distributary-mouth bar deposits. The dark material is transported organic debris. **J.** Alternating silty sand and clay layers common to the small overbank splays that cap the distributary-mouth bar deposits.

Fig. 19—Part of a continuously cored boring through the distributary-mouth bar sequence. This section represents the lower part of the distal bar deposit. Diameter of cores is 8 cm (3 in.).

Fig. 20—Part of a continuously cored boring through the distributary-mouth bar sequence. This section represents the lower part of the distributary-mouth bar deposit. Diameter of cores is 8 cm (3 in.).

Fig. 21—Part of a continuously cored boring through the distributary-mouth bar sequence. This section represents the upper part of the distributary-mouth bar deposits. Diameter of cores is 8 cm (3 in.).

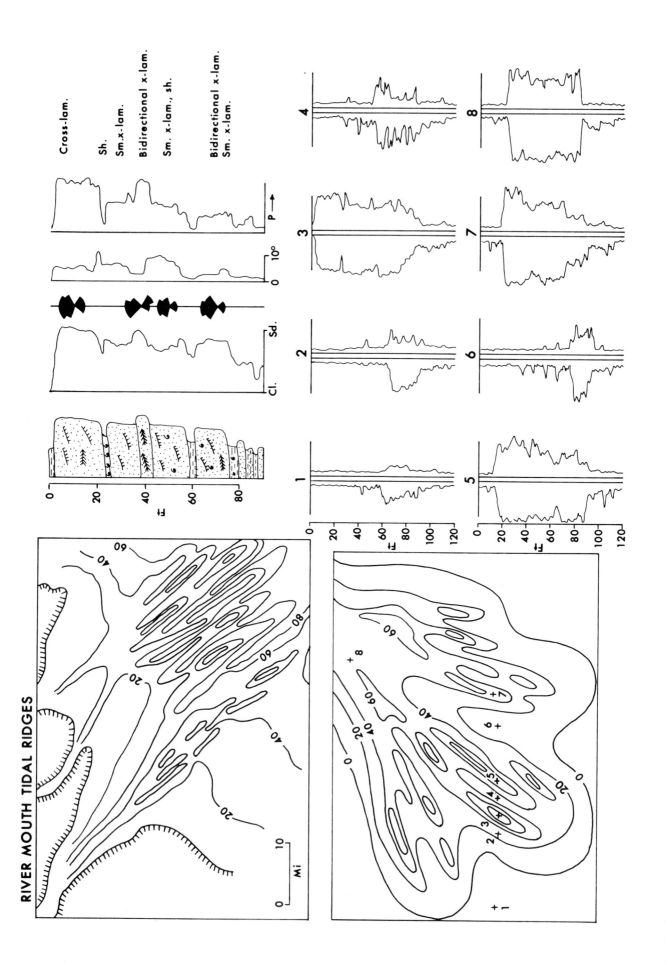

Fig. 22—Summary diagram illustrating the major characteristics of river-mouth tidal ridge deposits in the subaqueous delta plain.

Subaqueous Slump Deposits

Recent detailed marine geologic investigations on subaqueous parts of continental shelves seaward of many river deltas experiencing high depositional rates have revealed contemporary recurrent subaqueous gravity-induced mass movements as common phenomena worthy of consideration as an integral component of the normal deltaic process and marine sediment transport. Off river deltas such as the Mississippi, Magdalena (Columbia), Orinoco (Venezuela), Surinam (Surinam), Amazon (Brazil), Yukon (Alaska), Niger (Nigeria), Nile (Egypt), and Hwang-Ho (China), subaqueous slumping and downslope mass movement of sediments are common processes. Instabilities and mass movement of sediment in these regions generally display the following characteristics: (a) instability occurs on very low angle slopes (generally less than 2°); (b) large quantities of sediment are transported from shallow water to deeper water offshore along well-defined mudflow gullies (debris flows) and in a variety of translational slumps. Although individual mudflow features vary in size and frequency, they generally possess a source area consisting of subsidence and rotational slumping, an elongate, often sinuous chute or channel (mudflow gully), and a composite depositional area composed of overlapping lobes of remolded debris.

River deltas displaying an abundance of submarine landslides are generally characterized by high rates of sediment accumulation within both fine-grained and coarse-grained fractions. Sediments therefore have an extremely high water content and most commonly display excess pore fluid pressures.

The abundance of fine-grained organic material within fine-grained clays is also subjected to rapid degradation by biochemical processes and produces large accumulations of sedimentary gas (primarily methane and carbon dioxide). The basic conditions for failure exist when stresses exerted on the sediment are sufficient to exceed its strength. This can be due to stress increases, strength reduction, or a combination of the two.

The Mississippi River delta and adjacent shelf region have been sites of active investigation for decades. The past decade has seen substantial advances in the systematic utilization of various techniques for marine geological exploration. Application of side-scan sonar and high-resolution seismic techniques has allowed substantial improvements in documentation and mapping of subaqueous flowslides off the delta. Essentially the entire subaqueous part of the Mississippi delta has been covered with overlapping side-scan sonar imagery and high-resolution seismic lines on a grid spacing of 250 m. Using these techniques, and aided by a large number of offshore foundation borings, Coleman et al (1974), Coleman (1976), Coleman and Garrison (1977), and Prior and Coleman (1978a, b; in press) have identified a wide variety of slope deformational features. These data form the basis for the following discussion of subaqueous mass-movement deposits.

The main types of slope and sediment instability mapped in 5- to 300-m water depths are illustrated schematically in Figure 23, which shows their distribution around a single distributary. Similar spatial organization can be identified around the entire periphery of the modern river delta. Although a large number of types have been identified, the major types having geological significance include peripheral slumping, elongate retrogressive slides and mudflow gullies (and their associated overlapping depositional lobes), and large shelf-edge arcuate slumps and contemporaneous faults.

Peripheral slumps are associated primarily with immediate distributary-mouth-bar deposits. Bottom slopes in this region range from 0.2 to 0.6°. Peripheral slumps display abrupt stair-step scarps on the seafloor, the heights of the scarps ranging from 3 to 8 m. Tensional cracks are often present upslope from the major scarp, and frequently small mud vents are associated with these scarps.

Sizes of the slump blocks vary considerably, ranging from only a few hundred meters to well over 3 km. Although the features are referred to as rotational types of slumps because of the extremely low-sloping bottom gradients, the major movement is translational downslope. Displacements of the sediments begin as shallow rotational slumps, continuing over basal shears as predominantly translational movement.

Depth of the shear plane, and hence the thickness of the block, varies, but rarely exceeds 35 m. Movement rates are hard to determine, but repeated surveys over a 1-year period display movements ranging from a few hundred meters to nearly 1,000 m. This type of block slumping results essentially in downslope movement of sediment from shallow-water environments to deeper offshore and outer continental shelf water depths. Since the features originate in and near the distributary-mouth bar, they are frequently responsible for carrying coarse distributary-mouth bar sands farther offshore into deeper waters.

Instability of a second major type consists of elongate, retrogressive mudflow gullies and their overlapping depositional lobes. These features extend radially seaward from each major distributary and occur in water depths from 10 to 100 m. Depositional lobes extend farther seaward to water depths as great as 300 m. Each feature possesses a long, narrow chute or channel linking a depressed, hummocky source area on the upslope end to composite overlapping depositional lobes of fans on the seaward end. Figure 24 is a side-scan sonar mosaic of several of these mudflow gullies. Source areas are normally bowl shaped and bounded by distinct scarps (A, Fig. 24), with the interior of the depression normally characterized by extremely hummocky, chaotically arranged blocks of clasts in a matrix of highly fractured, flowed sediments.

Narrow chutes or gullies (B, Fig. 24) extend downslope at approximately right angles to the regional depth contours and achieve lengths exceeding 8 to 10 km. They are rarely straight, and in plan view they display high sinuosity, with alternating narrow constrictions and wide bulbous sections. Widths of the gullies range from 20 to 50 m at the narrow section to 600 to 800 m where gullies are widest. Gully floors are generally depressed from a few meters to 20 m below the adjacent intact bottom. Slopes of side walls range from 1° to highs of 15°, and small rotational side slumps are often apparent.

During failure and movement of sediments, the material is apparently viscous enough to occasionally be ejected out of the narrow channel, forming overbank or natural-levee-type splays (C, Fig. 24). On seaward ends of the elongate chutes, broad overlapping composite depositional lobes composed of debris discharged from the gullies are present. Depositional lobes display extremely irregular bottom topography characterized by crenulated blocky, disturbed debris and often abundant mud vents and volcanoes. Seaward mud nose scarps range in height from a few meters to more than

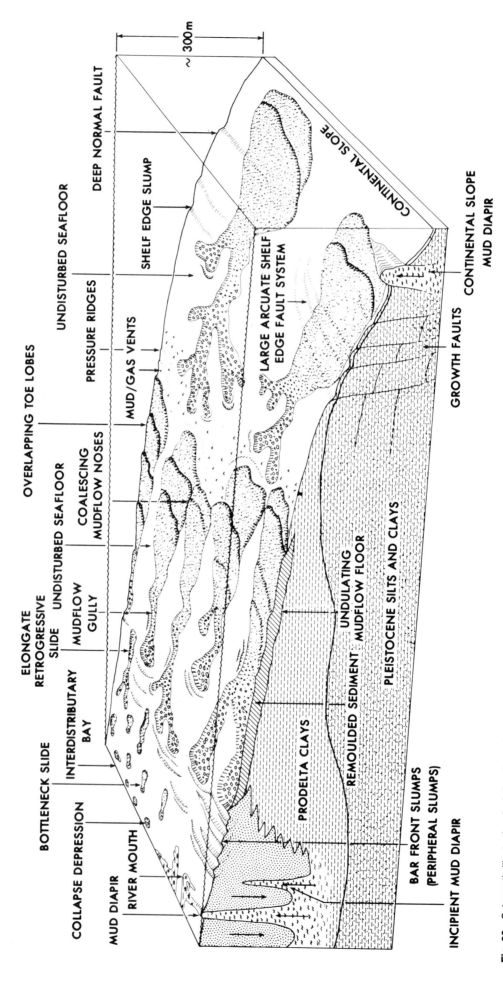

Fig. 23—Schematic illustration depicting the major types of submarine landslides, diapirs, and contemporary faults in the Mississippi River delta.

25 m. In plan view, scarps are curved and adjacent lobes often coalesce, forming an almost continuous complex sinuous frontal scarp possibly extending for distances of 20 to 25 km more or less parallel with the bathymetric contours.

Depositional areas are composed of several overlapping lobes owing to periodic discharge events, and each discharge is associated with its own distinctive nose. Seaward of the edge of the lobes, extensive small-scale pressure ridges are arranged sinuously and parallel. Extensive fields of mud vents and volcanoes emitting gas, water and fluid mud are found associated with the lobes and directly seaward of the noses; these undoubtedly result from rapid loading of underlying sediment as well as consolidation processes within the debris itself. Thicknesses of the lobes are difficult to determine, but each distinct lobe is normally 20 or so meters thick, and because of overlapping, the total thickness of mudflow can often approach 50 to 60 m. In one area of the Mississippi River delta, in water depths of approximately 200 to 250 m, depositional lobes cover approximately 770 sq km with discharged debris volume of 11.2 x 10^6 cu m.

A third major type of sediment instability of significant geological importance is arcuate rotational slump and growth fault. This type commonly occurs on the outer continental shelf in front of the advancing or prograding deltaic system. Large, arcuate-shaped families of shelf-edge slumps and deep-seated contemporaneous faults tend to be active along the peripheral margins of delta fronts. In most instances, these large-scale features tend to cut the modern sediment surface, often forming localized scarps on the seafloor. These surface scarps provide localized areas for accumulation of downslope mass-moved shallow-water sediment. In many places shelf-edge slumps tend to give a stairstepped appearance to the edge of the continental shelf and are highly reminiscent of rotational peripheral slumps higher on the continental shelf, near the mouths of the modern distributaries. However, these features generally occur on a much larger scale and cut a column of sediment ranging from 50 to 150 m in thickness.

Lateral continuities of individual slump scarps range from a few kilometers to as much as 8 to 10 km, and scarps on the seafloor produced by this slumping process may have heights of 30 m. A similar type of slump is commonly referred to as a contempora-

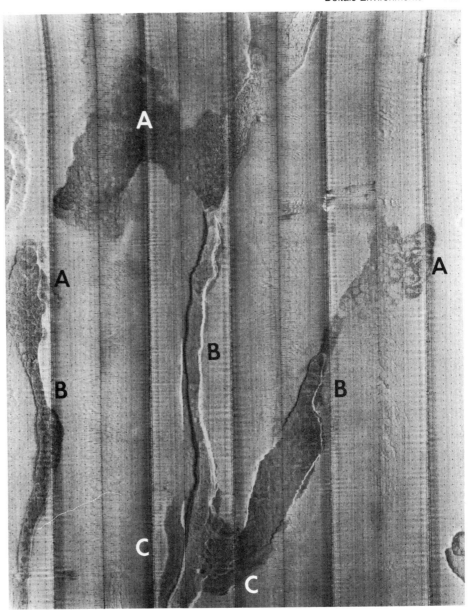

Fig. 24—Side-scan sonar mosaic of subaqueous landslide gullies in the Mississippi River delta. The width of the mosaic is 1.5 km, and the superimposed grid is a 25-m square. The slope is from top (approximately 10-m water depth) to bottom (water depth 60 m).

neous or growth fault and is the feature moving continuously along the shear plane with deposition. Hence with time and continued movements, offsets of individual marker beds increase with depth, and thickness of these beds increases abruptly across the fault.

Figure 25 illustrates active growth faults in the Mississippi River delta seaward of the mouth of South Pass. Figure 25A shows a large mudflow lobe upslope from an active growth fault. This fault extends from the surface to depths beyond the bottom of the record. Sparker data run simultaneously indicate that the fault extends 700 to 750 m below sea bottom before merging into a bedding-plane fault. Offsets in the uppermost units are generally 5 to 10 m, while at depth (400 m), offsets of marker beds approach 70 to 80 m.

Note the increased thickness of the sediment units on the downthrown side of the fault and the small rollover anticline or reverse-drag characteristic of this type of fault.

The amorphous zone upslope represents a surface mudflow that has progressed to and slightly beyond the limits of the faults. As surface mudflow crosses the fault zone, it thickens. It is highly possible that increased thicknesses on the downthrown sides of these faults result from movement of subsurface mudflows across the fault zone. As the fault zone is blanketed by a large mass of rapidly introduced mass-moved sediment, surface scarps on the seafloor are eliminated. Continued movement along the fault, however, will cause a new scarp, and given enough time, another mudflow will

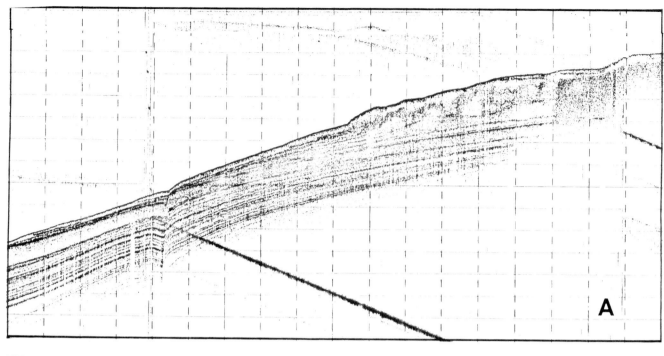

A

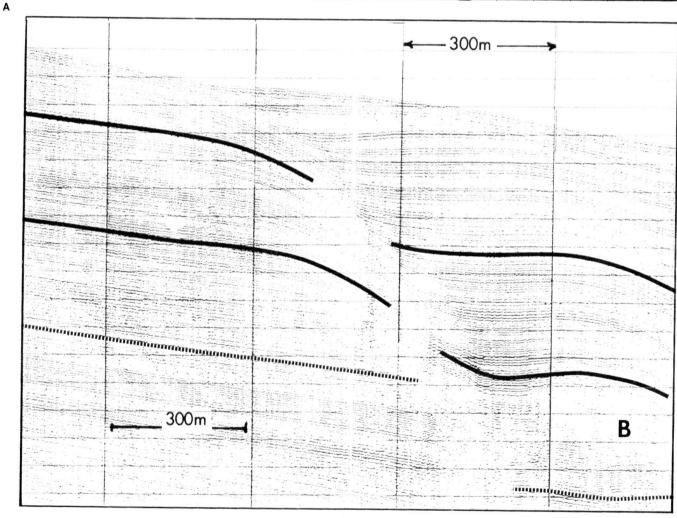

B

Fig. 25—High-resolution seismic record run across an active growth fault seaward of the mouth of South Pass, Mississippi River delta. **A.** Seismic line showing active growth fault seaward of a large upslope mudflow. Note the increased thickness of sediment on the downthrown side of the fault. Horizontal scale is 300 m between shot points and vertical scale is 25 milliseconds per time line, or 19 m (62.5 ft). **B.** Detailed subbottom seismic record run across an active growth fault. Note the presence of a rollover structure and the increased accumulation of sedimentation on the downthrown side of the fault. Horizontal scale is 300 m between shot points and vertical scale is 10 milliseconds per time line, or 7.6 m (25 ft).

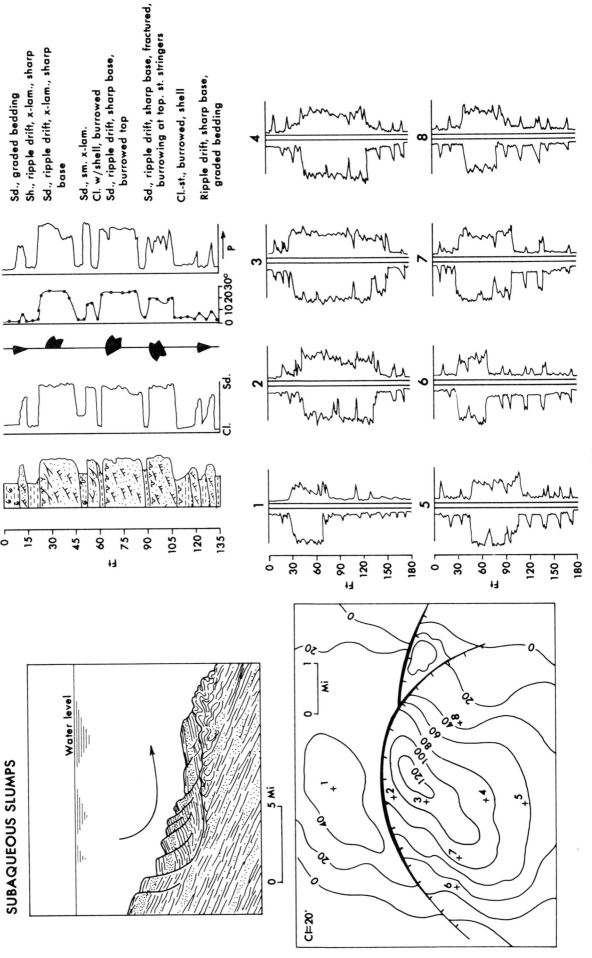

Fig. 26—Summary diagram illustrating the major characteristics of slump deposits in the subaqueous delta plain.

then move across the feature, adding increased amounts of sediment on the downthrown side.

This type of interaction between surface mudflow movements and contemporaneous faults may quite possibly play a large role in maintaining the continuing movement along these fault planes. Figure 25B shows details of the upper levels of a growth fault in the same vicinity.

A feature commonly associated with growth faults and of extreme importance to petroleum trapping is the association of rollovers or reverse drag with the downthrown side of a growth fault (Fig. 25A, B). These features are common on the contemporaneous faults presently active in the delta. Rollover structures tend to form soon after deposition of sediment on the downthrown side and do not require a considerable amount of overburden and weighting to form. Mass-moved material flowing downslope from higher levels on the delta front (sands, silts and clays) contains high water and gas contents. It is speculated that, as sediment accumulates slightly more thickly on the downthrown side of the fault, early degassing and dewatering associated with movement along the fault take place. Pore waters and pore gases are permitted to escape upward in the zone of movement associated with the fault, thereby decreasing the volume of sediment and allowing an early change in density to occur nearly contemporaneously with the fault. As greater and greater amounts of sediment are added and overburden pressures become increasingly larger, this feature is then amplified and becomes more pronounced with time and depth.

Figure 26 illustrates through a summary diagram some of the major characteristics associated with subaqueous slump deposits. Although boring control and core control are limited in deeper offshore waters, enough foundation borings have penetrated some of the sequences to give a fairly good indication of the deposit types accumulating offshore on the downthrown sides of some slump fault features. In addition, numerous articles on Gulf Coast Tertiary sequences indicate the type of deposition associated with slump deposits. The upper right-hand diagram illustrates a vertical sequence commonly associated with offshore slump deposits. The first striking characteristic is the extreme variations in grain size. Sandy deposits generally occur as distinct isolated blocks showing both sharp bases and sharp tops. Grain size

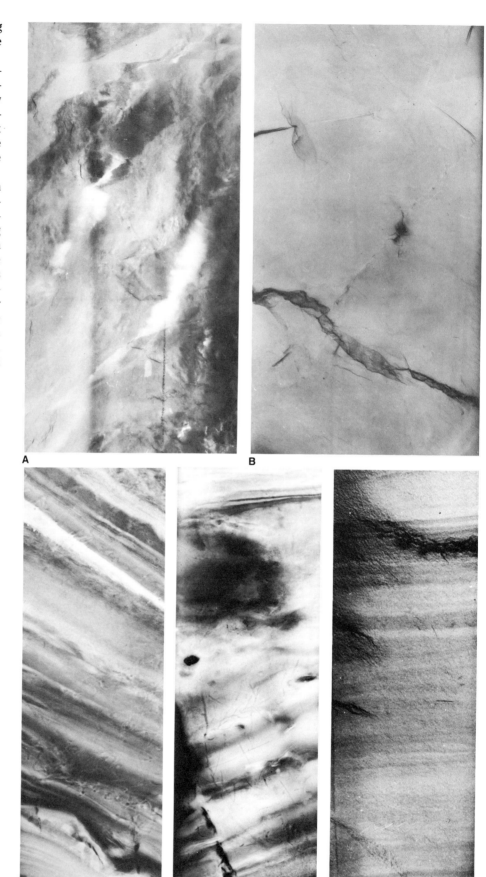

Fig. 27—Core photographs of subaqueous slump deposits. Diameter of cores A, B, and E is 13 cm (5 in.) and of cores C and D 8 cm (3 in.). **A.** X-ray radiograph of highly distorted clay layers in marine deposits beneath the slump block. **B.** X-ray radiograph of multiple fracturing in clays in the shear plane zone. **C.** X-ray radiograph of silt and sand core in the slump block. Note that bedding is preserved with only minor fracturing but is tilted at angles of 20 to 30°. **D.** X-ray radiograph of disturbed structures in mudflow deposit that caps a slump block. **E.** X-ray radiograph of core in normally deposited marine clays, which often cap the slump deposits. Note the lack of disturbance in these deposits.

Fig. 28—Part of a cored boring through a slump block composed of distributary mouth bar sands. This slump block is off Southwest Pass, Mississippi River delta, and is underlain and capped by marine clays. Diameter of cores is 8 cm (3 in.).

depends on the source of the slump material, and in such a deltaic setting, sources are commonly distributary-mouth-bar deposits trapped on the downthrown sides of these slump features.

Thus many sedimentary structures are the same as those described for distributary-mouth-bar deposits. Having been mass moved downslope, however, they lie on entirely marine clay deposits and thus normally have a sharp lower bounding surface. The upper surface is also usually extremely sharp and generally is characterized by a high degree of intensive burrowing on the top of the sand body. Because most of the deposits are mass moved, depositional dips increase significantly, and high-angle dips of 10 to 25° are not uncommon in these beds. Fracturing and localized faulting and slump structures are also abundant in most of the sand bodies.

Figure 27A-E illustrates specific types of stratification often encountered in finer grained sequences separating slumped sand blocks. Both normal bedded marine clay deposits (Fig. 27E) and highly distorted marine clays (Fig. 28A, B, D) are common within the finer grained sequences.

Figure 28 shows part of a cored boring through a slump block composed of distributary-mouth-bar sands. The core was taken in a buried slump block off Southwest Pass, Mississippi River delta. The sand unit is capped above and below by marine clays having faunal content that indicates middle to outer continental shelf depths. The sand body itself contains no fauna except for an occasional shallow-water microfossil. The thickness of the slump block cored is approximately 15 m, of which approximately 10 m is illustrated in Figure 28. The lowermost part of the sand body is characterized by extremely distorted sedimentary structures, with little or no primary stratification preserved. However, within the same slump block unit, well-preserved small-scale stratification exists as isolated blocks. There is a tendency for excessive dips to be present in the deposits, and fracturing systems such as those shown in the first two cores are also common within the sequences. The lateral extent of the sand body is relatively unknown, though additional borings in the area tend to show a minimum lateral continuity of at least 1 km. It seems highly probable that there is reservoir continuity across this particular sand body.

Figure 29 represents a part of a core taken through a mudflow lobe that moved down from shallower waters in the delta front across the continental shelf and on to outer continental slope water depths. This particular mudflow consists primarily of fine-grained silty clays of the distal bar and prodelta clay deposits. As is evident in the core, a wide variety of distorted laminations are present. Numerous fractures exist throughout the entire deposit; however, this particular depositional lobe has moved downslope from water depths of approximately 20 m and now lies at the edge of the continental shelf, where water depths are about 130 m (downslope movement of approximately 3 to 4 km). Even with this magnitude of movement, primary stratification is still preserved within individual blocks, as shown in the cored boring in Figure 29.

The lower two diagrams in Figure 26 attempt to illustrate a sand isopach map associated with a growth fault offshore, and accumulation of slump-block material on the downthrown side of a growth fault. This type of isopach is common in many of the published papers discussing Tertiary depocenters. Variations in electric-log response in various parts of the isopach map are shown in the lower right-hand diagram of Figure 26. One of the most characteristic features of sand bodies deposited by slumping processes is the extremely blocky character of the electric-log response. Sands generally tend to be sharp based, producing rather uniform log response. Correlation of individual kicks on the sand body is extremely tentative simply because of the nature of the slumping process. It is the writers' belief that many of the sand bodies associated with growth fault systems represent shallow-water sand bodies that have slumped downslope into deeper water and become trapped on the seafloor near growth fault systems.

In other instances, normal progradation of deltaic sequences across the growth fault zone results in thicker accumulations of mass-moved sediment on the downthrown side of the fault systems. Mass-moved sediment on the downthrown side is most common near the lower, more marine parts of the delta sequence.

Literature on subaqueous slumps and their influence on sediment distribution is extensive. Major articles dealing with processes of slumping and characteristics of the slumped deposits include those by Shepard (1955, 1973), Fisk (1956), Moore (1961), Ocamb

(1961), Morgan et al (1968), Bea (1971), Bruce (1972), Woodbury (1973, 1977), Coleman et al, (1974), Garrison (1974), Hedberg (1974), Busch (1975), Whelan et al (1975), Coleman (1976), Edwards (1976), Roberts et al (1976), Coleman and Garrison (1977), Embley and Jacobi (1977), Molnia et al (1977), Carlson (1978), Prior and Coleman (1978a, b; in press), Rider (1978), and Coleman and Prior (in press).

CONCLUSIONS

Delta environments have a wide variety of individual depositional facies within the overall delta sequence. This complexity results from the following factors: (a) modern deltas exist in a wide range of geographic settings, ranging in climatic regimes from arctic to temperate to tropical to arid, with basin tectonics ranging from rather stable basins to extremely actively subsiding basins; (b) deltas form primarily in the zone of interaction between freshwater and marine processes, one of the most complex process settings in all coastal environments; (c) deltas carry large volumes of sediment, ranging in grain size from gravel to clay, and deposit these sediments both overbank and into the marine environment through distributary channels; (d) rapid rates of deposition often result in formation of extremely weak foundations, with a wide variety of mass-movement processes resulting in complex redistribution of the deltaic sediment. Thus sand bodies within deltas display a variety of geometries and vertical-sequence characteristics.

The complexity of environmental settings under which deltas exist results in a variety of vertical sequences that can form within the delta facies. Delta types range from river dominated to tide dominated and wave-current dominated (Coleman, 1976). From the standpoint of petroleum accumulation, however, river- and tide-dominated deltas are probably the most important. In these two delta settings, reservoir-quality rocks are often deposited in close proximity to potential source beds, contemporaneous structure which forms major trapping potentials is common, and most deltas exist in rapidly subsiding basins, allowing thick deltaic sequences to develop over a rather short time framework. The highly wave-reworked delta sequences are often devoid of major source rock deposits and often do not form in structure settings that result in major trapping characteristics of the deposits.

Fig. 29—Part of a core taken through a mudflow lobe on the continental shelf off the modern Mississippi River delta.

Deposits described represent the most common sand-body types found in river- and tide-dominated delta sequences.

Funds for support of this research were provided by the Geography Programs, Office of Naval Research, Arlington, Virginia 22217. The Marine Geology Branch of the United States Geological Survey supported the work on mass-movement processes in the Mississippi River delta. Discussions with colleagues H. H. Roberts and L. E. Garrison are acknowledged with gratitude. The authors would also like to extend appreciation to various members of the petroleum industry and consulting firms for access to data not normally available and for many useful and stimulating discussions.

REFERENCES CITED

Allen, J. R. L., 1965a, Fining upward cycles in alluvial succession: Liverpool, Manchester, Geol. Jour., v. 4, p. 229-246.

——— 1965b, A review of the origin and characteristics of Recent alluvial sediments: Sedimentology, v. 5, p. 88-191.

——— 1965c, Late Quaternary Niger delta and adjacent areas: AAPG Bull., v. 49, p. 547-600.

——— 1970, Sediments of the modern Niger Delta: in J. P. Morgan, ed., Deltaic sedimentation: modern and ancient: SEPM Spec. Pub. 15, p. 138-151.

Are, H. T., 1963, The braided stream depositional environment: PhD dissertation, Univ. Wyoming, 205 p.

Arnborg, L. E., 1948, The delta of the Angerman River: Geog. Ann., Stockholm, v. 30, p. 673-690.

Arndorfer, D. J., 1973, Discharge patterns in two crevasses of the Mississippi River delta: Marine Geol. v. 15, p. 269-287.

Axelson, V., 1967, The Laitaure Delta, a study of deltaic morphology and processes: Geog. Ann., Stockholm, v. 49, p. 1-127.

Bagans, B. P., J. C. Horne, and J. C. Ferm, 1975, Carboniferous and Recent Mississippi lower delta plains: Gulf Coast Assoc. Geol. Soc. Trans., v. 25, p. 183-191.

Bates, C. C., 1953, Rational theory of delta formation: AAPG Bull., v. 37, p. 2119-2162.

Bea, R. G., 1971, How sea floor slides affect offshore structures: Oil and Gas Jour., v. 69, no. 48, p. 88-92.

Belt, E. S., 1975, Scottish carboniferous cyclothem patterns and their paleoenvironmental significance: in M. L. Broussard, ed., Deltas, models for exploration: Houston Geol. Soc., p. 427-449.

Bernard, H. A., et al, 1970, Brazos alluvial plain environment: Texas Bureau of Econ. Geol. Guidebook No. 11.

Bluck, B. J., 1971, Sedimentation in the meandering river Endrick: Scot. Jour. Geol., v. 7, p. 93-138.

Borichansky, L. S., and V. N. Mikhailov, 1966, Interaction of river and sea water in the absence of tides: in Problems of the humid tropic zone deltas and their implications: UNESCO, Proc. Dacca Symp., p. 175-180.

Bradley, W. H., 1964, Geology of Green River Formation and associated Eocene rocks in southwestern Wyoming and adjacent parts of Colorado and Utah: U. S. Geol. Survey Prof. Paper 496A, 86 p.

Bruce, C. H., 1972, Pressured shale and related sediment deformation: a mechanism for development of regional contemporaneous faults: Gulf Coast Assoc. Geol. Soc. Trans., v. 22, p. 23-31.

Busch, D. A., 1975, Influence of growth faulting on sedimentation and prospect evaluation: AAPG Bull., v. 59, p. 217-230.

Cant, D. J., and W. G. Walker, 1976, Development of a braided-fluvial facies model for the Devonian Battery Point Sandstone: Can. Jour. Earth Sci., v. 13, p. 102-119.

Carlson, P. R., 1978, Holocene slump on continental shelf off Malaspina Glacier, Gulf of Alaska: AAPG Bull., v. 62, no. 12, p. 2412-2426.

Chien, N., 1961, The braided stream of the lower Yellow River: Sci. Sinica, Peking, v. 10, p. 734-754.

Clapp, F. G., 1922, The Hwang Ho, Yellow River: Geogr. Review, v. 12, p. 1-18.

Coleman, J. M., 1966, Ecological changes in a massive freshwater clay sequence: Gulf Coast Assoc. Geol. Soc. Trans., v. 16, p. 159-174.

——— 1969, Brahmaputra River: channel processes and sedimentation: Sedimentary Geology, v. 3, p. 129-239.

——— 1976, Deltas: Processes of deposition and models for exploration: Continuing Education Publ. Co., Champaign, IL (now available from Burgess Publishing Co., 7108 Ohms Lane, Minneapolis, MN 55435), 102 p.

——— and S. M. Gagliano, 1964, Cyclic sedimentation in the Mississippi River deltaic plain: Gulf Coast Assoc. Geol. Socs. Trans., v. 14, p. 67-80.

——— ——— 1965, Sedimentary structures—Mississippi delta plain: in G. V. Middleton, ed., Primary sedimentary structures and their hydrodynamic interpretation—a symposium: SEPM Spec. Pub. 12, p. 133-148.

——— and L. E. Garrison, 1977, Geological aspects of marine slope instability, northwestern Gulf of Mexico: Marine Geotechnology, v. 2, p. 9-44.

——— and C. L. Ho, 1968, Early diagenesis and compaction in clays: Proc. Symp. on Abnormal Subsurface Pressures, Louisiana State Univ., Baton Rouge, p. 23-50.

——— and D. B. Prior, in press, Contemporary gravity tectonics—an everyday catastrophe? in Uniformitarianism—a contemporary perspective: Am. Assoc. Petroleum Geologists-Soc. of Econ. Paleon. and Mineral.

——— and L. D. Wright, 1975, Modern river deltas: variability of processes and sand bodies: in M. L. Broussard, ed., Deltas, 2nd ed:, Houston Geol. Soc., p. 99-150.

——— S. M. Gagliano, and W. G. Smith, 1970, Sedimentation in a Malaysian high tide tropical delta: in J. P. Morgan, ed., Deltaic sedimentation: modern and ancient: SEPM Spec. Pub. 15, 312 p.

——— ——— and J. E. Webb, 1964, Minor sedimentary structures in a prograding distributary: Marine Geol., v. 1, p. 240-258.

——— et al, 1974, Mass movements of Mississippi River delta sediments: Gulf Coast Assoc. Geol. Soc. Trans., v. 24, p. 49-68.

Collinson, J. D. 1970, Bedforms in the Tana River, Norway: Geog. Ann., Stockholm, v. 52, p. 31-56.

Dapples, E. C., and M. E. Hopkins, 1969, Environments of coal deposition: Geol. Soc. America Spec. Paper no. 114, 204 p.

Doeglas, D. J., 1962, The structure of sedimentary deposits of braided rivers: Sedimentology, v. 1, p. 167-190.

Donaldson, A. C., 1967, Deltaic sands and sandstones: in Guidebook, p. 31-62h, Symposium on recently developed geologic principles and sedimentation of the Permo-Pennsylvania of the Rocky Mountains: Wyoming Geol. Assoc. 20th Ann. Field Conf., Casper, Wyoming, 1966.

Edwards, M. B., 1976, Growth faults in Upper Triassic deltaic sediments, Svalbard: AAPG Bull., v. 60, p. 341-355.

Embley, R. W., and R. Jacobi, 1977, Distribution and morphology of large submarine sediment slides and slumps on Atlantic continental margins: Marine Geotechnology, v. 2, p. 205-228.

Fahnestock, R. K., 1963, Morphology and hydrology of a braided stream: U.S. Geol. Survey Prof. Paper 422A, p. 1-70.

——— and W. C. Bradley, 1973, Knik and Matanuska Rivers, Alaska: a contrast in braiding: in M. Morisawa, ed., Fluvial geomorphology: Publications in Geomorphology, State Univ. of New York, Binghampton, 314 p.

Ferm, J. C., 1970, Allegheny deltaic deposits: in J. P. Morgan, ed., Deltaic sedimentation: Modern and ancient: SEPM Spec. Pub. 15, 312 p.

——— 1977, Allegheny deltaic deposits: in F. B. van Houten, ed., Ancient continental deposits: Benchmark Papers in Geology: Stroudsburg, Pa., Dowden, Hutchinson and Ross, 384 p.

Fisher, W. L., and L. F. Brown, 1972, Clastic depositional systems—a genetic approach to facies analysis: Bur. of Econ. Geol., Univ. of Texas at Austin, 211 p.

——— and J. H. McGowen, 1967, Depositional systems in the Wilcox Group of Texas and their relationship to oil and gas: Gulf Coast Assoc. Geol. Soc. Trans., v. 17, p. 105-125.

Fisk, H. N., 1944, Geological investigation of the alluvial valley of the lower Mississippi River: U.S. Army Corps of Engr., Mississippi River Commission, Vicksburg, Miss.

——— 1952, Geological investigation of the Atchafalaya Basin and problems of Mississippi River diversion: U.S. Army Corps of Engineers, Mississippi River Commission, Vicksburg, Miss., p. 1-145.

——— 1955, Sand facies of Recent Mississippi delta deposits: 4th World Petroleum Cong. Proc., Rome, sec. 1, p. 377-398.

——— 1956, Nearshore sediments of the continental shelf off Louisiana: Proc. 8th

Texas Conf. on Soil Mech. and Foundation Eng., p. 1-23.

———— 1961, Bar-finger sands of the Mississippi delta: in J. A. Peterson and J. C. Osmond, eds., Geometry of sandstone bodies: AAPG Spec. Pub., p. 29-52.

———— and E. McFarlan, Jr., 1955, Late Quaternary deltaic deposits of the Mississippi River—local sedimentation and basin tectonics: in A. Poldervaart, ed., Crust of the earth, a symposium: Geol. Soc. America Spec. Paper 62, p. 279-302.

———— et al, 1954, Sedimentary framework of the modern Mississippi delta: Jour. Sed. Petrology, v. 24, p. 76-99.

Frazier, D. E., and A. Osanik, 1961, Point-bar deposits, Old River Locksite, Louisiana: Trans. Gulf Coast Assoc. Geol. Soc., v. 11, p. 121-137.

Garrison, L. E., 1974, The instability of surface sediments on parts of the Mississippi delta front: U.S. Geol. Survey Open File Rept., Corpus Christi, Texas, 18 p.

Harms, J. C., D. B. MacKenzie, and D. G. McCubbin, 1963, Stratification in modern sands of the Red River, Louisiana: Jour. Geology, v. 71, p. 556-580.

Hedberg, H. D., 1974, Relation of methane generation to undercompacted shales, shale diapirs, and mud volcanoes: AAPG Bull., v. 58, no. 4, p. 661-673.

Hobday, D. K., and David Mathews, 1975, Late Paleozoic fluviatile and deltaic deposits in the northeast Karroo Basin, South Africa: in M. L. Broussard, ed., Deltas, models for exploration, 2nd ed:, Houston Geol. Soc., p. 457-469.

Horne, J. C., et al, 1978, Depositional models in coal exploration and mine planning in Appalachian region: AAPG Bull., v. 62, no. 12, p. 2379-2411.

Houbolt, J. J. H. C., 1968, Recent sediments in the southern bight of the North Sea: Geol. en Mijnbouw, v. 47, p. 245-273.

Humphreys, Matthew, and G. M. Friedman, 1975, Upper Devonian Catskill deltaic complex in north-central Pennsylvania: in M. L. Broussard, ed., Deltas, models for exploration, 2nd ed: Houston Geol. Soc., p. 369-379.

Jackson, R. G., 1975, A depositional model of point bars in the lower Wabash River meander belt: PhD thesis, Illinois Univ., Urbana, 269 p.

Johnston, W. A., 1922, The character of the stratification of the sediments in the recent delta of the Fraser River delta, British Columbia, Canada: Jour. Geology, v. 30, p. 115-129.

Jopling, A. V., 1963, Hydraulic studies on the origin of bedding: Sedimentology, v. 2, p. 115-121.

Keller, G. H., and A. F. Richards, 1967, Sediments of the Malacca Straits, southeast Asia: Jour. Sed. Petrology, v. 37, p. 102-127.

Klein, G. de V., 1970, Depositional and dispersal dynamics of intertidal sand bars: Jour. Sed. Petrology, v. 40, p. 1095-1127.

Kolb, C. R., and J. R. Van Lopik, 1966, Depositional environments of the Mississippi River deltaic plain, southeastern Louisi-

ana: in M. L. Shirley and J. A. Ragsdale, eds., Deltas: Houston Geol. Soc., p. 17-62.

Krigstrom, A., 1962, Geomorphological studies of sandur plains and their braided rivers in Iceland: Geog. Ann., no. 44, p. 328-346.

Kruit, C., 1955, Sediments of the Rhone delta, 1, Grain size and microfauna: Kon. Nederlands Geol. Mijnb. Gen. Verhand., v. 15, p. 397-499.

LeBlanc, R. J., 1972, Geometry of sandstone reservoir bodies: in T. D. Cook, ed., Underground waste management and environmental implications: AAPG Mem. 18, p. 133-190.

Leopold, L. B., and M. G. Wolman, 1957, River channel patterns: braided, meandering, and straight: U.S. Geol. Survey Prof. Paper 282-B, p. 39-85.

———— 1960, River meanders: Geol. Soc. America Bull., v. 71, p. 769-794.

Lewin, J., 1976, Initiation of bedforms and meanders in coarse-grained sediment: Geol. Soc. America Bull., v. 87, p. 281-285.

Ludwick, J. C., 1970, Sand waves and tidal channels in the entrance to Chesapeake Bay: Old Dominion Univ., Inst. Oceanography, Tech. Rept. 1, 79 p.

Maldonado, Andres, 1975, Sedimentation, stratigraphy, and development of the Ebro Delta, Spain: in M. L. Broussard, ed., Deltas, models for exploration, 2nd ed: Houston Geol. Soc., p. 311-338.

McEwen, M. C., 1963, Sedimentary facies of the Trinity River delta: PhD dissertation, Rice Univ., Houston, Texas, 113 p.

McGowen, J. H., and L. E. Garner, 1970, Physiographic features and stratification types of coarse-grained point bars: modern and ancient examples: Sedimentology, v. 14, p. 77-111.

Meckel, L. D., 1975, Holocene sand bodies in the Colorado Delta, Salton Sea, Imperial County, California: in M. L. Broussard, ed., Deltas, models for exploration, 2nd ed: Houston Geol. Soc., p. 239-265.

Mikhailov, V. N., 1966, Hydrology and formation of river mouth bars: in Scientific problems of the humid tropic zone deltas and their implications: UNESCO, Proc. Dacca Symp., p. 59-64.

Molnia, B. F., P. R. Carlson, and T. R. Bruns, 1977, Large submarine slide in Kayak Trough, Gulf of Alaska: in Landslides: Geol. Soc. America Rev. Eng. Geol., v. 3, p. 137-148.

Moore, D. G., 1961, Submarine slides: Jour. Sed. Petrology, v. 31, p. 343-357.

Moore, G. T., 1970, Role of salt wedge in bar-finger sand and delta development: AAPG Bull., v. 54, p. 326-333.

Morgan, J. P., 1967, Ephemeral estuaries of the deltaic environment: in G. H. Lauff, ed., Estuaries: Am. Assoc. Adv. Sci. monograph, p. 115-120.

———— ed., 1970, Deltaic sedimentation, modern and ancient: SEPM Spec. Pub. 15, 312 p.

———— J. M. Coleman, and S. M. Gagliano, 1968, Mudlumps: diapiric structures in Mississippi Delta sediments: AAPG Mem. no. 8, p. 145-161.

Muller, G., 1971, Sediments of Lake Constance: Guidebook, 8th Inter. Sedimentology Cong., p. 237-252.

NEDECO, 1959, River studies and recommendations on improvement of Niger and Benue: Amsterdam, North Holland Pub. Co., 1000 p.

Ocamb, R. D., 1961, Growth faults of south Louisiana: Gulf Coast Assoc. Geol. Soc. Trans., v. 2, p. 139-175.

Off, Theodore, 1963, Rhythmic linear sand bodies caused by tidal currents: AAPG Bull., v. 47, p. 324-341.

Oomkens, E., 1967, Depositional sequences and sand distribution in a deltaic complex: a sedimentological investigation of the post-glacial Rhone delta complex: Geologie en Mijnbouw, v. 46, no. 7, p. 265-278.

———— 1974, Lithofacies relations in the Late Quaternary Niger delta complex: Sedimentology, v. 21, p. 195-221.

Picard, M. D., and L. R. High, Jr., 1972, Criteria for recognizing lacustrine rocks: in J. K. Rigby, and W. K. Hamblin, eds., Recognition of ancient sedimentary environments: SEPM Spec. Pub. 16, p. 108-145.

Prior, D. B., amd J. M. Coleman, 1978a, Disintegrating retrogressive landslides on very-low-angle subaqueous slopes, Mississippi Delta: Marine Geotechnology, v. 3, no. 1, p. 37-60.

———— 1978b, Submarine landslides on the Mississippi River delta-front slope: Geoscience and Man, v. XIX, p. 41-53: School of Geosciences, Louisiana State Univ., Baton Rouge.

———— ———— 1979, Submarine landslides — geometry and nomenclature: Zeitschrift fur Geomorphologie, v. 23, no. 4, p. 415-426.

Reeves, C. C., Jr., 1968, Introduction to paleolimnology: in Developments in sedimentology, v. II: Amsterdam, Elsevier, 228 p.

Reineck, H. E., and I. B. Singh, 1967, Primary sedimentary structures in the Recent sediments of the Jade, North Sea: Marine Geol., v. 5, p. 227-235.

———— 1973, Depositional sedimentary environments: New York, Springer-Verlag, 439 p.

Rider, M. H., 1978, Growth faults in Carboniferous of Western Ireland: AAPG Bull., v. 62, no. 11, p. 2191-2213.

Roberts, H. H., D. Cratsley, T. Whelan, III, 1976, Stability of Mississippi delta sediments as evaluated by analysis of structural features of sediment borings: Eighth Ann. Offshore Tech. Conf., Houston, Texas, May 3-5, 1976, p. 9-28.

Russell, R. J., 1936, Physiography of the lower Mississippi River delta: in Reports on the geology of Plaquemines and St. Bernard Parishes: Louisiana Dept. Cons. Geol. Bull. 8, p. 3-193.

Rust, B. R., 1972, Structure and process in a braided river: Sedimentology, v. 18, p. 221-245.

Schumm, S. A., 1972, River morphology: Dowden, Hutchinson and Ross, Stroudsburg, Pa., 448 p.

Scruton, P. C., 1955, Sediments of the eastern Mississippi Delta: *in* Finding ancient shorelines: SEPM Spec. Pub. 3, p. 21-51.

———— 1956, Oceanography of Mississippi Delta sedimentary environments: AAPG Bull., v. 40, p. 2864-2952.

———— 1960, Delta building and the deltaic sequence: *in* F. P. Shepard et al., eds., Recent sediments, northwest Gulf of Mexico: AAPG Spec. Pub., p. 82-102.

Shelton, J. W., and R. L. Noble, 1974, Depositional features of braided-meandering stream: AAPG Bull., v. 58, no. 4, p. 742-752.

Shepard, F. P., 1955, Delta front valleys bordering the Mississippi distributaries: Geol. Soc. America Bull., v. 66, p. 1489-1498.

———— 1956, Marginal sediments of the Mississippi delta: AAPG Bull., v. 40, p. 2537-2623.

———— 1973, Sea floor off Magdalena delta and Santa Marta area, Colombia: Geol. Soc. America Bull., v. 84, p. 1955-1972.

———— F. B. Phleger, and T. H. van Andel, eds., 1960, Recent sediments, northwest Gulf of Mexico: AAPG Spec. Pub., 394 p.

Shirley, M. L., ed., 1966, Deltas: Houston Geol. Soc., 251 p.

Smith, N. D., 1970, The braided stream depositional environment: Comparison of the Platte River with some Silurian clastic rocks, north-central Appalachians: Geol. Soc. America Bull., v. 81, p. 2993-3014.

Steinmetz, R., 1967, Depositional history, primary sedimentary structures, cross bed dips, and grain size of an Arkansas River point bar at Wekiwa, Oklahoma: Rept. F67-G-3, Amoco Production Co.

Sundborg, A., 1956, The River Klaralven: a study of fluvial processes: Geog. Ann., v. 38, p. 125-136.

Thompson, R. W., 1968, Tidal flat sedimentation on the Colorado River delta, northwestern Gulf of California: Geol. Soc. America Mem. 107, 133 p.

Twenhofel, W. H., 1950, Principles of sedimentation: New York, McGraw-Hill, 673 p.

van Houton, F. B., 1964, Cyclic lacustrine sedimentation, Upper Triassic Lockatong Formation: *in* D. F. Merriam, ed., Symposium on cyclic sedimentation: Kansas Geol. Bull. 169, p. 497-531.

Visher, G. S., 1965, Fluvial processes interpreted from ancient and recent fluvial deposits: *in* G. V. Middleton, ed., Primary sedimentary structures and their hydrodynamic interpretation: SEPM Spec. Pub. no. 12, p. 133-148.

———— 1972, Physical characteristics of fluvial deposits: *in* J. K. Rigby and W. K. Hamblin, eds., Recognition of ancient sedimentary environments: SEPM Spec. Pub. 16, p. 84-97.

———— S. B. Ekebafe, and James Rennison, 1975, The Coffeyville Formation (Pennsylvanian) of northeastern Oklahoma, a model for an epeiric sea delta: *in* M. L. Broussard, ed., Deltas, models for exploration, 2nd ed: Houston Geol. Soc., p. 381-397.

Welder, F. A., 1959, Processes of deltaic sedimentation in the lower Mississippi River: Tech. Rept. 12, Coastal Studies Inst., Louisiana State Univ., Baton Rouge, 90 p.

Whelan, Thomas, III, et al, 1975, The geochemistry of Recent Mississippi River delta sediments: gas concentration and sediment stability: Preprints 7th Ann. Offshore Tech. Conf., Houston, Texas, May 5-8, 1975, p. 71-84.

Williams, P. F., and B. R. Rust, 1969, The sedimentology of a braided river: Jour. Sed., Petrology, v. 39, p. 649-679.

Woodbury, H. O., et al, 1973, Pliocene and Pleistocene depocenters, outer continental shelf, Louisiana and Texas: AAPG Bull., v. 57, p. 2428-2439.

———— J. H. Spotts, and W. H. Akers, 1977, Movement of sediment on Gulf of Mexico continental slope and upper continental shelf: 9th Offshore Tech. Conf., Proc., v. 1, p. 59-68.

Wright, L. D., and J. M. Coleman, 1971, Effluent expansion and interfacial mixing in the presence of a salt wedge, Mississippi River delta: Jour. Geophys. Research, v. 76, no. 36, p. 8649-8661.

———— 1974, Mississippi River mouth processes: effluent dynamics and morphologic development: Jour. Geology, v. 82, p. 751-778.

———— and B. G. Thom, 1975, Sediment transport and deposition in a macrotidal river channel: Ord River, Western Australia: Estuarine Research, v. II: New York, Academic Press, p. 309-321.

Wright, M. D., 1959, The formation of cross-bedding by a meandering or braided stream: Jour. Sed. Petrology, v. 29, p. 610-615.

Estuarine Deposits

H. Edward Clifton
U. S. Geological Survey
Menlo Park, California

INTRODUCTION

An estuary is a semi-enclosed marginal-marine body of water in which salinity is measurably diluted by fluvial discharge (Fairbridge, 1968). Sediments deposited in this setting are influenced by a complex combination of tides and tidal currents, oceanic waves, locally generated waves, river discharge, precipitation, temperature, and local flora and fauna. These factors differ markedly among the world's estuaries, and accordingly the sedimentary facies produced vary widely. Depositional facies have been examined in relatively few estuaries, mostly in temperate climates (Howard and Frey, 1975; Klein, 1977; Lauff, 1967). The extent to which results of these studies can be generalized is uncertain. Nonetheless, features found are very likely recognizable in many ancient estuary deposits.

This paper is based primarily on analysis of depositional facies in Willapa Bay, a mesotidal, temperate-climate estuary on the southwestern coast of Washington. This bay location hosted repeated estuary development during the Pleistocene epoch, and is enclosed on three sides by Pleistocene terrace deposits composed of ancient estuarine sediment. The ancient deposits provide not only a basis for comparing ancient and modern depositional facies, but also a model of internal stratigraphy in a substantial accumulation of estuary deposits.

The specific character of estuarine facies depends on a combination of environmental setting, processes operating therein, and sediment texture. The following section considers the effect of the morphology of the setting and the character of processes. A subsequent section deals with textural distribution, controlled to a degree by availa-ble sources which may be unrelated to process or location within a bay.

MORPHOLOGY AND PROCESSES

Tidal effects tend to dominate the pattern of sedimentation within an estuary. Commonly, the range of the tide is amplified within an embayment (Komar, 1976). Accordingly, tidal influence is likely to be significant in bays on even a microtidal coast (tidal range less than 2 m, Davies, 1972).

The tides shape the interior of most estuaries into a series of tidal flats and channels (Figs. 1, 2). Two types of channels exist: tidal channels, which extend well below the position of the lowest tides; and runoff channels, which are perched atop the flats and graded approximately to the lowest low tide level. Both types focus the ebb and flood of the tide and are thereby dominated by tidal currents.

Tidal channels within an estuary may differ greatly in cross sectional profile and complexity. Near the inlet, where sediment is generally sand, bars and tidal ridges are common (Ludwick, 1974) and channels may form a complicated and frequently shifting network. Within the estuary, tidal channels may contain either longitudinal (Wright et al, 1975) or oblique bars (Phillips, 1979). In the upper reaches of the estuary where sediment is "muddier," channels become markedly asymmetric in cross-section, and longitudinal bars are replaced by simple point bars similar to those formed in a meandering river.

The geometry of runoff channels also depends on sediment texture. On sandy intertidal flats, channels are fairly straight, have relatively few tributaries, and are more or less symmetric in cross section (Fig. 1), whereas on muddy flats, channels are dendritic, more sinuous, and have well-developed point bars (Fig. 2).

On the tidal flats, tidal currents are relatively weak, and wave effects predominate. Tidal flats include two environments: intertidal flats, which are inundated by astronomical tides; and supratidal flats, which are inundated only by a combination of astronomical and meteorological (storm surge or river flood) tides. Wave effects are somewhat less important on supratidal flats than on intertidal flats.

TEXTURE

Estuarine sediment typically consists of well-sorted fine sand and mud, two very different types of material. The sand may be introduced mostly from the ocean, while the mud is contributed primarily by river discharge. Commonly, the mud and well-sorted sand are interlayered in sharply contrasting strata, although intense bioturbation may mix the components into a muddy sand or sandy mud.

In addition to source locations, a number of other factors contribute to the tendency for sediment to become progressively finer in an "up-estuary" direction. The winnowing effects of waves and tidal currents typically diminish toward the upper reaches of an estuary. Moreover, the finest material may be selectively transported landward by the tides (Van Straaten and Kuenen, 1957). In any event, sediment near the inlet tends to consist pre-

Many people have assisted in this analysis of modern and ancient estuary deposits at Willapa Bay, Washington, including R. L. Phillips, R. J. Anima, J. E. Scheihing, J. C. Posada, J. W. Hedenquist, Perry Howard, Linda Fano, Ross Miller, Richard Spicer, and James Crouch.

Bruce Richmond assisted in the field and provided much help in preparing the final copy of this manuscript.

Fig. 1—Oblique aerial photograph of a sandy intertidal runoff channel, Willapa Bay, Washington. Flats in the foreground are grass-covered; those bordering the main channel in the background are covered with fields of sand waves and dunes.

Fig. 2—Oblique aerial photograph of muddy intertidal flats crossed by runoff channels, Willapa Bay, Washington. The channels show the dendritic drainage pattern typical of muddy intertidal flats; Palix River in foreground.

dominantly of sand, whereas that in upper parts of an estuary is mostly mud.

Topographic position imparts a second textural trend to estuary sediments. The bottoms of channels typically contain the coarsest clasts available, including logs and smaller wood fragments, shells, mud clasts, and pebbles. Likewise, sediment within tidal channels tends to be somewhat coarser than that on adjacent intertidal flats. The supratidal flats, in contrast, consist predominantly of clay and silt deposited from suspension during episodes of unusually high water.

SEDIMENTARY STRUCTURES

The widely variable combination of processes and textures within an estuary produces a broad range of sedimentary structures. Structures are produced by both physical processes and biological activity. The following discussion describes some of the more common structures likely to occur in ancient estuarine deposits.

Bedforms are nearly ubiquitous wherever sand occurs. In the tidal channels, dunes and straight-crested sand waves ranging from 4 to 400 cm in height cover large parts of the floor (Fig. 3). Current ripples form almost everywhere the channel bed consists

of sand. Depending on their size and location, bedforms may reverse their direction of migration under alternating ebb and flood flow. Intensities of ebb and flood currents relative to each other differ within many tidal channels (Ludwick, 1975). If one of these strongly predominates, it will direct movement of the larger bedforms. The shape of the bedform may be modified by the weaker counterflow, but the bedform as a whole moves only in the direction of dominant flow (Fig. 4). In contrast, smaller bedforms such as ripples respond more abruptly to changing flow directions and readily reverse their movement. The variability induced by spring and neap tides also influences bedform migration. Some structures migrate only under stronger spring tidal currents and are inactive otherwise.

Because of the abundance of bedforms, cross-bedding and ripple lamination are common in tidal channel sands. Both tabular and trough sets occur. The cross-bedding may locally show "herringbone" reversals thought to typify tidal systems, but more commonly, cross-bedding in one specific part of the deposit is unidirectional. A reverse bimodal pattern is more likely to show up in rose diagrams that summarize many parts of the deposit. Ripple lamination, in contrast, is more likely to show a herringbone pattern in response to changing tidal flows.

Reversal of tidal flow commonly induces a characteristic feature within tidal cross-beds. In an area where either the ebb or flood current predominates, countercurrent commonly modifies the slip face, eroding some previously deposited foresets and depositing this sand in a small sand wave

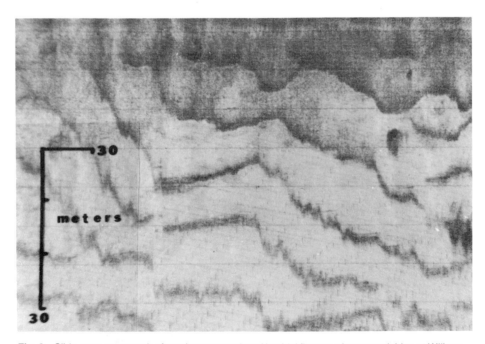

Fig. 3—Slide-scan sonograph of sand waves produced by tidal flow on a large sand ridge at Willapa Bay, Washington. The current direction from bottom to top as indicated by the dark lines on lower half of sonograph (reflections from slip faces) and light lines on upper half (reflection shadows over slip faces). Courtesy of R. L. Phillips.

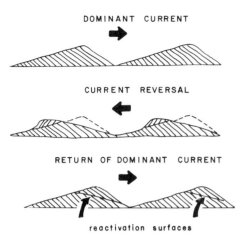

DOMINANT CURRENT

CURRENT REVERSAL

RETURN OF DOMINANT CURRENT

reactivation surfaces

Fig. 4—Schematic diagram showing formation of reactivation surfaces produced during a current reversal, a common feature of the tidal estuarine system (modified after Klein, 1970).

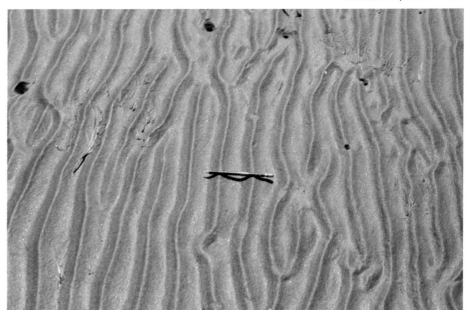

Fig. 5—Symmetrical wave-formed ripples on sandy intertidal flats.

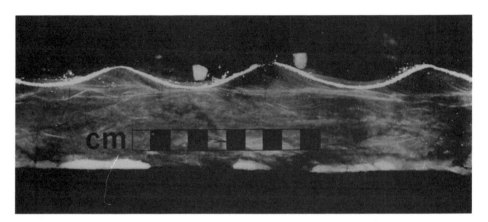

cm

Fig. 6—X-ray radiograph through symmetrical wave-formed ripples of a sandy intertidal flat, similar to that shown in Figure 5. Ripple cross-lamination is directed in a shoreward direction (i.e., to the right). Note the bioturbated sand beneath the ripple lamination.

Fig. 7—Thin, concave-up mud lenses in a predominantly sandy deposit from Pleistocene terrace exposures at Willapa Bay, Washington.

on the large structure's crest (Fig. 4). The subsequent flow of predominant current builds foresets over the erosional surface, preserving it in the set as a reactivation surface (Klein, 1970). Reactivation surfaces are not restricted to tidal deposits—they can form in fluvial or open marine crossbeds (McCabe and Jones, 1977). Where they recur at regular intervals within the set and are the site of mud accumulation, they strongly suggest tidal conditions.

Sandy intertidal flats are typically covered by long-crested symmetrical ripples formed by locally generated waves (Fig. 5). These ripples, up to one centimeter in height and spaced 5 to 10 cm apart, tend to parallel topographic contours because of wave refraction. Current ripples on intertidal areas tend to be restricted to runoff channels. In mesotidal estuaries (tidal range 2 to 4 m) such as Willapa Bay sand waves and dunes occur only on the lowermost intertidal flats, adjacent to larger tidal channels. But in macrotidal bays (tidal range greater than 4 m), such as the Bay of Fundy, these structures are common over much of the intertidal areas (Dalrymple et al, 1978).

Although the typical wave-formed ripples on intertidal flats are symmetrical, they tend to move sporadically in a shoreward direction. The internal structure of symmetrical ripples is shoreward-directed ripple lamination (Fig. 6). This structure is not likely to be preserved inasmuch as it typically occurs in a thin layer above thoroughly bioturbated sand. Even cross-bedding produced by intertidal sandwaves and dunes may not survive faunal mixing. *Callinassia* burrows penetrate the slip faces of large active dunes, and troughs of the dunes are underlain by structure-

Fig. 8—Mud and sand in approximately equal concentrations. The mud forms anastomosing layers around sand ripples.

Fig. 9—"Starved" ripples and sand beds in a mud-rich deposit.

Fig. 10—Mud foresets, in the center of the photograph, are fairly regularly spaced in a cross-bedded sand.

less sand.

Because tidal flow is so varied, the sedimentary bed is mobile only part of the time. Accordingly, silt and clay can accumulate over bedforms or in their troughs during those periods when the bed is immobile (during slack water or if the current is sufficiently weak). This is most likely to occur after floods or storms when bay water is highly charged with suspended fine sediment. Mud can also accumulate in megaripple troughs as fecal pellets of filter-feeding invertebrates (Pryor, 1975). These pellets behave as individual particles of low density and collect in depressions where currents are less intense.

Layers of silt and clay in otherwise clean, well-sorted sand constitute one of the most characteristic features of tidally deposited sand. Referred to as flaser bedding (Reineck and Wunderlich, 1968), the actual character depends on the ratio between sand and mud. Where sand greatly predominates, mud typically occurs in thin concave-up lenses of a few centimeters to more than a meter in width (Fig. 7). Where mud and sand are more or less equal in abundance, mud takes the form of anastomosing layers around sand ripples (Fig. 8). Where sand is rare, it occurs as isolated "starved" ripples (Fig. 9) or in discrete planar laminae. Mud also commonly overlies reactivation surfaces, where it forms regularly spaced foresets within a cross-bedded unit (Fig. 10).

Where the floor of the estuary is muddy, it tends to be flat unless disrupted by faunal activity. The internal structure in this case consists of more or less planar alternating laminations of silt, clay, carbonaceous detritus, or very fine sand.

The nature of lamination in estuarine mud can change within a short distance. Within tidal channels (Fig. 11), subtidal accretionary banks are underlain by thinly laminated mud, somewhat disrupted by faunal activity (Fig. 12). On intertidal banks of the same channels the laminae consists of broader (0.5 to 2 cm) alternations of fine sand, silt and clay (Fig. 13). The sandier layers are storm-winnowed lag deposits. On the accretionary banks of runoff channels, the mud is very finely and irregularly laminated and contains numerous paper-thin layers of carbonaceous detritus (Fig. 14). The finest stratification will record individual ebb and flood currents (Reineck, 1967). Supratidal flats are underlain by almost rhythmically alternating layers of silt (or fine sand) and clay a few millimeters thick (Fig. 15) deposited under storm or flood conditions.

Deformational (slump) structures are fairly common in estuarine deposits, particularly in muddier sediment. Structures range in size from small slumps several centimeters thick on the banks of runoff channels (Fig. 16) to large structures tens of centimeters or more thick and many meters in length (Fig. 17). The typical form consists of a glide plane underlying a section of sediment with laminations rotated to an inclination opposite the initial dip (Figs. 16, 17). In some places the deformation is more intense, producing a mass of highly folded and faulted sediment.

Estuaries typically contain a highly prolific fauna that leaves a marked imprint on the sediment. The fauna disrupt the substrate by burrowing through it or digging into its surface. By feeding on the sediment, their feces, as mentioned above, can produce discrete layers of mud. Large animals such as crabs and fish modify the fabric of the coarser material by overturning convex-up disarticulated pelecypod

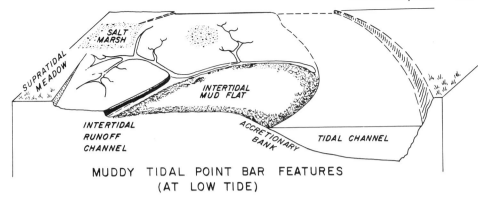

MUDDY TIDAL POINT BAR FEATURES
(AT LOW TIDE)

Fig. 11—Typical physiographic features formed on muddy substrate in the upper estuary (at low tide).

valves in searching for food (Emery, 1968; Clifton, 1971).

The bioturbation rate depends on a number of factors including substrate texture, salinity, and degree of subaerial exposure. Generally, fauna mix sandy sediment more rapidly than muddy sediment in the same setting. The sand tends to house "sand-swimming" forms such as errant polychaetes that produce intrastratal trails, whereas mud tends to contain fauna that reside primarily in discrete, open burrows. The rate of bioturbation de-

creases in the upper, less saline reaches of an estuary and tends to be quite low in areas rarely inundated, such as supratidal flats.

The degree of bioturbation in sediment depends on the rate of faunal mixing relative to that of physical reworking or accumulation. In topographically flat areas such as muddy channel floors or on intertidal flats where sedimentation rates are low, sediment tends to become thoroughly mixed. On accretionary banks (both intertidal and subtidal), sediment accu-

Fig. 12—X-ray radiograph of thinly laminated, predominantly muddy (dark) sediment on a subtidal accretionary bank, from the Palix River, Willapa Bay, Washington.

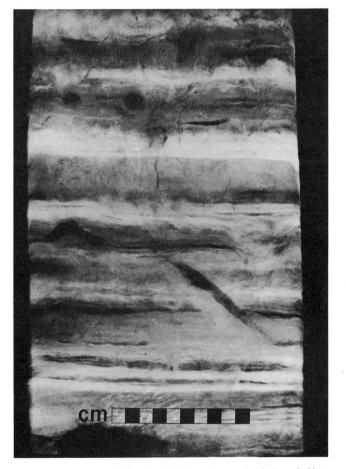

Fig. 13—X-ray radiograph of an intertidal accretionary bank deposit. Note the increase in sand (light material) and thickness of laminations relative to the subtidal deposit (Fig. 12).

Fig. 14—X-ray radiograph showing the very fine carbonaceous layers, and the general irregular nature of stratification on the accretionary bank of the intertidal runoff channel shown in the center of Figure 2.

Fig. 15—X-ray radiograph of supratidal flat deposits which exhibit alternating very fine layers of silt, sand, and clay, and root or rhizome structures.

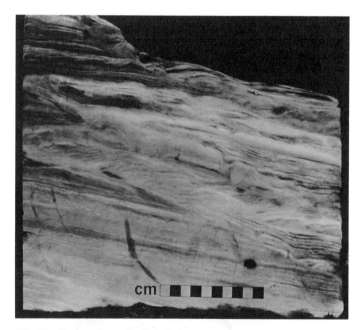

Fig. 16—X-ray radiograph of small slump structures on the accretionary bank of a small intertidal runoff channel.

Fig. 17—Large-scale slumps exposed in Pleistocene terrace deposits of estuarine origin, Needle Point, Willapa Bay, Washington.

mulates more rapidly and physical structures more likely are preserved. Stratification also likely persists within active sandwave or dune fields.

Although a number of different trace fossils exist within estuarine sediment, none have been recognized as diagnostic of this environment. Moreover, trace fossils are not yet useful for distinguishing between specific estuarine environments such as intertidal and subtidal settings.

In contrast, fossil root and rhizome structures are most helpful for identifying subaerially exposed (or nearly so) deposits. These features may be difficult to recognize. Not only do they resemble burrow structures, but it is often difficult to distinguish between roots or rhizomes and carbonaceous detritus. Additionally, the roots of modern plants may extend into poorly lithified ancient deposits and be taken for fossil forms.

Fortunately, subsurface structures of many estuarine angiosperms are fairly distinctive (Fig. 18). Otherwise, the best evidence for fossil roots or rhizomes is branching carbonaceous cylinders or threads that cut across bedding planes and are truncated by internal erosional surfaces or, in poorly lithified deposits, lie in a position where modern roots are unlikely to occur. Such features are particularly convincing where associated with supratidal stratification (Fig. 15) or with peaty beds.

VERTICAL SEQUENCES

Tidal channels in an estuary tend to wander back and forth through time. As they do, a vertical sequence is produced that, if complete, is capped by supratidal muds. Identification of vertical sequences is a key part of recognizing ancient estuary deposits. In this section three types of sequences are described—that produced by a sandy tidal channel, a muddy tidal channel, and one of mixed sand and mud.

The base of a sequence produced by a migrating sandy tidal channel (Fig. 19) typically consists of a lag deposit composed mostly of shells. The lag may or may not be bioturbated. It is typically overlain by either cross-bedded or ripple-bedded sand, depending on the velocity of the tidal currents. If cross-bedded, the thicknesses of the sets of cross-beds are likely to decrease upsection. Thickness of the cross-bedded part of the deposit depends on depth of the channel. The cross-bedded sand is likely overlain by

structureless or burrowed sand of the intertidal flat. This sand grades upsection into mud that in a complete sequence may show supratidal stratification or root or rhizome structures.

A migrating muddy channel produces a sequence (Fig. 20) in some respects similar to that of a sandy tidal channel. Both are floored by a lag deposit and culminate in supratidal facies. Muddy tidal channels tend to be more prevalent in upper reaches of an estuary, and a basal lag deposit is likely to contain more wood and pebbles than that of a sandy tidal channel. The bioturbated lag underlies laminated mud. If exposure is adequate, the progradational nature of the laminations may be evident. Laminated mud is either overlain by cross-bedded mud, produced by migrating runoff channels (Fig. 21), or it grades up into broadly stratified or bioturbated mud (muddy intertidal flats). This in turn grades up into supratidal facies.

Between the clearly sandy and clearly muddy parts of an estuary, a mixture of sand and mud floors the tidal channels. In this area, the part of the section formed by the accretionary bank consists of much interbedded sand and mud and flaser bedding, some laminated mud, and some cross-bedded sand. The part of the sequence deposited on intertidal flats may be bioturbated sand or mud. Otherwise it resembles the other sequences.

As the foregoing discussion implies, a trend exists within an estuary where

in the vertical sequence becomes increasingly muddy in an up-estuary direction. The lower reaches, dominated by strong tidal currents, winnowing by waves and an oceanic sand supply, produce sandy vertical sequences. The sequence changes systematically in an up-estuary direction through the mixed sand-mud to the muddy sequence. The lateral extent of the sequences depends on the size and physical character of the estuary. If the bay is small or tidal circulation is restricted, the muddy sequence is likely to predominate except near the estuary inlet.

ESTUARY FILL COMPLEXES

Estuarine deposits rarely consist of a single fill formed at one stand of the sea. Instead, they tend to occur in combination in an area where an estuary persisted off and on for a significant period of time (such as several hundred thousand years). During this time, relative changes of sea level induced by tectonism or eustatic events are likely. Each of these changes produces attendant changes in the environmental setting, by increasing or decreasing the size of the estuary and its tidal prism.

Estuaries are sites of rapid deposition and aggrade relatively quickly to a horizontal surface approximating the adjacent sea level. The resulting deposit is considered a stillstand unit, a body of sediment produced at a specific position of sea level. A stillstand unit includes all sediment deposited in a particular estuary, from the base of

Fig. 18—Distinctive root structure of the upper intertidal and supratidal angiosperm *Triglochin maritima* exposed in Pleistocene terrace deposits at Willapa Bay, Washington. *Triglochin* is abundant in the upper reaches of the modern estuary.

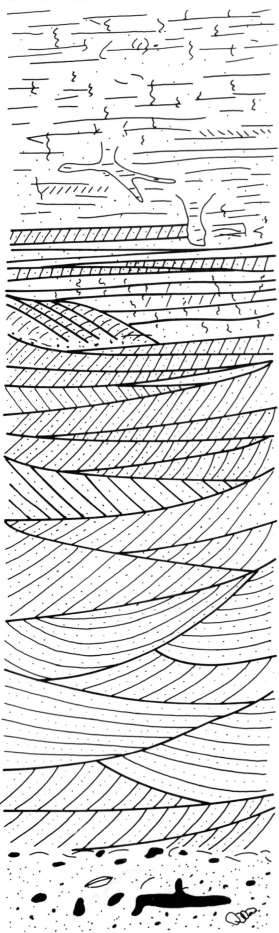

Supratidal and upper intertidal deposits. Generally interlaminated fine sand, silt, and clay often associated with rhizome structures.

Intertidal sand flat deposit. Intensely bioturbated with little physical structure preserved. May contain intertidal runoff channel deposits which consist of cross- or ripple-bedded sand.

Channel bank deposits. Composed predominantly of sand-sized material. The thickness of foreset units usually decreases upsection. Small to medium scale cross-stratification may reflect either floor or ebb currents, depending on location within the channel. Bioturbation can be intense locally.

Channel bottom deposit. Generally contains abundant shell debris often associated with gravel, sand, wood fragments and eroded mud clasts. Bioturbation is common only in the smaller, lower velocity channels.

Fig. 19—Schematic diagram of the vertical sequence produced by a migrating sandy tidal channel. For approximate scale, channel bottom deposit can be considered 1 meter thick.

the channel fills to the overlying intertidal-supratidal flats, and from the inlet to the adjacent rivers.

Delineation of stillstand units is a key to interpreting complexes of ancient estuarine deposits. Unfortunately, the bounding surfaces of stillstand units may be difficult to define. Most tidal channels are capable of eroding not only into sediment previously deposited during the same stand of the sea, but also into older subjacent stillstand units. It is rarely clear from the contact alone whether an erosional surface lies within a stillstand unit or separates superjacent units.

Mapping of stillstand units depends primarily on recognizing evidence for different sea-level positions within the deposit. Evidence of subaerial exposure, such as roots and rhizomes, mud-cracks, or recognition of vertical sequences or tidal runoff channels becomes significant in the interpretation of an ancient estuary-fill complex.

The nature of the sea-level change in part controls the geometry of the stillstand unit. In a complex where sea level has risen continuously, stillstand units are relatively tabular (Fig. 22). In a complex where sea level has fluctuated, deposits become dissected and more channelized (Fig. 23).

ECONOMIC ASPECTS

Ancient estuary-fill complexes should have excellent oil and gas source potential. Sand deposits are typically very well sorted and should have excellent reservoir characteristics. They are associated with muds rich in organic products; estuaries are among the most biologically productive sedimentary environments known (Lauff, 1967). The intercalation of supratidal and other mud with sand provides impermeable barriers necessary to the development of stratigraphic traps. A set of typical cross-sections through a filled estuary is shown in Figure 24.

RECOGNITION OF ESTUARY DEPOSITS

Estuarine deposits should typically be of limited geographic extent. Within this constraint, the best geologic evidence suggesting an estuarine environment is probably brackish-water fauna. This criterion becomes progressively less reliable in increasingly older deposits, however, where paleoecologic parameters are less certain. Moreover, many modern estuaries are occupied by fauna that also live in the open ocean. In such a case, the only clue may be the more restricted faunal assemblage in a bay relative to that of the open sea (Weyl, 1970).

Probably no diagnostic physical criterion exists for estuarine deposition. Nonetheless, several features occur that are characteristic of estuarine facies and serve to delineate it from adjacent environments seaward and landward.

An open-ocean deposit may show many of the same fauna as a bay and consist of fine well-sorted sand that, as in an estuary, contains reversing or highly variable cross-bedding (Clifton et al, 1971; Davidson-Arnott and Greenwood, 1976). However, an estuary deposit is likely to contain much more silt and clay. Mud drapes and flaser bedding, although present in some open marine deposits, are much more abundant in estuarine sediment. Channeling is also likely to be more common in estuary deposits, as is large-scale cross-bedding produced by migrating channel banks.

The fining-upward character of estuary fill resembles that of fluvial deposits (Miall, 1977). Fluvial deposits, however, rarely contain the marine or brackish fauna found in an estuary, nor do they show the intensity of variability of bioturbation and burrowing of an estuary deposit. Fluvial cross-bedding is generally unidirectional rather than bimodally opposed as in tidal sands.

In summary, estuary deposits are fairly distinctive, but may resemble other depositional facies. Evidence of tidal flow, runoff channels, the vertical sequences described herein, as well as faunal characteristics, serve to distinguish ancient bay facies from other types of sediments. Deposits are likely to form an estuary fill complex produced at a number of stages of relative sea level. Delineation of stillstand units within the complex provides a rationale for interpreting vertical sequences and lateral trends.

REFERENCES CITED

Boothroyd, J. C., 1978, Mesotidal inlets and estuaries, *in* R. A. Davis, ed., Coastal sedimentary environments: New York, Springer-Verlag, p. 287-360.

Clifton, H. E., 1971, Orientation of empty pelecypod shells and shell fragments in quiet water: Jour. Sed. Petrology, v. 41, p. 671-682.

———— R. E. Hunter, and R. L. Phillips, 1971, Depositional structures and processes in the non-barred high-energy nearshore: Jour. Sed. Petrology, v. 41, p. 651-670.

Dalrymple, R. W., R. J. Knight, and J. S.

Supratidal (meadow) deposits. Laminated silt and clay with abundant root traces.

Mud breccia formed at base of wave-cut scarps into supratidal deposits.

Salt marsh deposits characterized by rhizomes of *Triglochin* and other grasses.

Mud flat deposits. Commonly completely bioturbated, except where algal mats generate a broad alternation of silt and clay.

Intertidal creek accretionary bank deposit. Well-bedded, commonly finely laminated sand, silt, clay and carbonaceous debris. Strata generally form muddy cross-strata in units up to one meter thick. Slumped strata relatively common.

Intertidal creek channel deposit. Generally disorganized due to slumping from side banks, may contain mud clasts and concentrations of shells and wood fragments.

Accretionary bank deposit. Generally well-bedded gently inclined, alternations of silt (or fine sand) or clay. Silty layers may show intense small scale bioturbation. Deposits at similar depths tend to become progressively finer in an up-river direction.

Channel bottom deposit. Intensely bioturbated, disorganized shell lag wood fragments common. Toward river mouths, channel may contain interbedded gravel or coarse sand and mud.

Fig. 20—Vertical sequence produced by lateral migration of a muddy tidal point bar. Nature of individual facies may differ as a function of location within the estuary. For approximate scale, channel bottom deposit can be considered to be 1 meter thick.

Fig. 21—Cross-stratified muds produced by migrating muddy runoff channels in Pleistocene terrace deposits, Willapa Bay, Washington.

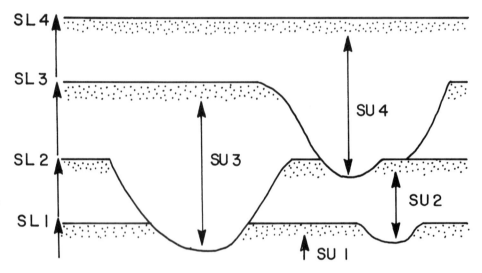

Fig. 22—Schematic diagram illustrating a hypothetical estuarine-fill complex deposited under continuously rising sea level (SL1, SL2, SL3, SL4 indicate progressive levels). The related stillstand units (SU1, SU2, SU3, SU4) are relatively tabular and extend from channels cut in the previous unit (channel fill) to the upper intertidal or supratidal (shown with x's).

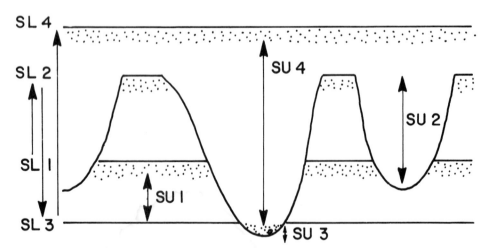

Fig. 23—Schematic diagram illustrating a hypothetical estuarine-fill complex deposited under fluctuating sea level conditions. Under conditions of sea level 3, stillstand units 1 and 2 may be highly dissected as depicted here, or depending on position within the estuary, they may be completely eroded. Stillstand unit 3 may only be preserved as localized channel fill deposits.

Lambiase, 1978, Bedforms and their hydraulic stability relationships in a tidal environment, Bay of Fundy, Canada: Nature, v. 275, p. 100-104.

Davidson-Arnott, R. G. D., and B. Greenwood, 1976, Facies relationships on a barred coast, Kouchibouguac Bay, New Brunswick, Canada, in Beach and nearshore sedimentation: SEPM Spec. Pub. 24, p. 149-168.

Davies, J. L., 1972, Geographical variation in coastal development: London, Longman, 204 p.

Emery, K. O., 1968, Positions of empty pelecypod valves on the continental shelf: Jour. Sed. Petrology, v. 41, p. 1264-1269.

Fairbridge, R. W., 1968, Estuary, in Encyclopedia of geomorphology: New York, Reinhold, p. 325-330.

Howard, J. D., and R. W. Frey, 1975, Estuaries of the Georgia coast, USA, Sedimentology and Biology II. Regional animal-sediment characteristics of Georgia estuaries: Senkenbergiana Maritima, v. 7, p. 33-103.

Klein, G. deV., 1970, Depositional and dispersal dynamics of intertidal sand bars: Jour. Sed. Petrology, v. 40, p. 1095-1127.

——— 1977, Clastic tidal facies: Champaign, Ill., CEPCO, 149 p.

Komar, P. D., 1976, Beach processes and sedimentation: Englewood Cliffs, N. J., Prentice-Hall Inc., 429 p.

Lauff, G. H., ed., 1967, Estuaries: Am. Assoc. Adv. Sci. Spec. Pub. 83, 757 p.

Ludwick, J. C., 1974, Tidal currents and zig-zag sand shoals in a wide estuary entrance: Geol. Soc. America Bull., v. 85, p. 717-726.

——— 1975, Tidal currents, sediment transport, and sand banks in Chesapeake Bay entrance, Virginia, in L. E. Cronin, ed., Estuarine Research, Vol. II: New York, Academic Press, p. 365-380.

McCabe, P. J., and C. M. Jones, 1977, Formation of reactivation surfaces within superimposed deltas and bedforms: Jour. Sed. Petrology, v. 47, p. 707-715.

Miall, A. D., ed., 1977, Fluvial Sedimentology: Canadian Soc. Petrol. Geologists, Mem. 5, Calgary.

Phillips, R. L., 1979, Bedforms and processes on an estuarine tidal current ridge, Willapa Bay, Washington: AAPG Abs. with programs, Ann. Mtg., p. 145.

Pryor, W. A., 1975, Biogenic sedimentation and alteration of argillaceous sediments in shallow marine environments: Geol. Soc. America Bull., v. 86, p. 1244-1254.

Reineck, H. E. and F. Wunderlich, 1968, Classification and origin of flaser and lenticular bedding: Sedimentology, v. 11, p. 99-104.

Van Straaten, L. M. J. U., and Ph. H. Kuenen, 1957, Accumulation of fine grained sediments in the Dutch Wadden Sea: Geol. en Mijnbouw, v. 19, p. 329-354.

Weyl, P. K., 1970, Oceanography, an introduction to the marine environment: New York, John Wiley and Sons, 535 p.

Wright, L. D., J. M. Coleman, and B. G. Thom, 1975, Sediment transport and deposition in a macrotidal river channel: Ord River, Western Australia, in L. E. Cronin, ed., Estuarine Research, Vol. II: New York, Academic Press, p. 309-321.

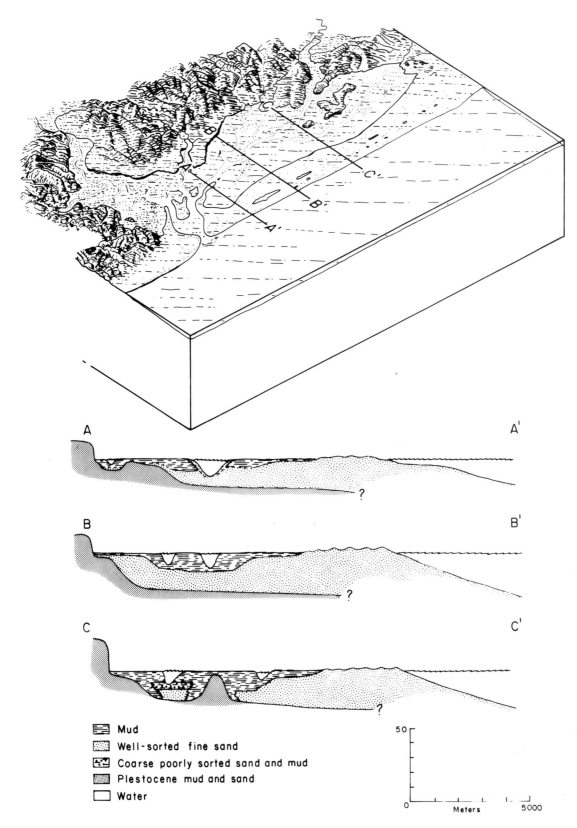

A

A'

B

B'

C

C'

▤ Mud

▨ Well-sorted fine sand

▧ Coarse poorly sorted sand and mud

▦ Plestocene mud and sand

☐ Water

Meters   5000

Fig. 24—Schematically drawn cross-section through a filled estuary.

Tidal Flats and Associated Tidal Channels

Robert J. Weimer
Colorado School of Mines
Golden, Colorado

James D. Howard
Skidaway Institute of Oceanography
Savannah, Georgia

Donald R. Lindsay
Occidental Geothermal, Inc.
Bakersfield, California

INTRODUCTION

Tidal flats occur on open coasts of low relief and relatively low energy and in protected areas of high-energy coasts associated with estuaries, lagoons, bays, and other areas lying behind barrier islands. Conditions necessary for formation of tidal flats include a measurable tidal range and the absence of strong wave action.

Extent of tidal flats on present-day coasts varies greatly and includes small, locally restricted areas of several hundred square meters or regional features extending over hundreds of square kilometers, (coast of the Netherlands, Germany, and Denmark and salt marshes along the Atlantic coast of the United States). Occasional confusion in terminology occurs because tidal flats may carry local geographic names such as "lagoon, bay, salt marsh, etc." Although the term tidal deposit has become popular in recent years, in this chapter we primarily describe tidal flats and genetically related tidal channels (dominantly subtidal) representing two specific types of tidal deposits.

Early detailed studies on tidal flats were carried out on the North Sea coast by German and Dutch geologists, and these classic studies have become models for subsequent investigations and interpretation of ancient tidal flat deposits. There is, however, considerable variation in tidal flats, depending on sediment types and availability, presence or absence of vegetation, tide range, and coastal energy and morphology.

Tidal flats are subdivided into intertidal and subtidal environments which control facies distribution. Parts of the tidal flat lying between high and low tide range, the intertidal zone, make up the major areal extent of the tidal flat. If a noticeable variation in sediment type is present throughout the flat, for example muds and sand, the intertidal area commonly possesses alternating layers of both textures.

Subtidal areas of tidal flats are important to understanding of this environment because they represent that part of the tidal flat most likely to be preserved. Most tidal flat deposition results from lateral accretion in association with progradation of the flat and point bar associated with meandering tidal channels. Hence, a major portion of the sedimentary record for most tidal flat sequences includes features associated with channel fill and tidal point bars.

Within these two principal facies, many subdivisions are possible based on sediment texture, morphology, sedimentary structures, biogenic aspects or special interests of the investigators. Furthermore, degree of subdivision increases as more information is gathered and analyzed.

Animals and plants play a significant role in tidal flat environments. They are influential in trapping sediment, creating sedimentary particles as feces or pseudofeces, and forming biogenic sedimentary structures and bioturbation as a result of the processes of feeding, dwelling, and moving. Tidal flats are excellent sedimentary environments for the preservation of trace fossils as well as physical sedimentary structures because of the alternating layers of sand and mud enhancing the expression of the structures. In addition to traces, zones of shells are found, sometimes with shells still articulated and in growth position.

Intertidal and subtidal parts of tidal flats are continuously affected by tidal currents as well as wind-induced wave currents. Current velocity can be highly variable in different settings or in the same area under different conditions. Regional and local geomorphology, height and stage of the tide, and strength and direction of local winds are factors controlling wave and current processes. For example, consider the differences expected between currents in the Bay of Fundy tidal flats, where a normal tide range is greater than 10 m, and flats on the Gulf Coast of the United States where tides are less than 0.5 m.

A wide variety of physical sedimentary structures occurs in response to wave and current activity. Especially important in many tidal flats is the fluctuation of energy and tremendous variety in strength of the transporting currents common not only in a single tidal cycle, but over monthly or even seasonal cycles, or under varying wind conditions. This range of energy in combination with a sediment supply—which includes sands and suspended mud—is expressed by a continuum of sedimentary structures. Particularly important is the possible formation of bidirectional structures in response to ebb and flood currents.

Tidal flats developed under regressive or prograding conditions are characterized by a fining upward sequence, consisting of coarse sediments at the base and progressively finer sediments toward the top in an uninterrupted, vertical sequence. This reflects decreasing energy in a progression from subtidal to intertidal parts of the tidal flat. Commonly, this occurrence is represented by: (1) a dominantly sandy subtidal zone of channel-fill, point bar and shoal sediments; (2) a mixed sand and mud intertidal flat deposit; or (3) a muddy upper intertidal flat or salt marsh deposit. Decrease in grain size may also be seen within an individual subfacies.

Despite the rich and varied suite of sedimentary structures commonly developed in the tidal flat environment, no structures appear restricted only to this facies. Hence, recognition of tidal flats requires examination of individual features and consideration of the main relationship to vertical and lateral facies.

Specific examples of structures are illustrated in subsequent sections. Megaripples and large-scale cross-bedding, and associated small-scale current ripples and flaser bedding, are common in channel-fill material. The intertidal flat displays a variety of intertidal sand and mud layers, including flaser, wavy, and lenticular bedding. The upper-intertidal flat surfaces are commonly bioturbated or contain slightly laminated muds with thin sand lamina. All of these structures, however, can be modified or completely destroyed by intense burrowing, bioturbation, or profuse tracks and trails.

Tidal flats are biologically rigorous environments because of subjection to extremes in currents, water depth, salinity, temperature, desiccation, erosion, and rapid deposition. Indigenous organisms are generally well adapted forms capable of dealing with the stresses of the environment. The number of species is therefore generally small, but the number of individuals is large. Areas of slow sedimentation such as intertidal flats may be highly bioturbated, whereas areas of rapid progradation, such as point bars in rapidly meandering tidal creeks or channels, may have only a few biogenic structures, or may contain thin layers of dense biogenic reworking. Tidal flats, as in shallow marine facies, have an inverse relationship between physical and biogenic sedimentary structures. As the influence of physical processes increases, the influence of biogenic processes decreases and vice versa.

APPLICATION TO THE GEOLOGIC RECORD

Study of modern tidal flats for purposes of interpretation of similar deposits in the stratigraphic record demands caution to assure that information derived from modern de-

Fig. 1—Intertidal flat, Jade Bay, Germany. The vast expanses of gently sloping intertidal flats overlie the less obvious but geologically significant tidal channel deposits. Photo by J. D. Howard.

posits is pertinent to interpretaton of the ancient environments. A cautious approach is obviously necessary in applying all modern analogs, but especially for tidal flats, as the facies most likely to be preserved (channel-fill deposits) are the least obvious when looking at the surface of a modern tidal flat. Much of its record is overlain by facies (intertidal flats) which are less likely to be preserved (Fig. 1). When preserved, deposits of intertidal flats will often be represented as only a thin veneer overlying the channel-fill sequence. Probably no other coastal facies is so potentially misleading in interpreting an ancient sequence.

Within the stratigraphic record, examples exist of depositional sequences of significant thickness possessing features interpreted to represent intertidal flats that lack an obvious significant channel-fill record. Explanation for the origin of such deposits is a mystery, with no modern analogs yet proposed except perhaps the fascinating tidal sequence associated with the Colorado River delta in the northwest Gulf of California. Even this, as pointed out by Thompson (1975), would, in an ancient sequence, probably be interpreted as part of an inland basin.

Our initial purpose is to examine modern examples of tidal flats and illustrate the important physical and biogenic sedimentary structures. Examples of modern tidal flats are described from different geographic and depositional settings and their characteristics are summarized. In the second part of the chapter, examples from the ancient record are presented that are interpreted to have resulted from tidal flats and tidal channel deposition. Two of the ancient sequences are petroleum productive from tidal channel sandstones.

MODERN TIDES AND TIDAL FLATS

Modern tidal flats, as they relate to their fossil counterparts, can be subdivided into three zones: (1) normally subtidal; (2) normally lower intertidal; and, (3) upper intertidal. These subdivisions are useful because of the manner tides fluctuate and the way tidal flats have been described in the literature.

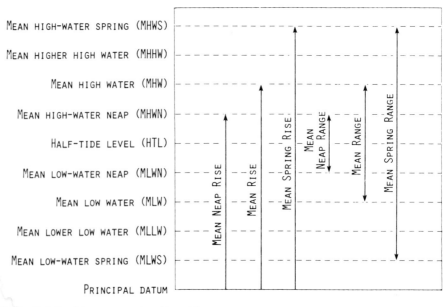

Fig. 2—Relative tide ranges and positions. This diagram shows the positions the tide may reach under different conditions and the way in which ranges vary. Winds may cause additional variations (modified from Frey and Basan, 1978).

Figure 2 illustrates the levels of tides under varying conditions. While this may be too specific to relate to in ancient sediments, it is necessary to have some understanding of these tidal ranges in order to apply the "Principle of Uniformitarianism." The normally subtidal zone (subsequently referred to as subtidal) is considered to have as its upper boundary the mean low water line; and, the normally intertidal zone (subsequently referred to as intertidal zone), the area between mean low water and mean higher high water. The upper intertidal zone is the area above mean higher high water (includes supratidal of some investigators). These subdivisions for consideration of those areas of tidal flats which are, for the majority of the time, subjected to certain types of physical process. Boundaries are gradational because tide levels change gradually from day to day throughout the lunar month. Furthermore, these levels are raised or lowered to some extent nearly every day in response to unpredictable onshore and offshore winds and, occasionally, due to heavy rainfall.

Subtidal zone deposits primarily result from lateral accretion and are affected primarily by tide-induced currents and secondarily by wave processes. The intertidal zone is a mixed zone of vertical and lateral accumulation, normally affected by wave processes with secondary effects by currents. The upper intertidal zone is only marginally affected by currents and is less often subjected to wave action. In some areas, this is a salt marsh environment where vegetation serves to dampen wave energy. The upper intertidal zone is incorrectly, but frequently, referred to as the supratidal zone. This causes confusion when trying to compare different tidal flats. Supratidal means above tidal influence, but the upper intertidal zone is within tidal range even if the tides covering this area do not occur daily. Furthermore, unusually high tides have a relatively greater impact on this environment, which normally remains unflooded for days, than the effects of unusually low tides have on the subtidal environment, because the latter floods to some extent on every tide.

Based on size, two types of tidal channels associated with modern flats are observed. The major channels supplying water and sediment to the tidal flat are called tidal channels or estuaries. These are of sufficient size and energy to produce "reservoir rock" for petroleum. Smaller distributary chan-

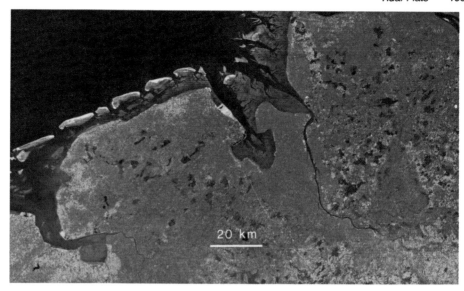

Fig. 3—ERTS false-color imagery of part of the North Sea coast of Germany. The North Sea subtidal area is blue, the mainland red, and the broad intertidal area gray. The large northwest-southeast re-entrant is Jade Bay and the adjacent river the Weser River.

nel patterns associated with tidal flat deposits are called tidal creeks. Because of their shallow depth and low energy, resulting deposits are thin, poorly sorted, silty, clayey sandstones which are "nonreservoir" for petroleum. Deposits of tidal flats, away from major channels, are also generally "nonreservoir deposits."

North Sea

The principal example of a modern tidal flat cited here is from the North Sea coast of Europe (Figs. 3, 4). These are the most thoroughly studied tidal flats in the world and can indeed be considered the "type locality" for tidal flats. Discussion of these tidal flats is derived from various papers by Graham Evans, Hans-Erich Reineck, and L. M. J. U. van Straaten, which are included in the reference list.

Subtidal Facies—Channel-fill deposits are, from a geologic standpoint, the most important tidal facies. Examination of a tidal flat surface at any instant in time usually shows tidal channels occupying less than 50% of the total surface area. However, tidal creeks and major tidal channels are dynamic features that continually shift position. Commonly, the migration rate is measurable in mm/day or cm/day and may be as much as m/day, yet channel form and dimension remains approximately the same (Fig. 5A, B). Thus, this is a meandering system of erosional occurrence on one bank and depositional occurrence at channel margins, or on point bars (Fig. 5C, D, E).

Whereas the thickness of intertidal

Fig. 4—Aerial view of North Sea tidal flat at low tide showing a typical relation of tidal creek to total intertidal flat area at one instant of time. However, the positions of tidal channels change and thus most of the intertidal flat area is underlain by tidal channel-fill deposits. Photo by J. Dorjes.

deposits is roughly equal to the tide range across the flat surface, the thickness of channel-fill deposits, except in the most landward parts of the flat, will equal the depth of the channel. In the North Sea tidal flats, this may exceed 15 m for major tidal channels. Through time the intertidal deposits build across channel-fill sequences but in comparison to the channel-fill material, they represent only a thin capping veneer (Fig. 6A). Furthermore, if the tidal flat sequence is partially eroded (as during a transgression) channel deposits will most likely be preserved (Fig. 6B). An excavation into Holocene sediments at Velsen, The Netherlands, described by Van Straaten (1957), illustrates some excellent examples of

A

B

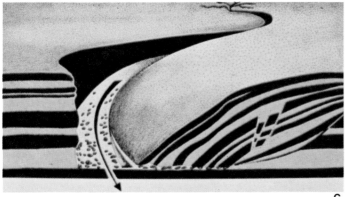

C

D

E

Fig. 5—Tidal channel meandering and channel-fill deposits. **A** and **B** illustrate meandering of a small tidal channel on the tidal flat at Ameland, The Netherlands. The two photos by L. M. J. U. van Straaten were taken a few weeks apart. Height of stakes is approximately 50 cm. **C.** Diagram illustrating the process of tidal channel migration. As channel migrates to the left, the intertidal flat deposits are eroded. On the depositional side of the channel, point bar and longitudinal foreset beds are deposited (drawing by Reineck, 1967). **D.** Geologists slogging through the soft sediments on rapidly deposited longitudinal beds on depositional side of meandering tidal creek, Jade Bay, Germany. Vertical face of eroding intertidal flat deposits seen in the background. Photo by H. -E. Reineck. **E.** Meandering tidal creek illustrating point bar and longitudinal cross-bedding. Photo by G. Evans from the Wash, England.

Fig. 6—Principal tidal flat deposits. **A.** Thin sequence of intertidal flat deposits overlying channel-fill sediments. **B.** Exhumed and truncated channel-fill deposits of a channel that migrated from left to right (truncation of dipping strata at surface). Flaser bedding within these strata are the dominant small-scale sedimentary structure. Photos from the Jade Bay, Germany, by F. Wunderlich.

channel fill preservation (Fig. 7).

Longitudinal cross-beds, megaripples, and ripple laminae are major bedding types in the subtidal deposits. Whereas the latter two bed forms are usually current-formed features in this facies and, therefore, indicative of current flow direction at the time of deposition, the longitudinal crossbeds form on the channel margin at approximate right angles to current flow. Both ebb- and flood-oriented structures may be present in subtidal deposits, but one or the other is usually dominant. An examination of the sinuousity of tidal creeks (channels) suggests that some variation in paleocurrent flow patterns should be anticipated.

Interbedded sand and mud, resulting in lenticular and flaser bedding, are common in subtidal facies and result from flucutations of energy. Sand is deposited during tidal and wave induced current flow, and mud is deposited during slack water periods. These alternations of textures can occur in thicknesses ranging from a few cm to less than 1 mm. What is important is the fluctuation of energy indicated by such interbedded sequences. Principle small-scale sedimentary structures (Fig. 8) are a form of ripple bedding and relative sand-mud percentages result in ripple structures as described by Reineck and Wunderlich (1968).

Intertidal Flat Facies—The area lying approximately between mean low and mean high high water is referred to as intertidal flat. Thickness and width of this environment in a modern setting is directly related to the tide range. However, because of subsidence, compaction, rate of progradation, or erosion, this would not necessarily be true of a preserved sequence. Intertidal areas of the North Sea tidal flats are gently sloping surfaces with widths of several kilometers perpendicular to the shoreline (Figs. 3, 4). Lateral continuity of tidal flats parallel with the shoreline varies depending on coastal morphology. For instance, tidal flats of the Netherlands, Germany and Denmark are continuous for over 400 km whereas intertidal flats of the Wash on the east coast of England are restricted to an embayment a few tens of km wide.

North Sea tidal flats point out the significance of variations in the vertical sequence. This reflects decreasing energy from the lower to upper parts of the intertidal flat and is recorded by textures, types of sedimentary structures, and the transition from dominantly physical to dominantly biogenic sedimentary structures.

Deposition on the intertidal part of tidal flats is primarily by vertical accretion, except in areas of small laterally migrating tidal creeks. In a normal prograding sequence this means relatively slow rates of sedimentation and greater opportunity for biogenic sediment. Furthermore, on the intertidal flat a greater likelihood exists for bioturbation preservation in topographically higher parts of the flat than in lower areas. Reasons for this tendency center around the lower part of the flat, which is more frequently subjected to wave reworking, and migrating tidal creeks, which may deeply incise the lower flat. Creeks gradually shallow, become less abundant, and have lower energy with less reworking of sediment in higher areas of the flat.

Preservation of bioturbation, however, is dependent on more than just the possibility of tidal creek reworking. Availability of nutrients, sedimentary texture, and salinity plays a key role. Indeed, incorporated into a vertical sequence anywhere on the flat there may be units of muddy sand more highly bioturbated than overlying or underlying interbedded sand and mud units. Examples of this can be seen in flats which are densely populated by the deposit-feeding worm *Arenicola* that effectively disrupts primary physical sedimentary structures to a depth of nearly 30 cm. Another general characteristic of the intertidal area is a de-

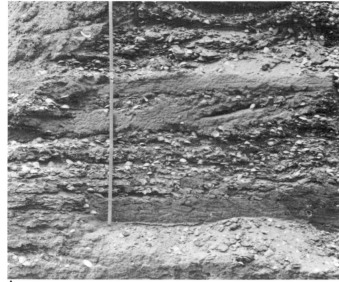

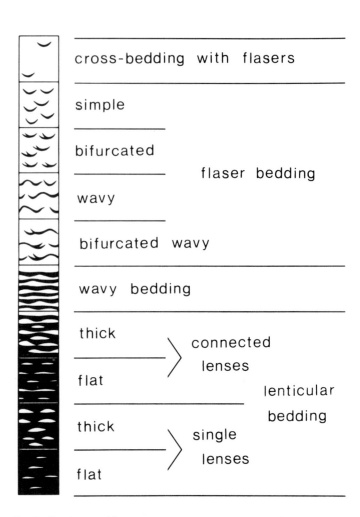

cross-bedding with flasers

simple

bifurcated

wavy

flaser bedding

bifurcated wavy

wavy bedding

thick

connected
lenses

flat

lenticular
bedding

thick

single
lenses

flat

Fig. 8—Continuum of flaser, wavy, and lenticular bedding. Figure illustrates the relationship of sand (white) and mud or shale (black) found as small-scale physical sedimentary structures in many subtidal and intertidal tidal flat deposits. Although characteristic of tidal flats, they are not restricted to that environment. (After Reineck and Wunderlich, 1968).

Fig. 7—Channel-fill sediments at Velsen, The Netherlands. A. Channel-floor deposits containing abundant transported shells and mud pebbles. Meter stick for scale. B. Small sand-filled abandoned channel. Meter stick for scale. C. Erosional contact of sandy tidal channel against muddy tidal flat beds. Shovel for scale. Photos from van Straaten (1957).

Fig. 9—Megaripples on sand flats from the Wash, England. Current flow from left to right. Small-scale symmetrical wave ripples superposed on larger forms. Shells and shell fragments are concentrated in megaripple troughs. Heights of bedforms are approximately 50 cm. Photo by M. E. Pugh.

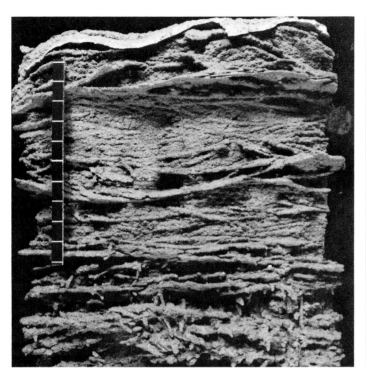

Fig. 10—Sedimentary structures from sand flats, Jade Bay, Germany. **A.** Ripple bedding in upper part. In lower part, former ripple bedding has been nearly destroyed by burrows of polychaete worms. **B.** Ripple bedding in upper part and cross-bedding in lower part. The pelecypod *Cardium edula* is in growth position in the upper left. Transported pelecypod valves in lower center are oriented along cross-beds. Both A and B are epoxy impregnated relief casts. Scales are in cm. Photos by F. Wunderlich.

crease in grain size in a landward direction. This also relates to the gradual decrease in current and wave energy across the intertidal surface.

Reineck (1975) proposed the terms sand flats, mixed flats, and mud flats to subdivide respectively the lower, middle, and upper parts of the area between low tide and high tide on the intertidal flats. These descriptive textural terms are relative, however, and mud, for instance, does not mean that only silt and/or clay is present—minor sand may also be present. Thus, intertidal flats in different settings may contain different proportions of textures depending on both the material available to form the flats and the energy range. Presence or absence and relative abundance of both physical and biogenic sedimentary structures also depend on the local hydrographic conditions and coastal morphology and the biological regime. Nevertheless, the energy gradient gradually changes across the flat and, therefore, textures and structures change, providing the basis for the Reineck subdivision.

Sand flats occupy areas closest to low tide line and are subjected to the strongest wave and current action. As a result they may contain both wave and current ripple laminae (Fig. 9). Muddy lenses and layers, usually present as flasers (ripple laminated sand with mud-filled troughs) and wavy bedding, enhance the definition of bedding. Locally, however, units lacking interbedded muds may be found, with laminated sand also present. In general, sand flats should be less bioturbated than other intertidal areas (Fig. 10), although beds that are completely bioturbated may be present.

Mixed flats contain physical sedimentary structures of intertidal flats which include flaser bedding (Fig. 11A), wavy bedding (Fig. 11B), lenticular bedding (muddy beds with discontinuous lenses of sand, Fig. 11C), and interbedded sand and mud (Fig. 11D). Both symmetrical and asymmetrical ripple laminae are found. This area of intertidal flat may be highly bioturbated by deposit-feeding invertebrate organisms. Because of bioturbation, the mud fraction of the substrate, which is thoroughly mixed with sand, will be less apparent than in lesser bioturbated units where sands and muds are interbedded.

Mud flats, despite the name, usually contain significant amounts of sand as thin ripple-laminated, or lenticular beds, or incorporated into highly bioturbated beds. In addition, algal horizons may be preserved in this subfacies as well as roots of plants. Roots from plants in the overlying upper-intertidal marsh may also penetrate this facies if it is buried by prograding deposition. Generally, this is the muddiest (Fig. 12) of the intertidal flats and usually contains the most bioturbation (Fig. 13).

Intertidal Flat—Marsh Contact— On the North Sea coast, the contact between marsh and the intertidal flats may be gradual or distinct. In the vertical profile a change from highly bioturbated intertidal sediments to the less bioturbated, crudely laminated marsh sediments takes place (Fig. 14A). Increase in physical sedimentary structures is not, in this case, because of increasing physical energy but rather because the marsh environment is not favorable to benthic invertebrate organisms which, in the intertidal zone, are able to destroy primary physical structures.

A second reason to expect a recognizable break is that this is commonly an erosional contact. All accounts of tidal flat sequences from the North Sea coast indicate the presence of a small erosional cliff, generally 1 m or less in height along a significant percentage of the salt marsh edge (Fig. 14B, C). Through time, as the salt marsh progrades, this wave-cut cliff is smoothed as vegetation slowly colonizes the upper intertidal surface.

*Salt Marsh—*This area is flooded only by storms forcing water above maximum high tide position. Hence, the area is completely covered by wa-

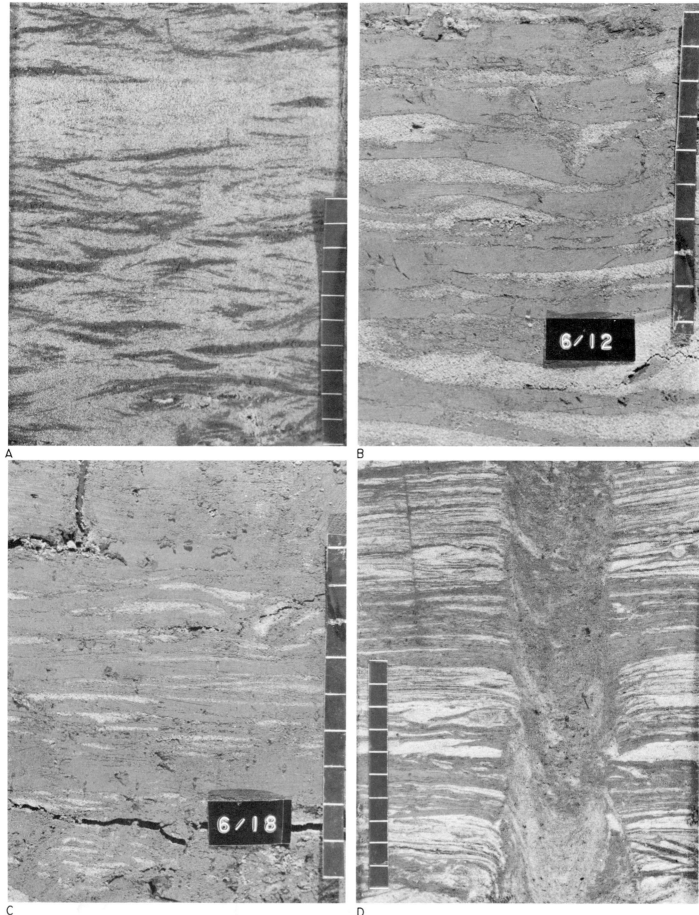

C

D

Fig. 11—Primary physical sedimentary structures from mixed flats, Jade Bay, Germany. Light layers are sand, dark layers are mud. **A.** Flaser bedding in which mud lenses were deposited in ripple troughs. **B.** Wavey bedding. **C.** Lenticular bedding. **D.** Interbedded sand and mud. In the center of D is the trace left (escape structure) by the pelecypod *Mya* *arenaria* as it moved upward to keep pace with sedimentation. Compare this photograph with Figure 8. All photos by H. -E. Reineck. Scale in cm.

Fig. 12—Typical mud flat of the Jade Bay, Germany. As the mud-flat tidal creeks meander, channel bottom becomes layered with eroded shells of *Mya arenaria* and *Scrobicularia plana*. Tidal cracks are about 1 m deep. Photo by J. Dorjes.

ter infrequently and may be subjected to extended periods of high temperatures and desiccation. In winter months, it may experience periods of ice cover. These conditions of intermittent flooding, variations in salinity, and temperature extremes make the salt marsh one of the most biologically rigorous environments of any coastal environment.

In the North Sea the salt marsh is an area vegetated by salt tolerant plants. Because of the physical extremes of this environment, only a few species of plants and benthic organisms are present. This results in relatively little biogenic reworking of the substrate by benthic organisms. Most bioturbation existing here results from plant roots and, therefore, crude stratification is usually preserved and enhanced by the presence of alternating layers of fine sand and mud (Fig. 15A). However, this layering is not the well-defined lamination found in other, especially subtidal, parts of tidal flats. Most of the salt marsh surface has numerous irregularities because of the presence of plants, plant debris, and even mud cracks. Sand layers are more abundant

near tidal creeks and along the seaward margin of the marsh. Farther landward in the salt marsh, they become very thin, sometimes present only as discontinuous layers. In dominantly muddy areas, the thin sand laminae result in parting planes between thicker, finely laminated mud layers. Coarse material, especially shells and shell layers, are also present and result primarily from deposition associated with storms and are most abundant along the seaward margin of the marsh. The salt marsh is a low energy environment, thus, coarse material once introduced is not likely to be transported again. Abundant organic detritus derived form the salt marsh vegetation is dispersed throughout the clays.

Even though burrows and bioturbation made by benthic organisms are uncommon in the salt marsh facies, an initial examination of these sediments may easily give the impression that they exist because of disturbance by plant roots disrupting laminae (Fig. 15B,C). Equally confusing may be structures formed by infilling of decayed plant roots by sand or iron hydroxide.

Fig. 13—Bioturbation in mud flats. **A.** Interbedded sand and mud in which bioturbation has almost completely destroyed primary physical sedimentary structure, Jade Bay, Germany. Scale in cm. Photo by H. -E. Reineck. **B.** Complete biogenic reworking of interbedded sand and mud. U-in-U structures (sprite) are present in upper and lower left, and sand-filled, mud-lined burrow at upper right. Center of core. 1.5 m below mud flat surface, Tjummarum, The Netherlands. Photo by L. M. J. U. van Straaten.

A

B

C

Fig. 14—Intertidal flat and upper intertidal flat (marsh) contact. **A.** Erosional exposure of stratified upper intertidal (salt marsh) sediments overlying highly bioturbated intertidal flat sediments in the Somme Estuary, France. Vertical sequence approximately 150 cm. Photo by L. M. J. U. van Straaten. **B.** Figure shows the erosional cliff which commonly is present at the marsh edge and the colonization of the intertidal flat by vegetation. Scarp height approximately 1 m, Lauwerszee, The Netherlands. Photo by L. M. J. U. van Straaten. **C.** Similar scarp of less relief which is being healed by deposition on the intertidal flat surface. Note the presence of an algal mat on the surface at right. Small plants are *Salicornea herbacea* (pickelweed) a common marsh halophyte. Mellum Island, Jade Bay, Germany. Plants are 15 cm high. Photo by H. -E. Reineck.

In a sense, structures created by roots are burrows and differentiation between the two is certainly difficult but important in facies interpretation.

Unvegetated areas referred to as barrens are locally present. These subfacies lack features commonly associated with the surrounding salt marsh.

Sediments may be highly bioturbated or contain specific biogenic structures such as those formed by crabs and insects. Contacts between planes of sand and mud can contain tracks and trails of vertebrate and invertebrate organisms. During periods of flooding, it is possible for wave and current ripples to form on these unvegetated surfaces and, during periods of drying and subaerial exposure, mud cracks (Fig. 16) and salt crusts can form. Surfaces of these barren areas are sometimes covered with algal mats. All of these features may be preserved as sedimentary structures.

A

B

C

Fig. 15—Sedimentary structure of upper intertidal (marsh) sediments. North Sea coast tidal flats. **A.** Interbedded sand and mud

showing crude stratification, Jade Bay, Germany. Photo by H. -E. Reineck. **B.** and **C.** Poorly stratified, root mottled and slightly

burrowed beds of sand and clay from salt marsh near Zoutkamp, The Netherlands. Photos by L. M. J. U. van Straaten.

Fig. 16—Desiccation structures in salt marsh, Jade Bay, Germany. During extended periods of time when the upper intertidal (salt marsh) zone in not flooded, mud cracks can develop. Note the presence of young *Salicornea* plants and tracks of birds on surface. Centimeter scale in center foreground. Photo by F. Wunderlich.

Fig. 17—Salt marsh creek. View of meandering tidal creek in upper intertidal zone during ebbtide. Rates of channel migration are less than in the lower intertidal flat because of influence of vegetation. Schiermonnikoog, The Netherlands. Photo by L. M. J. U. van Straaten.

Salt marshes are dissected by branching tidal creeks and it is through these channel ways that tidal waters move into the marsh area (Fig. 17). On nearly every tide, waters flood and ebb to some extent within these channels but only rarely do they overflow their banks. Hence flows in these creeks are generally sluggish and sediments deposited here are generally fine grained and become finer grained in a landward direction. Whereas channels in lower parts of the tidal flat commonly migrate rapidly, those in the salt marsh are more permanent due to the stabilizing influence of vegetation and

to the less frequent periods of strong tidal flow.

Animal-Sediment Relationships— Distribution of benthic invertebrate organisms on tidal flats is not random. They tend to occur within certain zones controlled by salinity, wave or current energy, nutrients, substrate type and depth. Thus, across a tidal flat surface the zones of habitation generally parallel the depositional strike and the limits of these zones are perpendicular to strike. This is important because the preservable record of an organisms' existence, i.e., fossil shells or more likely, in the case of tidal flats, the trace fossils, would make changes in a

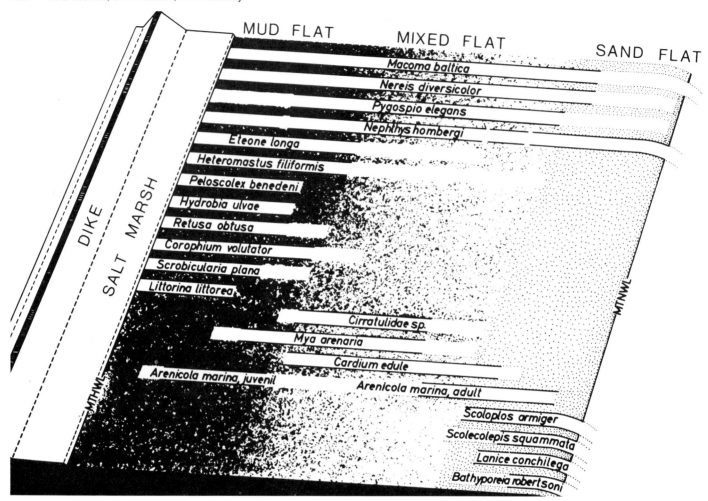

Fig. 18—Zonation of benthic organisms in Jade Bay, Germany, tidal flats. Some forms extend across almost all of the intertidal flat areas. Others are narrowly restricted due to texture; wave, and current energy; degree of flooding; etc. Modified from Dorjes (1978).

Fig. 19—Shell accumulations, Jade Bay, Germany. Widespread beds of living *Mytilus* furnish abundant shell material to the tidal creek channel floor near Mellum Island, Germany. Photo by F. Wunderlich.

vertical sequence similar to those of the physical sedimentary structures and textures.

One example of this zonation from the German tidal flat described by Dorjes (1978) is shown in Figure 18. This example illustrates how zonations occur across the flat in an orderly and gradational way. Some forms are obviously more restricted than others and some forms may extend completely across the flat. In addition, not all hard parts or biogenic structures are equally likely to be preserved. Beause so many factors affect the specific organism, it would be purposeless to attempt to set forth specific forms to look for in the process of recognizing ancient intertidal flat sequences. It is important, however, to be aware that the potential for such zonation exists.

Biogenic structures are more important in intertidal flats than elsewhere in the tidal flat sequence. As pointed out, the salt marsh represents a difficult environment for survival of burrowing organisms. The subtidal environment often features high rates of erosion and deposition, and although numerous or-

Fig. 20—Tidal flat shell accumulations. **A.** Dense accumulation of articulated and disarticulated shells. This disorganized shell accumulation indicates transport has occurred but the numerous articulated valves suggest very short transport distance. **B.** Accumulation of single valves of the clam *Mya arenaria* in an imbricated pattern. Both photographs from tidal flat area south of Elbe River, Germany by F. Werner.

Fig. 21—Biogenic sedimentary structures in North Sea tidal flats. **A.** Dense accumulations of sandy fecal strings of *Arenicola marina* on ripple marked surface of mixed flat surface. **B.** Close-up view of fecal castings of *Arenicola marina*. **C.** Vertical and horizontal view of muddy flat sediments which have been bioturbated primarily by the crustacean *Corophium volutator*. Trench is 60 cm deep. **D.** Detailed view of burrows and bioturbation by *Corophium volutator*. A, B, and C, are from Jade Bay, Germany. D is from tidal flats in Solway Firth, Scotland. All photos by J. D. Howard.

ganisms are present, the likelihood of preservation of trace fossils is greatly reduced. Shell lags, however, are important characteristic features of subtidal channels and the abundance of shelled forms on tidal flat surfaces (Figs. 19, 20), plus erosion of migrating tidal creeks, readily accounts for these shell lags (Fig. 12). If shells are preserved, they are useful facies indicators because protected environments such as this are characterized in part by certain forms not found in open shelf deposits.

In most detrital sand deposits, shell preservation is poor and trace fossils, used here in its broadest sense to mean tracks, trails, burrows, borings, and bioturbation, are especially important (Fig. 21). Trace fossil structures have a good chance of preservation in this environment because so much of the facies is characterized by interbedded sand and mud. Such alternations of tex-

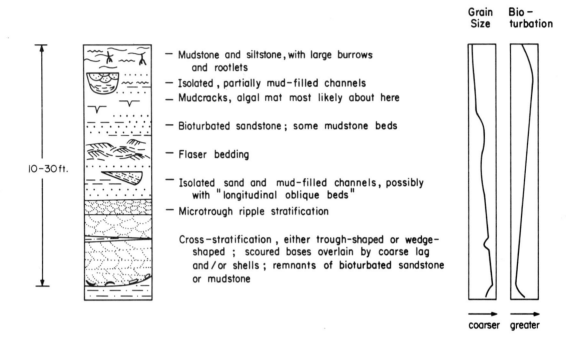

- Mudstone and siltstone, with large burrows and rootlets
- Isolated, partially mud-filled channels
- Mudcracks, algal mat most likely about here

- Bioturbated sandstone; some mudstone beds

- Flaser bedding

- Isolated sand and mud-filled channels, possibly with "longitudinal oblique beds"
- Microtrough ripple stratification

Cross-stratification, either trough-shaped or wedge-shaped; scoured bases overlain by coarse lag and/or shells; remnants of bioturbated sandstone or mudstone

10-30 ft.

Grain Size   Bio-turbation

coarser   greater

Fig. 22—Model of inferred vertical sequence in prograding intertidal sand flat deposits based on interpretation of four examples from the Holocene (from MacKenzie, 1968).

tures serve to enhance and emphasize biologically produced sedimentary structures. This is true of both those structures oriented in the vertical plane, where burrows cut through interbedded sediments, and those which are predominantly in a horizontal or subhorizontal plane produced by organisms that burrow or feed at sediment interfaces. Even surface markings stand a relatively greater chance for preservation, as when, for example, tracks and trails are made on a muddy surface which is subsequently covered by a thin sand layer.

Model of Vertical Sequence Applicable to Ancient Sequences

MacKenzie (1968) compiled data from studies of the North Sea tidal flats into a model for a vertical sequence in prograding intertidal sand flat deposits. The model (Fig. 22) shows a deposit 10 to 30 ft (3 to 9 m) in thickness which has an upward decrease in energy indicated by changes in sedimentary structures, fining upward grain size, and a general upward increase in bioturbation. The model was used by MacKenzie (1968) to describe Early Cretaceous tidal flats near Denver, Colorado, but may be applied as well to interpretation of other ancient sequences. The vertical profile may be modified by utilizing details of the sedimentary structures and textures presented in the previous discussion.

Georgia Coast, USA

A different type of tidal flat complex is observed along the Georgia coast; one which may initially cause some confusion when compared with the North Sea example. These are tidal flats which, through much of their lateral extent, are covered by halophytes (plants which grow in saline soils). Many kilometers of coastline throughout the world are characterized by vast expanses of intertidal salt marsh. Although thoroughly studied as biological habitats, salt mash tidal flats have not been studied extensively by geologists. Of the numerous possible examples, the Georgia coast (Fig. 23A) is cited because of the writers' studies of this area and because of the detailed studies by Robert W. Frey and his students at the University of Georgia.

If the coastal setting and scale of the Georgia coast is compared with that of the North Sea coast, some similarities are apparent. A barrier island coast is backed by a vast intertidal flat dissected by estuaries, tidal rivers, and creeks. The most obvious difference is that the flats landward of the North Sea barrier are unvegetated except on the landward fringe, whereas those landward of the Georgia barriers are highly vegetated by marsh grasses (Fig. 23A, B, C). To aid in the discussion, these board expanses of salt marsh are referred to as vegetated tidal flats in contrast to those of the North Sea referred to as unvegetated tidal flats.

An important aspect of vegetated flats is the relationship of flooded area to tidal range. In salt marsh tidal flats, a majority of the areal extent of the flat lies in the upper half of the tide range

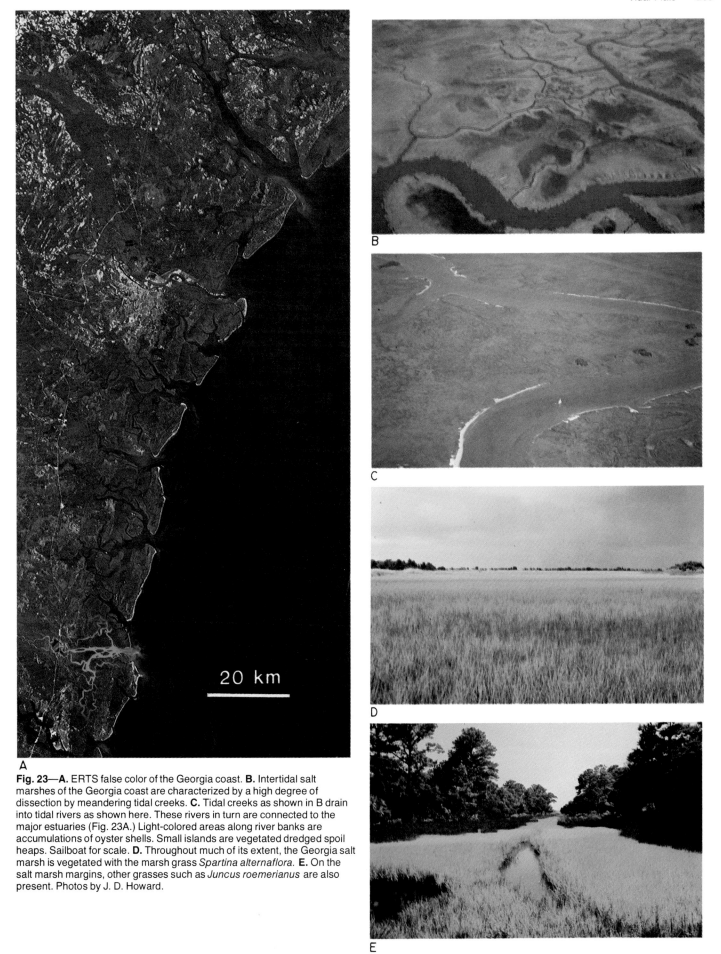

Fig. 23—A. ERTS false color of the Georgia coast. **B.** Intertidal salt marshes of the Georgia coast are characterized by a high degree of dissection by meandering tidal creeks. **C.** Tidal creeks as shown in B drain into tidal rivers as shown here. These rivers in turn are connected to the major estuaries (Fig. 23A.) Light-colored areas along river banks are accumulations of oyster shells. Small islands are vegetated dredged spoil heaps. Sailboat for scale. **D.** Throughout much of its extent, the Georgia salt marsh is vegetated with the marsh grass *Spartina alterniflora.* **E.** On the salt marsh margins, other grasses such as *Juncus roemerianus* are also present. Photos by J. D. Howard.

Fig. 24—Lower part of intertidal zone, Georgia coast, at low tide. **A.** Tidal creek margin along a tidal creek at Sapelo Island, Georgia. **B.** Soft, muddy creek bank. This zone represents most of the lower intertidal zone. Grass in both views is *Spartina alternaflora.* Photos by R. W. Frey.

zone whereas the lower half of the intertidal zone is only a narrow band bordering the tidal creek (Fig. 24). Compare this to the North Sea example where the salt marsh occurs in the upper intertidal of the tide range and the intertidal surface slope is more constant across the broad intertidal area.

Despite differences in these two tidal flat settings, records which may be preserved would show some similarities. In vegetated tidal flat areas, as in the unvegetated example, the most important facies to be preserved is the subtidal, lower intertidal, and channel-fill sediment formed in response to meandering tidal creeks and larger tidal channels. Indeed, cores from estuaries and tidal creeks of the Georgia estuaries show structures similar to those from the subtidal areas in the North Sea example (Howard and Frey, 1975).

While unvegetated flats have been shown to possess vertical zonation of physical and biogenic sedimentary structures and textural changes showing a general fining upward, vegetated flats of the Georgia coast contain fewer physical sedimentary structures. Bioturbation is intense but specific diagnostic burrows are few. Textures fine upward initially but may become coarser near the top of the intertidal sequence because of washover of sand from barrier beaches, small splays of sand from tidal creeks, or erosion of Pleistocene barrier island remnants.

In the narrow unvegetated lower part of the intertidal zone, bioturbation is common though primary physical sedimentary structures are also preserved. This interval may contain longitudinal crossbedding which dips at low angles from vegetated salt marsh surfaces into subtidal zones of the tidal creeks. In these beds, the uppermost part is highly disturbed by crab burrows and plant roots but grades downward into interbedded sand and mud containing flaser bedding, wavy bedding, and lenticular bedding (Fig. 25). Bioturbation in the vegetated zone is 100% and, other than finding peat layers or fossilized plant remnants, it is usually impossible to distinguish disturbance by plants from bioturbation by benthic organisms.

Hence, the vertical sequence in a vegetated tidal flat is characterized by channel-fill sediments including longitudinal crossbedding, trough crossbedding, and large and small ripple lamination with flaser bedding and interbedded sands and mud representing the subtidal channel-fill sequence. This sequence is in many ways similar to the North Sea tidal flats. Upward in this sequence, increasing bioturbation results from benthic organisms. The uppermost part of the sequence, commonly bioturbated by plant roots and organisms, is a sandy mud rich in organic matter. In some deposits this could include layers of peat, although they are not present in the Georgia coast. The top of the sequence can show an increase in sand and even grade to washover or dune deposits.

Gulf of California

Tidal flats of the Gulf of California occur in a region where evaporation far exceeds precipitation. These tidal flat deposits, because of their depositional setting, contain few tidal channels and hence, the significance of channel-fill sediments and sedimentary structures emphasized previously are of minor importance here. Robert W. Thompson conducted a very thorough study of these tidal flats and his results are the major source of information for this discussion.

Depositional setting for these tidal flat deposits is somewhat different than in examples previously described. They lie in the upper reaches of the Gulf of California (Fig. 26) and have been formed by deposition of fine-grained sediments carried by the Colorado River. The river has built a large delta into an embayment which, due to its configuration, has a relatively large tidal range (normally 5 m with maximum range of 8 m on spring tides) and low wave energy. Elements of deposition are: (1) a high supply rate of silt and clay; (2) high tidal range; (3) low wave energy; and (4) arid climate. These conditions occur over an approximately 2,000 sq km area and include a vertical sequence of nearly 16 m. Settings such as this, whether arid or not, may help explain some of the ancient tidal flats that lack significant channel-fill deposits.

Based on morphology, three major subdivisions can be recognized. High flats lie at and above the level of spring higher high water (this is a nearly flat surface which can be considered a supratidal zone). The intertidal zone is between spring higher high water and spring lower low water. The subtidal zone lies below the level of spring lower low water. Together these environments include the area from 5 m above mean sea level to a depth of 11 m below mean sea level.

Subtidal Facies—The subtidal zone, comprising the maximum thickness of the vertical sequence (1 to 11 m), is made up predominantly of poorly laminated silty clay in which little bioturbation is present, which is quite different from previous examples. This facies has formed in response to a very rapid influx of fine sediment from the Colorado River combined with low wave energy. The general lack of biogenic re-

Fig. 25—X-ray radiograph prints of cores of intertidal sedimentary structures. **A.** Upper part of sequence in which plant roots and benthic organisms have highly disturbed the substrate. Dark fragments are plant fragments. X-ray radiographs by R. W. Frey. **B.** Interbedded sand and mud from the lower part of intertidal zone.

working is considered due to rapid sedimentation.

Lower Intertidal Facies—Between mean high and mean low water, (approximately 4 m) is the principal zone of biogenic reworking of silty clays. These sediments are almost totally bioturbated, and any evidence of primary physical sedimentary structures is extremely rare. The biogenic reworking of the sediment is by two types of organisms and a much more restricted fauna than in intertidal parts of other tidal flats.

In the lower part, from approximately neap high level to neap low, bioturbation is accomplished by clams especially adapted to muddy substrates. The upper part of the zone is bioturbated primarily by fiddler crabs burrowing as deep as 30 cm (Fig. 27A). This species of fiddler is adapted to the rigorous environment, and during neap tides the area may lie exposed to long periods of desiccation, (to the point that the surface becomes mud-cracked and salt encrusted). During this time the fiddlers remain in the depths of their burrows only to reappear on the tidal flat surface at the onset of spring tides.

Locally, accumulation of shell and coarse sand may lie at the top of the lower intertidal zone (at the contact between lower and upper inter-tidal zones, i.e. overlying the fiddler burrow zone). These deposits form low amplitude beach ridges that derive their shell material from the lower intertidal zone (Fig. 27B).

Upper Intertidal Facies—The upper intertidal zone which lies approximately betwen mean high water and spring high water is composed primarily of laminated silt, and to a lesser extent, clayey silt. In this facies of approximately 1 m thickness, laminations become more regular upward and occasionally cross laminae are present. In the upper part, however, laminated silts are highly disturbed due to shrinkage and crystallization of evaporites during times of desiccation, which is most of the time for this facies (Fig. 27B). Suspended silts and clays deposited across the upper intertidal surface during spring tides are capable of infilling and preserving any tracks (for instance vertebrate tracks, especially birds), and mud cracks made when the flat is subaerially exposed. It is often difficult to distinguish mud cracks from root structures and burrows, especially in cores.

The uppermost intertidal zone (high flats) and the supratidal zone are either

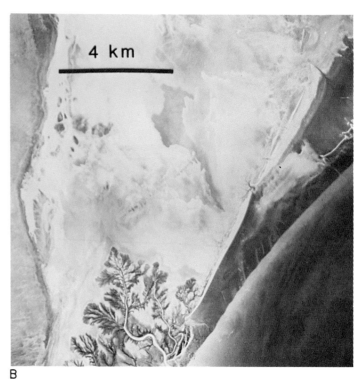

Fig. 26—A. ERTS photo of north coast of Gulf of California. **B.** Aerial view of Gulf of California tidal flats. Triangular shaped area in the southeast part of the view is the subtidal area. Dark area to the west is the intertidal zone. The light gray area is the contact point in the uppermost intertidal area or high flats. Tidal channels are seen in the lower center part of the view. Photos by L. A. Hardie.

completely lacking in lamination or bioturbation (Fig. 27F, lower part) or the lamination present is lenticular with gypsum crystals filling desiccation cracks. This is a facies of evaporites and silts highly disturbed due to desiccation. Thompson refers to these as choatic muds.

Tidal Channel Facies—Recalling again the significance placed on tidal creek and tidal channel deposits in most tidal flat sequences, their absence in this setting deserves emphasis. Channels are present in the laminated clayey silts of the upper intertidal facies (Fig. 27G), but they are erosional features developed only where thin sand and shell deposits form low relief beach ridges. These are very shallow channels and maximum scour depth is only 2 m below sea level. In an ancient sequence such channels would proba-

bly constitute a very minor part of the stratigraphic sequence.

Overall, the tidal flat example from the Gulf of California is distinctly and uniquely different from a "typical" tidal flat sequence. Deposits contain an overall uniformly fine grain size, a relative absence of major channels, an absence of marsh deposits, and highly bioturbated sediments occurring only in very restricted parts of the sequence. Although desiccation features are abundant in the upper part of the sequence, they are preserved as disturbed bedding which might be confused with biogenic activity. Thompson suggests that some of the western interior red bed sequences should be examined using this example as a model. Many sequences of this type may appear in the stratigraphic record but are not considered as tidal flats because

the usual concept of a tidal flat is derived from a different type of model.

Bay of Fundy

No discussion of tidal flat sequences would be complete without mention of the Bay of Fundy. Like the Gulf of California, the setting is a large coastal embayment of rapid depositional progradation (Fig. 28). However, no river provides a sediment source, the climate is temperate, and much of the locally derived sediment is sand. Tidal channels are important and the spectacular tidal range has a mean of 11.5 m and spring fringe of 15 m.

Geologic studies of the Bay of Fundy area tying it to the interpretation of the stratigraphic record began with studies by George de Vries Klein and continue now by students at McMaster University under the direction of Gerard V.

Fig. 27—Gulf of California. **A.** View of lower intertidal zone. Surface is covered by dense accumulation of fiddler crab burrows. **B.** Beach ridge formed by shells. Ridge separates the lower and upper intertidal zones. Shells are derived from the lower intertidal zone. **C.** Channel erosion of laminated silts in the lower part of the upper intertidal zone. **D.** Close-up view of mud crack structures developed in laminated silts. **E.** Surface view of uppermost flats. **F.** Impregnated core from the high flats. Chaotic muds at base of core. Disrupted silt layers and gypsum-filled mud cracks in the upper part. Core is 8 cm in length. **G.** Tidal channels eroding into tidal flat sediments. **H.** Undercutting by erosional tidal channels of algal-covered flat. Photography by L. A. Hardie.

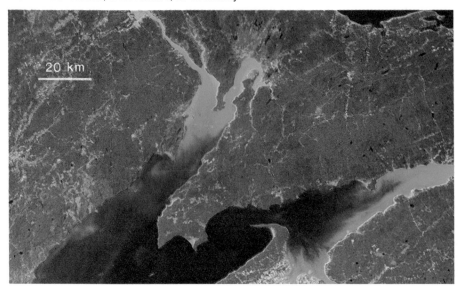

Fig. 28—ERTS view of Bay of Fundy.

Middleton (physical processes) and Michael J. Risk (biological processes).

Most detailed studies to date have been in Cobequid Bay, an on-trend extension of the Bay of Fundy. Here the vertical sequence thickness is 15 m and overlies glacial gravels or Triassic bedrock. The basal 9 m is well-sorted medium-grained sands. These tidal deposits result from depositional processes associated with extensive bars exposed at low tide in the channels. Wave processes in this restricted embayment are of relatively minor importance. At high tide the bars are covered by 5 to 15 m of water. Thus, in a vertical sequence of deposits resulting from bar migration, the lower and middle part of the intertidal zone also are represented.

Internal features of the Holocene deposits are interpreted by studying processes associated with the bars. Bed forms on the surface include megaripples (features also referred to as dunes) and sand waves (Figs. 29, 30). The largest features are sand waves with wavelengths of up to 215 m and wave heights as much as 3.4 m. Megaripples possess wave lengths of <1 to 25 m and wave heights of 10 to 70 cm, and ripples have less than 30 cm wavelengths with wave heights of less than 5 cm. These impressive bed forms are the major features of the bar.

In addition, associated sedimentary structures more common in other facies and other tidal flat examples are present. These include plane beds, interference ripples, reactivation surfaces, and evidence of scour and deposition during falling tide. Reactivation surfaces indicate an interruption in flow, erosion, and then continued deposition. In tidally controlled deposition and erosion, these reactivation surfaces commonly represent instances of current reversal. However, such features can also occur in response to current or depth changes under unidirectional flow. With strong tidal action the opportunity for both ebb- and flood-oriented structures occurs. Bidirectional cross-stratification, though present (Fig. 30B), is not common in the Bay of Fundy bars. For the most part, channel axes preserve ebb-oriented structures whereas bars are characteristically dominated by flood-oriented features.

Bioturbation is definitely not a characteristic of the bar facies even though benthic organisms are abundant in adjacent facies. Their absence can be attributed to strong tidal currents and frequent shifting of the substrate that eliminate preservation of most burrows. Channel-fill sequences are locally present in bars as small drainage ways cutting perhaps 2-3 m into bar surfaces. Large-scale channel-fill deposits have not been documented in this sequence but probably occur. Recognition of specific channels in equivalent ancient deposits would be difficult except in well-exposed outcrops because channels would be filled with cross stratified sand similar in appearance to deposits resulting from bar migration.

In a prograding sequence, bar sands would be overlain by a 2 m thick unit sand which today is infilling the head of Cobequid Bay and which Swift and McMullen (1968) compared to bars in a braided river. Physical sedimentary structures of these "braid bars" are similar to, but of somewhat smaller scale than, those in the underlying facies described above. Especially im-

portant in this sequence is the presence of flaser bedding, wave ripples and, locally, biogenic structures. Laterally this sequence could be replaced by shoals, bars and tidal delta sediments where rivers are present. In the Bay of Fundy setting this sequence would constitute approximately 2 m of the vertical section. Above the "braid bar," a thin beach sequence of perhaps 1 m in thickness should be found. Recognition of this unit as a beach might be easily overlooked in the stratigraphic record.

Sandy and muddy flats and salt marsh (Fig. 31) make up the upper 4 m of this sequence. This is the upper intertidal zone with the salt marsh referred to as the uppermost intertidal because it is inundated only during spring tides.

Sand and mud flats are dissected by tidal channels (Fig. 31A, B). The flats contain silt, clay or sand and, although the surface is commonly ripple-marked or bioturbated (Fig. 31C), internal structures are principally thinly laminated fine sands and muds (Fig. 31D). The dense pattern of tidal channels on the flats suggest that channel-fill sequences would make up a significant part of the depositional record. Locally, the substrate may be highly bioturbated and biogenic reworking is present throughout the flats. Preliminary studies indicate that diagnostic vertical zonation of organisms and hence biogenic structures occur in these flats, similar to that described previously for the German tidal flats.

The upper 1.5 m of the sequence is made up of salt marsh (Fig. 31E, F), which is flooded only at spring and storm tides. This facies is slightly stratified due to disturbance by plant roots. Channels are present in the marsh but probably meander very little due to the resistance to erosion by the vegetation.

The sequence described above is hypothetical in the sense that a complete vertical sequence has yet to be established. Observations are predicated on the assumption of how sediments would accumulate if basin filling occurred.

Wind-Tidal Flats

Wind-tidal flats are those flooding during times of strong onshore winds; hence, the flooding is unpredictable and such flats occur only in the supratidal zone. The example cited here, studied by James A. Miller, is part of the Laguna Madre which lies landward of Padre Island barrier on the Texas coast (Fig. 32A, B). This area, similar to topographically higher parts of the Gulf of

Fig. 29—Sedimentary structure, Bay of Fundy area. **A.** Ebb-oriented megaripples, East Bar, Economy Point. **B.** Ebb-oriented megaripples with surface cover of smaller-scale current ripples. **C.** Field of megaripples grading into mudflat in the background. **D.** Aerial view of sand waves on Selmah bar. Photos A, B, C, and D are by R. W. Dalrymple. **E.** Aerial view of Big bar, Minas Basin. Wave length of dunes is 60 m. **F.** Reactivation surface and cross-stratification in dunes. West bar, Economy Point, Minas Basin. Photos E and F by G. de Vries Klein.

California flats, is one in which evaporation exceeds precipitation. Wind-driven tides are less than 0.5 m.

This depositional sequence is made up primarily of interbedded clays, algal layers, gypsum, and sand (Fig. 32C). Bedding is well developed but irregular and consists primarily of alternating clay and algal laminae. Algal layers became active when the surface is flooded and clays suspended in tidal waters settle out and are caught in the algal mat (Fig. 32D). Gypsum is present in several forms and precipitated below the surface from interstitial brines.

Sand layers are minor, and are derived from sand flowing across flats when they are not flooded (Fig. 32E). Carbonate minerals are present but in minor amounts. Significant disturbance of bedding results from desiccation, mud cracks (Fig. 32F), crystal growth, and to a lesser extent from burrowing of insects, worms, and by plant roots.

Facies such as these may not be abundant in the stratigraphic record but are important environmental indicators when found associated with beach and back barrier deposits. The Texas wind-tidal flat deposits (gener-

ally less than 1 m thick) are underlain by lagoon deposits and overlain by washover fans and eolian dunes. This facies contains no significant channel-fill deposits.

ANCIENT TIDAL FLATS AND TIDAL CHANNELS

Lithologic features and lateral facies changes observed in modern intertidal settings have been compared with similar features in ancient rocks to establish tidal-controlled sedimentation. Three ancient examples are presented in the second part of this chapter. Two

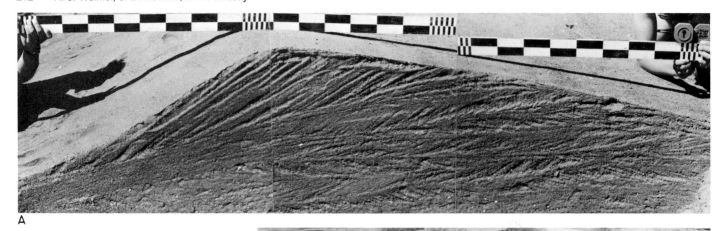

A

B

Fig. 30—Trenches through sand waves and megaripples. **A.** Photomosaic of trench through ebb-asymmetric sand wave. Gently inclined surfaces separated by steeper dipping cross-beds are reactivation surfaces. **B.** Trench through megaripple. This area is flood dominated (flow from left to right) but ebb structures are also present. Cross-bed sets are separated by reactivation surfaces. Photo by R. W. Dalrymple.

examples are from the Cretaceous of the Western Interior of the U.S.A. and one is from the Mississippian of the Illinois basin. Petroleum production has resulted in tidal channels from all 3 sequences but, in each case, the finer grained, lower energy, tidal-flat deposits have been nonproductive, or shown poor reservoir qualitites. Therefore, to describe petroleum production associated with tidal deposits on ancient coastal plains, a description of tidal channels genetically associated with the tidal flats is needed. These concepts are illustrated by discussion of the Almond Formation of the West Desert Springs and Patrick Draw fields, Wyoming and the Aux Vases Formation, Rural Hill field, Illinois.

Dakota Group (Cretaceous) Sedimentary Structures, Textures, and Facies

Characteristic suites of sedimentary structures and related textures, as previously described, have been used by many investigators to identify ancient tidal flats and tidal channels. The genetic association of inorganic and biogenic structures is most diagnostic in environmental interpretation, although deformational structures can also be important. Both vertical and lat-

eral associations of structures and textures must be considered in reconstructing the processes characteristic of tidal sedimentation on ancient coastal plains.

Rather than provide a glossary of photos of all structures commonly observed in ancient tidal flat and channel environments, three case histories are presented as "typical" in order to illustrate outcrop and subsurface data. Outcrop photos are from one of the best documented tidal flat sequences in the Lower Cretaceous of the Western Interior, the upper Dakota Group exposed at Alameda Avenue west of Denver, Colorado.

The Dakota Group stratigraphy along the Front Range in Colorado has been the subject of many published papers, but only work of the past 15 years has emphasized the use of sedimentary structures, textures, and facies in reconstructing the environments of deposition. The Alameda Avenue section (Fig. 33) is unique because road construction allows observation of structures in both section and bedding plane views. MacKenzie (1968, 1971), Weimer and Land (1972), and Chamberlain (1976) have described in detail the textures and structures found in a 230-ft (70.1m) section of the South Platte For-

mation, the upper formation of the Dakota Group (Fig. 34). Because of the classic bedding plane exposures of the fine-grained units, descriptions of the section have emphasized tidal flat environments found in units 12, 13, 19, and 23 as labeled on Figure 34. These units represent thin fine-grained tidal flats that cap coarser-grained and thicker major channel deposits. The most diagnostic structures of the tidal flats and associated strata are ripple modified tabular sets of cross strata, dinosaur footprints, root zones, abundant trace fossils, and small channels, either sand or partially mud filled.

General lithology, textural variations, inorganic and biogenic structures, and porosity are plotted as a stratigraphic column (Fig. 34) with graphic details shown in the explanation. Unit numbers and the interpreted environment of deposition are indicated on the right side of the lithology column; grain size variation is shown by maximum and average plots on the left side of the column.

The lowermost marine strata, associated with the transgression related to a water deepening cycle in the Albian, are included in the Plainview Sandstone Member. Units 12 and 13 of this member constitute a 17 ft thick (5.2 m)

Fig. 31—Bay of Fundy tidal flat facies. **A.** Aerial view of mud flats showing dense dendritic pattern of tidal channels. **B.** Detailed view of tidal channel on muddy tidal flat. **C.** Highly burrowed tidal flat surface. Burrowing is primarily by the clam *Macoma* and the crustacean *Corophium*. **D.** Parallel and wavey laminated silts in intertidal mud flat. Scale in meters. **E.** View of salt marsh and meandering tidal channel. **F.** Truncated margin of salt marsh showing weakly laminated bedding. Scale is 1 m. Compare with similar feature from North Sea tidal flats in Figure 15A. Photos by R. J. Knight.

burrowed sandstone and shale sequence (Fig. 35A) which shows many characteristics of prograding tidal flat sedimentation (Figs. 21A, 35). Unit 11 represents the first tidal flat deposit on the overall transgressive cycle. Bed thickness and grain size in units 12 and 13 decrease upward. At the base are 1-ft thick (0.3 m) beds of fine-grained sandstone which show alternating sets of trough and tabular cross-stratification and subparallel lamination (Fig. 35B). The upper few feet are very thin-bedded, very fine-grained sandstone with thin interbeds of gray, silty shale. Sandstones in the upper part are ripple cross-stratified and exhibit flaser bedding (Fig. 35D). Groove casts and other current markings on the base of some beds may represent tidal runoff structures.

Tidal flat deposits are locally burrowed and contain casts or trails and other organic markings on the base of beds. Trace fossils are *Planolites, Skolithus, Trichichnus* and roots. On the south side of the Alameda Avenue cut, unit 12 contains an interesting sequence of burrow structures which repeats in the vertical section several times. Two to five cm-thick intensely burrowed zones alternate with 5 to 10 cm-thick "clean" sandstone layers (Fig.

Fig. 32—Wind tidal flats, Texas coast. **A.** Wind tidal flooding of flat. **B.** Supratidal flat west of Padre Island. **C.** Trench on wind tidal showing algal and clay laminations. **D.** Drying of algal mats, Buena Vista, Texas. **E.** Wind tidal flat with windblown sand dunes on surface. **F.** Mud cracks developed on the surface of Buena Vista, Texas, flat. Photos by J. A. Miller.

35C). The "clean" sandstones contain numerous vertical to oblique, *Skolithos*-burrows 2 to 5 mm in diameter. Smaller tubular burrows have light gray clay walls, while others appear to be completely filled with light gray clay.

The J Sandstone interval is the upper 135 ft (41.4 m) of the Dakota Group at Alameda Avenue and is the approximate equivalent of the J Sandstone in the Denver basin to the east and south of Denver. The J Sandstone is the principal petroleum-producing formation in the basin with the main subsurface reservoir rock interpreted as channel sandstone similar to units 16, 18, 20 and 22.

Typical structures observed in tidal flats capping the channels in the J Sandstone are shown in Figures 36, 37, 38 and 39. Unit 19 is a tidal flat sequence containing long-crested flat-topped ripples, interference or ladder ripples, tracks, burrows, and mud cracks (Fig. 36B, C, D). The orientation of ripple crests shows polymodal current directions. The underlying unit 18 (Fig. 36A), interpreted as a beach deposit, is parallel laminated fine-grained well sorted friable sandstone with *Skolithos*, *Arenicolites*, and *Planolites* trace fossils.

The best assemblage of trace fossils is found in the tidal flats of unit 23. Chamberlain (1976) summarized the overall fauna and vertical changes in Figure 39. Trace fossils and other structures observed in unit 23 are shown on Figures 35 and 36.

Channel sandstones are 10 to 40 ft (3.05 to 12.2 m) thick and are present throughout the J interval. Most are medium-grained and become finer upward, with fine- to very fine-grained clayey sandstone in the upper few feet.

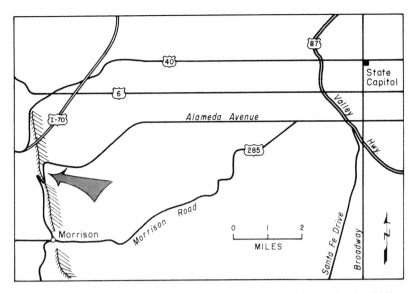

Fig. 33—Index map showing location of Alameda Avenue road cut (after MacKenzie, 1968).

The sandstone is friable and porous, constituting the best potential reservoir rock in the J interval. Channels have sharp scour bases, and at some localities, several feet of relief can be observed within short distances. Clay clasts, concentrations of coarse quartz grains, and wood imprints are present locally at the base of the units. A few log imprints show impressions of borings by organisms (tentatively identified as *Teredo*, Fig. 38B).

The most common type of stratification is 1 to 3 ft (0.3 to 0.9 m) thick trough sets where foresets dip 15 to 25° and become tangential at the base (Fig. 36E). Transport directions within the channels vary, but most fall into the 90° quadrant from northeast to southeast. Thin, carbonaceous clay or shale beds in the middle and upper parts of some units are interpreted as clay drapes.

Scattered small, nondescript tubular burrows are present in the middle and upper parts of a few channels; burrows and other organic markings are present on the base of some beds.

Channel sandstones on Figure 34, interpreted as estuary-tidal, are those exhibiting features suggestive of brackish water and/or tidal currents, such as burrowing, bored wood, and ripples on the bounding surface of sets of cross-strata. Concentrations of fine carbonaceous detritus on foresets of cross-strata suggest periodic fallout of suspended material—possibly during periods of slack current flow, such as at high or low tide.

The channels, interpreted as being distributaries, are those showing unidirectional transport. They are not appreciably burrowed, contain abundant plant fragment casts along bedding surfaces, and have fining-upward textures with a scour base.

Upper Almond Formation (Cretaceous), West Desert Springs and Patrick Draw Fields, Wyoming

Geologic Setting—Stratigraphically trapped petroleum was found on the east flank of the Rock Springs uplift in 1958. Lenticular sandstones in the upper 100 ft (30.5 m) of the Almond Formation (Upper Cretaceous) produce oil and gas at the Desert Springs, West Desert Springs, and Patrick Draw fields. Although major production is primarily related to porous and permeable sandstones of ancient shoreline deposits (Weimer, 1966, Van Horn, 1979), minor oil and gas production has been established in a tidal channel and tidal flat sequence at West Desert Springs and the north end of Patrick Draw (Fig. 40). This production is the subject of this discussion.

Sandstones in the Upper Almond Formation occur at different stratigraphic levels and have been given in descending order the informal designation of UA1 to UA6 sands (UA means upper Almond). UA1 to UA3 sandstones are exposed at the surface, whereas UA4 to UA6 are subsurface units (Fig. 41).

Recent outcrop and subsurface work by Van Horn (1979) has established a sedimentation model for the upper Almond Formation of tidal-dominated deposits in a shoreline and coastal plain setting. Excellent exposures in the outcrop area (Fig. 42) pro-

vide criteria for establishing the model and these data are applied to interpreting well and core data for fields to the east.

The UA6 sandstone production is found in the West Desert Springs field and the north end of the Patrick Draw field (Arch Unit) from depths of 3,600 to 4,800 ft (1,097 to 1,463 m). In both fields, production is found from sandstones other than UA6. The strata dip approximately 3° to the east as indicated on the structure contour map (Fig. 43).

Although oil is the principal production from UA6 sandstone, small areas of gas production suggest isolated reservoirs within overall sandstone trends. Water is found in the UA6 sandstone interval in a low structural position along the east side of the fields.

Origin and distribution of UA6 Sandstone—An electric log cross section (Fig. 44) shows the stratigraphic position and lateral lenticularity of the UA6 sandstone. Location of the cross section and postulated productive trends for the main sandstone of UA6 are shown on Figure 43. Production from a lower thin sandstone in UA6 is not shown on the map.

The stippled pattern on the structure contour map (Fig. 43) shows areas where tidal channel sandstones range in thickness from 5 to 15 ft (1.5 to 4.6 m). Between this pattern and the zero isopach, sandstone is generally less than 5 ft (1.5 m) thick and may or may not contain an economic thickness of porous and permeable sandstone. These thin sandstones are thought to have been deposited in shallow tidal creeks, or lower sand flats, or both. The area beyond the zero isopach is interpreted as middle (mixed) to upper (mud) tidal flats where thin flaser-bedded sandstone and mudstone lack porosity and permeability. This facies provides the trap for petroleum in tidal channel sandstones.

The stratigraphic model for the reservoir and trap in UA6 sandstone is shown on Figure 45. Production is found in the landward edge of tidal channels where reservoir characteristics in the channel and lower sand flat sandstones are lost by facies changes to shales and siltstones of the mixed and mud tidal flats, and the salt and freshwater marsh, or possibly lagoonal shales. The main shoreline zone during UA6 deposition is postulated to be from 6 to 12 mi (9.6 to 19.3 km) to the east.

Following UA6 deposition, the shoreline transgressed into the area of

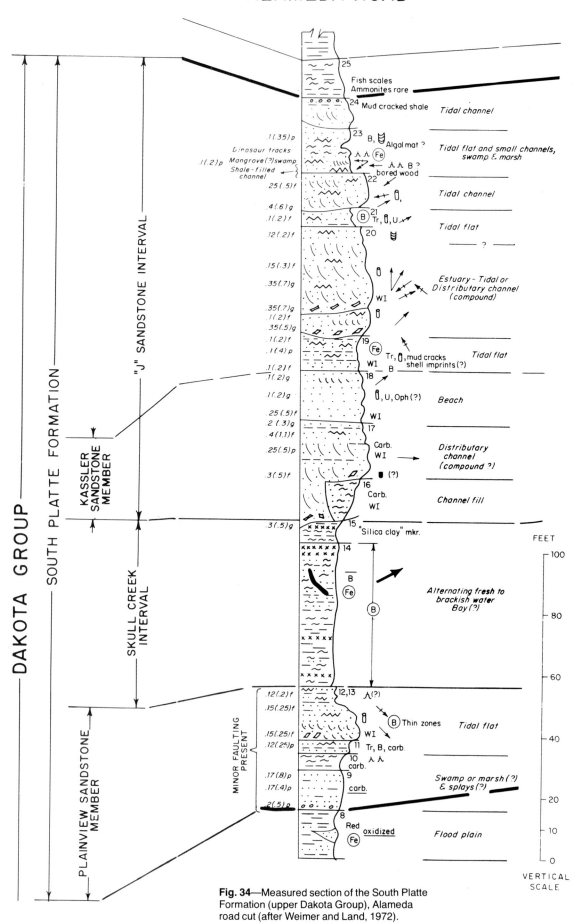

Fig. 34—Measured section of the South Platte Formation (upper Dakota Group), Alameda road cut (after Weimer and Land, 1972).

EXPLANATION

SEDIMENTARY STRUCTURES

 Cross-stratified sandstone (large scale)

 Ripple-stratified sandstone

 Direction of transport indicated by large scale cross-stratification

 Direction of transport indicated by ripple-stratification

x x x x x x   Thin white Kaolinite bed

 Locally derived clasts

carb   Carbonaceous material

WI   Small wood and plant fragment imprints

Log(s)   Log or large wood imprints

 Root imprints

(Fe)   Local concentrations of iron oxide cement

⌣   Load casts

 Slump deformation with direction of movement shown

 Leaf imprints

TRACE FOSSILS

Oph. Ophiomorpha

⌷   Simple vertical burrow

U   Simple U-shaped burrow - oriented normal to bedding

⩄   U-shaped burrow with spreiten - oriented normal to bedding

 U-shaped burrow with spreiten - oriented at low angles or parallel to bedding

B   Unoriented, nondescript burrows

Ⓑ   Bioturbation by organisms

Tr   Winding trails or burrows on sandstone bedding surface

B̄   Horizontal burrows or organic markings on base of sandstone bed

▮   Blade-shaped burrows on base of sandstone bed

LITHOLOGY

 Sandstone - average and maximum grain sizes and sorting shown at left. g = good, f = fair, p = poor.

 Conglomeratic sandstone

Shale or claystone

Siltstone

Coal bed

Fig. 35—Vertical tidal flat sequence (from Weimer and Land, 1972). **A.** Plainview Sandstone Member at Alameda Avenue, units 11, 12, 13. Section is locally offset 1 to 4 ft (0.3 to 1.3 m) by several small faults which strike approximately N 70°E. P=Plainview Member, SC=Skull Creek interval. **B.** One-foot-thick (0.3 m) set of trough cross-strata in Plainview Sandstone Member, unit 12. View is of surface parallel to axis of trough. Location of photo is shown by arrow on photo A. **C.** Thin zones of bioturbation (B) and alternating beds of "clean" sandstone with vertical burrow (X). Plainview Sandstone Member, unit 12. Photo is of exposures along trail southwest of road. **D.** Alternating very fine grained sandstone and shale beds showing irregular bedding and minor flaser structure. Upper part of Plainview Sandstone Member, unit 13.

Fig. 36—Beach capped by tidal flats, units 18 and 19, Figure 34 (from Weimer and Land, 1972). **A.** Alameda Avenue, unit 18. Note even bedding of very fine-grained, well sorted sandstone (beach stratification). **B.** Symmetrical, flat-topped ripples and several other types of ripples unit 19. Arrow points to location of photo C. **C.** Symmetrical, flat-topped ripples. Location shown by arrow on photo B. **D.** Mud cracks on rippled bedding surface, unit 19. **E.** Unit 20. Three-foot-thick (1 m) set of tabular cross-strata showing north transport. Lower bounding surface is at man's feet. **F.** Contact of units 21 and 22 (arrow). Scour surface with several feet of relief is present at base of unit 22.

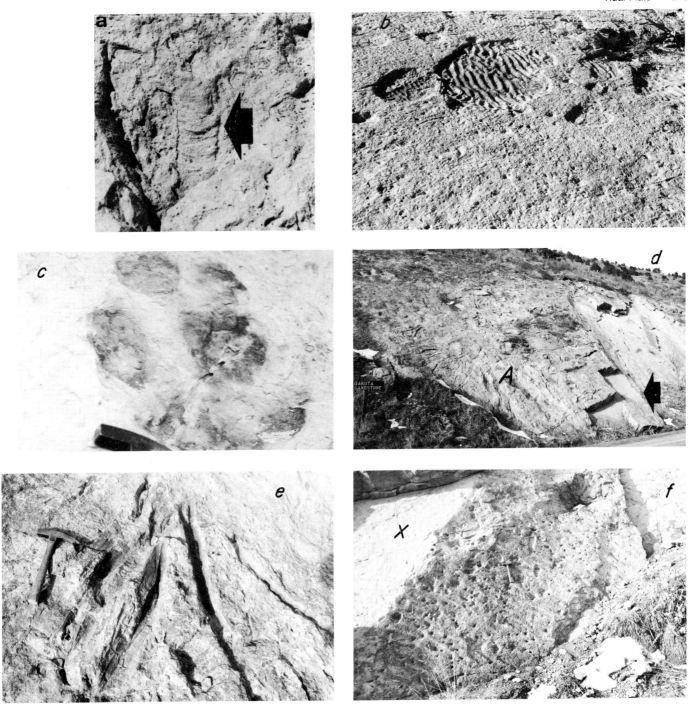

Fig. 37—Physical and biogenic structures (from Weimer and Land, 1972). **A.** U-burrow with spreiten-*Diplocriterion* (arrow), unit 23 (Fig. 34). View is of surface perpendicular to bedding. Burrow is approximately 2.5 cm wide. **B.** Bedding surface with possible algal mats with trails and burrows. Depression with ripple-marked surface may represent area where algal mat was removed, unit 23. **C.** Three-toed dinosaur footprint (compound ?), unit 23. **D.** Bedding surface with large log imprints (A) representing in-place destruction of mangrove-like swamp, unit 23. Scour channel at right cuts 10 ft (3 m) into underlying beds. Arrow points to general location of Figures 37F, 38A and B. **E.** large log imprints at A on photo D represent collapse of trees. Surface is underlain by root zone. **F.** Markings of *Skolithus* burrows on channel margin at location shown by arrow on photo D. Location of Figure 38A shown by "X," unit 23.

Fig. 38—Trace fossils, unit 23 (from Weimer and Land, 1972). **A.** *Skolithus* burrows on bedding surface. Location of photo shown of Figure 37F, unit 23. **B.** Bored log imprint. Location shown on Figure 37D.

R 99 W to initiate deposition of UA4 and UA5 sandstones. These shoreline sandstones are fine- to medium-grained, with higher energy deposits than fine-grained UA6 sandstones.

Figures 46 and 47 illustrate the lithologies of the UA6 interval. Photographs and interpretations of environments are taken from Van Horn (1979).

The productive tidal channel sandstone facies is gray very fine- to fine-grained cross-stratified porous and permeable sandstone with a minor amount of burrowing, minor slump structures, and clay drapes. Laminae of detrital carbonaceous clay are common in steeply dipping sets of cross-strata. Because only a few inches of each foot of core was kept for the Forest Arch Unit 63, certain details of the lithology are not present in Figure 46. Other features commonly observed in the tidal channel sandstones are a sharp scour base, a fining-upward texture, abraded shell fragments near the base, and clay clasts. Generally, core data do not permit determination of bi-model transport directions from cross-stratification. The UA6 sandstone is at a depth of 4,514 ft to 4,528 ft (1,376 to 1,380 m) and 432 b/d was produced through a 1/4 in. choke from a notch at 4,520 ft (1,378 m). The UA5 sandstone is from 4,480 to 4,500 ft (1,365 to 1,372 m). The UA6 is capped by a 14-ft (4.3 m) interval containing shale, siltstone, and coal interpreted as marsh and swamp deposits and tidal flats.

Cyclic sedimentation marginal to the major tidal channels is illustrated in core photographs of the Texas National UPRR 28A well (Fig. 47A, B, C). The stratigraphic position of the cored interval is shown on the electric log cross section (Fig. 44). Van Horn (1979, p. 70) stated:

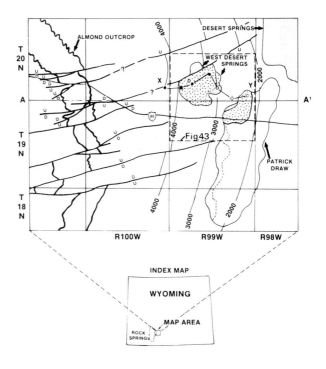

Fig. 40—Index map with location of producing areas from UA6 Sandstone in West Desert Springs and Patrick Draw fields (stippled areas). Location of lines of cross section A-A' and X-Y are indicated. Gas area in fields is indicated by diagonal ruling. Structure contours are of top of Almond Formation.

"The core clearly exhibits several symmetrical sedimentary cycles. Complete cycles contain an onlapping and offlapping sequence of coal, shale, siltstone and sandstone. Each sequence or half cycle, from the coal-shale contact to the midpoint of sandstone, is interpreted to represent sedimentation in environments that existed between high and low tide. The sequences range from 4 to 7 feet in thickness indicating a possible tidal range of 6

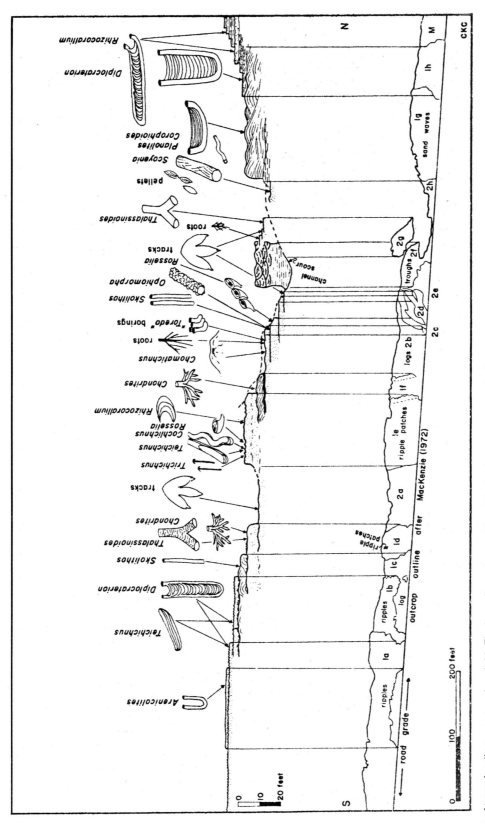

Fig. 39—Details of trace-fossil occurrences in unit 23. The lower figure is an outline of bedrock exposed by the roadcut with some physical characteristics cited to help locate beds; the outline is after the photographic mosaic of the area by MacKenzie (1968). The upper figure is a rotation back to horizontal of the tilted beds with lines showing corresponding points to the exposure outline (from Chamberlain, 1976).

to 10 feet when considering a 2 to 1 compaction factor for shaly intervals."

The coals represent freshwater marshes or swamps and are important reference horizons. In a typical cycle, a thin coal 1 to 4 in. thick (2.54 to 10.2 cm) is overlain by a salt marsh deposit of dark gray to brown shales containing carbonaceous plant material and brackish water pelecypods identified as *Corbula undifera* and *Modiolus sp.*, and the pelecypod *Durcella insculpta*. The central part of the cycle is light gray claystone, siltstone and very fine-grained sandstone that may be highly burrowed suggesting a lagoonal deposit, or ripple laminated and slightly burrowed layers suggesting tidal flat deposits. Overlying these layers are salt marsh shales and thin coals of freshwater marsh, or swamps which represent a regressive progradational part of the cycle.

Because of the shaly nature, the UA6 interval is nonproductive in the Texas National well. However, since the tidal flat and saltwater marsh are genetically related to tidal channels, the well developed cycles in this borehole might serve as a proximity indicator of a nearby channel sequence. This would aid significantly in exploration and development work in the area of the fields now incompletely developed.

SHORELINE AND TIDAL CHANNEL SANDS, AUX VASES (MISSISSIPPIAN) FORMATION, RURAL HILL FIELD, ILLINOIS

Sandstone reservoirs in the oil productive Aux Vases (Mississippian) zone in the Rural Hill field, Illinois, are composed of two distinct genetic sand types. These are interpreted to be: (1) shoreline sands deposited in linear buildups parallel with the regional northeast-southwest Aux Vases depositional trend; and, (2) tidal channel sands cutting across the shoreline trend and occurring in elongate confined belts. Channel sands occur in association with impermeable tidal flat and marsh deposits, and the zone is overlain and underlain by shallow marine oolitic limestones.

The two sandstone types have distinctively different textures, depositional structures, cementation, and reservoir geometries, and hence their identification is important in the prediction of reservoir trends and fluid flow properties within the zone. This discussion describes characteristics of these deposits and summarizes the criteria for their recognition.

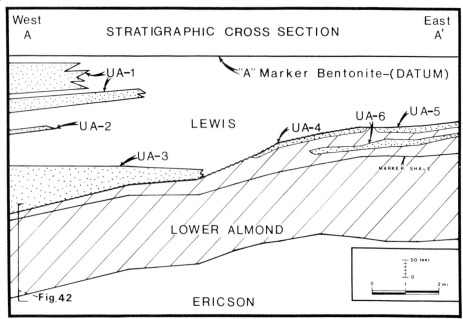

Fig. 41—Stratigraphic cross section with the position and terminology of the upper Almond sandstones in the outcrop and subsurface (UA1 through UA6). Stratigraphic interval shown on Figure 42 is indicated (from Van Horn, 1979). Location of section is along north line of T19 N, R99 and 100 W.

Fig. 42—Outcrop of approximately 200 ft (62 m) of lower Almond in SW NE Sec. 11, T19 N, R101 W. Upper 30 ft (9.1 m) of section (approximately equivalent to UA6 in subsurface) is tidal flats, salt and freshwater marsh deposits (above white sandstone). Lower lenticular sandstones and interbedded shale are interpreted as dominantly deposits of distributary channels, levees, swamps, and lakes (from Van Horn, 1979).

Geologic Setting and Data Control

Rural Hill field, in the southern part of the Illinois basin, was a region of slow subsidence surrounded by weakly positive areas during the Paleozoic Era (Fig. 48). The Aux Vases zone is 40 ft (12.2 m) thick and occurs at an average depth of 3,150 ft (960 m). The Aux Vases Formation is the basal member of the Upper Mississippi Chester Series and is interbedded between the Renault and Ste. Genevieve oolitic limestones (Fig. 49).

This study was based on examination of electric logs from 167 wells, and 49 cored intervals in the Aux Vases zone. Wells were drilled on 10-acre (4 ha.) spacing in the 1,954-acre (791 ha.) Rural Hill Water Flood Unit (RHFU).

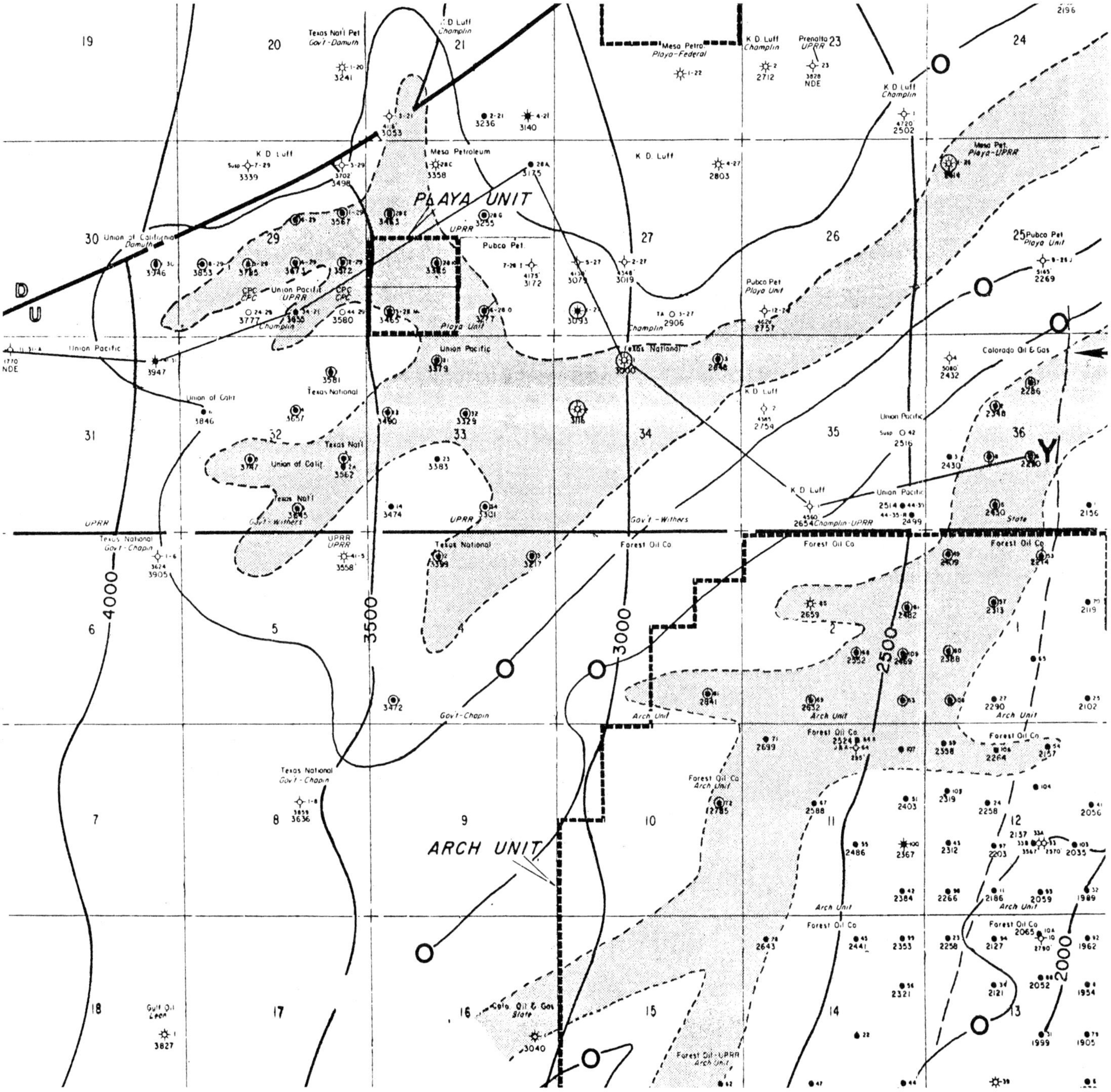

Fig. 43—Structure contour map on top of Almond Formation. Two tidal channel systems for the main body of UA6 sandstone are shown in stippled patterns. Productive wells from interval are circled. The pattern is where sandstone is interpreted as dominantly tidal channel and net pay is 5 to 15 ft (1.5-4.6 m) thick. Between 0 and the stippled pattern, net pay is less than 5 ft (1.5 m) and is interpreted as tidal creeks or the lower tidal sand flats. Compiled and interpreted from maps made available by Champlin Petroleum Company.

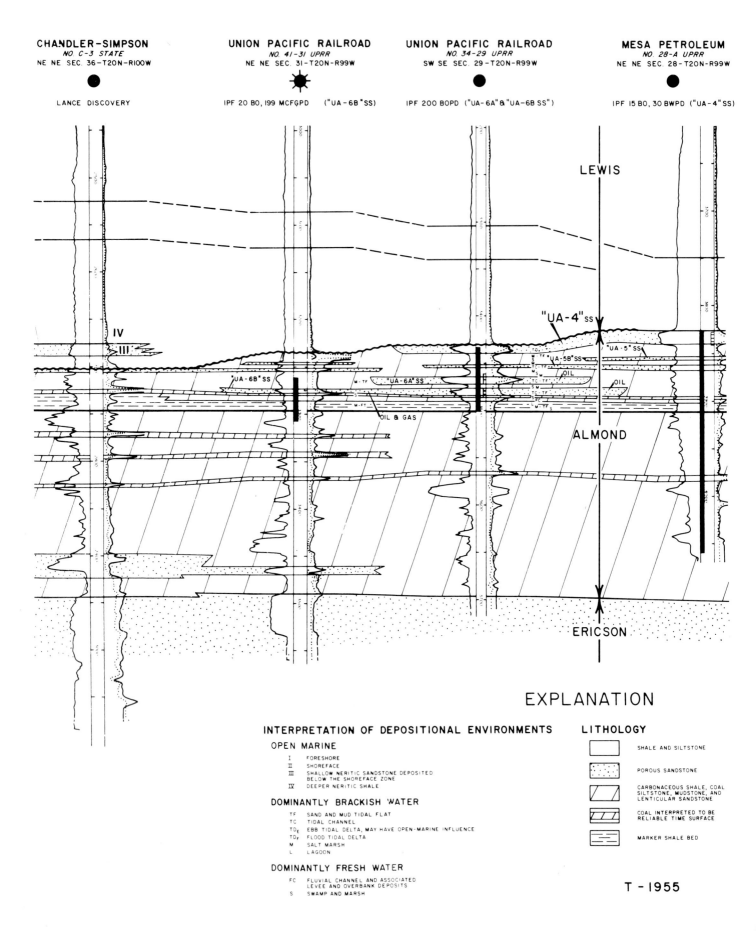

CHANDLER-SIMPSON
NO. C-3 STATE
NE NE SEC. 36-T20N-R100W

LANCE DISCOVERY

UNION PACIFIC RAILROAD
NO. 41-31 UPRR
NE NE SEC. 31-T20N-R99W

IPF 20 BO, 199 MCFGPD ("UA-6B"SS)

UNION PACIFIC RAILROAD
NO. 34-29 UPRR
SW SE SEC. 29-T20N-R99W

IPF 200 BOPD ("UA-6A" & "UA-6B SS")

MESA PETROLEUM
NO. 28-A UPRR
NE NE SEC. 28-T20N-R99W

IPF 15 BO, 30 BWPD ("UA-4"SS)

LEWIS

"UA-4"SS

"UA-5"SS

"UA-5B"SS

OIL

"UA-6B"SS

"UA-6A"SS

OIL

OIL & GAS

ALMOND

ERICSON

EXPLANATION

INTERPRETATION OF DEPOSITIONAL ENVIRONMENTS

OPEN MARINE

I   FORESHORE
II   SHOREFACE
III SHALLOW NERITIC SANDSTONE DEPOSITED BELOW THE SHOREFACE ZONE
IV   DEEPER NERITIC SHALE

DOMINANTLY BRACKISH WATER

TF   SAND AND MUD TIDAL FLAT
TC   TIDAL CHANNEL
TDe EBB TIDAL DELTA, MAY HAVE OPEN-MARINE INFLUENCE
TDf FLOOD TIDAL DELTA
M   SALT MARSH
L   LAGOON

DOMINANTLY FRESH WATER

FC   FLUVIAL CHANNEL AND ASSOCIATED LEVEE AND OVERBANK DEPOSITS
S   SWAMP AND MARSH

LITHOLOGY

SHALE AND SILTSTONE

POROUS SANDSTONE

CARBONACEOUS SHALE, COAL SILTSTONE, MUDSTONE, AND LENTICULAR SANDSTONE

COAL INTERPRETED TO BE RELIABLE TIME SURFACE

MARKER SHALE BED

T-1955

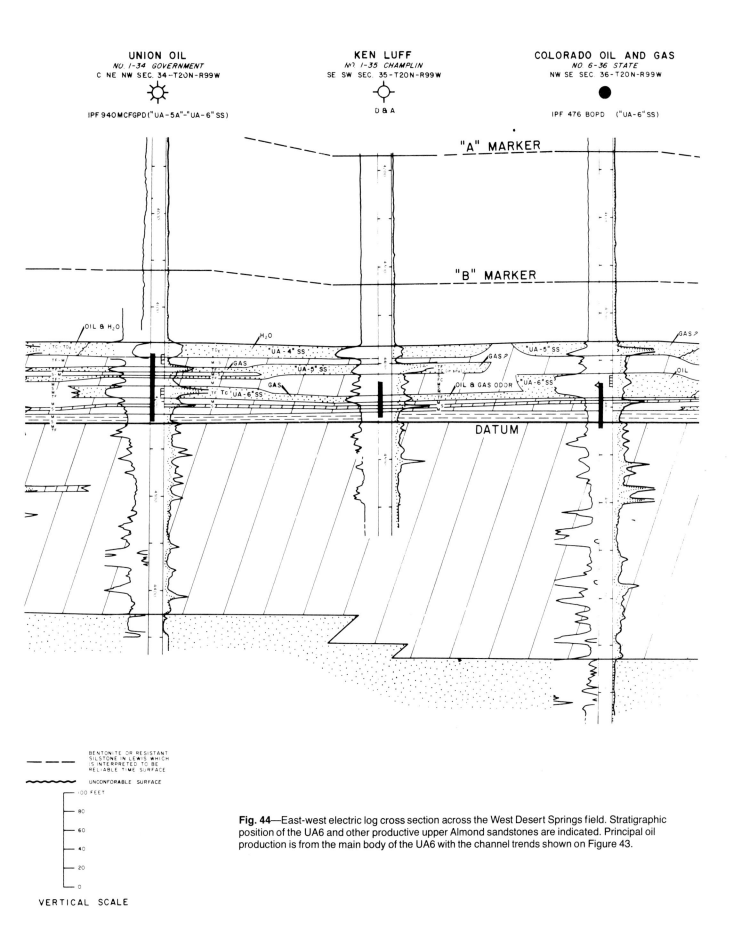

Fig. 44—East-west electric log cross section across the West Desert Springs field. Stratigraphic position of the UA6 and other productive upper Almond sandstones are indicated. Principal oil production is from the main body of the UA6 with the channel trends shown on Figure 43.

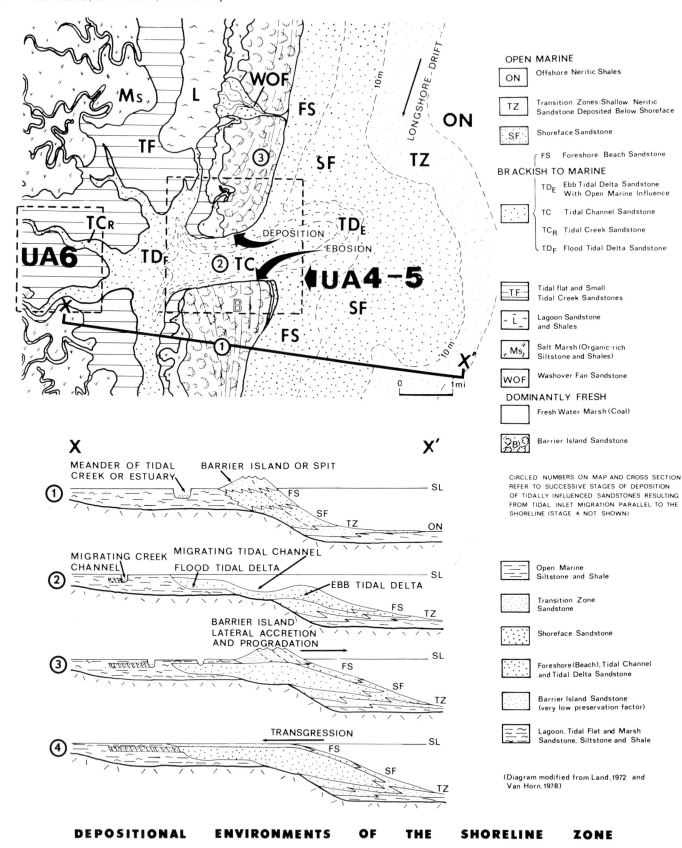

DEPOSITIONAL ENVIRONMENTS OF THE SHORELINE ZONE

Fig. 45—Depositional environments of the shoreline zone summarized from modern environments and modified to represent stratigraphic model for petroleum-productive Almond sandstones. UA6 sandstone at West Desert Springs is believed representative of tidal channel, tidal creeks, and lower tidal sand flats inland from main shoreline zone. UA4 and UA5 sandstones are tidal channels of main shoreline sand zone.

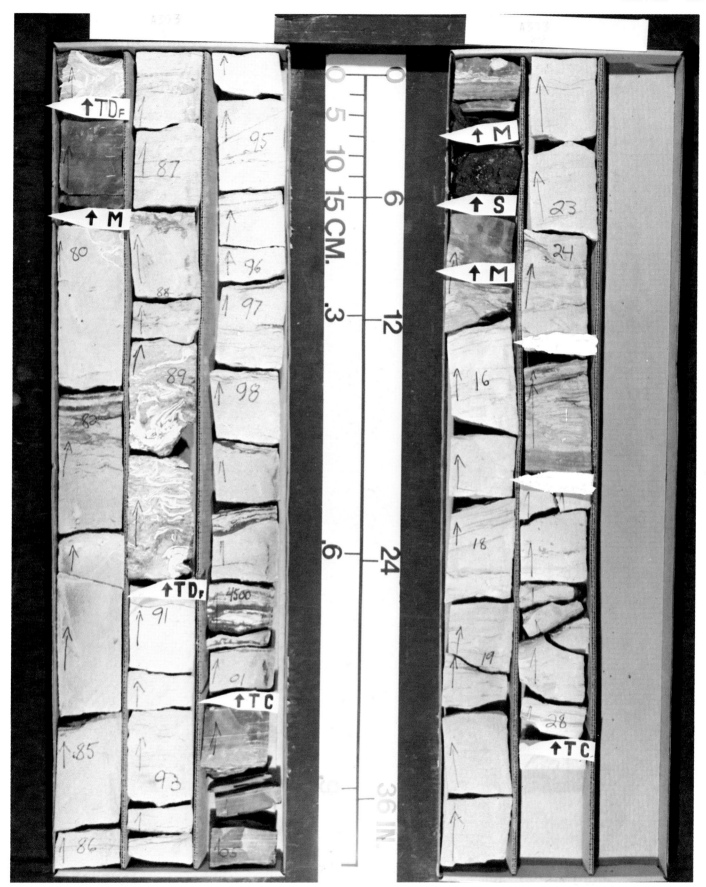

Fig. 46A—Core photo from Forest Arch Unit 63, C SE SE, Sec. 2, T19 N, R 99 W interval 4,479 to 4,528 ft (1,365 to 1,380 m). UA5 sandstone from 4,479 to 4,501 ft (1,365 to 1,372 m); UA6 sandstone from 4,516 to 4,528 ft (1,377 to 1,380 m). Only 2 to 4 in. (5 to 10 cm) representative of each foot of core was saved for analysis. Symbols are same as on Figure 44.

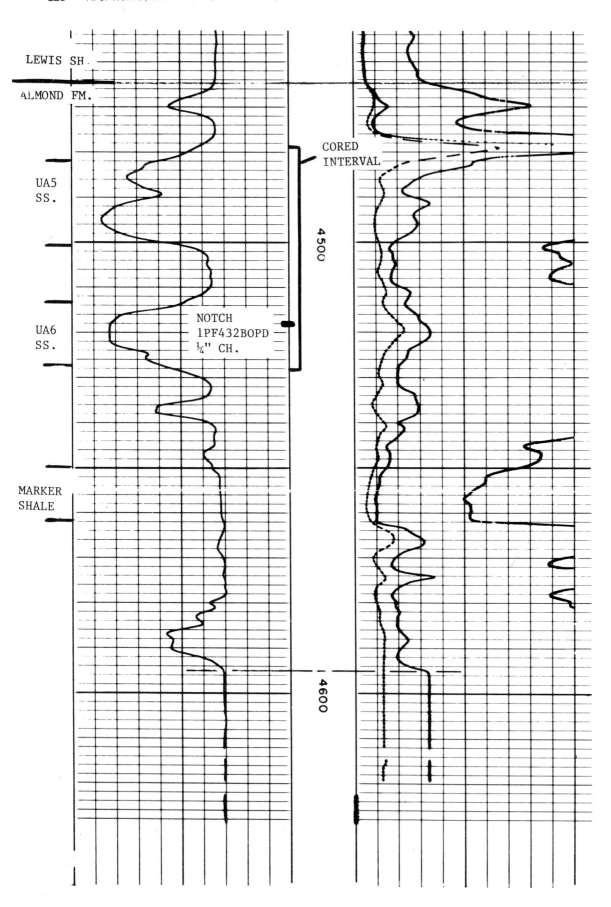

Fig. 46B—Electric log of Forest Arch Unit 63. Cored interval, sandstone intervals, and marker shale are designated.

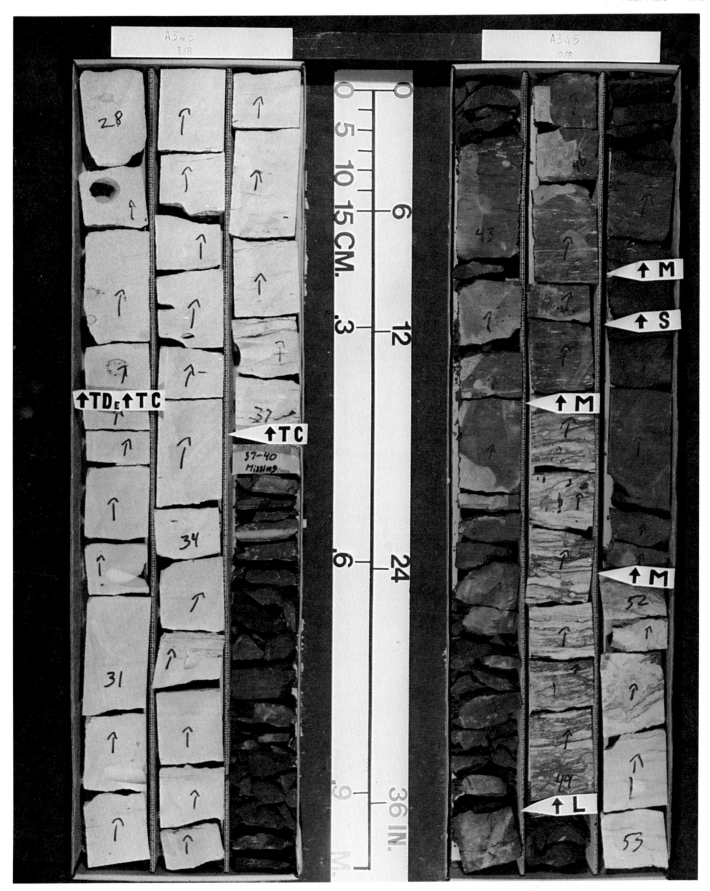

Fig. 47A—Core photo from Mesa UPRR 28A, NE NE Sec. 28, T20 N, R99 W, interval 3828-3853 ft (1167 to 1175 m). Core interval and symbols are indicated on Figure 44. Minor oil production was from UA4 sandstone from 3828-3837 ft (1167 to 1175 m) and is interpreted as dominantly a tidal channel deposit. Log of cored interval is on Figure 44. Core is stored at U.S. Geological Survey core library, Denver Colorado.

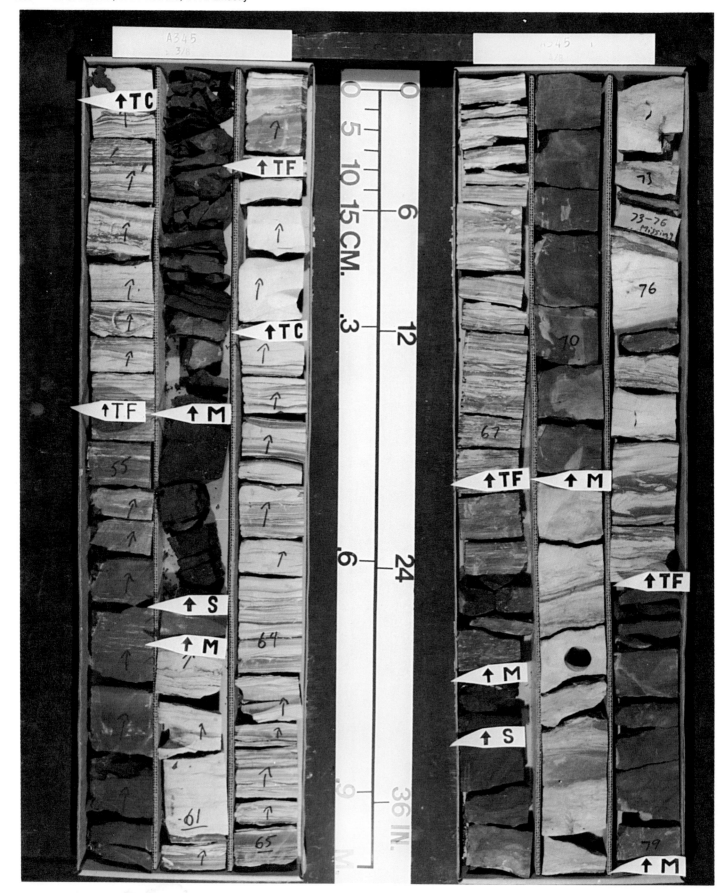

Fig. 47B—Core photo from Mesa UPRR 28A, interval 3,853 to 3,879 ft (1,174 to 1,182 m). Upper UA6 interval is from 3,868 to 3,879 ft (1,179 to 1,182 m) and consists of nonreservoir shales, siltstones, and sandstones interpreted as tidal flat, marsh, and swamp deposits.

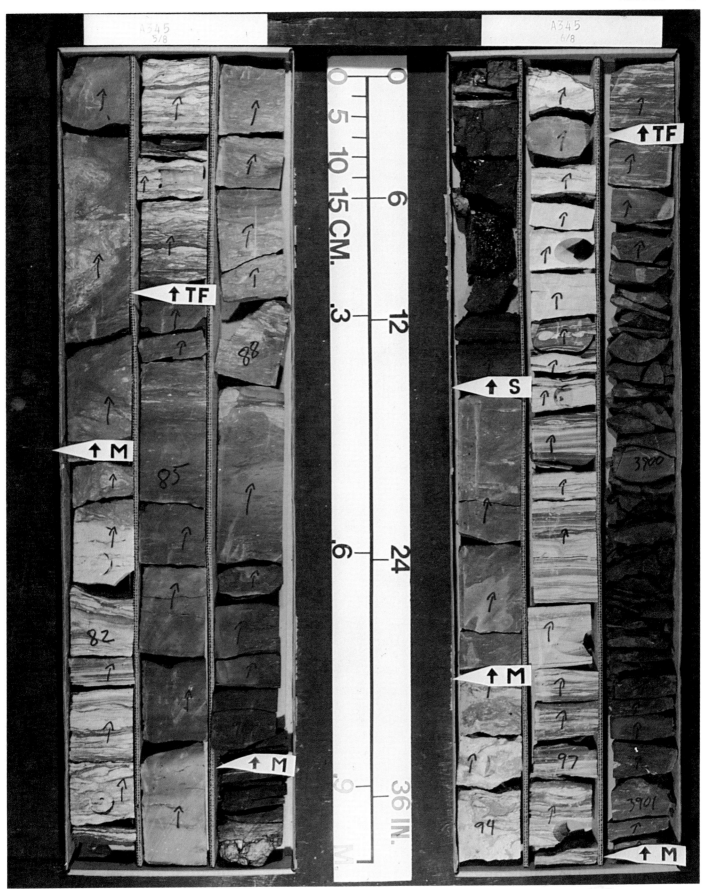

Fig. 47C—Core photo from Mesa UPRR 28A, interval 3,879 to 3,902 ft (1,182 to 1,189 m). Lower UA6 interval is from 3,879 to 3,890 ft (1,182 to 1,186 m) and is interpreted as tidal flat, marsh, and swamp deposits.

General Characteristics of the Aux Vases Reservoir Sandstones

Shoreline and tidal channel sandstones have some similar characteristics. Both are mature orthoquartzites, typically clean and well sorted, containing less than one percent clay by volume. Both contain two to three percent fossils consisting of oolitically coated skeletal fragments of crinoids, brachiopods, and other marine macrofauna, and small amounts of calcite cement which tends to be localized in the more fossiliferous parts of the sandstones.

Two sandstone types can be distin-guished by different sedimentary struc-tures, average grain size, and cementa-tion. Shoreline sandstones are characteristically horizontally to nearly horizontally laminated and fine grained (Fig. 50). In contrast, channel sandstones are cross-bedded and very fine grained. Furthermore, shoreline sandstones invariably contain silica ce-ment which does not occur in the chan-nel deposits. Grain size of the shoreline sandstones has been slightly increased by authigenic silica overgrowths, but the sands as originally deposited were fine grained and, therefore, coarser than the channel sands.

Petrophysical Properties and Log Expression

Figure 51 illustrates that pore-filling silica cement has reduced porosity and permeability of shoreline sandstones. Core analyses indicate the average ef-fective porosity in shoreline sand-stones is 15% as opposed to 21% in the channel sandstones. Air permeabilities rarely exceed 100 millidarcys in shore-line sandstones, but range up to several hundred millidarcys in channel sand-stones.

As a result of larger average grain size and porosity reduction by addition of silica cement, electric log resistivity values are contrastingly higher oppo-site shoreline sandstones. The reason for selective silica cementation of the shoreline sands is not known, but the resistivity contrast is useful in identify-ing the two sandstone types on electric logs of uncored wells and mapping their extent and distribution.

Reservoir Zonation and Correlation

The Aux Vases is divided into lower and upper subzones (Fig. 52). The lower subzone is a composite of shore-line and tidal channel sandstones, while the upper subzone consists of tidal channel sandstones separated lat-erally (and sometimes vertically) by impermeable tidal flat and marsh de-posits. In some wells, the two subzones are separated vertically by a thin bed consisting of variously impermeable shaly siltstone, tightly cemented sand-stone, or limestone. All well correla-tions in this study were made using a thin, widespread lime mudstone in the Ste. Genevieve limestone as datum.

Sequence and Distribution of Genetic Deposits

Shoreline Sandstones—A core pho-tograph of a typical shoreline sand-stone sequence is shown in Figure 53. The sequence consists almost entirely of fine-grained, clean, well sorted quartz sand with occasional thin silt and clay laminae. Grain size is nearly uniform with minor fluctuations, and decreases at the top and bottom of the interval. The most characteristic sedi-mentary structures are the nearly hori-zontal laminations and low-angle cross laminations. Higher angle festoon cross-beds occur but are rare. Bur-rowed structures and small scale ripple cross-beds are present at the base and top of the sequence. The maximum net permeable sandstone thickness in an Aux Vases shoreline deposit is 20 ft (6.1 m).

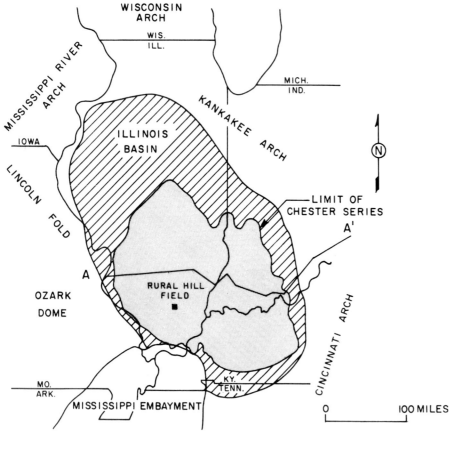

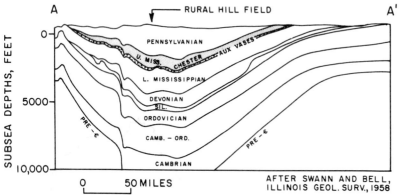

Fig. 48—Map and cross section showing the location and geologic setting of the Rural Hill field, Illinois.

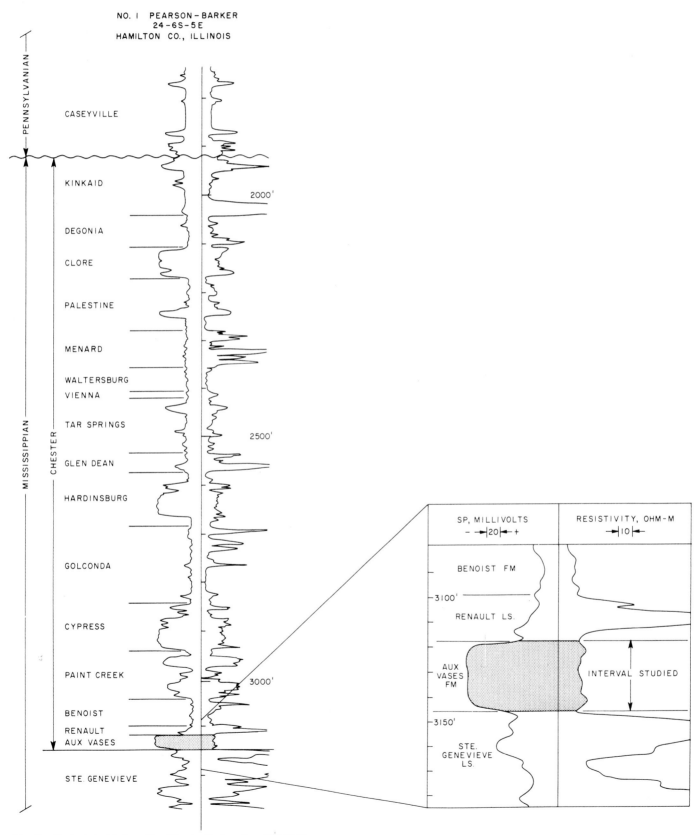

Fig. 49—Typical electric log of the Chester Series, Rural Hill Field.

Fig. 50—Photographs of typical Aux Vases shoreline and channel sandstones. **A.** Shoreline sandstone—horizontally to nearly horizonally laminated, fine grained, well sorted, silica cemented, Shell, RHFU 128-W, 3,123 ft (952 m). **B.** Channel sandstone—cross bedded, very fine grained, well sorted, not silica cemented, Shell, RHFU 4, 3,118 ft (950 m).

The overall sequence records a cycle of increasing and then decreasing current strength. Finer grained, shaly, laminated to ripple-bedded sandstones and siltstones near the top and bottom indicate deposition by relatively low energy currents, whereas coarser, well winnowed sandstones in the middle part indicate higher current strength.

Shoreline sandstones occur only in the lower Aux Vases subzone. The distribution and thickness of shoreline and associated tidal channel sandstones are illustrated in Figures 54 and 55. Shoreline sands are interpreted as having been deposited by wave generated currents.

The fossils and gradational contact with oolitic limestones indicate a shallow marine depositional environment. Well-sorted sand textures and the virtual absence of clay indicate strong winnowing currents, while horizontal laminations and low-angle cross-laminations are characteristic of modern shore deposits. The occasional festoon cross-beds indicate directed currents which could result from either longshore or tidal currents. These characteristics suggest a shoreface to foreshore beach depositional environment.

Tidal Channel Sandstones—A cored tidal channel sandstone sequence (Fig. 56) consists almost entirely of very fine-grained, well sorted, clean quartz sandstone. Sandstones are mostly cross-bedded, with small- (<3 in.; 7.6 cm) to medium-scale (3 in. to 3 ft; 7.6 cm to 0.91 m in thickness) cross-laminated beds truncated at the top and tangential at the base. Some burrows occur, as at interval 3,104 to 3,105 ft (946.1 to 946.4 m) in the photograph. Maximum thickness of a channel sandstone sequence in the Aux Vases reservoir (upper and lower subzones combined) is 40 ft (12.2 m). The clean, well sorted, cross-bedded sandstones indicate deposition by high energy, winnowing, strongly directed currents.

Alternating scour and deposition is shown by truncated bedding, and a shallow marine sand source is indicated by fossil fragments. Since the avalanche slope deposits of migrating ripple cross-beds face downcurrent, the dip directions of the laminations serve as indicators of current direction. In aligned core segments from the Aux Vases tidal channel sandstones, it can be seen that dip directions of the cross strata in places are 180° opposed. This is illustrated in Figure 57, which shows the interval 3,119 to 3,120 ft (950.7 to 950.9 m) from Figure 56 in detail. These sands were deposited by opposed currents interpreted to be flood and ebb tidal currents.

In the lower Aux Vases subzone, the intertonguing of shoreline and tidal channel sandstone indicates contemporaneity of the two depositional processes. The slightly coarser grain size of shoreline deposits suggests that as those sands were concentrated along the shore by waves, finer sands were transported into tidal inlets during flood tides. The upper subzone consists of confined, elongate belts of tidal channel sandstones separated laterally by tidal flat and marsh deposits (Fig. 58). Because modern tidal channel sands characteristically die out

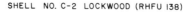

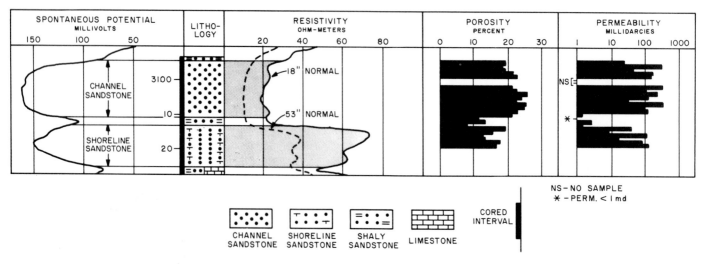

Fig. 51—Typical porosity and air permeability values, and electric log response to the reservoir sandstone types. Because of variation in silica and calcite cementation, SP does not reflect grain size variation.

landward in tidal creeks, the Aux Vases channel sandstones probably thin and terminate in a landward (northwesterly) direction (Fig. 54). Superposition of tidal flat deposits on underlying shorelines sands indicates a regressive depositional Aux Vases sequence, followed by transgression and deposition of the Renault oolitic limestones.

Tidal Flat and Marsh Deposits—Tidal flat and marsh strata (Fig. 59) consist of interlaminated very fine sandstone, siltstone, and claystone, characterized by an upward decrease in grain size and increase in clay content. Thin, horizontal laminations and very fine ripple beds are present throughout most of the sequence, indicating deposition by weak, fluctuating currents. Clay-rich sediments at the top are poorly bedded and extensively reworked by burrowing organisms. Color changes upward from gray and green to predominantly red, indicating oxidation of the upper part of the sequence.

The overall sequence is interpreted to consist of tidal flat deposits grading upward into marsh sediments. The same types of marine fossils as occur on the shoreline and channel sandstones are present in the lower part of the tidal flat sequence, but diminish upward and are absent in the marsh beds. Dessication cracks and root marks occur in the upper part of some tidal flat and marsh sequences (Fig. 60).

Criteria for Recognition of the Genetic Sandstone Bodies

Taken individually, sedimentary features of the Aux Vases reservoir sandstones can also be observed in other different genetic types of sand deposits. For example, horizontal laminations may occur in alluvial point bar deposits as well as in shoreline sands, and festoon cross-beds characterize many different types of sands bodies deposited by directed currents. Therefore, interpretations will depend on combinations of related observations indicating a particular depositional system. Distinguishing features of the Aux Vases shoreline and tidal channel sandstone deposits are summarized in Table 1.

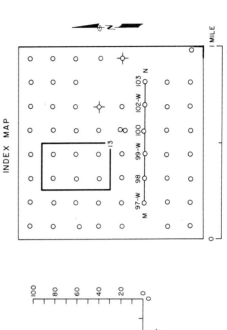

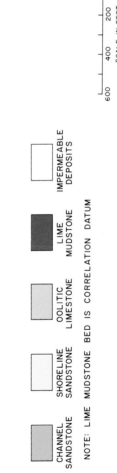

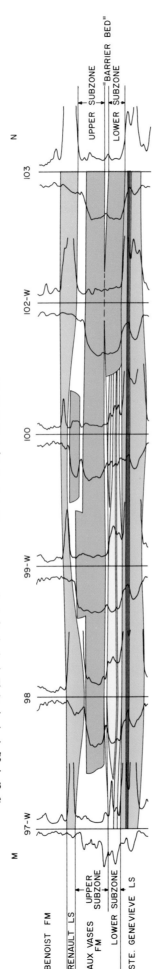

Fig. 52—Correlation section showing Aux Vases upper and lower subzones.

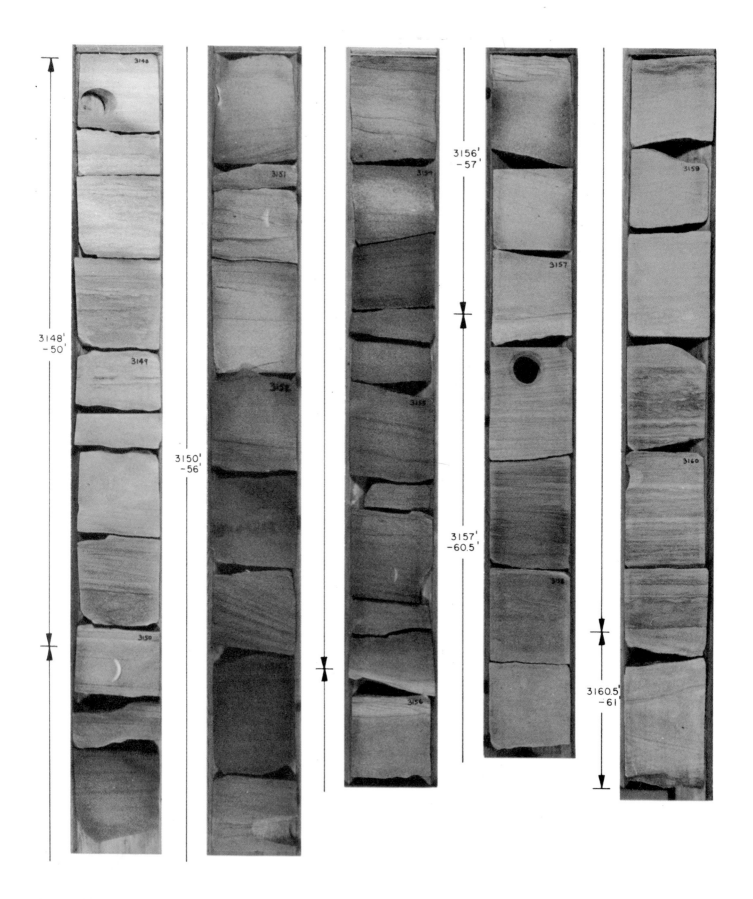

Fig. 53—Core photograph of a shoreline sandstone sequence. Shell, RHFU 109, 3,148 to 3,161 ft (960 to 963 m).

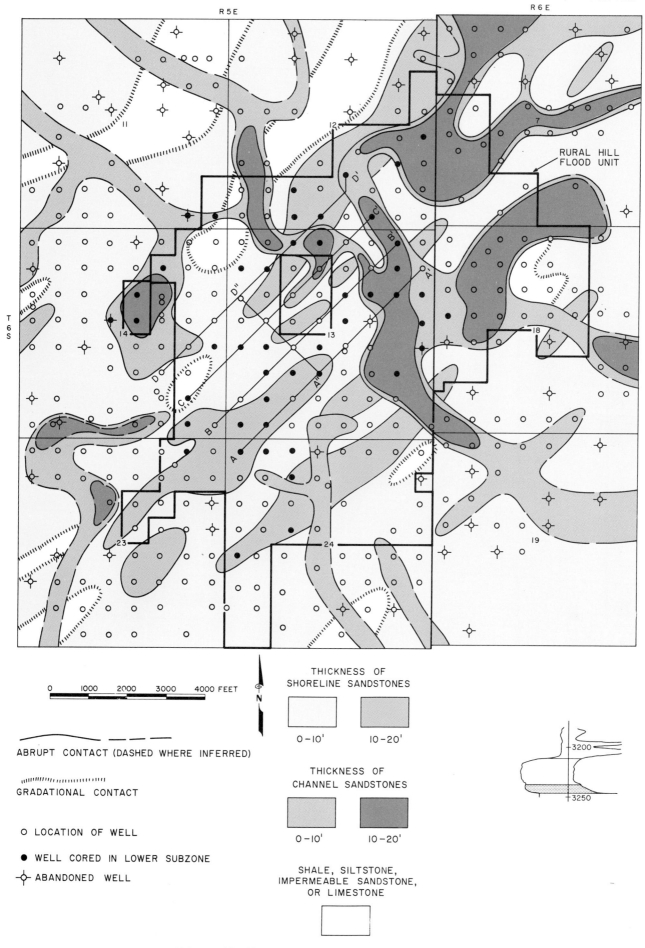

R 5 E

R 6 E

RURAL HILL
FLOOD UNIT

T 6 S

0   1000   2000   3000   4000 FEET

N

THICKNESS OF
SHORELINE SANDSTONES

0–10'   10–20'

THICKNESS OF
CHANNEL SANDSTONES

0–10'   10–20'

SHALE, SILTSTONE,
IMPERMEABLE SANDSTONE,
OR LIMESTONE

ABRUPT CONTACT (DASHED WHERE INFERRED)

GRADATIONAL CONTACT

O LOCATION OF WELL

● WELL CORED IN LOWER SUBZONE

✛ ABANDONED WELL

3200

3250

Fig. 54—Map showing distribution and thickness of Aux Vases shoreline and tidal channel sandstones in the lower subzone.

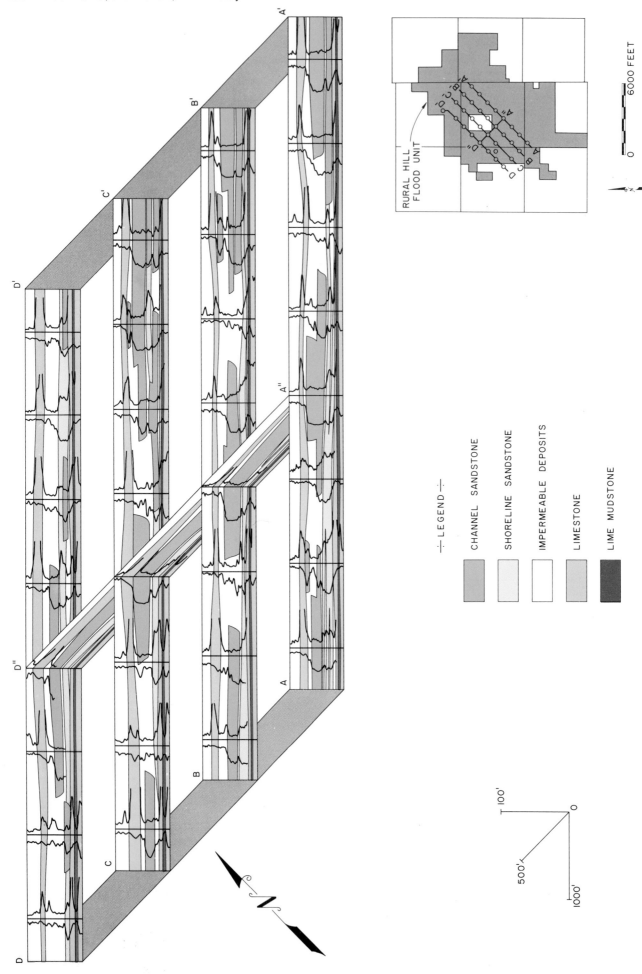

Fig. 55—Cross sections showing vertical and lateral distribution of Aux Vases shoreline and tidal channel sandstone reservoirs. The impermeable deposits in the upper subzone consist primarily of tidal flat and marsh sequences.

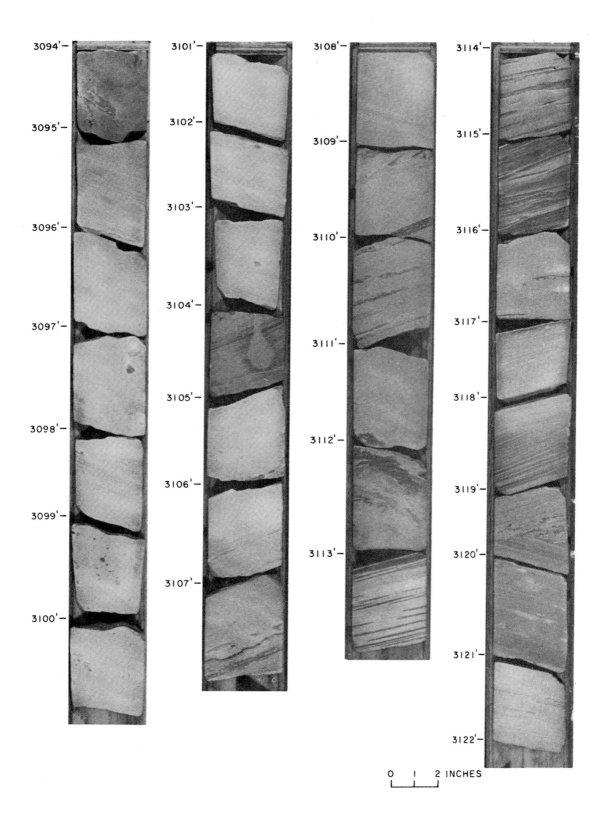

3094'—
3095'—
3096'—
3097'—
3098'—
3099'—
3100'—

3101'—
3102'—
3103'—
3104'—
3105'—
3106'—
3107'—

3108'—
3109'—
3110'—
3111'—
3112'—
3113'—

3114'—
3115'—
3116'—
3117'—
3118'—
3119'—
3120'—
3121'—
3122'—

0   1   2 INCHES

Fig. 56—Core photograph of a tidal channel sandstone sequence. Shell, RHFU 4, 3,094 to 3,122 ft (942 to 952 m).

DIP VIEW

INFERRED CURRENT DIRECTIONS

0   1   2 INCHES

STRIKE VIEW

INFERRED CURRENT DIRECTIONS

AWAY FROM OBSERVER
—————————————————
TOWARD OBSERVER

Fig. 57—Opposed cross-beds in a tidal channel sandstone. Shell, RHFU 4, 3,119 to 3,120 ft (951 m).

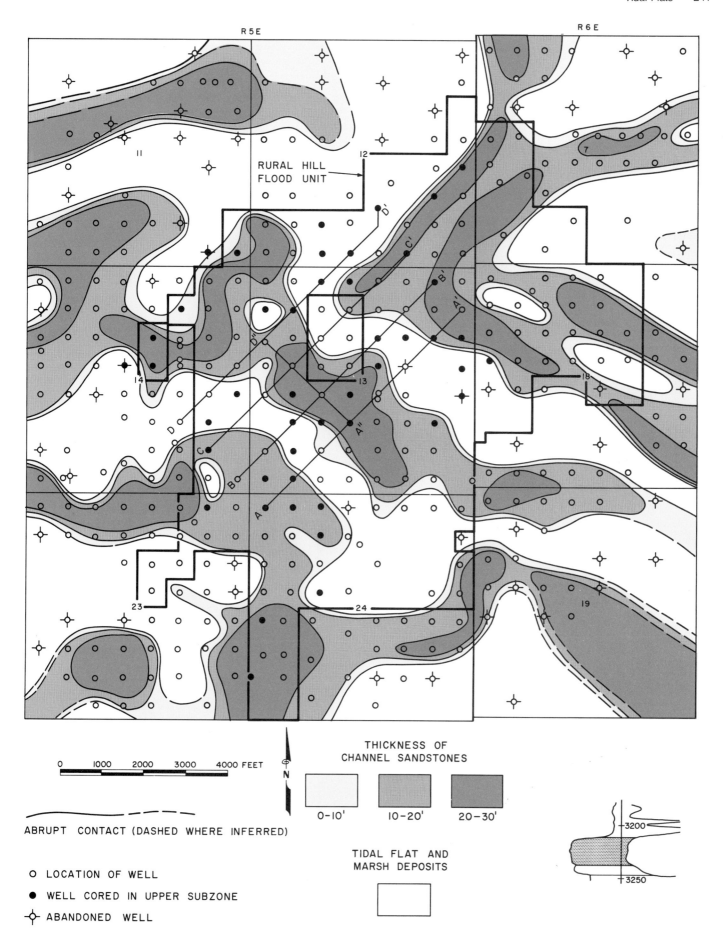

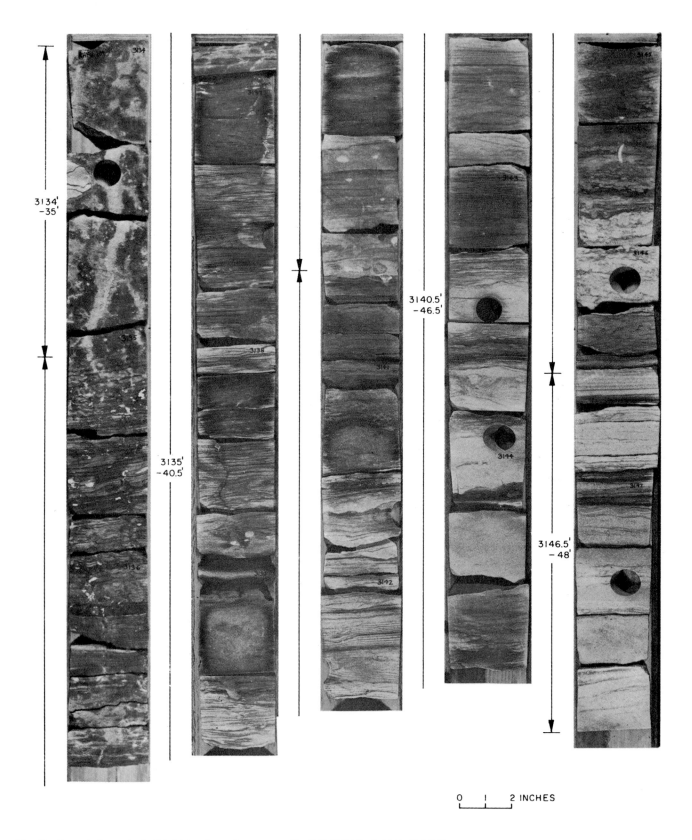

Fig. 59—Core photograph of a tidal flat and marsh sequence. Shell, RHFU 109, 3,134 to 3,148 ft (955 to 960 m). (Note: This sequence directly overlies the shoreline sandstone in Fig. 53).

A

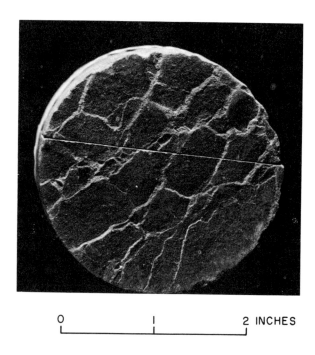

0   1   2 INCHES

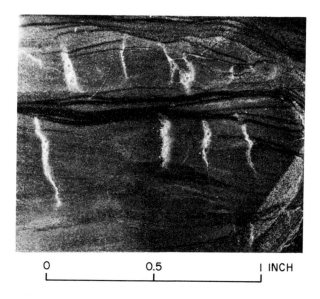

0   0.5   1 INCH

B

Fig. 60—Photographs of desiccation cracks and root marks in some tidal flat and marsh deposits. **A.** Desiccation cracks in a thin clay laminae. Cracks are filled with inwashed silt. Shell 1 Crabtree (RHFU 94), 3,128 ft (953 m). **B.** Calcite-filled root marks extending downward from clay laminae in siltstone. Shell 2, Ventress (RHFU 22), 3,164 ft (964 m).

	Shoreline Sands	Tidal Channel Sands
Sand textures	Clean, well sorted, fine grained.	Clean, well sorted, very fine grained.
Vertical sequence	Nearly uniform grain size, minor vertical fluctuations.	Nearly uniform grain size, variable within a narrow range.
Sedimentary structures	Horizontal to slightly inclined laminations; small low-angle cross-laminations, occasional small-scale festoon cross-beds and ripple beds, burrows rare.	Small- to medium-scale, high-angle cross-beds; some opposed cross-bedding directions, burrows rare.
Faunal content	Reworked skeletal fragments of contemporaneous shallow marine fossils; many fragments oolitically coated, some well preserved.	
Sand trends and distribution	Buildups parallel to depositional strike.	Elongate confined belts nearly perpendicular to depositional strike.
Boundary relations	Gradational except where cut by tidal channel.	Sharp and erosional at sides and base, gradational to abrupt at top.
Association with other rock types	Conformably overlies oolitic limestone, grades upward into tidal flat deposits, breached by tidal channels.	Situated laterally between tidal flat and marsh deposits or shoreline sand buildups, interbedded with oolitic limestones.

Table 1. Distinguishing features of Aux Vases shoreline and tidal channel sandstones.

REFERENCES CITED

General

Ginsburg, R. N., 1975, Tidal deposits: A casebook of recent examples and fossil counterparts: Springer-Verlag, New York, 428 p.

Reineck, H. E., 1973, Bibliographic geolgischer Arbeiten uber rezente und fossile Kalk- und Silikatwatten. (Bibliography of recent and ancient tidalites): Cour. Forsch. Inst. Senckenberg 6, Frankfurt a.M., 57 p.

——— and F. Wunderlich, 1968, Classification and origin of flaser and lenticular bedding: Sedimentology, v. 11, p. 99-104.

North Sea

Dorjes, J., 1978, Das Watt als Lebenstraum, in H. E. Reineck, ed., Das Watt: Kromer Frankfurt a.M., p. 107-143.

Evans, G., 1965, Intertidal flat sediments and their environments of deposition in the Wash: Geol. Soc. London Quart. Jour., v. 121, p. 209-245.

Larsonneur, C., 1975, Tidal deposits, Mont Saint Michel Bay, France, in R. N. Ginsburg, ed., Tidal deposits: A casebook of recent examples and fossil counterparts: Springer-Verlag, New York, p. 21-30.

Reineck, H. E., 1967, Layered sediments of tidal flats, beaches, and shelf bottom of the North Sea, in G. H. Lauff, ed., Estuaries: Am. Assoc. Adv. Sci., Pub., Washington, D.C., p. 191-206.

——— 1972, Tidal flats, in J. K. Rigby and W. K. Hamblin, eds., Recognition of ancient sedimentary environments: SEPM Spec. Pub. 16, p. 146-159.

——— 1975, German North Sea tidal flats, in R. N. Ginsburg, ed., Tidal deposits: A casebook of recent examples and fossil counterparts: Springer-Verlag, New York, p. 5-12.

Straaten, L. M. J. U. van, 1954, Composition and structure of recent marine sediments in the Netherlands: Leidse Geol. Meded XIX, p. 1-110.

——— 1957, The Holocene deposits: The excavation at Velsen: Verhand Kon. Ned. Geol. Mijnb. Gen. Geol. Ser. XVII, p. 158-183.

——— 1959, Minor structure of some recent littoral and neritic sediments: Geologie en Mijnbouw, v. 21, p. 197-216.

——— 1961, Sedimentation in tidal flat areas: Alberta Soc. Petrol. Geol. Jour., v. 9, p. 203-226.

Georgia

Basan, P. B., and R. W. Frey, 1978, Actualpalaeontology and neoichnology of salt marshes near Sapelo Island, Georgia, in T. P. Crimes and J. C. Harper, eds., Trace Fossils 2: Geol. Jour. Spec. Pub. 9, p. 41-70.

Edwards, J. M., and R. W. Frey, 1977, Substrate characteristics within a Holocene salt marsh, Sapelo Island, Georgia: Senchenbergiana marit., v. 9, p. 215-259.

Frey, R. W., and P. B. Basan, 1978, Coastal salt marshes, in R. A. Davis, Jr., ed., Coastal sedimentary environments: Springer-Verlag, New York, p. 101-169.

Howard, J. D., and R. W. Frey, 1975, Estuaries of the Georgia coast, U.S.A.: Sedimentology and biology. II. Regional animal-sediment characteristics of Georgia estuaries: Senchenbergiana merti., v. 7, p. 33-103.

Gulf of California

Thompson, R. W., 1968, Tidal flat sedimentation on the Colorado River delta, northwest Gulf of California: Geol. Soc. America Mem. 107, 133 p.

——— 1975, Tidal flat sediments of the Colorado delta, northwestern Gulf of California, in R. N. Ginsburg, ed., Tidal deposits: A casebook of recent examples and fossil counterparts: Springer-Verlag, New York, p. 57-65.

Bay of Fundy

Dalrymple, R. W., R. J. Knight, and G. V. Middleton, 1975, Intertidal sandbars in Cobequid Bay (Bay of Fundy), in L. E. Cronin, ed., Estuarine Research, Vol. III: Academic Press, New York, p. 293-307.

——— R. J. Knight, and J. J. Lambiase, 1978, Bedforms and their hydraulic stability relationships in a tidal environment, Bay of Fundy, Canada: Nature, v. 275, p. 100-104.

Klein, G. de V., 1970, Depositional and dispersal dynamics of intertidal sandbars: Jour. Sed. Petrology, v. 40, p. 1095-1127.

Knight, R. J., and R. W. Dalrymple, 1975, Intertidal sediments from the southshore of Cobequid Bay, Bay of Fundy, Nova Scotia, Canada, in R. N. Ginsburg, ed., Tidal deposits: A casebook of recent examples and fossil counterparts: Springer-Verlag, New York, p. 47-55.

Swift, D. J. P. and R. M. McMullen, 1968, Preliminary studies of intertidal sand bodies in the Minas Basin, Bay of Fundy, Nova Scotia: Canadian Jour. Earth Sci., v. 5, p. 175-183.

Wind-Tidal Flats

Miller, J. A., 1975, Facies characteristics of Laguna Madre wind-tidal flats, in R. N. Ginsburg, ed., Tidal deposits: A casebook of recent examples and fossil counterparts: Springer-Verlag, New York, p. 67-73.

References For Ancient Examples

Campbell, C. V., and R. Q. Oaks, Jr., 1973, Estuarine sandstone filling tidal scours, Lower Cretaceous, Fall River Formation, Wyoming: Jour. Sed. Petrology, v. 43, no. 3, p. 765-778.

Chamberlain, C. K., 1976, Field guide to trace fossils of the Cretaceous Dakota Hogback, along Alameda Avenue, West of Denver, Colorado: School of Mines Prof. Contrib. 8, p. 242-250.

Land, C. B., Jr., 1972, Stratigraphy of Fox Hills Sandstone and associated Formations, Rock Springs Uplift and Wamsutter Arch area, Sweetwater County, Wyoming: a shoreline-estuary sandstone model for the Late Cretaceous: Colo. School of Mines Quart., v. 67, no. 2, 69 p.

McCubbin, D. G., and M. J. Brady, 1968, Depositional environment of the Almond reservoirs, Patrick Draw field, Wyoming: Mtn. Geologist, v. 6, no. 1, p. 3-26.

MacKenzie, D. E., 1963, Depositional environments of Muddy Sandstone, Western Denver Basin, Colorado: AAPG Bull., v. 49, p. 186-206.

———— 1968, Studies for students: sedimentary features of Alameda Avenue Cut, Denver, Colorado: Mtn. Geologist, v. 5, no. 1, p. 3-13.

———— 1971, Post Lytle Dakota Group on west flank of Denver basin, Colorado: Mtn. Geologist, v. 8, no. 3, p. 91-131.

Mitchell, G. C., 1976, Grieve oil field, Wyoming: a Lower Cretaceous estuarine deposit: Mtn. Geologist, v. 13, no. 3, p. 71-87.

Oomkens, E., 1974, Lithofacies relations in the Late Quaternary Niger delta complex: Sedimentology, v. 21, p. 195-222.

Spearing, D. R., 1974, Summary sheets of sedimentary deposits: Geol. Soc. America MC-8, 7 sheets.

Swann, D. H., and A. H. Bell, 1958, Habitat of oil in the Illinois basin in L. G. Weeks, ed., Habitat of Oil: AAPG Spec. Pub., p. 447-472.

Van Horn, M. D., 1979, Stratigraphy of the Almond Formation, east-central flank, Rock Springs uplift, Sweetwater County, Wyoming—a mesotidal-shoreline model for the Late Cretaceous: master's thesis, Colo. School of Mines, Golden, 150 p.

Waage, K. M., 1955, Dakota Group in the northern Front Range foothills: U.S. Geol. Survey Prof. Paper 274-B, p. 15-51.

Weimer, R. J., 1966, Time-stratigraphic analysis and petroleum accumulations Patrick Draw Field, Sweetwater County, Wyoming: AAPG Bull., v. 50, no 10, p. 2150-2175.

———— and C. B. Land, 1972, Field guide to Dakota Group (Cretaceous) stratigraphy, Golden-Morrison area, Colorado: Mtn. Geologist, v. 9, nos, 2-3, p. 241-267.

ACKNOWLEDGMENTS

Our purpose in this study has been to summarize and synthesize. Different workers in different geographic settings have used different approaches to describe modern tidal flat depositional sequences. Because of this we have, in trying to make comparisons, taken the liberty to sometimes use different terms or even to make interpretations different than those used by the original author. We hope this statement will serve as an apology to anyone whose work we have tampered with and likewise as a warning to those who, in seeking additional detail, may find some discrepancies as they examine the original papers of various authors.

The references for this chapter are not exhaustive. Rather we include some of the main papers which deal with the examples we have cited and additional papers which present an overview on the subject of tidal flats.

We asked numerous colleagues to share some of their prized photographs of the various tidal flats where they have worked. The response was overwhelming and we thank them for their generosity. Photocredits are indicated throughout the chapter but we wish specifically to thank the following who spent considerable time going through their files for photographs and who gave us valuable assistance in other ways: George P. Allen; Robert Dalrymple; Jurgen Dorjes; Graham Evans; Robert W. Frey; Lawrence A. Hardie; George de Vries Klein; R. John Knight; Claude Larsonneur; Gerard V. Middleton; James A. Miller; Mary E. Pugh; Hans-Erich Reineck; Michael J. Risk; L. M. J. U. van Straaten; Friedrich Werner; Friedrich Wunderlich.

We thank Rick Brokaw of Skidaway Institute of Oceanography for printing most of the black and white photographs and Robert W. Frey for reviewing the manuscript.

We are grateful to Michael Van Horn and the Champlin Petroleum Company for making available unpublished material in the West Desert Springs and Patrick Draw areas.

The Aux Vases Sandstone part of the study is based on a study done by Lindsey at Shell Development Company, Houston, Texas, during 1961-1963. Appreciation is expressed to Shell for permission to publish this material. Particular thanks are extended to Robert M. Sneider for inspiration and guidance during the original study.

Barrier-Island and Strand-Plain Facies

Donald G. McCubbin
Marathon Oil Company
Denver Research Center

INTRODUCTION

Barriers and strand plains are prominent depositional features of many modern coasts, and sandstone bodies of similar origin are represented in the stratigraphic record. In contrast to river deltas, which result from interaction of fluvial and marine processes, barriers and strand plains are supplied and molded almost entirely by marine processes. For purposes of this discussion, barriers are defined as sandy islands or peninsulas elongate parallel with shore and separated from the mainland by lagoons or marshes. Barriers are transitional in character to strand plains, which are wider in a land-sea direction and generally lack well developed lagoons and inlets. Chenier plains are a type of strand plain consisting of coastwise sandy ridges, separated by coastal mudflat deposits.

Some of the major environments and facies associated with barriers and strand plains are shown in Figure 1. The environments of sand deposition include: (1) beach and shoreface environments on the seaward side of barriers and strand plains; (2) inlet channels and tidal deltas, separating barriers laterally; and (3) washover fans on the landward or lagoonward side of barriers. Seaward or longshore migration of these environments results in facies sequences constituting much of the volume of many coastal sand bodies. For example, the emergent parts of many barriers and strand plains are underlain by progradational beach and shoreface sequences. Barrier sand bodies may also include sequences formed by coastwise or seaward migration of tidal inlets and tidal deltas.

Studies of modern coastal environments and sediments in a variety of settings show that these major facies have recognizable characteristics but also show some significant variations, depending mainly on local wave conditions and tidal range. Distribution of facies, external geometry of sand bodies, and nature of associated facies are also variable and depend on sediment supply, relative sea-level changes, and other aspects of setting and history of individual examples. A useful approach in stratigraphic reconstruction is to identify major facies, determine their lateral distribution and relationships, and use any available information on local and regional setting for interpretation.

The main objective of this review is to summarize and illustrate the characteristics of barrier and strand-plain sandstone facies judged to be most useful for stratigraphic interpretation from outcrops, cores, and logs. Emphasis is on sedimentary structures and textures and, particularly, on their vertical sequence. A second objective is to illustrate, using three case-study examples, some important variations in geometry of coastal sandstone bodies and in the nature of associated facies.

Nomenclature used here for stratification types and bed forms is generally that of Harms et al (1975), and the reader should refer to that source for more detailed discussion of stratification types and their hydrodynamic interpretation. Some of the more common stratification types in coastal deposits are briefly described and defined in Figure 2.

BEACH AND SHOREFACE DEPOSITS

Sequences formed by seaward progradation of beach and shoreface (nearshore) deposits account for a major part of the volume of many Holocene barriers and strand plains. Two of the best known and most frequently cited examples (Figs. 3-7) are Galveston Island, on the Texas coast (Bernard, LeBlanc, and Major, 1962; and Bernard and LeBlanc, 1965), and the strand plain north of San Blas, on the Pacific coast of the State of Nayarit, Mexico (Curray and Moore, 1964; Curray et al, 1969). Other Holocene shoreline sand complexes are partly progradational, as shown by accretionary beach-dune ridge topography, vertical sequence of facies, or radiocarbon ages (e.g. Wilkinson, 1975). Many well documented examples of ancient shoreline sandstone bodies, including those discussed in this review, were formed mostly by seaward accretion of beach and shoreface deposits.

Studies of sediments on modern beach-to-offshore profiles, using large undisturbed samples obtained by box coring or by vibrocoring, have resulted in a relatively good understanding of sedimentary structures and other features of sediments in these environments. The areas studied show a considerable range in setting, especially in terms of tidal range and wave energy (Reineck, 1963, 1976; Reineck and Singh, 1971; Clifton, Hunter, and Phillips, 1971, 1972; Howard and Reineck,

Generalizations and specific examples summarized in this review are based on published literature, work at Marathon Oil Company, and discussions with colleagues both inside and outside the Company. I am particularly indebted to D. L. Wiegand and others who did much of the work on the La Ventana sandstones included here as an example; to J. D. Howard and H.-E.

Reineck for allowing me to include data from their study of the Oxnard-Ventura, California area, and to H.-E. Reineck and J. R. Curray who provided originals of some photographs used here. Illustration work was done by Barbara Steele, and support for production of the manuscript was provided by Marathon Oil Company.

MAP

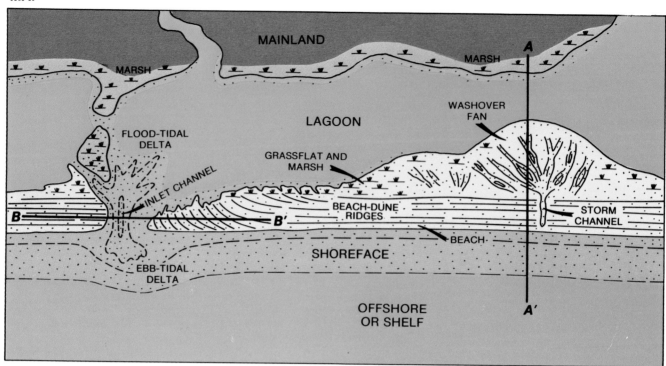

SECTION PERPENDICULAR TO SHORE

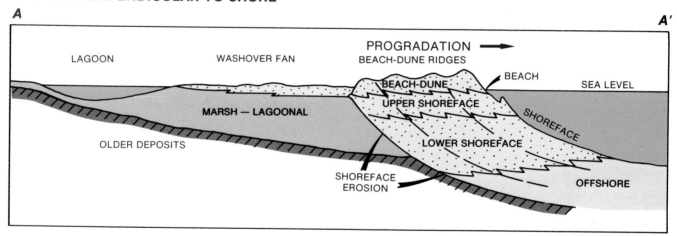

SECTION PARALLEL TO SHORE

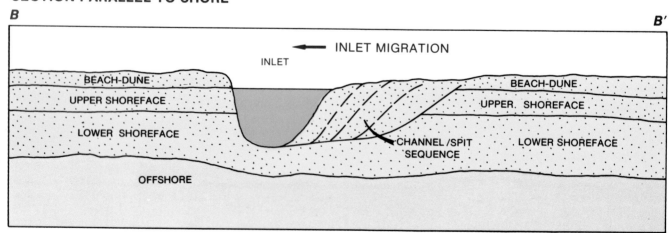

Fig. 1—Generalized map and cross sections showing major environments and facies of a barrier island-lagoonal system. Similar environments are associated with a strand-plain system, except that the progradational beach-ridge plain generally is wider and inlet-lagoonal environments are less well developed on strand plains.

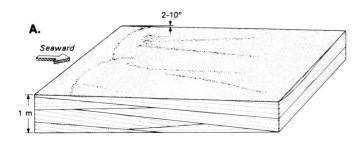

SWASH CROSS STRATIFICATION. Low-angle (2°-10°) cross stratification, subparallel to bases of wedge-shaped sets. Stratification and set boundaries are formed parallel to changing slope of beachface and dip generally seaward.

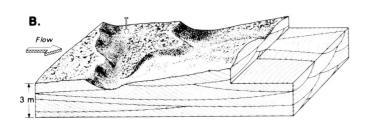

LARGE-SCALE TROUGH CROSS STRATIFICATION formed by subaqueous dunes or "megaripples". High-angle (25°-30°) cross stratification, tangential to bases of trough-shaped sets Cross strata dip parallel to flow direction.

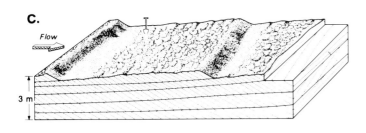

TABULAR CROSS STRATIFICATION FORMED BY MIGRATING SAND WAVES. High-angle (near 30°) cross stratification in tabular sets. Cross strata are planar and angular to bases of sets where flow is steady but may be tangential under some conditions.

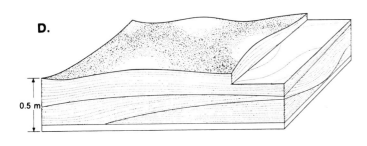

HUMMOCKY CROSS STRATIFICATION. Low-angle (less than 15°) cross stratification, subparallel to smooth, undulatory lower boundaries of sets. Similar appearance in all vertical orientations. Commonly associated with wave ripples.

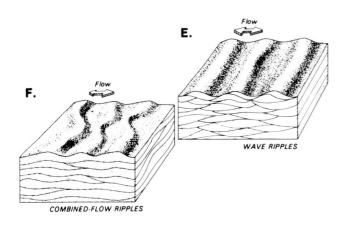

WAVE RIPPLES. Ripple-trough profiles are symmetrical and rounded, and stratification dips in both directions of oscillatory flow.

COMBINED-FLOW RIPPLES. Formed by superimposed wave and current action or by shoaling waves. Small-scale cross strata are curved and tangential, dipping in direction of dominant flow.

Fig. 2—Block diagrams showing some common stratification types and associated bed forms in coastal-marine sands. Variations in form related to grain size, changing flow conditions, and rates of deposition are discussed by Harms et al, 1975.

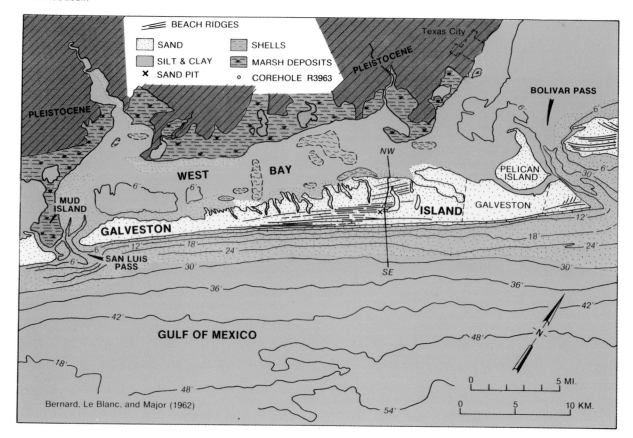

Fig. 3—Map of Galveston Island, Texas, showing morphology and environments.

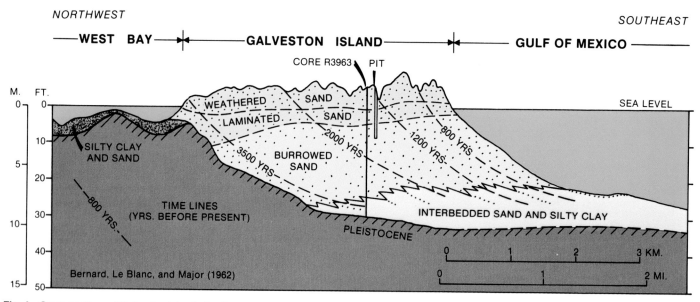

Fig. 4—Cross sections of Galveston Island, showing progradational origin of the main beach-dune-ridge part of the island. Location of cross section is shown on Figure 3.

1972, 1979; Davidson-Arnott and Greenwood, 1976; and Hill and Hunter, 1976). These studies, and comparisons with ancient examples, allow inferences about the vertical sequence preserved in prograding beach-to-offshore environments.

Studies of sediments on modern beach-to-offshore profiles, using large undisturbed samples obtained by box coring or by vibrocoring, have resulted in a relatively good understanding of sedimentary structures and other features of sediments in these environments. The areas studied show a considerable range in setting, especially in terms of tidal range and wave energy (Reineck, 1963, 1976; Reineck and Singh, 1971; Clifton, Hunter, and Phillips, 1971, 1972; Howard and Reineck, 1972, 1979; Davidson-Arnott and Greenwood, 1976; and Hill and Hunter,

1976). These studies, and comparisons with ancient examples, allow inferences about the vertical sequence preserved in prograding beach-to-offshore environments.

Beach-to-Offshore Morphology and Processes

Beach and shoreface (nearshore) environments are commonly described in

Fig. 5—Aerial photo of Galveston Island, looking northeast. West Bay (lagoon) is to the left, Gulf of Mexico is to the right, and part of San Luis Pass (inlet) is at the bottom.

Fig. 7—Aerial photo of Nayarit strand plain; Pacific Ocean visible to the left. The swales between the mangrove-covered beach ridges are partially submerged during flood season. Prominent discontinuity in beach-ridge orientation indicates that the coastline was reoriented before accretion of the younger ridges to the left. (Photo from Curray and Moore, 1964).

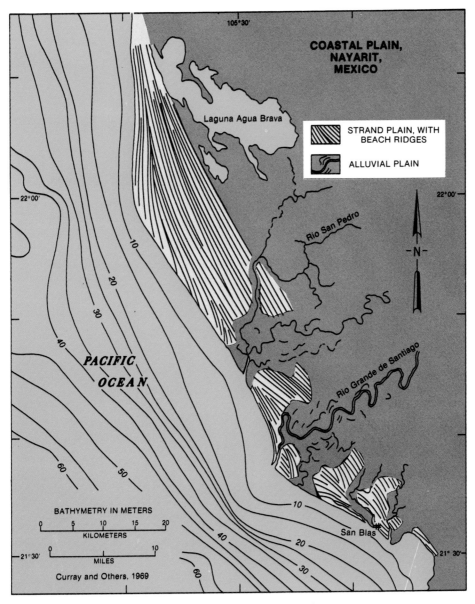

Fig. 6—Map of strand plain and alluvial-delta plains of the Rio Grande de Santiago and Rio San Pedro, State of Nayarit, Mexico. The beach ridges of the strand plain are generally parallel with the present coast but locally are truncated, indicating changes in shoreline orientation.

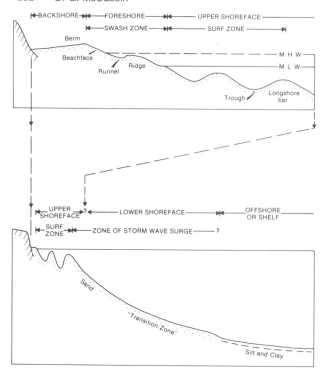

Fig. 8—Generalized beach-to-offshore profiles showing terminology applied to principal morphological features and zones.

Fig. 9—Beach on Plum Island, Massachusetts, showing ridge and runnel, berm, and seacliff cut in vegetated dune ridge. Purple area on landward margin of beach is a garnet-rich beach placer.

Fig. 10—Beach foreshore in Galveston Island, Texas, showing breakers (on skyline), surf, and swash zone (wet area). The low relief features in the swash zone are antidunes formed by backwash.

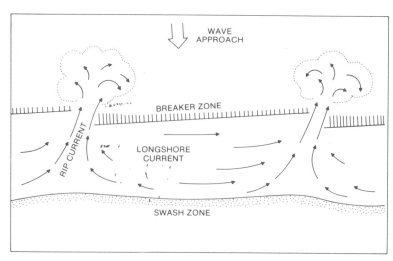

Fig. 11—Diagram showing longshore currents and rip currents generated in the surf zone by obliquely approaching breakers.

Fig. 12—Wedge-shaped set of cross stratification in the seaward face of a welded berm, Plum Island, Massachusetts. Purple color is due to garnet concentrated in the backshore by wind and wave action.

Fig. 13—Stratification on gently sloping backshore, Sapelo Island, Georgia. Thin sets of low-angle cross stratification formed by antidunes occur only at the top and are rarely preserved.

terms of their morphological features, which both control and reflect the dominant processes in these environments. These morphological features and associated processes are similar on mainland coasts and on seaward coasts of barriers. A generalized profile from beach to offshore, with the terminology used here, is shown in Figure 8. The processes of beach and nearshore environments are described in more detail in a recent review article by Davis (1978).

The beach is sometimes divided into a backshore, which consists of a nearly level berm, and a foreshore which slopes seaward from the berm edge or crest (Fig. 9). The foreshore includes the beachface and, on some beaches, one or more elongate bars and intervening troughs called ridge and runnel. The shoreface, as the term is used here, extends from the beach offshore to a depth of 6 to 20 m (20 to 65 ft), where there is commonly a change of gradient from the gently sloping shoreface to the nearly level shelf (Price, 1954; Shepard, 1960). The upper shoreface commonly, but not always, has one or more bars elongate parallel or slightly oblique to the shore. The lower shoreface is relatively smooth and generally concave upward in profile.

Sediment transport on the beach and shoreface is dominated by waves and wave-induced currents, although tidal currents may be important near inlets and estuaries. As waves move toward shore, they begin interacting with the seafloor near the base of the shoreface, become oversteepened, and collapse to form breakers in the upper shoreface. Plunging breakers are transformed into translation waves or bores in the surf zone. As the surf runs up the beachface, it forms a thin uprush of water called the swash, followed by an even thinner return flow called the backwash (Fig. 10). As breakers and onshore winds pile water against the beach, they generate a system of currents including both bidirectional oscillatory motions and unidirectional longshore currents and rip currents. When the waves approach the beach obliquely, one direction of longshore currents predominates (Fig. 11).

Beach and nearshore morphology and associated sediment-transport systems vary from one coast to another and also vary with time on the same coast, depending mainly on wave conditions. Evidently, both the absolute magnitude and degree of variation or range in wave energy are important (Wright et al, 1979). On coasts with

Fig. 14—Cross section of a foreshore ridge on Plum Island, Massachusetts, showing tabular set of high-angle cross stratification formed by landward migration of the ridge into the adjacent runnel. Migration occurred over several tidal cycles, with some modification of the slipface of the ridge during rising or falling tides.

large seasonal variations in wave energy, such as some southern California beaches (Winant et al, 1975), the beach is eroded during periods of high wave energy and sediment moves offshore to form a relatively broad shallow nearshore zone with longshore bars. With decrease in wave energy, much of the sediment returns to the beach by shoreward migration of the bars. If low energy conditions persist long enough, a relatively steep, nonbarred profile may be developed nearshore, and a wide, high berm may be built on the beach. In areas of relatively little seasonal variation of wave energy, as on east-facing shores in the Gulf of St. Lawrence (Greenwood and Davidson-Arnott, 1975; Owens, 1977), beach and nearshore zones show some morphological changes related to erosion and deposition by individual storms, but longshore bars persist throughout the year. A high rate of sediment supply also helps maintain a relatively shallow nearshore zone with longshore bars.

Wave energy depends on exposure to oceanic swell, generated by distant storms, and on size and frequency of storm waves (Davies, 1964). In the northern hemisphere, south of the temperate-zone storm belt, wave energy consists mostly of southwesterly and northwesterly swell generated in the southern and northern storm belts. On west-facing coasts in this area, such as those of California, wave energy is generally much higher in the winter months than in the summer because of the seasonal nature of the northwest-

erly swell. On coasts within the northern hemisphere storm belt and in areas protected from westerly swell, coastal processes are dominated by storm waves. In these areas, wave energy depends on frequency of storms, shoreline orientation relative to wind direction, and fetch distance. Where storm frequency is high, such as in the Gulf of St. Lawrence, wave energy is high but seasonal variations are small, especially on east-facing coasts. Coasts farther from the northern storm belt and protected from westerly swell, such as the middle and southern U.S. Atlantic coast and the Gulf of Mexico, have less frequent storms and generally lower wave energy.

Modern Beach-to-Offshore Facies

Beach Sands — Characteristics of beach sediments are quite well known because of process-oriented sedimentological studies in areas with a wide range of tides, wave energy, and nearshore morphology (e.g. Thompson, 1937; Hayes et al, 1969; Wunderlich, 1972; Berg, 1977). The dominant process in the beach environment is wave swash, which runs up the foreshore and occasionally overtops the berm during high tides. The main effect of tides is to shift the swash and surf zones laterally and vertically on the beach and nearshore profile.

Wave swash produces planar, nearly horizontal stratification on the berm, and planar, seaward-dipping stratification on the beachface. The angle of dip on the beachface depends on grain size,

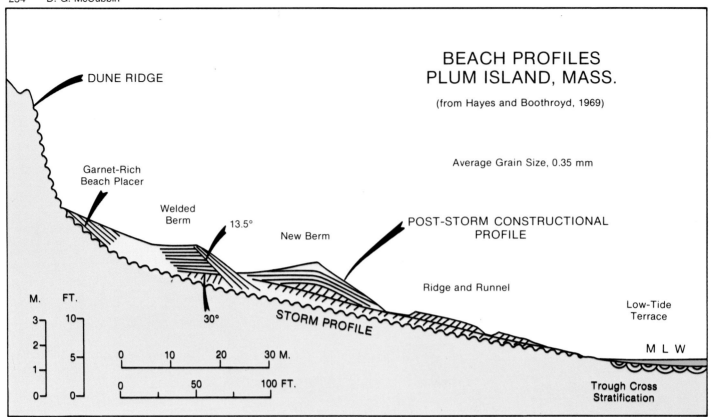

**BEACH PROFILES
PLUM ISLAND, MASS.**

(from Hayes and Boothroyd, 1969)

Average Grain Size, 0.35 mm

DUNE RIDGE

Garnet-Rich
Beach Placer

Welded
Berm   13.5°

New Berm

POST-STORM CONSTRUCTIONAL
PROFILE

Ridge and Runnel

Low-Tide
Terrace

M. FT.

30°   STORM PROFILE

M L W

Trough Cross
Stratification

Fig. 15—Profiles of beach on Plum Island, Massachusetts, immediately after a major "northeaster" (lower, concave-upward profile) and three weeks later (upper profile with ridge and runnel, new berm, and welded berm). The wedge of sediment deposited during fair-weather conditions may be partly or completely removed by later storms.

Fig. 16—Beach foreshore and adjacent nearshore on south end of Sapelo Island, Georgia, at low tide. Dunes or megaripples partially exposed on low-tide terrace have been modified by wave action as the tide receded.

Fig. 17—Planed-off, near-horizontal surface on bar off the Island Norderney, Germany, showing trough-shaped sets of cross stratification formed by migrating dunes or megaripples. Compass is 10 cm square. (Photo from Reineck, 1963, 1967).

but is usually 2 to 10° in fine- to medium-grained sand. Because the slope of the beachface changes somewhat with varying wave conditions, swash stratification on the foreshore commonly occurs in wedge-shaped sets, bounded by low-angle surfaces of truncation (Figs. 2, 12). Antidunes are sometimes formed during backwash (Fig. 10), depositing thin lenses with very gently dipping cross stratification

(Fig. 13), but these lenses are usually eroded as conditions change. Swash stratification is marked by variations in grain size of sand or by concentrations of heavy minerals. Clay, plant fragments, and other hydraulically light particles are sparse in these sands, compared with associated upper and lower shoreface deposits.

High-angle, landward-dipping cross stratification formed by migration of

ridge and runnel is also a common sedimentary structure on some modern beaches (Fig. 14). This type of stratification is superficially similar to that formed by migration of sand waves under unidirectional flow (Fig. 2). It usually differs in detail, however, because of its origin within the intertidal zone over several tidal cycles. Deposition of foresets near the angle of repose occurs during high tide, when sand is carried up the seaward slope of the ridge by swash and is deposited by avalanching down the steep slope into standing water of the runnel. These high-angle foresets may be modified by changing water level, by small wind waves, or by unidirectional currents in the runnel. High-angle cross-stratified sands deposited by migrating ridge and runnel may be expected to contain more fecal mud pellets and hydraulically light particles than associated swash-stratified sands.

On beaches showing distinct cycles of erosion and deposition related to individual storms or seasonal changes in wave conditions, planar to concave-upward erosional profiles develop during periods of large waves. On the New England coast, where beach erosion is related to northeasterly storms, subsequent accretion occurs by migration of ridge and runnel toward the backshore. In some cases, a mature, convex-upward profile with a well developed berm is restored within 2 to 6 weeks after the storm (Hayes and Boothroyd, 1969). Sands deposited during the constructional part of the cycle (Fig. 15) may have some high-angle, landward-dipping cross stratification in the lower part and low-angle, seaward-dipping or nearly horizontal stratification underlying the restored beachface and berm. Although part or all of this sequence may be eroded during subsequent storms, deposits of the upper foreshore and berm will likely be preserved during net seaward progradation. Similar cycles of erosion and deposition occur on beaches undergoing seasonal changes in wave energy, such as some California beaches.

Because of the active physical processes, sands in modern foreshore and backshore environments rarely are bioturbated completely, although discrete burrows are formed by *Callianassa* and other filter feeders in the lower foreshore and by burrowing crabs in the backshore of some beaches. Vegetation may also become established in the backshore, resulting in disturbance of the stratification and preservation of root traces.

Upper Shoreface Deposits — Sediment transport in the upper shoreface or surf zone is dominated by bidirectional, shore-normal oscillatory motion related to primary incident waves and associated breakers and by secondary unidirectional longshore currents and rip currents. Subaqueous dunes (also called scour megaripples or lunate megaripples) are sometimes exposed on the low-tide terrace and on crests of

nearshore bars during unusually low tides (Fig. 16). Medium- to large-scale trough cross stratification is formed by migration of these dunes or megaripples (Figs. 2, 17).

Observations by divers and studies of box cores indicate that high-angle, tangential cross stratification in trough-shaped sets is the dominant stratification type in longshore troughs and rip-current channels (Davidson-Arnott and Greenwood, 1976; Hunter et al, 1979). Planar, nearly horizontal stratification is formed by shore-normal oscillatory motion on the crests and seaward slopes of longshore bars in some areas (Davidson-Arnott and Greenwood, 1976; Hill and Hunter, 1976).

On many barred coasts, sands of the nearshore zone or upper shoreface are coarser than those of the associated beach and lower shoreface environments. Some grains of mica and plant material may be present in sands with high-angle cross stratification, but are generally less abundant than in sediments of the lower shoreface. Sand-sized pellets of fecal mud may be abundant, and mud layers occasionally are deposited and preserved. Distinct burrows formed by *Callianassa* and other filter feeders are often present, but complete bioturbation is relatively uncommon (Fig. 18A).

The nearshore zone or upper shoreface, like the associated beach environments, shows cycles of erosion and deposition related to changing wave conditions. Generally, the upper shoreface is eroded as sand moves onto the beach during periods of relatively low wave energy and is built up during periods of higher wave energy (Fig. 19). On southern California shores, where these changes in morphology are seasonal, longshore bars may be completely removed during summer and restored during winter (Winant et al, 1975). In nearshore areas on the east coast of Canada and New England, where erosion and deposition are related to individual storms, nearshore bars may oscillate landward and seaward but are never completely removed (Greenwood and Davidson-Arnott, 1975; Owens, 1977). In some areas, longshore troughs are deepest and widest during periods of highest wave energy and become shallower and narrower as bars migrate landward during periods of declining wave energy (Wright et al, 1979). Because of the lateral migration of the bars and troughs, deposits of the nearshore zone probably include nearly-horizontal ero-

sion surfaces overlain by cross-stratified sands deposited in trough and rip-current environments.

Lower Shoreface Deposits — The lower shoreface environment, on a barred coast, extends seaward from the outermost bar to the break in slope between the shoreface toe and the relatively flat shelf. In some modern settings, the outer margin of the shoreface is marked by a change in sediment type from the finer sands deposited under present conditions to coarser palimpsest sediments partly reflecting earlier conditions. In other settings, where very fine sediments accumulate on the shelf, sands and silts of the lower shoreface grade offshore to shelf muds. Extensive box coring in many areas shows that biogenic structures are dominant seaward from the surf zone off many coasts (Fig. 18B, C) but that physical sedimentary structures are preserved in some. Physical processes in the lower shoreface environment are mostly inferred from characteristics of the sediments themselves, rather than from direct observation.

Stratification types on seaward slopes of offshore bars are mostly small-scale cross stratification formed by predominantly landward migration of ripples, and planar stratification dipping seaward at low angles. Farther offshore, toward the base of the shoreface and onto the shelf, grain size generally decreases and interbedded silts and muds become more abundant. Preserved stratification in the sands is mostly planar and nearly horizontal. Some individual beds of fine sand or coarse silt are characterized by low-angle cross stratification with both upward-concave and upward-convex components nearly parallel with smooth, curved surfaces of erosion (Figs. 2, 20). This distinctive style of cross stratification was first recognized in ancient shoreface deposits (Campbell, 1966) and later termed hummocky cross stratification by Harms et al (1975). In some cases, these planar- to hummocky-stratified beds have basal lag deposits of shells or mud clasts and are capped by thin intervals with wave-ripple stratification, suggesting that each was deposited by a single, high-energy (storm?) event (Hayes, 1967; Reineck, 1976; Kumar and Sanders, 1976).

Off the Oxnard-Ventura coast of California, at depths exceeding 9 m (30 ft) below sea level, "parallel-laminated" beds of fine sand and coarse silt are interbedded with bioturbated muds (Howard and Reineck, 1979). These

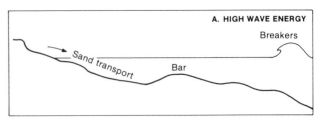

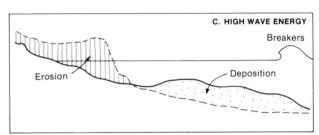

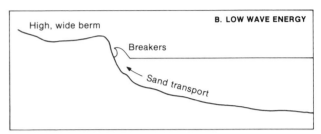

Modified from Wright and others, 1979

Fig. 19—Generalized profiles of beach and nearshore zone on coasts showing a large temporal variation in wave energy. During seasons or shorter term periods of low energy, sand is eroded from the nearshore and a high, wide berm is built on the beach. During periods of high wave energy, the beach is eroded and sand is deposited to form longshore bars and troughs in the surf zone or upper shoreface.

Fig. 18—Relief casts of box-core samples from off the coast of the Gulf of Gaeta. **A.** Medium- and small-scale trough cross stratification, formed by the migration of dunes or megaripples in 0-2 m water depth. **B.** Nearly horizontal stratification (top) and some bioturbated sand (below) in 2-4 m water depth. **C.** Bioturbated sand in a 4-7 m water depth. Each sample is about 12 cm wide. (Photos from Reineck and Singh, 1971).

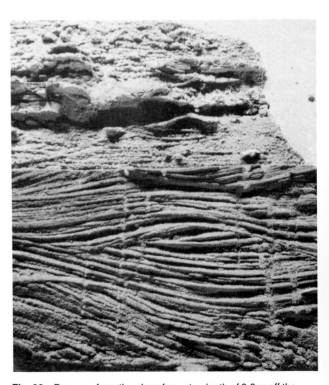

Fig. 20—Box core from the shoreface at a depth of 3.8 m off the seaward coast of Norderney, Germany, showing sand with medium-cale hummocky cross stratification. (Photo from Reineck, 1976).

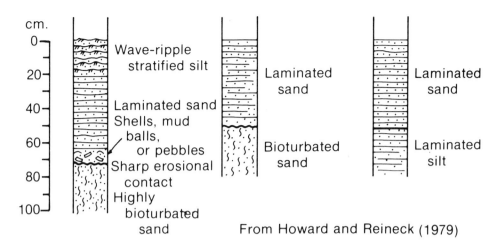

From Howard and Reineck (1979)

Fig. 21—Examples of laminated to burrowed beds of fine sand and silt observed in vibrocores from the lower shoreface or "transition zone" off the Ventura-Oxnard coast, California. These beds are interpreted as the products of storm-wave activity.

sand and silt beds have sharp, erosional lower contacts, and some have concentrations of shells or mud clasts at their bases; wave-ripple stratification is sometimes preserved at the tops (Fig. 21). Individual beds decrease in average thickness from 74 cm seaward of the mouth of the Santa Clara River to 45 cm away from the river. Howard and Reineck (1979) interpret these beds as the result of a two-part process, with sediment initially brought to nearshore areas by river floods and then redistributed offshore by storm-induced waves.

Another characteristic of "storm layers" consistent with their origin by rapid deposition after a single high-energy event, is the reported abundance of plant material, mica flakes, and other hydraulically light particles

SEQUENCE ON GALVESTON ISLAND

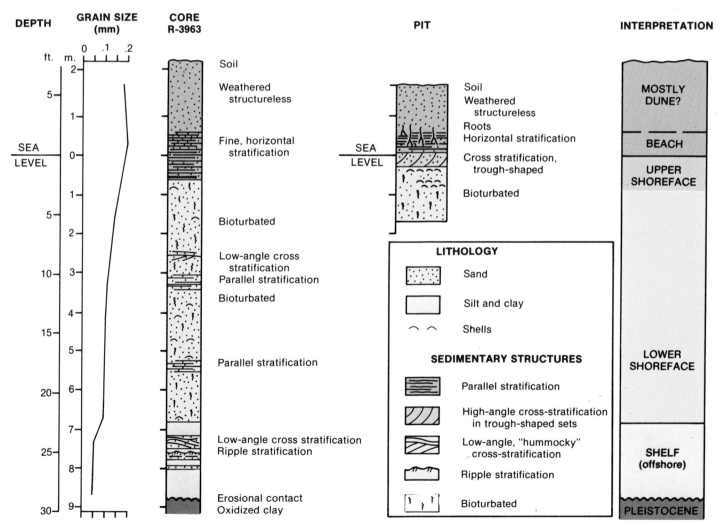

Modified from Bernard, LeBlanc, and Major (1962)

Fig. 22—Vertical sequence observed in a long core and in a sand pit on Galveston Island, Texas. In this setting, characterized by low wave energy (except for hurricanes), the trough cross stratification of the upper shoreface is thin or only locally present. Locations of corehole and pit are shown on Figure 3.

in both modern and ancient examples. Bioturbation of interbedded muds and tops of the sand layers suggests slow sedimentation between episodes of sand deposition. In normal marine environments, the abundance and variety of trace fossils may be a distinctive characteristic of deposits of the lower shoreface (Howard, 1972).

Vertical Sequence Models

The vertical facies sequence formed by seaward progradation of beach and shoreface deposits on Galveston Island has been observed and described in detail through studies of cores and sand pits (Fig. 22). Vertical sequence models have been inferred for other areas from studies of sediments of modern beach-to-offshore environments. These models generally assume that the vertical sequence from top to bottom will approximate the lateral sequence observed on modern profiles from beach to offshore (Howard and Reineck, 1979). Some models also take into consideration the effects of erosional and depositional cycles on the preservation of beach and nearshore facies during progradation (Hunter et al, 1979).

Significant differences are observed on modern beach-to-offshore profiles in the depth range of the major facies. Howard and Reinech (1979) suggested that these differences are related to the average wave energy of the environment. On low-energy coasts, such as those of Texas, Georgia, and the Gulf of Gaeta, the facies of the surf zone or upper shoreface extends offshore to depths of only 1 to 2 m (3 to 6 ft). On high-energy coasts, such as that of the Oxnard-Ventura area (Fig. 23), a similar facies extends offshore to depths of as much as 9 m (30 ft). Presumably, these differences would be reflected in a greater thickness of upper shoreface deposits in high-energy progradational systems as compared with low-energy systems (Fig. 24).

The nature and thickness of the sequence formed during progradation are also affected by changes in beach and nearshore morphology related to individual storms or seasonal variation in wave conditions. During periods of high wave energy, the beach is eroded and sand is deposited in the nearshore zone with well developed longshore bars; during periods of low wave energy, bars migrate landward and a wide, high berm is constructed on the beach. Because of these cyclic changes in morphology and process, beach and nearshore sediments that accumulate over long periods of time likely will consist mostly of low energy beach deposits and high energy nearshore deposits with numerous erosional surfaces sloping gently seaward. The depth range or thickness of beach and nearshore facies will reflect extreme conditions and may be greater than that observed over short time periods.

Vertical sequence models for progradation on both barred and nonbarred coasts have been proposed by Hunter et al (1979), based on studies of Oregon

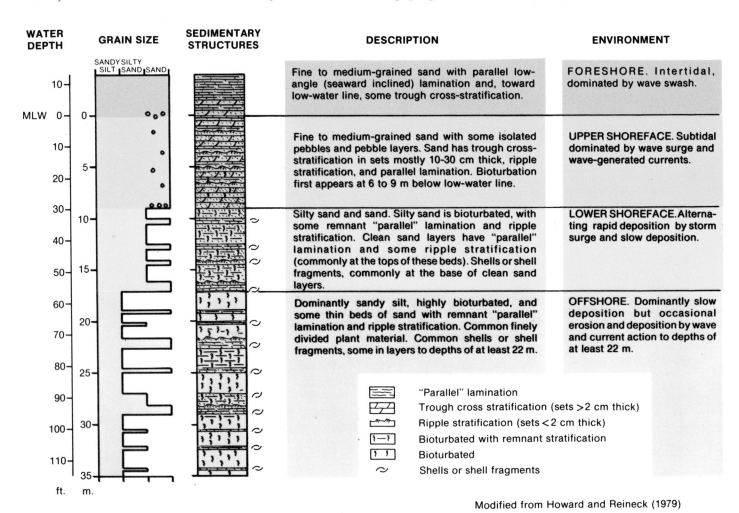

WATER DEPTH	GRAIN SIZE	SEDIMENTARY STRUCTURES	DESCRIPTION	ENVIRONMENT
			Fine to medium-grained sand with parallel low-angle (seaward inclined) lamination and, toward low-water line, some trough cross-stratification.	FORESHORE. Intertidal, dominated by wave swash.
			Fine to medium-grained sand with some isolated pebbles and pebble layers. Sand has trough cross-stratification in sets mostly 10-30 cm thick, ripple stratification, and parallel lamination. Bioturbation first appears at 6 to 9 m below low-water line.	UPPER SHOREFACE. Subtidal dominated by wave surge and wave-generated currents.
			Silty sand and sand. Silty sand is bioturbated, with some remnant "parallel" lamination and ripple stratification. Clean sand layers have "parallel" lamination and some ripple stratification (commonly at the tops of these beds). Shells or shell fragments, commonly at the base of clean sand layers.	LOWER SHOREFACE. Alternating rapid deposition by storm surge and slow deposition.
			Dominantly sandy silt, highly bioturbated, and some thin beds of sand with remnant "parallel" lamination and ripple stratification. Common finely divided plant material. Common shells or shell fragments, some in layers to depths of at least 22 m.	OFFSHORE. Dominantly slow deposition but occasional erosion and deposition by wave and current action to depths of at least 22 m.

"Parallel" lamination
Trough cross stratification (sets >2 cm thick)
Ripple stratification (sets <2 cm thick)
Bioturbated with remnant stratification
Bioturbated
Shells or shell fragments

Modified from Howard and Reineck (1979)

Fig. 23—Beach-to-offshore sequence off the Ventura-Oxnard coast, California, shown as a vertical sequence relative to present sea level. The depth zones referred to here as upper and lower shoreface were called "nearshore" and "transition" zones by Howard and Reineck (1979). High-angle cross stratification in small-to large-scale trough-shaped sets occurs to depths of 9 m (30 ft), and sands of this zone are fine to medium grained with some pebbles. The Ventura-Oxnard coast has "high wave energy" and is subjected to large storm waves during the winter months (Howard and Reineck, personal commun.).

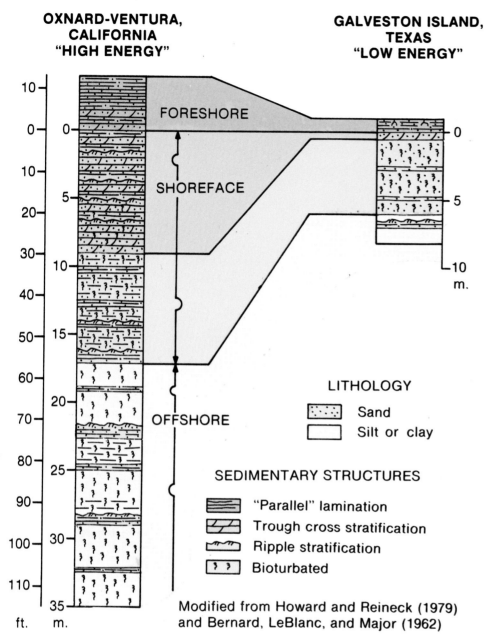

OXNARD-VENTURA, CALIFORNIA "HIGH ENERGY"

GALVESTON ISLAND, TEXAS "LOW ENERGY"

FORESHORE

SHOREFACE

OFFSHORE

LITHOLOGY

Sand

Silt or clay

SEDIMENTARY STRUCTURES

"Parallel" lamination

Trough cross stratification

Ripple stratification

Bioturbated

Modified from Howard and Reineck (1979) and Bernard, LeBlanc, and Major (1962)

Fig. 24—Comparison of the beach-to-offshore facies sequence and thickness from a low-wave-energy coast (Galveston Island) with that inferred for a high-wave-energy coast (Oxnard-Ventura area).

beach and nearshore areas and incorporating effects of temporal change in coastal morphology. The models for a high-energy barred coast proposed by Hunter et al are similar to that proposed by Howard and Reineck (Fig. 23), except that Hunter et al inferred the existence of subhorizontal erosional surfaces. One model incorporates a widespread scour surface formed by lateral migration of rip-current channels, overlain by relatively coarse cross-stratified sands recording generally seaward flow directions. Another model suggests a scour surface formed by onshore and offshore migration of longshore troughs, overlain by relatively coarse cross-stratified sands deposited in the troughs and recording

flow generally parallel with the troughs. A third model, based on studies of a nonbarred, high-energy Oregon coast, proposes planar-stratified sand of the lower shoreface overlain by high-angle, tangential cross stratification in trough-shaped sets, recording dominantly landward transport in the nearshore zone. These models differ mainly in the directional aspects of cross stratification formed by sand transport in the upper shoreface or nearshore zone.

Ancient Progradational Sequences

Vertical sequence models based on studies of modern beach-to-offshore deposits seem to adequately explain many of the features observed in an-

cient progradational sequences. The upper Tertiary Cohansey Sand of New Jersey (Carter, 1978) has characteristics expected for a progradational nearshore-to-beach sequence deposited on a relatively high-wave-energy coast, but without clear evidence indicating whether it was barred or nonbarred. Progradational, offshore-to-beach sequences in the middle Miocene Caliente Formation of California have erosional surfaces overlain by sandstones with seaward-dipping cross strata and are interpretable as the deposits of high-wave-energy coasts with oblique bars and rip-current channels (Clifton, 1967, 1973; Hunter et al, 1979). An upward-shoaling, offshore-to-beach sequence in the upper Miocene Canos Formation of Spain has wave-ripple and planar stratification overlain by high-angle, tangential (trough) cross stratification indicating offshore and onshore flow; this sequence may be an example of progradation on a high-wave-energy, nonbarred coast (Roep et al, 1979).

Many Upper Cretaceous shoreline sandstone units of the Rocky Mountain region (Campbell, 1971; Howard, 1972; Ryer, 1977) are progradational shoreface and beach sequences, evidently deposited on coasts with moderate to high wave energy. The Gallup Sandstone, New Mexico, sequence (McCubbin, 1972) is typical in many respects, although thicker and coarser than some. In this sequence (Fig. 25), the uppermost facies unit consists of fine-grained, well sorted sandstone showing nearly horizontal, planar stratification with some low-angle truncation surfaces (Fig. 26). The middle facies division consists of fine- to medium-grained sandstone with high-angle cross stratification in trough-shaped sets (Fig. 27), some thin intervals with horizontal, planar stratification, and some wave ripples preserved beneath thin shale or siltstone beds. The lower facies division consists of fine- to very fine-grained sandstone in tabular beds ranging in thickness from a few centimeters to as much as a meter. In the thinner beds, stratification is mostly subhorizontal and planar (Fig. 28); thicker, slightly coarser beds commonly have low-angle, undulatory, hummocky cross stratification (Fig. 29). The base of each bed is sharp, truncating sedimentary structures below, and the top is either burrowed or has small-scale cross stratification formed by migration of wave ripples (Fig. 30). These three facies divisions are interpreted as deposits of beach, upper shoreface, and lower shoreface environments on a prograding coast with relatively high wave energy.

In the Gallup Sandstone, at the locality shown in Figure 25, the shoreline orientation during deposition was northwest-southeast and the offshore direction was to the northeast, as shown by the attitude of northeastward-sloping, heavy-mineral-rich beach placers (Fig. 31), as well as by regional control. Flow directions recorded by high-angle cross stratification in the middle (upper shoreface) division are bidirectional, with a dominant mode to the southeast and a secondary mode to the northwest (Fig.

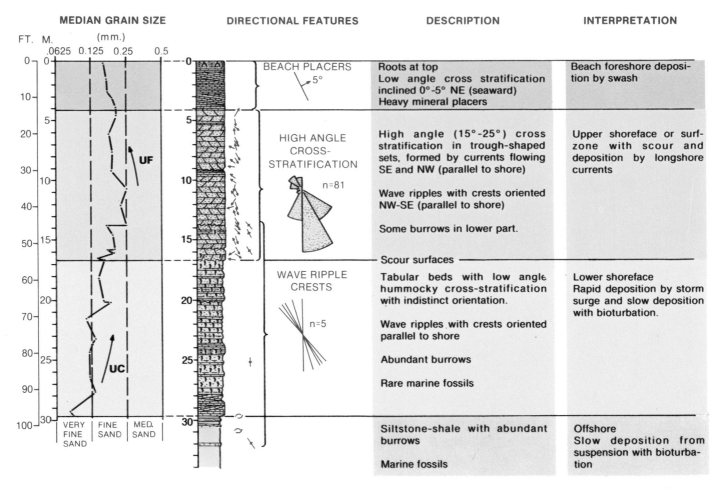

MEDIAN GRAIN SIZE	DIRECTIONAL FEATURES	DESCRIPTION	INTERPRETATION

BEACH PLACERS — 5°

Roots at top
Low angle cross stratification inclined 0°-5° NE (seaward)
Heavy mineral placers

Beach foreshore deposition by swash

HIGH ANGLE CROSS-STRATIFICATION   n=81

High angle (15°-25°) cross stratification in trough-shaped sets, formed by currents flowing SE and NW (parallel to shore)

Wave ripples with crests oriented NW-SE (parallel to shore)

Some burrows in lower part.

Upper shoreface or surf-zone with scour and deposition by longshore currents

Scour surfaces

WAVE RIPPLE CRESTS   n=5

Tabular beds with low angle hummocky cross-stratification with indistinct orientation.

Wave ripples with crests oriented parallel to shore

Abundant burrows

Rare marine fossils

Lower shoreface
Rapid deposition by storm surge and slow deposition with bioturbation.

Siltstone-shale with abundant burrows

Marine fossils

Offshore
Slow deposition from suspension with bioturbation

Fig. 25—Vertical facies sequence in part of Gallup Sandstone, northwestern New Mexico. The vertical sequence of sedimentary structures and textures formed by progradation is similar to that observed from beach to offshore off the high-wave-energy Ventura-Oxnard coast (Fig. 23).

Fig. 26—Nearly horizontal stratification in well sorted, fine-grained sandstone near the top of the Gallup sequence. This is interpreted as swash stratification formed on the beachface or possibly the beach berm.

Fig. 27—High-angle, tangential cross stratification in fine- to medium-grained sandstone of the middle facies of the Gallup sequence. The cross stratification occurs in bidirectional trough-shaped sets and was formed by longshore migration of dunes or megaripples in the upper shoreface. Rod shows scale in feet and tenths.

Fig. 28—Tabular beds of fine- to very fine-grained sandstone with mostly subhorizontal stratification in the lower facies (lower shoreface) of the Gallup Sandstone. The top of each bed is bioturbated.

Fig. 29—Fine-grained sandstone bed with low-angle, hummocky cross stratification from basal part of the Gallup Sandstone sequence. The characteristics of this bed and its position near the base of a progradational beach-to-offshore sequence suggests deposition by unusually large storm waves.

Fig. 30—Tabular, fine-grained sandstone bed from lower part of Gallup sequence with nearly horizontal stratification and capped wtih wave-ripple stratification and profiles of widely spaced wave ripples.

Fig. 31—Outcrop of Gallup Sandstone near location of Figure 25, showing dark, heavy-mineral-rich layers in upper part (near top of cliff). These layers, sloping gently northeastward (to the right), are interpreted as beach placers formed during progradation to the northeast. Cliff is about 25 m (80 ft) high.

25). These flow directions are sub-parallel with the inferred shoreline orientation and perpendicular to the direction of wave approach, as indicated by ripple-crest orientations (Fig. 25). This strongly suggests that the high-angle cross stratification was formed by longshore currents in a nearshore zone with well developed longshore bars and troughs. Northeastward- or seaward-dipping cross strata are sparse or absent, suggesting that rip-current channels were absent or that their deposits were not preserved. Low relief, subhorizontal erosional surfaces at and near the base of the facies with high-angle cross stratification (Fig. 32) may be due to onshore-offshore migration of longshore troughs. These surfaces are commonly overlain by thin lag-like deposits of relatively coarse grains.

Other characteristics of this progradational sequence include a general upward increase in average grain size from thin siltstones in the offshore facies to the top of the lower shoreface division, and a trend toward upward-decreasing grain size from the upper shoreface through beach divisions (Fig. 25). A general upward decrease in the amount of detrital mica and carbonaceous fragments also is observed, especially through the lower shoreface sequence, with these particles being very sparse in the upper shoreface and beach divisions. In this sequence and in similar sequences of the Book Cliffs, Utah, described by Howard (1972), there is an upward decrease in the amount and variety of burrowing; sandstones of the lower shoreface have burrows formed by a variety of grazing and filter-feeding animals, whereas sand-

stones of the upper shoreface have structures formed by deep-burrowing filter feeders (Ophiomorpha).

Mechanical Log Response

Some aspects of the vertical sequence in progradational sandstones can be interpreted from common mechanical logs, aiding in recognition where regional control from cores or outcrops is also available. Figure 33 is an example of an induction-electrical log of the Gallup Sandstone, compared with the vertical sequence in outcrop a few miles away along the depositional strike. In the subsurface sequence interpreted as a progradational sandstone unit, the spontaneous potential (SP) trace shows a general upward deflection toward more negative values (leftward), compared with the marine siltstone and shale below. This results

Fig. 32—Scour surface between middle, cross-stratified facies and lower, tabular-bedded, burrowed facies of Gallup Sandstone. Cross strata overlying this surface are bidirectional, indicating flow directions generally parallel with the inferred shoreline orientation. Vertical section shown is about 2 m.

from an upward decrease in the amount of clay, probably both as thin clay-rich beds and as matrix clay in the sandstone. A gamma-ray trace would show a similar pattern, with an upward decrease in gamma-ray count because of the decrease in the proportion of clay. Electrical resistivity traces show a general upward increase in resistivity through the same interval, probably in response to both an upward decrease in relatively conductive clay and an upward increase in invasion of more permeable sandstone by relatively resistive drilling-mud fluids. The upward-increasing separation between resistivity measured by the shallower-reading 16″-normal and the deeper-reading induction devices also indicates upward increase in mud-fluid invasion and permeability.

This example is believed to be reasonably typical of log response expected for progradational beach-to-offshore sequences where hydrocarbon saturation is zero and formation-water salinity and drilling-mud resistivity are within "normal" ranges. Variations in fluids and also variations

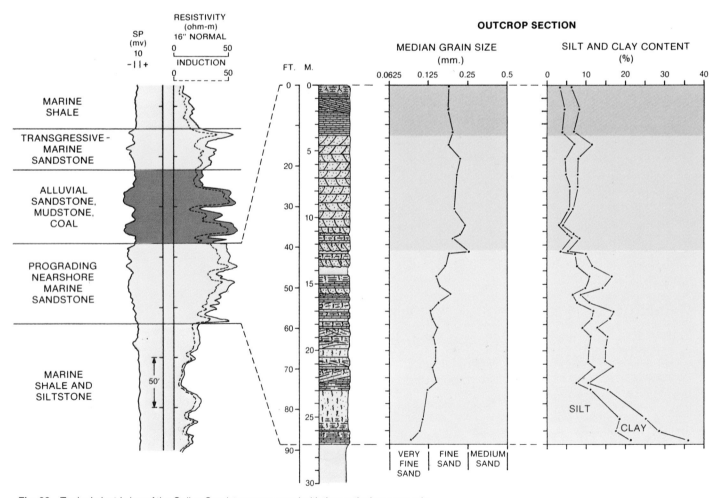

Fig. 33—Typical electric log of the Gallup Sandstone compared with the vertical sequence in outcrop a few miles away along depositional strike.

in mineral cements within sandstones can significantly alter log responses. Additional examples of mechanical log response in progradational sequences and associated facies are shown in a later section on "case-study examples."

Lateral Variation in Facies Sequence

An idealized cross-sectional model of onshore to offshore facies changes in a prograding shoreline sandstone body, based on studies of Rocky Mountain Cretaceous examples, is shown in Figure 34. Near the landward margin, where shoreline sandstone lies with disconformable contact on older nonmarine or lagoonal deposits, the vertical sequence consists of cross-stratified upper shoreface deposits overlain by beach deposits (location A, Fig. 34). Farther seaward, lower shoreface deposits were laid down before upper shoreface and beach deposits, and the complete sequence is present (location B, Fig. 34). In the most seaward parts of individual shoreline sandstone bodies, beach and upper shoreface deposits are lacking, and sands deposited in lower shoreface environments constitute the entire thickness of the unit (location C, Fig. 34). In some cases these seaward areas were probably beyond the beach and upper shoreface following progradation, but the seaward extent of these facies may have been modified in other cases by shoreface erosion during a succeeding transgression. Examples of these variations in vertical sequence are shown in the later section on "case studies." In-

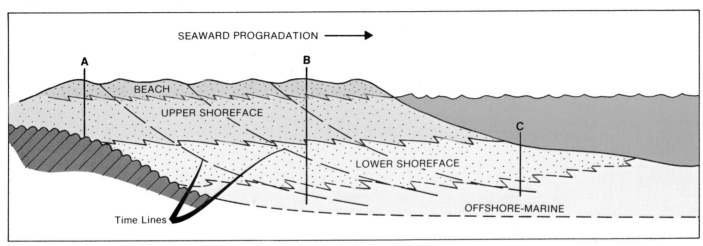

Fig. 34—Idealized cross section of progradational beach-to-offshore sandstone and shale, illustrating variation in the vertical facies sequence in a landward to seaward direction.

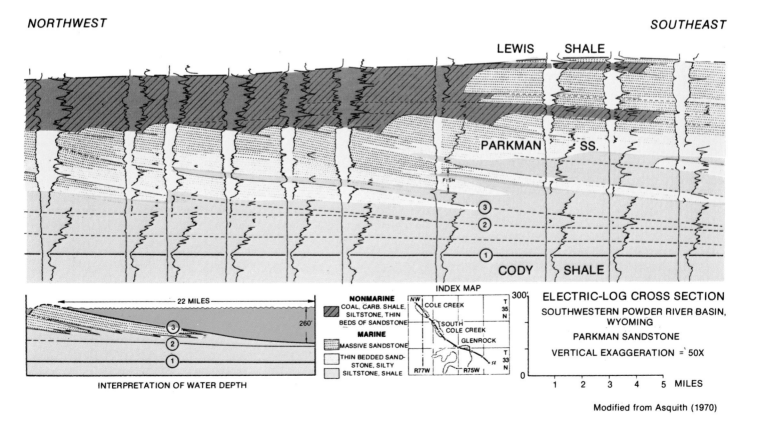

Fig. 35—Log correlation section of Parkman Sandstone, Wyoming, showing correlation markers sloping southeastward relative to top of sandstone and to marker 1 below. This indicates progradation of beach-to-offshore sequence toward the southeast.

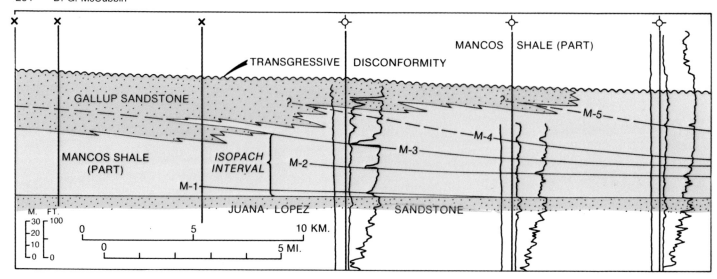

Fig. 36—Outcrop to subsurface correlation section showing north-eastward progradation of Gallup Sandstone and underlying marine shale. Location of section is shown on Figure 37.

complete sequences described by Clifton (1967, 1973) and by Ryer (1977) may also be related to the position of these sequences in a land-to-sea direction.

Because a prograding beach-to-offshore sequence is an offlap sequence, time lines slope seaward (in the direction of progradation) through successively lower facies and into marine shales that underlie the sandstone body (Fig. 34). In some deposits, this is recorded by seaward-sloping surfaces marked by heavy-mineral placers, surfaces of erosion, or clay-rich partings. The attitude of these surfaces provides a measure of the local direction and amount of offshore slope during deposition. These surfaces are relatively steep in beach foreshore deposits and very gentle in the more offshore deposits. Heavy-mineral placers in the upper part of the prograding shoreline sequence of the Gallup Sandstone slope seaward (northeastward) at an angle of 4.5° relative to the top and base of the unit (Fig. 31); this is similar to the slope of present-day beach foreshores made up of sand of comparable grain size. On the basis of electric log correlations (Fig. 35), clay-rich beds in the Parkman Sandstone of the southwestern Powder River basin, Wyoming, slope seaward (southeastward) at about 2.84 m per km (15 ft per mi) relative to markers in underlying marine shale (Asquith, 1970). The attitude of these beds is similar to that of time lines in the Galveston Island sand body (Bernard et al, 1962) and to offshore profiles of modern wave-dominated coasts (Shepard, 1960).

Mapping of thickness, lateral variation in vertical facies sequence, and attitude of time lines can all be useful in determining the distribution and trend of sandstone bodies formed by seaward progradation of beach and shoreface environments. In the Gallup Sandstone, for example, the interval between a widespread bentonite marker, M-3, and the top of the underlying Juana Lopez unit thins seaward (northeastward; Fig. 36). An isopach map of this interval shows that depositional strike was northwest-southeast, parallel with the trend of the overall Gallup shoreline (Fig. 37).

TIDAL-INLET DEPOSITS

Tidal inlets are more or less permanent passages between barrier islands that allow tidal exchange between the open sea and lagoons, bays, and tidal marshes behind the islands (Figs. 1, 38). Inlet channels are generally deepest between tips of the islands and shoal onto tidal deltas both seaward and landward. Relatively flat channel-margin platforms may occur on one or both sides of the main inlet channel but commonly are best developed on the side adjacent to the barrier that is growing by spit accretion. The main channels range in maximum depth from 4.5 m (15 ft) to 40 m (130 ft), depending largely on the amount and duration of tidal exchange (Allen, 1967). The frequency or spacing of inlets on present barrier-island coasts also depends on tidal range (Hayes, 1975). Morphology of associated flood-tidal and ebb-tidal deltas relates both to tidal range and to the amount of wave energy (Hayes, 1976). Ebb-tidal deltas are poorly developed on coasts where wave energy dominates over tidal cur-

rents, but may form extensive offshore shoals in other areas (Fig. 39).

Modern Inlet Deposits

The processes and characteristics of modern tidal inlets and associated environments are summarized in a recent review article by Boothroyd (1978). Knowledge is good for intertidal and shallow subtidal sedimentary environments but is much more limited for deep inlet channels, except in a few areas where box-coring studies and diving observations have been made. Processes are complex because of the reversing tidal flow through inlets and the influence of waves and wave-generated currents in shallower environments. It is generally agreed, however, that ebb-current (seaward) sand transport is dominant in many deep inlet channels and ebb-tidal deltas, but that flood-current (landward) transport is dominant on channel-margin platforms and flood-tidal deltas. The processes and sedimentary characteristics of the beach and berm of the recurved spit on the associated barrier island are similar, except for their orientation, to those of ocean-facing beaches. Parts of the flood-tidal deltas may eventually become inactive and subject to the processes of marsh environments.

The thickness of sediments deposited by migration of tidal inlets and associated environments may be as great as the depth of the inlet, but the facies sequence depends on the direction of inlet migration. Some modern inlets have migrated laterally or along shore during historical time by spit accretion on one side of the inlet and net erosion on the opposite side. The direction of lateral inlet migration is determined by the dominant direction of longshore sediment transport and is probably somewhat variable over long periods of time. Some modern inlets associated with prograding barrier islands are presumably migrating seaward, along with the barriers. Because the depth of the inlet channel may exceed the depth of

sand deposition in the associated environments, inlet-fill deposits may locally replace all other facies of the barrier complex, as at Sapelo Island, Georgia (Hoyt and Henry, 1967), and Fire Island, New York (Kumar and Sanders, 1974).

A vertical sequence model for lateral or longshore migration of tidal inlets

was proposed by Kumar and Sanders (1974) and is redrawn here (Fig. 40). This model is based on morphological and box-core studies in Fire Island Inlet and is supported to some degree by long cores from the part of Fire Island that has accreted by inlet migration during the past 115 years. Deposits of this area consist of medium- to coarse-

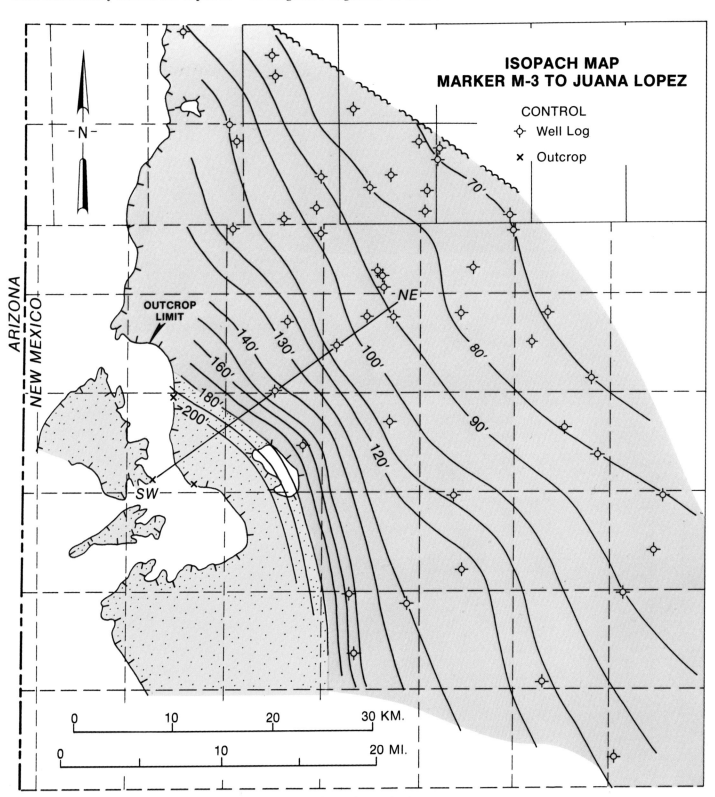

Fig. 37—Isopach map of interval from bentonitic marker M-3 to top of Juana Lopez showing thinning northeastward, in a seaward direction, and strike northwest, parallel with the Gallup shoreline. Interval mapped is shown on Figure 36.

Fig. 38—Aerial view of flood-tidal delta and tidal inlet, San Luis Pass, Texas. Looking south toward Gulf of Mexico.

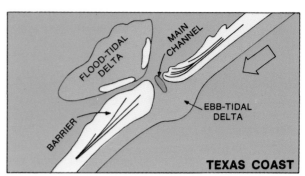

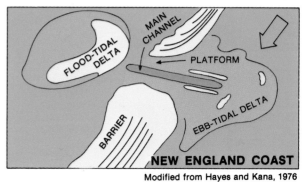

Modified from Hayes and Kana, 1976

Fig. 39—Variation in form of inlet channels and tidal deltas. Large arrows show direction of dominant longshore transport. New England coast has higher tidal range and better developed ebb-tidal deltas than does the Texas coast.

SEDIMENTARY STRUCTURES	DESCRIPTION	ENVIRONMENT AND PROCESSES
	Medium sand, with subhorizontal to low-angle (2°-5°) stratification.	Spit berm. Deposition by swash and overwash.
	Medium sand, with low-angle (5°-10°) and some high-angle cross stratification.	Spit beachface. Deposition by swash.
	Medium sand; with flood-oriented, large- and small-scale, trough cross stratification at top; longshore-dipping foreset strata in middle; and ebb-oriented, tabular-planar and trough cross strata below.	Spit platform or channel - margin platform. Deposition by flood-tidal currents (in shallow water), by lateral (longshore) accretion of spit-platform slope, and by ebb-tidal currents (in deeper water).
	Medium sand, with plane, parallel stratification.	Shallow channel.
	Medium sand; with ebb-oriented, high angle (15°-25°), planar cross stratification, in sets at least 10 cm thick; reactivation surfaces; some shell fragments.	Deep channel. Deposition dominated by ebb-oriented tidal currents with minor erosion of sand waves by flood-oriented tidal currents. Sand waves 0.5 to 2.0 m high.
	Lag gravel of large shells and pebbles.	Erosion by channel migration.

Modified from Kumar and Sanders, 1974

SEA ⟶

Fig. 40—Vertical facies sequence formed by longshore migration of main inlet channel and longshore accretion of barrier island on the updrift side. Based on studies of Fire Island Inlet and cores from laterally accreting part of Fire Island, New York.

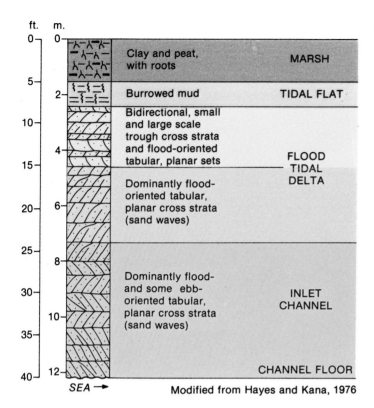

ft.	m.		
0	0	Clay and peat, with roots	MARSH
5	2	Burrowed mud	TIDAL FLAT
10	4	Bidirectional, small and large scale trough cross strata and flood-oriented tabular, planar sets	FLOOD TIDAL DELTA
15		Dominantly flood-oriented tabular, planar cross strata (sand waves)	
20	6		
25	8	Dominantly flood-and some ebb-oriented tabular, planar cross strata (sand waves)	INLET CHANNEL
30			
35	10		
40	12		CHANNEL FLOOR

SEA →   Modified from Hayes and Kana, 1976

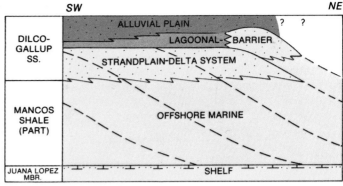

Fig. 42—Diagrammatic cross section showing three major depositional systems represented in Gallup Sandstone, northwestern New Mexico.

Fig. 41—Vertical facies sequence formed by seaward migration of inlet channel and flood-tidal delta. Inferred from studies of modern U.S. Atlantic coast areas with a moderate tidal range and generally fine-grained sand.

grained sand with some pebbles. The sequence is divided into several facies representing shoaling upward from the channel floor to the berm of the spit on the laterally accreting barrier. In this model, it is suggested that sands deposited in the deep channel, below about 4.5-m water depth, are characterized mainly by ebb-oriented, tabular, planar cross strata, with reactivation surfaces formed by flood-oriented tidal currents. Sands of the channel-margin platform or "spit platform" have trough-shaped sets of cross strata formed by both ebb- and flood-oriented megaripples, as well as large-scale foresets formed by migration of the steep margin of the spit platform. The sediments of the beachface and berm of the associated spit have swash stratification, similar to that of the ocean-facing beach except for its variable but generally longshore dip direction. A sequence model based on studies of inlet channel to spit profiles in other areas was proposed by Hayes (1976, Fig. 81). This model is generally similar to that based on Fire Island except that the deep-channel facies includes alternating ebb-oriented and flood-oriented cross strata with only a slight dominance of the ebb direction; also, Hayes' model does not include the thick foresets of the spit-platform or channel-margin-platform facies.

The vertical sequence formed by seaward migration of inlets and associated tidal deltas is likely to differ from that formed by longshore migration. Deep-channel deposits will be overlain by flood-tidal delta deposits and capped by lagoonal or mudflat and marsh deposits. A generalized model for such a sequence was proposed by Hayes (1976, Fig. 74) and is reproduced here (Fig. 41). In this sequence, sands of the deep inlet channel are dominated by large-scale, ebb-oriented, tabular-planar cross stratification, with some smaller scale, bidirectional, trough cross stratification. As in the case of the sequence formed by lateral migration of the tidal inlet, however, flood currents may be recorded by reactivation surfaces rather than by landward-dipping cross strata. Facies deposited by flood-tidal deltas are inferred to consist of sands with small-scale, bidirectional, trough cross stratification and medium-scale, flood-oriented, tabular-planar cross stratification. The sequence is capped by burrowed mud of the tidal flat and peat of the salt-marsh environment.

Other characteristics proposed for inlet-fill sequences include the presence of a mixed open-marine and lagoonal or marsh fauna (Shepard, 1960; Kumar and Sanders, 1974) and some bioturbation or distinct burrows. Because of fluctuating energy levels, some mud drapes may be preserved, as well as internal scour surfaces. Auger-hole samples from inlet-fill deposits underlying some modern barrier islands also show a general fining-upward sequence (Moslow and Heron, 1978).

Ancient Inlet Deposits

Many authors have suggested that tidal-inlet deposits should be well represented in the stratigraphic record because of their preservation potential, but relatively few examples have been recognized. One possible example, from the Cretaceous Blood Reserve-St. Mary River Formations, southern Alberta, is described by Reinson (1979). In this example, the inlet-channel deposit is about 7 m (22 ft) thick and consists of fine- to medium-grained sandstone with: bidirectional, small- to medium-scale trough cross stratification; some tabular, planar cross stratification; and rare horizontal plane beds. Abundant *Ophiomorpha* burrows indicate marine influence. The sandstone is overlain by coal lenses and shale with oysters and oyster-coquina beds, interpreted as marsh and subtidal lagoonal deposits. The total sequence is similar to the model shown in Figure 41, suggesting overall seaward migration of the inlet channel, rather than lateral or longshore migration.

The Cretaceous Gallup Sandstone of northwestern New Mexico includes a barrier island-lagoonal system, overlying a strand plain-deltaic system (Fig. 42). A probable inlet-fill sequence in the barrier-island system is shown in Figure 43. This sequence is about 9 m (30 ft) thick and consists mostly of fine- to medium-grained sandstone with some thin shale and siltstone beds. The sandstones have mostly tabular, planar cross stratification formed by migration of sand waves (Fig. 44). Dip directions of the cross strata are dominantly seaward (northeastward) in the lower part, but many are reversed, especially in the upper part. Interbedded siltstones are highly bioturbated, and burrows (including *Ophiomorpha*) are common in sandstones (Fig. 44). The sandstones are overlain by bioturbated mudstones interpreted as lagoonal deposits, and the sequence is capped by a thin coal bed with underlying root traces suggesting very shallow water (marsh?) environments. Northwest and southeast, the interval equivalent to this sequence consists of progradational beach and shoreface sandstones interpreted as contemporaneous barrier-island deposits. Southwestward, the inlet-fill sequence grades into bioturbated sandstones and foraminifer-bearing shales interpreted as lagoonal deposits. Again, the nature of the inlet-fill sequence and associated facies suggests seaward migration of the inlet channel.

PALEOFLOW DIRECTIONS	DESCRIPTION	INTERPRETATION
	Lignite	Marsh
	Clayey siltstone, bioturbated	Tidal flat or lagoonal mud
	Sandstone, fine to medium-grained, with tabular, planar cross strata in sets up to .6 to .9 m thick.	Upward-shoaling sequence from main inlet channel (below) to tidal delta and lagoonal (above). Alternating flood-current and ebb-current sand waves.
	Wave and wave-current ripples.	Some wave influence.
	Abundant burrows in sandstone and interbedded siltstone.	Abundant marine infauna.
	Sandstone, mostly medium-grained, with tabular, planar cross strata, ebb-current dominated.	Inlet channel floor with ebb-current sand waves and dunes.
	Few wave ripples.	Very little wave influence.
	Burrows in sandstone and siltstone.	Some marine infauna.
		Scour, related to tidal currents.
		Strand plain deposits, related to earlier period of progradation.

SHORELINE ORIENTATION
N
n=41

Fig. 43—Vertical facies sequence interpreted as an upward-shoaling inlet channel to lagoonal and marsh sequence, in part of the barrier island-lagoonal system of the Gallup Sandstone.

Fig. 44—Thick set of high-angle, planar, tabular cross stratification in inlet-fill sequence of Gallup Sandstone. Also note burrows in sandstone. Cross stratification like this was formed by migration of large sand waves both lagoonward and seaward under the influence of reversing tidal currents (Fig. 43).

WASHOVER-FAN DEPOSITS

Washover fans are subaerial, fan-shaped landforms occurring on the landward or lagoonward side of some barrier islands (Fig. 1). Washovers are formed where storm tides overtop and erode channels through beach-dune ridges fronting the open sea and transport sand landward from these channels. Relatively small washover lobes or fans occur on many barrier coastlines, including the U.S. Atlantic coast (Schwartz, 1975; Deery and Howard, 1977), but large washover fans are most common on coasts with a low tidal range, such as the Texas coast (Hayes, 1967; Andrews, 1970; Dickinson et al, 1972).

Small washovers on the Outer Banks, North Carolina (Schwartz, 1975), occur on barrier flats landward from washover passes or channels cut through the foredune ridge (Fig. 45). Individual washover units, deposited during a single storm, can be 1 m thick and extend landward a few hundred meters. Most common stratification types are planar, "horizontal" stratification where deposition is by shallow flow above normal high tide, and medium-scale foreset stratification where the washover fan or lobe builds into shallow standing water such as a pond on the barrier flat (Fig. 46). Foreset stratification may show sigmoidal topset-foreset-bottomset forms, or foresets may be truncated by overlying topsets.

Large washover fans, such as those on the Texas coast, have a maximum width in a land-sea direction of as much as 6.4 km (4 mi). These large fans are composite features, deposited by many individual storms during late Holocene time. The large fan on St. Joseph Island (Andrews, 1970) has numerous broad, shallow distributary channels, radiating lagoonward from the main channel through the higher seaward part of the island. Deposits of the distributary channels consist of thin (8 to 38 cm) sand units with planar, subhor-

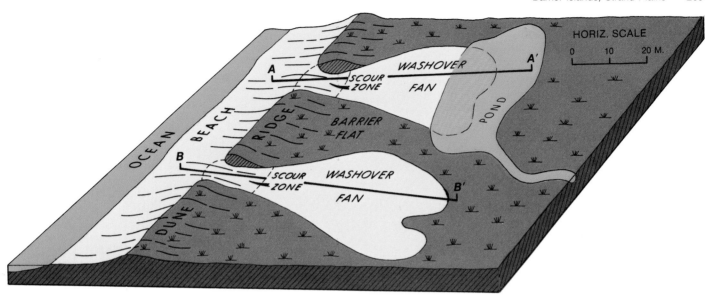

Fig. 45—Small washover fans on barrier flats landward from washover channels cut through foredune ridge, Outer Banks, North Carolina.

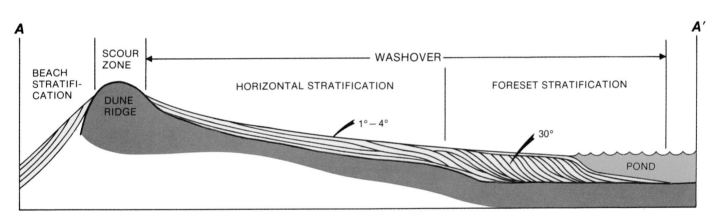

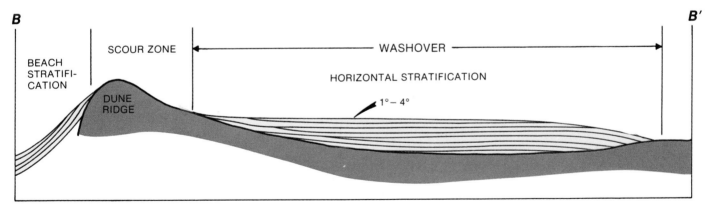

Fig. 46—Diagrammatic cross sections showing stratification in washover-fan units deposited by a single storm. Locations of sections are shown on Figure 45.

izontal stratification, commonly becoming burrowed toward the top of the unit, and in some cases separated by burrowed muddy sediments deposited during periods when the channel was not active. Low elongate mounds between channels are underlain by eolian sands, mostly disturbed and disrupted by plant roots. The sandy washover-fan sequence has a maximum thickness of

at least 1.3 m (4 ft) near its apex, decreases in thickness toward the lagoon, and is underlain by muddy lagoonal sediments.

Other characteristics of washover deposits include the presence of transported shells, mud clasts, and other debris from the open sea, concentrated at the base of each depositional unit or decreasing in abundance and average

size upward in the unit (Hayes, 1967; Andrews, 1970). Transported shells are dominantly open-marine forms, in contrast to the restricted-marine fauna of interbedded muddy pond deposits or underlying lagoonal deposits. Burrows in both the washover sands and interbedded and underlying muds are formed by lagoonal infauna or supratidal animals such as crabs, and may be recognizably different from those in contemporaneous open-marine deposits.

Examples of sandstones interpreted as washover deposits on the basis of their characteristics and relationships to other facies include parts of the Lower Silurian of southwest Wales (Bridges, 1976), the Upper Carboniferous of northwest England (Elliott, 1975), the Carboniferous of eastern Kentucky and southern West Virginia (Horne and Ferm, 1978), and the Pleistocene of South Africa (Hobday and Jackson, 1979). In the Upper Cretaceous La Ventana Tongue of northwestern New Mexico, described in more detail later, some thin sandstones extending landward at least 1.5 km from associated barrier-island facies may be washover deposits. These sandstones range up to 4 m (14 ft) thick and are interbedded with carbonaceous mudstones containing a restricted-marine foraminiferal fauna and other evidence of a lagoonal origin. The thinner sandstones have planar, horizontal stratification. Thicker sandstones have a lower part with landward-dipping

tangential foresets as much as 1.5 m (5 ft) thick and an upper part with horizontal stratification. Marine burrows *(Ophiomorpha)* occur in the lower cross-stratified part, and root traces occur at the top, indicating marine influence and very shallow to emergent conditions.

In the subsurface, washover deposits might be recognizable only on the basis of their geometric relationships and association with barrier sandstones and lagoonal deposits. Possible examples include parts of the Oligocene Frio of south Texas (Boyd and Dyer, 1964) and the Cretaceous La Ventana (this paper).

PRESERVATION DURING TRANSGRESSION

In preceding sections of this review, emphasis has centered on depositional sequences formed by coastal progradation under conditions of more or less constant sea level. The record formed by progradation is fairly well understood, and many ancient examples have been recognized. The record

formed by transgression, or landward migration of the strand line, is more controversial. In particular, disagreement exists concerning the potential for preservation of barriers and other coastal sand bodies during a relative rise of sea level.

Shoreface Retreat Versus In-Place Drowning

From studies of Holocene history of modern shelves and shorelines, two general mechanisms of transgression have been proposed. Landward retreat by shoreface erosion (Fig. 47, A and B) is exemplified by Holocene transgressive barriers such as those of the U.S. middle Atlantic coast (Kraft, 1971) and has been elaborated into a general theory by coastal geomorphologists (Bruun, 1962; Schwartz, 1967; Swift, 1975). According to this theory, a relatively steep concave-upward shoreface profile is maintained but translated upward and landward during a relative rise of sea level. To maintain the profile, net erosion occurs in the upper shoreface and beach zones and net deposition occurs in more landward and more seaward areas. If transgression continues by this process, preservation of beach and upper shoreface deposits is unlikely. The record would include a transgressive disconformity or "ravinement" (Swift, 1968) overlain by sediments deposited in lower shoreface or shelf environments. Preservation of back-barrier facies below the disconformity would depend on the vertical component of shoreface translation relative to the slope of the surface being transgressed (Swift, 1975).

Another general hypothesis of shoreline response during transgression is that of in-place "drowning" (Fig. 47, C and D). According to this concept (Sanders and Kumar, 1975), barriers or other coastal sand bodies can be submerged during transgression when the surf zone "jumps" from the former shoreline to a position farther inland. Presumably, this is more likely when the rate of subsidence or sea level rise is relatively high and the general slope of the surface being transgressed is low. A low slope results in rapid lateral movement of the surf zone and dissipation of wave energy over a relatively large offshore area (Keulegan and Krumbein, 1949).

Holocene deposits of the U.S. Atlantic shelf evidently record transgression by both of these general mechanisms. In some areas, Holocene lagoonal and marsh deposits are overlain by sands interpreted as the product of erosional

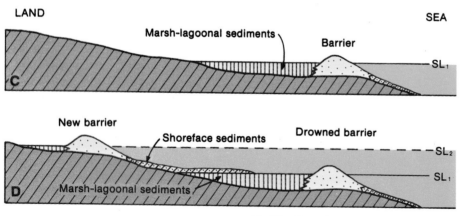

Modified from Sanders and Kumar (1975)

Fig. 47—Diagrammatic sections showing inferred effects of relative rise of sea level on barrier-island coasts. **A.** and **B.**—Barrier retreat by shoreface erosion and offshore deposition under conditions of relatively slow and steady rise of sea level. **C.** and **D.**—Barrier "drowning" in place, as the surf zone jumps landward across the flat marsh-lagoonal area under conditions of relatively rapid rise of sea level.

shoreface retreat (Swift, 1975; Field and Duane, 1976). Other submerged sand ridges with lagoonal deposits on the landward side and shoreface deposits on the seaward side are interpreted as drowned Holocene barriers (Sanders and Kumar, 1975). Some of the sands deposited during Holocene transgression, whether by continuous shoreface retreat or by stillstand and subsequent drowning, have evidently been reworked to various degrees by modern shelf processes.

Ancient Examples

The Cretaceous basal Niobrara (Tocito) sandstone, which forms significant oil reservoirs in northwestern New Mexico, is a possible example of the record of continuous shoreface retreat (McCubbin, 1969). Sandstones as much as 12 m (40 ft) thick onlap southwestward against a regional unconformity and interfinger northeastward into marine shale. Where examined in outcrop and in cores, these sandstones consist entirely of offshore-marine ("shelf") deposits. If contemporaneous beach and back-beach sediments were deposited during the transgression to the southwest, those sediments were evidently removed by subsequent shoreface erosion.

Intertonguing marine and non-marine rocks of middle Miocene age in the Caliente Range, California, include coastal-marine sandstones and conglomerates interpreted by Clifton (1967, 1973) as complete to incomplete progradational sequences separated by transgressive erosional surfaces. The erosional surfaces have low relief but are commonly overlain by thin (less than 0.3 m thick) conglomerates with shell fragments. In some places these thin conglomerates grade upward to bioturbated siltstones forming the basal, offshore facies of overlying progradational sequences. These surfaces may record shoreface erosion during transgressive episodes, but the amount of erosion is unknown. Some progradational sequences may be incomplete at the top because of their position in a

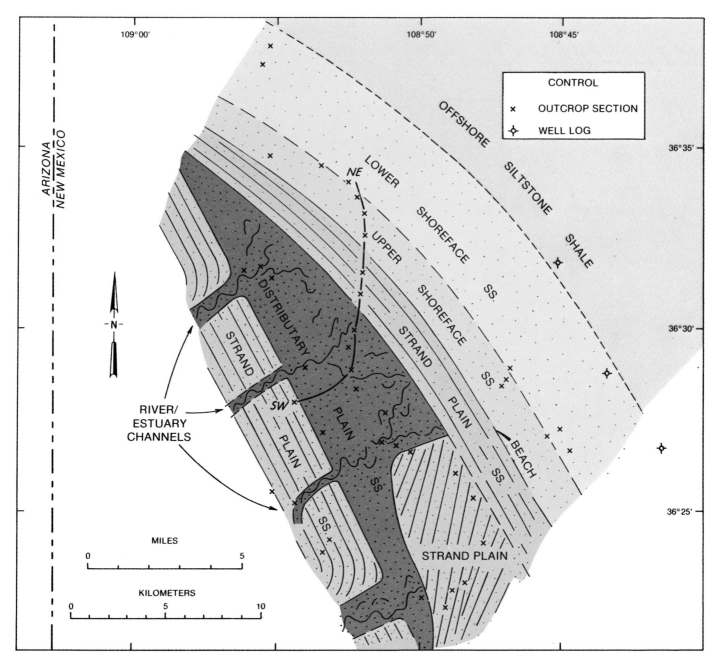

Fig. 48—Paleogeography after deposition of strand plain-deltaic system of Gallup Sandstone, northwestern New Mexico. Progradation to the northeast occurred by seaward accretion of a strand plain, then deposition of a deltaic distributary plain and, finally, accretion of a younger strand plain.

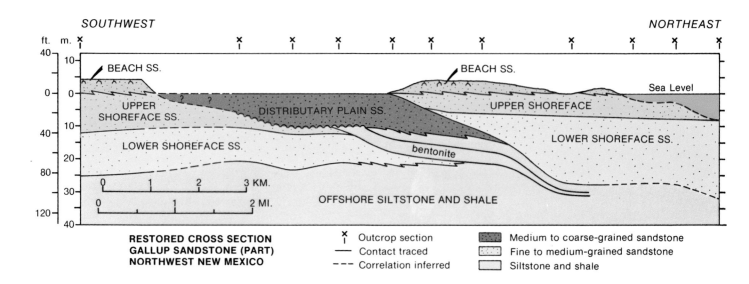

SOUTHWEST   *NORTHEAST*

RESTORED CROSS SECTION
GALLUP SANDSTONE (PART)
NORTHWEST NEW MEXICO

✗ Outcrop section
— Contact traced
‑‑‑ Correlation inferred

▓ Medium to coarse-grained sandstone
░ Fine to medium-grained sandstone
▒ Siltstone and shale

Fig. 49—Cross section showing facies distribution and relationships in strand plain-deltaic system of Gallup Sandstone. Location of cross section is shown on Figure 48.

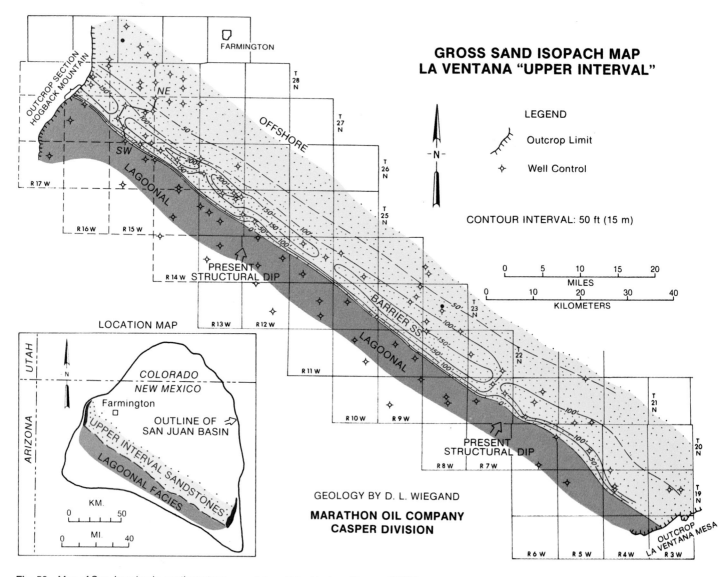

GROSS SAND ISOPACH MAP
LA VENTANA "UPPER INTERVAL"

LEGEND

⌣ Outcrop Limit

✦ Well Control

CONTOUR INTERVAL: 50 ft (15 m)

LOCATION MAP

GEOLOGY BY D. L. WIEGAND

MARATHON OIL COMPANY
CASPER DIVISION

Fig. 50—Map of San Juan basin, northwestern New Mexico, showing outcrop and subsurface distribution of sandstone in part of "upper interval", La Ventana Tongue, Cliff House Formation. Interval mapped is shown on Figure 52.

land-sea direction at the end of progradation (Fig. 34), rather than because of substantial shoreface erosion.

Two Cretaceous examples discussed later in this report, the upper La Ventana of New Mexico and upper Almond of Wyoming, are "transgressive" sequences in the sense that they consist of coastal deposits overlain by marine shale. The vertical and lateral sequences within sandstone bodies themselves, however, clearly show that they were formed by upward building and seaward progradation during a stillstand in the overall transgression. After progradation, the shoreline evidently jumped landward with little erosion of the shoreline sands and contemporaneous back-barrier facies. This transgression is recorded only by a sharp contact between the progradational sequence below and the offshore-marine siltstones and shales above.

CASE-STUDY EXAMPLES

Three examples of coastal sandstone bodies briefly described in this section illustrate some important variations in geometry and stratigraphic relationships. In each of these examples, shoreline sand deposition started during a period of stillstand or shoreline stabilization. Subsequent history differs in the relationship between the rate of submergence and rate of sediment influx. In one example, the shoreline prograded to form strand plains. In another example, progradation was interrupted by minor transgressions to form a series of stacked barriers and thick lagoonal deposits. In a third example, a prograding barrier and relatively thin lagoonal deposits were evidently localized over a submerging delta complex. In each case, comparison with Holocene strand plains and barriers helps in interpretation of geometry and stratigraphic relationships.

Gallup Sandstone—Strand-Plain Complex

The Gallup Sandstone of northwestern New Mexico is a nearshore-marine to coastal-plain deposit formed on the southwestern margin of the Late Cretaceous interior seaway. In the area discussed here, this sequence consists of three distinct depositional systems (Fig. 42). The lower one is interpreted as a strand plain-deltaic complex formed during a period of high sediment influx relative to subsidence or sea-level change (McCubbin, 1972). The depositional history of this part of the Gallup consisted of progradation of beach and shoreface deposits to form sandy strand plains, alternating with episodes of deltaic progradation (Fig. 48). Characteristics of the beach and shoreface deposits and their vertical sequence are shown on Figure 25.

The northeastern strand-plain unit (Fig. 48) thins abruptly to the southwest against deltaic sandstone deposits, with contemporaneous lagoonal sediments very thin or absent (Fig. 49). Evidently, this strand plain began as a mainland beach on the seaward margin of the abandoned delta. Some shoreface erosion occurred initially, partly truncating the distributary plain and an older strand plain along shore to the southeast (Fig. 48). The emergent part of the new strand plain, presumably with beach-ridge topography, prograded northeastward about

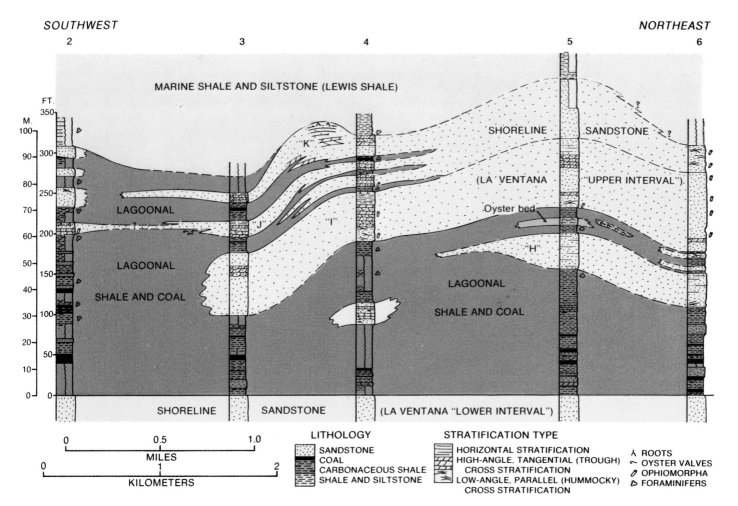

Fig. 51—Stratigraphic cross section in outcrop of "upper interval" of La Ventana Tongue at Hogback Mountain. Location of section is shown on Figure 50.

3.2 km (2 mi), and the submerged (shoreface) part extended offshore an additional 6.4 km (4 mi) where it graded into offshore silts and muds (Fig. 48). The gradually northeastward-thinning toe of the westernmost strand plain (Fig. 49) illustrates changes in a seaward direction.

The depositional setting of the strand plain-deltaic system of the Gallup (Fig. 48) probably paralleled that of the Holocene coastal plain north of the San Blas, State of Nayarit, Mexico (Fig. 6), as described by Curray and Moore (1964) and Curray, Emmel, and Crampton (1969). In the Nayarit area, relatively rapid progradation of the strand plain resulted from an abundant supply of sand from nearby rivers or distributaries. Because of the relatively high level of wave energy, the deltas were cuspate in form and much of the sand was transported along shore by wave-generated currents to form strand plains.

Although internally complex, the strand plain-deltaic system of the Gallup is an externally simple sandstone sheet, with a sharp, nearly flat top and a gradational base where underlain by offshore-marine siltstone and shale. Subsurface correlation of bentonitic beds, such as that overlying the seaward part of the westernmost strand plain (Fig. 49), shows that these beds diverge northeastward from the top and base of the overall sandstone unit, reflecting the history of progradation in this direction. Figure 37 is an isopach map of the interval between one of these bentonitic markers and the top of

the underlying Juana Lopez Sandstone. The thickness of this interval, which consists of sandstone and partly contemporaneous siltstone and shale, decreases northeastward and is uniform to the northwest and southeast parallel with general shoreline orientation and depositional strike.

The strand plain-deltaic unit of the Gallup produces no oil or gas on anticlines in the subsurface nearby and is probably flushed by meteoric water. In widespread coastal sandstone bodies of this type, stratigraphic traps formed by facies changes in a landward direction are unlikely because of the lateral continuity of potentially permeable sandstone facies. Where not exposed to flushing by meteoric water, excellent reservoirs may exist on structural closures. A knowledge of internal facies distribution might be useful to the exploitation geologist concerned with efficient development and recovery methods.

La Ventana Sandstone— Interdeltaic Barrier Chain

The upper part of the La Ventana Tongue of the Cliff House Formation, northwestern New Mexico (Fig. 50), illustrates outcrop and subsurface characteristics of stacked barrier-island sandstones and associated lagoonal deposits. These sandstones are exposed on the west and east margins of the present structural basin and occur along a relatively long, straight trend in the subsurface. The general deposi-

tional setting is believed to have resembled that of the Holocene barrier-island chain of the present Texas coast.

In outcrops along the west side of the basin, the La Ventana Tongue consists of a thick, regressive-marine sandstone ("lower interval"). The "upper interval" of La Ventana includes several correlatable sandstone units, three of which disappear to the south near the southern end of Hogback Mountain (units "I", "J", and "K", Fig. 51). Each of these sandstone units has a vertical and lateral sequence indicating deposition in upward-shoaling, shoreface and beach environments during northeastward progradation (Fig. 51). The characteristics of the lower shoreface, upper shoreface, and beach facies are similar to those of the Gallup Sandstone (Fig. 25). Marine origin of these sandstones is indicated by *Ophiomorpha* burrows, rare sharks' teeth and casts of pelecypods *(Cardium?)*, and long-spaced wave ripples.

Sandstone units of the "upper interval" are interpreted as barrier-island deposits (as opposed to strand-plain deposits) because they are replaced laterally to the southwest by fine-grained carbonaceous sediments believed to be lagoonal deposits (Fig. 51). These fine-grained sediments include dark, laminated shales with abundant fragmental plant material, but with only widely spaced horizons with carbonized plant roots. Interbedded coals, some at least 1.8 m (6 ft) thick, are not consistently underlain by root zones,

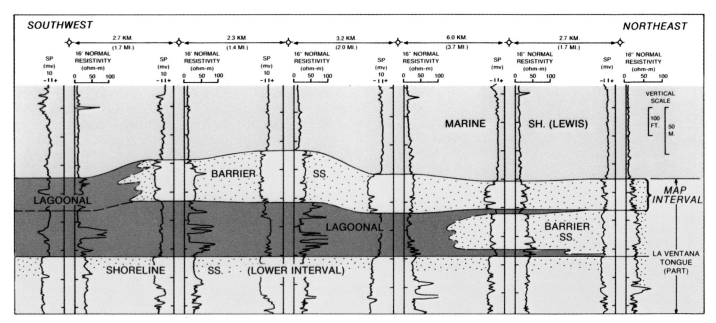

Fig. 52—Stratigraphic cross section in subsurface of La Ventana "upper interval," showing correlation and facies interpretation based on logs and nearby outcrop section. Location of section is shown on Figure 50.

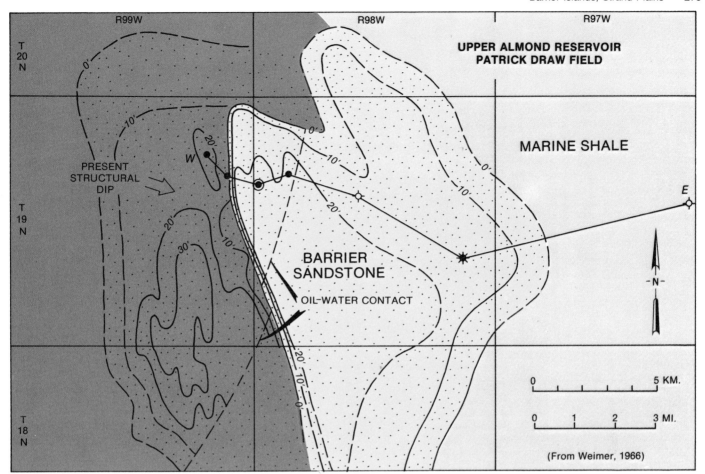

Fig. 53—Map showing distribution and thickness of upper Almond reservoir sandstone (UA-5), Patrick Draw field and nearby area, southwestern Wyoming, The eastern, somewhat younger, sandstone body is interpreted as a barrier-island deposit.

and probably represent transported plant material which settled out in quiet, ponded environments. Abundant arenaceous foraminifers occur in the carbonaceous shales, both intervening between sandstone units and south of their pinchout. The foraminifers include some forms (especially *Miliammina*) absent in closely associated open-marine shales of the Lewis and that are major components of present-day brackish, restricted (lagoonal) environments. Also present are some lenticular beds consisting mostly of oyster shells in a silty or sandy matrix. Thin sandstones in the lagoonward equivalents of units "J" and "I" probably represent washover-fan deposits.

In the nearby subsurface, similar stratigraphic relationships are recognized on the basis of mechanical log correlation and interpretation (Fig. 52). Here, as in the outcrop, barrier sandstones of the "upper interval" thin abruptly to the southwest and are replaced by shales and coals (the coals are recognized by their high resistivity and low density on mechanical logs). The uppermost interval of sandstone

shown on Figure 52 and mapped on Figure 50 probably consists of several individual units stacked vertically as they are in outcrop. The isopach map of sandstone in this interval shows an almost straight trend extending northwest-to-southeast across the entire structural basin (Fig. 50). The isopach map also shows that the sandstones thin rapidly to the southwest, where they are replaced by lagoonal deposits, and more gradually to the northeast, where they become interbedded with marine siltstone and shale. Farther northeast, beyond the contoured area on the map, they are replaced entirely by marine siltstone and shale.

Outcrop characteristics and subsurface distribution of this barrier-lagoonal system suggest that it was deposited in an "interdeltaic" setting, like that of the present Texas coast. Barriers were presumably supplied by longshore transport from contemporaneous river-deltaic sources, possibly in the outcrop area on the southeast margin of the present basin, where parts of the La Ventana have been interpreted as deltaic (Mannhard, 1976). Deposition of considerable thicknesses of bar-

rier sandstones and lagoonal shales resulted from an overall balance between rates of sediment accumulation and basin subsidence. Vertical and lateral distribution of facies within the sandstone units, however, shows that each unit, after an initial transgression over older barrier or lagoonal sediments, was deposited by upward building to the swash zone and then by seaward accretion of shoreface and beach deposits. After deposition of the uppermost barrier (unit "K" in outcrop), abrupt southwestward transgression resulted in drowning of this barrier and deposition of overlying offshore-marine siltstones and shales of the Lewis Shale (Figs. 51, 52).

Sandstone bodies of this type are potential stratigraphic traps for oil or gas, where sandstones are replaced updip by lagoonal shales and are overlain by marine shales. These particular sandstones appear to be relatively continuous laterally from the outcrop and may be partially flushed by meteoric water. If these or similar sandstones were found to be productive, however, reconstruction of the depositional facies, geometry, and trend would obviously be useful both in exploration and exploitation.

Almond Sandstone Reservoirs—Post-Deltaic Barrier

Sandstone units near the top of the Upper Cretaceous Almond Formation (Fig. 53) are important oil and gas res-ervoirs in the Patrick Draw and other nearby fields in southwestern Wyo-ming (Weimer, 1966). One of these sandstone units, the UA-5 of Weimer (1966), consists in part of a laterally discontinuous sandstone body in-terpreted as a barrier-island deposit (Weimer, 1966; McCubbin and Brady, 1969). This sandstone body (Fig. 53) overlies coal-bearing deltaic deposits near its landward margin, is replaced laterally to the west by oyster-bearing lagoonal siltstones and shales, and in-terfingers to the east with marine shales. The depositional setting of this barrier is believed to have been similar

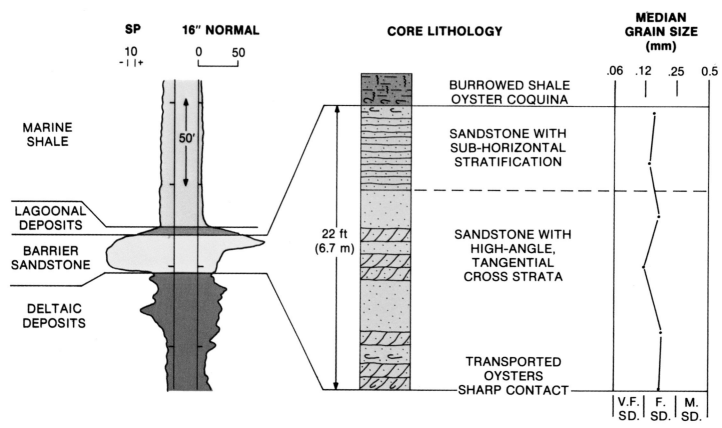

Fig. 54—Electric log and core sequence from near westward (lagoonward) margin of barrier-island sandstone body, upper Almond reservoir, Patrick Draw field. Interpreted as a progradational, upper shoreface to beach sequence with sharp contact on an upward-shoaling, marine to nonmarine (deltaic?) sequence. Location of well is shown by circled well symbol on Figure 53.

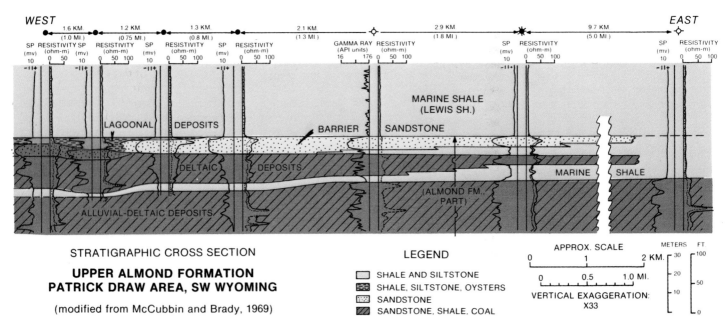

STRATIGRAPHIC CROSS SECTION

**UPPER ALMOND FORMATION
PATRICK DRAW AREA, SW WYOMING**

(modified from McCubbin and Brady, 1969)

Fig. 55—Stratigraphic cross section showing correlation and facies interpretation of part of the upper Almond reservoir sandstone unit. Location of cross section is shown on Figure 53.

to that of the modern Chandeleur Islands, along the margin of abandoned St. Bernard delta of the Mississippi River (Kolb and Van Lopik, 1958).

The vertical and lateral sequence of facies in the barrier-island sandstone indicates deposition by eastward progradation of beach, shoreface, and offshore deposits. Near the westward margin of this unit (Fig. 53), where it directly overlies coal-bearing deltaic deposits, the barrier sandstone consists of a lower facies with high-angle, tangential cross stratification interpreted as an upper shoreface or surf-zone deposit, and an upper facies with planar, nearly horizontal stratification interpreted as a beach or swash-zone deposit (Fig. 54). Farther eastward, the sandstone unit is separated from the deltaic sequence by a tongue of marine shale and siltstone interpreted as a partly contemporaneous offshore facies (Fig. 55). In this area, the sandstone consists of a facies with low-angle, hummocky cross stratification and is interpreted as a lower shoreface deposit.

West of the barrier sandstone are laterally equivalent, oyster-bearing, bioturbated shales and siltstones. These beds, which locally intertongue with the barrier sandstone (Fig. 55) and partly overlap it (Fig. 54), are interpreted as lagoonal deposits. The shales contain a sparse to abundant foraminiferal fauna indicating marine influence but not diagnostic of specific environments. The oyster valves are very abundant in one interval near the top, where they may represent oyster banks.

The sequence underlying the barrier-island sandstone and equivalents (Fig. 55) can be divided into three cycles, each of which consists of dark-gray marine shale grading upward into ripple-stratified siltstone and very fine-grained sandstone, which is in turn overlain by thin, very carbonaceous mudstone or siltstone. In most places, a coal bed forms the top unit of each cycle. These cycles are interpreted as thin, upward-shoaling sequences, deposited in low-energy, marginal-marine environments, and may record the progradation of small river deltas (McCubbin and Brady, 1969). The sandstone unit of one of these cycles is correlated with the UA-6 sandstone, which is a gas reservoir in the nearby West Desert Springs field. That sandstone was interpreted by Weimer (1966, p. 2,160) as a tidal-delta deposit, but the map distribution pattern for the reservoir (Weimer, 1966, Fig. 5) resembles a small eastward-building river delta.

The origin of the western sandstone body of the UA-5 unit (Fig. 53), between the coal bed of the uppermost cycle and the overlying lagoonal deposits (Fig. 55), is uncertain, but this sandstone also may represent a deltaic deposit. The original distribution of that sandstone and its facies equivalents to the east evidently has been modified by shoreface erosion prior to deposition of the barrier sandstone and its equivalents.

Deposition of the barrier-island sandstone began after marine transgression over the cyclic, deltaic sequence partially eroded that sequence but deposited only a thin basal lag of coarser sand and rounded shell fragments. Except for this basal lag deposit, initial deposition was in the upper shoreface or surf zone near the western margin of the barrier sandstone body, followed by upward shoaling to the beach or swash zone. The barrier sandstone body then prograded eastward by seaward accretion of outer shoreface and offshore-marine deposits. Deposition of lagoonal deposits occurred during both upward building and seaward accretion. Barrier-island and lagoonal deposition in this area ended with an abrupt transgression to the west, after which offshore-marine siltstone and shale of the overlying Lewis Shale were deposited.

The barrier-island sandstone of the UA-5 thins both south and north along depositional strike, but another sandstone body occurs in about the same stratigraphic position along trend to the north (Weimer, 1966). The geometry of the barrier-island sandstone and its stratigraphic relationships to other shoreline sandstone bodies along the regional trend may be related to an origin by deposition over laterally shifting deltaic plains, in much the same way the modern Chandeleur Islands were formed near the margin of the abandoned St. Bernard delta of the Mississippi River. If this interpretation is correct, reconstruction of the sequence and distribution of upper Almond deltas should aid in exploring for similar barrier-island sandstone bodies along the regional shoreline trend.

REFERENCES CITED

Allen, J. R. L., 1967, Depth indicators of clastic sequences: Marine Geology, v. 5, p. 429-446.

Andrews, P. B., 1970, Facies and genesis of a hurricane-washover fan, St. Joseph Island, central Texas coast: Bur. Econ. Geology, Univ. Texas, Austin, Rept. Invest. 67, 147 p.

Asquith, D. O., 1970, Depositional topography and major marine environments, Late Cretaceous, Wyoming: AAPG Bull., v. 54, p. 1184-1224.

Berg, J. H. van den, 1977, Morphodynamic development and preservation of physical sedimentary structures in two prograding Recent ridge and runnel beaches along the Dutch coast: Geologie en Mijnbouw, v. 56, p. 185-202.

Bernard, H. A., and R. J. LeBlanc, 1965, Resume of the Quaternary geology of the northwestern Gulf of Mexico province; in Quaternary of the United States, Wright, H. E., Jr., and D. G. Frey, eds.: Princeton Univ. Press, p. 137-185.

——— ——— and C. F. Major, 1962, Recent and Pleistocene geology of southeast Texas; in E. H. Rainwater, and R. P. Zingula, eds., Geology of the Gulf Coast and central Texas, guidebook of excursions: Houston Geol. Soc. p. 175-224.

Boothroyd, J. C., 1978, Mesotidal inlets and estuaries; in Davis, R. A., Jr., ed., Coastal sedimentary environments: Springer-Verlag, New York, p. 287-360.

Boyd, D. R., and B. F. Dyer, 1964, Frio barrier bar system of south Texas: Gulf Coast Assoc. Geol. Soc. Trans., v. 14, p. 309-322.

Bridges, P. H., 1976, Lower Silurian transgressive barrier islands, southwest Wales: Sedimentology, v. 23, p. 347-362.

Bruun, P., 1962, Sea level rise as a cause of shore erosion: Am. Soc. Civil Engineers Proc., Jour. Waterways and Harbors, v. 88, p. 117-130.

Campbell, C. V., 1966, Truncated wave-ripple laminae: Jour. Sed. Petrology, v. 36, p. 825-828.

——— 1971, Depositional model, Upper Cretaceous Gallup beach shoreline, Ship Rock area, northwestern New Mexico: Jour. Sed. Petrology, v. 41, p. 395-409.

Carter, C. H., 1978, Regressive barrier and barrier-protected deposit, depositional environments and geographic setting of the late Tertiary Cohansey Sand: Jour. Sed. Petrology, v. 48, p. 933-950.

Clifton, H. E., 1967, Cyclic facies of a middle Miocene littoral sandstone in the California Coast Ranges, U.S.A.: 7th International Sedimentological Congress Proc., Reading, England.

——— 1973, Marine-nonmarine facies change in middle Miocene rocks, southeastern Caliente Range, California; in Sedimentary facies changes in Tertiary rocks — California Transverse and southern Coast Ranges, SEPM Trip 2, 1973 Annual Meeting AAPG-SEPM-SEG. p. 55-57.

——— R. E. Hunter, and R. L. Phillips, 1971, Depositional structures and processes in the non-barred high-energy nearshore: Jour. Sed. Petrology, v. 41, p. 651-670.

——— ——— and ——— 1972, Depositional models from a high-energy coast (abs.): AAPG Bull., v. 56, p. 609.

Curray, J. R., F. J. Emmel, and P. J. S. Crampton, 1969, Holocene history of a

278   D. G. McCubbin

strand plain, lagoonal coast, Nayarit, Mexico; *in* Castanares, A. A., and F. B. Phleger, eds., Coastal lagoons, a symposium: Univ. Nac. Autonoma de Mexico, p. 63-100.

——— and D. G. Moore, 1964, Holocene regressive littoral sand, Costa de Nayarit, Mexico; *in* van Straaten, L. M. J. U., ed., Deltaic and shallow marine deposits, Developments in sedimentology, v. 1: Elsevier, p. 76-82.

Davidson-Arnott, R. G. D., and B. Greenwood, 1976, Facies relationships on a barred coast, Kouchibouguac Bay, New Brunswick, Canada; *in* Davis, R. A., Jr., and R. L. Ethington, eds., Beach and nearshore sedimentation: SEPM Spec. Pub. 24, p. 149-168.

Davies, J. L., 1964, Mophogenic approach to world shorelines: Zeit. fur Geomorphologie, v. 8, p. 127-142.

Davis, R. A., Jr., 1978, Beach and nearshore zone; *in* Davis, R. A., Jr., ed., Coastal sedimentary environments: Springer-Verlag, New York, p. 237-285.

Deery, J. R., and J. D. Howard, 1977, Origin and character of washover fans on the Georgia coast, U.S.A.: Gulf Coast Assoc. Geol. Soc. Trans., v. 27, p. 259-271.

Dickinson, K. A., H. L. Berryhill, Jr., and C. W. Holmes, 1972, Criteria for recognizing ancient barrier coastlines; *in* J. K. Rigby, and W. K. Hamblin, eds., Recognition of ancient sedimentary environments: SEPM Spec. Pub. 16, p. 192-214.

Elliott, T., 1975, Sedimentary history of a delta lobe from the Yoredale (Carboniferous) cyclothem: Proc. Yorks. Geol. Soc., v. 40, p. 505-536.

Field, M. E., and D. B. Duane, 1976, Post-Pleistocene history of the United States inner continental shelf; significance to origin of barrier islands: Geol. Soc. America Bull., v. 87, p. 691-702.

Greenwood, B., and R. G. D. Davidson-Arnott, 1975, Marine bars and nearshore sedimentary processes, Kouchibouguac Bay, New Brunswick; *in* Hails, J., and A. Carr, eds., Nearshore sediment dynamics and sedimentation: Wiley, p. 123-150.

Harms, J. C., et al, 1975, Depositional environments as interpreted from primary sedimentary structures and stratification sequences: SEPM Short Course No. 2, SEPM, Tulsa, 161 p.

Hayes, M. O., 1967, Hurricanes as geological agents; case studies of Hurricanes Carla, 1961, and Cindy, 1963: Bur. Econ. Geology, Univ. Texas, Austin, Rept. Invest. 61, 54 p.

——— 1975, Morphology of sand accumulations in estuaries; *in* Cronin, L. E., ed., Estuarine research, v. 2, Geology and engineering: Academic Press, New York, p. 3-22.

——— 1976, Lecture notes; *in* Hayes, M. O., and T. W. Kana, eds., Terrigenous clastic depositonal environments, some modern examples; AAPG Field Course: Univ. South Carolina, Tech. Rept. 11-CRD, p. I-1 to I-131.

——— F. S. Anan, and R. N. Bozeman, 1969,

Sediment dispersal trends in the littoral zone; a problem in paleogeographic reconstruction; *in* Coastal environments of northeastern Massachusetts and New Hampshire: Contri. 1-CRG, Geol. Dept., Univ. Mass., p. 290-315.

——— and J. C. Boothroyd, 1969, Storms as modifying agents in the coastal environment; *in* Coastal environments of northeastern Massachusetts and New Hampshire: Contri. 1-CRG, Geol. Dept., Univ. Mass., p. 245-265.

Hill, G. W., and R. E. Hunter, 1976, Interaction of biological and geological processes in the beach and nearshore environments, northern Padre Island, Texas; *in* Davis, R. A., Jr., and R. L. Ethington, eds., Beach and nearshore sedimentation: SEPM Spec. Pub. 24, p. 169-187.

Hobday, D. K., and M. P. A. Jackson, 1979, Transgressive shore zone sedimentation and syndepositional deformation in the Pleistocene of Zululand, South Africa: Jour. Sed. Petrology, v. 49, p. 145-158.

Horne, J. C., and J. C. Ferm, 1978, Carboniferous depositional environments, eastern Kentucky and southern West Virginia: Dept. Geol., Univ. South Carolina, 151 p.

Howard, J. D., 1972, Trace fossils as criteria for recognizing shorelines in stratigraphic record; *in* Rigby, J. K., and W. K. Hamblin, eds., Recognition of ancient sedimentary environments: SEPM Spec. Pub. 16, p. 215-225.

——— and H.-E. Reineck, 1972, Georgia coastal region, Sapelo Island, U.S.A., sedimentology and biology; IV. Physical and biogenic sedimentary structures of the nearshore shelf: Senckenbergiana Marit., v. 4, p. 81-123.

——— ——— 1979, Sedimentary structures of "high energy" beach-to-offshore sequence; Ventura-Port Hueneme area, California (abs.): AAPG Bull., v. 63, p. 468-469.

Hoyt, J. H., and V. J. Henry, Jr., 1967, Influence of island migration on barrier-island sedimentation: Geol. Soc. America Bull., v. 78, p. 77-86.

Hunter, R. E., H. E. Clifton, and R. L. Phillips, 1979, Depositional processes, sedimentary structures, and predicted vertical sequences in barred nearshore systems, southern Oregon coast: Jour. Sed. Petrology, v. 49, p. 711-726.

Keulegan, G. H., and W. C. Krumbein, 1949, Stable configuration of bottom slope in a shallow sea and its bearing on geological processes: Am. Geophysical Union Trans., v. 30, p. 855-861.

Kolb, C. R., and J. R. Van Lopik, 1958, Geology of the Mississippi River deltaic plain, southeastern Louisiana: U.S. Army Engineers, Waterways Experiment Station, Tech. Rept. 3-483, 120 p.

Kraft, J. C., 1971, Sedimentary facies patterns and geologic history of a Holocene marine transgression: Geol. Soc. America Bull., v. 82, p. 2131-2158.

Kumar, N., and J. E. Sanders, 1974, Inlet sequence, a vertical succession of

sedimentary structures and textures created by lateral migration of tidal inlets: Sedimentology, v. 21, p. 491-532.

——— ——— 1976, Characteristics of shoreface storm deposits; modern and ancient examples: Jour. Sed. Petrology, v. 46, p. 145-162.

Mannhard, G. W., 1976, Stratigraphy, sedimentology, and paleoenvironments of the La Ventana Tongue (Cliff House Sandstone) and adjacent formations of the Mesaverde Group (Upper Cretaceous), southeastern San Juan basin, New Mexico: PhD dissertation, Univ. New Mexico, Albuquerque, 182 p.

McCubbin, D. G., 1969, Cretaceous strike-valley sandstone reservoirs, northwestern New Mexico: AAPG Bull., v. 53, p. 2114-2140.

——— 1972, Facies and paleocurrents of Gallup Sandstone, model for alternating deltaic and strand-plain progradation (abs.): AAPG Bull., v. 56, p. 638.

——— and M. J. Brady, 1969, Depositional environment of the Almond reservoirs, Patrick Draw field, Wyoming: The Mountain Geologist, v. 6, no. 1, p. 3-26.

Moslow, T. F., and S. D. Heron, Jr., 1978, Relict inlets, preservation and occurrence in the Holocene stratigraphy of southern Core Banks, North Carolina: Jour. Sed. Petrology, v. 48, p. 1275-1286.

Owens, E. H., 1977, Temporal variations in beach and nearshore dynamics: Jour. Sed. Petrology, v. 47, p. 168-190.

Price, W. A., 1954, Dynamic environments; reconnaissance mapping, geologic and geomorphic, of continental shelf of Gulf of Mexico: Gulf Coast Assoc. Geol. Soc. Trans., v. 4, p. 75-107.

Reineck, H.-E., 1963, Sedimentgefuge im bereich der sudlichen Nordsee: Abh. Senckenberg. Naturforsch. Gessell., v. 505, p. 1-138.

——— 1967, Layered sediments of tidal flats, beaches, and shelf bottoms of the North Sea; *in* Lauff, G. H., ed., Estuaries: Pub. 83, Am. Assoc. Adv. Sci., Washington, p. 191-206.

——— 1976, Primargefuge, bioturbation und makrofauna als indikatoren des sandversatzes im seegebiet vor Norderney (Nordsee); I. Zonierung von primargefugen, und bioturbation: Senckenbergiana Marit., v. 8, p. 155-169.

——— and I. B. Singh, 1971, Der Golf von Gaeta (Tyrrhenisches Meer); III. Die gefuge von vorstrand- und schelfsedimenten: Senckenbergiana Marit., v. 3, p. 185-201.

Reinson, G. E., 1979, Facies models 14, barrier island systems: Geoscience Canada, v. 6, p. 51-68.

Roep, Th. B., et al, 1979, Prograding coastal sequence of wave-built structures of Messinian age, Sorbas, Almeria, Spain: Sedimentary Geology, v. 22, p. 135-163.

Ryer, T. A., 1977, Patterns of Cretaceous shallow-marine sedimentation, Coalville and Rockport areas, Utah: Geol. Soc. America Bull., v. 88, p. 177-188.

Sanders, J. E., and N. Kumar, 1975, Evidence of shoreface retreat and in-place "drowning" during Holocene submergence of barriers, shelf off Fire Island, New York: Geol. Soc. America Bull., v. 86, p. 65-76.

Schwartz, M. L., 1967, The Bruun theory of sea-level rise as a cause of shore erosion: Jour. Geology, v. 75, p. 76-92.

Schwartz, R. K., 1975, Nature and genesis of some storm washover deposits: U.S. Army, Corps of Engineers, Coastal Engineering Research Center, Tech. Memo. 61, Ft. Belvoir, Va., 69 p.

Shepard, F. P., 1960, Gulf Coast barriers; *in* Recent sediments, northwest Gulf of Mexico: AAPG Spec Pub., Tulsa, p. 197-220.

Swift, D. J. P., 1968, Coastal erosion and transgressive stratigraphy: Jour. Geology, v. 76, p. 444-456.

——— 1975, Barrier-island genesis; evidence from the central Atlantic shelf, eastern U.S.A.: Sedimentary Geology, v. 14, p. 1-43.

Thompson, W. O., 1937, Original structures of beaches, bars, and dunes: Geol. Soc. America Bull., v. 48, p. 723-752.

Weimer, R. J., 1966, Time-stratigraphic analysis and petroleum accumulations, Patrick Draw field, Sweetwater County, Wyoming: AAPG Bull., v. 50, p. 2150-2175.

Wilkinson, B. H., 1975, Matagorda Island, Texas; the evolution of a Gulf Coast barrier complex: Geol. Soc. America Bull., v. 86, p. 959-967.

Winant, C. D., D. L. Inman, and C. E. Nordstrom, 1975, Description of seasonal beach changes using empirical eigenfunctions: Jour. Geophys. Research, v. 80, p. 1979-1986.

Wright, L. D., et al, 1979, Morphodynamics of reflective and dissipative beach and inshore systems, southeastern Australia: Marine Geology, v. 32, p. 105-140.

Wunderlich, F., 1972, Georgia coastal region, Sapelo Island, U.S.A., sedimentology and biology; III. beach dynamics and beach development: Senkenbergiana Marit., v. 4, p. 47-79.

Continental Shelf and Epicontinental Seaways

Arnold H. Bouma
Henry L. Berryhill
U.S. Geological Survey
Corpus Christi, TX

Robert L. Brenner
University of Iowa
Iowa City, Iowa

Harley J. Knebel
U.S. Geological Survey
Woods Hole, Massachusetts

INTRODUCTION

Definition and Remarks on Facies

The continental shelf is that part of the sea floor between the shoreline and the shelf break, or upper edge of the continental slope. However, an examination of bathymetric charts shows that both the position of the shelf break and the shelf width vary. The shelf break may be as shallow as 18 m (10 fathoms) or as deep as 915 m (500 fathoms), and the shelf width may range from a few kilometers to more than 1,000 km (550 n. mi). On the average, the shelf break occurs at 124 m (67 fathoms), and the width is 75 km (40 n. mi).

Morphologic and sedimentary characteristics of the shelf also vary considerably. The shelf surface may be smooth, covered by a variety of bedforms, or may contain banks, islands, and shoals near its offshore edge. Likewise, the sedimentary characteristics change from one area to the next depending on differences in waves, currents, climatic conditions, and proximity to large sources of sediment. For example, muddy shelves may contain nearly homogeneous sediments except for the presence of layers formed during storms. Other shelves consist primarily of tillites dissected by valleys and then partly filled with either fine- or coarse-grained sediments. In many areas, the overall smooth shelf has a sand blanket that has been molded into a variety of small and large bedforms.

Ancient shelf deposits form important parts of the stratigraphic record. Sediments were deposited not only on ancient continental shelves, but also on depositional and structural shelves that existed within ancient epicontinental seaways. Such seaway deposits are recognized by associations of sandstone bodies in parts of ancient seaways that were located far from paleo-shorelines (e.g. Asquith, 1974; Spearing, 1976; Brenner, 1978a). Analogies appear to exist between processes constructing sand waves, ridges and related features on modern continental shelves, and those forming similar features preserved in ancient epicontinental seaway deposits (Brenner and Davies, 1974; Spearing, 1976; Brenner, 1978, 1979; Klein and Ryer, 1978).

Similarities and differences between continental shelves and epicontinental seaways are not completely understood; however, it appears that similar sedimentologic processes can operate in both types of environments (see also Brenner, 1980). The sedimentary results of these processes may differ because of influences of local features and tectonic settings. Sufficient data for presentation of in-depth discussions on the similarities and differences between continental shelves and epicontinental seaways are not available. In lieu of a general discussion of the relations between these two major domains, a few examples from the Mesozoic epicontinental seaway deposits of North America are presented.

Although descriptions of ancient offshore sandstone bodies are relatively rare, the few that have been published range in age from Precambrian (Hobday and Reading, 1972; Johnson, 1977) to Tertiary (Narayan, 1971; Nio, 1976). Examples described from epicontinental seas or seaways in North America include parts of the Ordovician St. Peter Sandstone of the Mid-continent (Pryor and Amaral, 1971; Dott and Roshardt, 1972), parts of the Jurassic Sundance Formation of Wyoming (Brenner and Davies, 1974), the Upper Cretaceous Shannon and Sussex Sandstone Members of the Steele Shale of

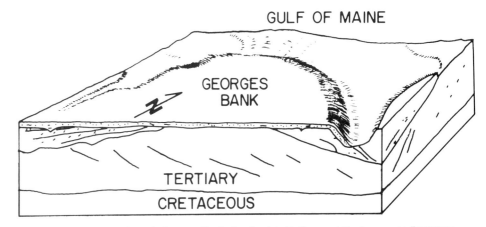

Fig. 1—Block diagram through Georges Bank showing late Tertiary and Quaternary sedimentary units and structures. Inferred geologic history is as follows: later Tertiary or Quaternary erosion leaves Tertiary coastal plain strata as a cuesta beneath central Georges Bank; lowlands that developed on the eastern and western flanks of the cuesta subsequently were filled with either prograded beds (east) or episodic accumulations of detritus (west); erosion planed all previous units; subaerial erosion with channel development followed late Pleistocene glacial and outwash deposits blanket the bank; these deposits reworked during and since the last transgression of sea level. From Lewis and Sylwester (1976, Fig. 6F).

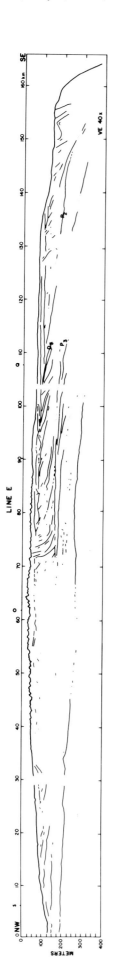

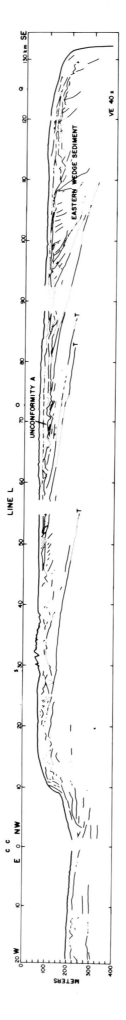

Fig. 2—Line drawings of high-resolution, seismic-reflection profiles across the western (Line E) and eastern (Line L) parts of Georges Bank. Depositional (D) and erosional (P) surfaces are indicated within the sedimentary sequence that was formed by episodic infilling of the lowland on the western flank of the bank (Line E). To the east, Tertiary coastal plain strata (T) are overlain by prograded, delta-like beds that partially fill the lowland on planation surface across the bank that subsequently was scoured by stream channels (Line L). From Lewis and Sylwester (1976, Plates 1, 2).

the Powder River basin (Spearing, 1976; Brenner, 1978), the Hygiene Sandstone Member of the Pierre Shale in Colorado (Porter, 1976), and the Woodside unit of the Upper Cretaceous Ferron Sandstone Member of the Mancos Shale in Utah (Cotter, 1975). In addition, basal deposits of some ancient pericontinental seaways, or paleocontinental shelves, can be observed in some forms of offshore bars. Examples may include the Cambrian Eriboll Sandstone of Scotland (Swett et al, 1971), the Exum's (1973) Upper Jurassic Cadeville Sandstone Tongue of the Schuler Formation in Louisiana, and some Tertiary sandstones of western Europe described by Nio (1976).

Diagnostic Criteria for Recognition

Sediments on modern continental shelves normally are not in complete equilibrium with present conditions. Complex interactions between such factors as tectonics, sea-level fluctuations, daily and seasonal wave and current dynamics, along with special events such as storms result in a mixture of relict and modern detrital sediments within a variety of microenvironments. As a consequence, benthic (animal) communities vary in time and space, precluding the identification of diagnostic fossils.

A complete set of diagnostic sedimentary criteria for recognizing shelf environments in the geologic record cannot be given because of data coverage that is both insufficient and spotty. Before diagnostic criteria can be given, large segments of continental shelves in different geographic, climatic, and tectonic areas must be studied uniformly by integrating various geologic and geophysical data. In this contribution, therefore, we can only present some insight into the complexities of the shelf environment. This is done by citing examples of sedimentary structures on a number of shelf areas around the United States.

SETTING OF THE CONTINENTAL SHELF IN GEOGRAPHIC, PHYSIOGRAPHIC AND TECTONIC SENSE

Introduction

The setting of the continental shelf can be discussed in its relation to lateral facies, in a depositional sense, or in a tectonic framework. However, all three types of settings are closely interrelated and, because sedimentary structures form the main emphasis of this publication, we will briefly describe a number of shelves under the heading of depositional relations.

Relation to Lateral Facies

The continental shelf is generally a composite of several sedimentary facies. On the one hand, the modern shelf is bounded landward by various coastal environments and seaward by the continental slope. Facies presently forming in these peripheral environments are the subject of other papers in this volume. On the other hand, relict deposits are common on most shelves. During low stands of sea level, several coastal and continental environments prevailed on the exposed shelf surface. These environments formed such features as barrier bars, river valleys, and even dunes atop former shelf sediments. During the ensuing transgressions of the sea, many of these features were abandoned, submerged, and eventually incorporated into the shelf sedimentary column. Some of these relict deposits are presently being reworked by currents and waves, into a facies that reflects several sedimentary environments.

Depositional Setting

Shepard (1977) divides the present continental shelves into six major categories based largely on tectonics and climates.

Glaciated shelves

At high latitudes, glaciers spread from land masses onto the continental shelf, forming distinctive features and deposits. Glaciers not only scoured the substrate directly but also eroded shelf sediments by meltwater channels. Some glacial troughs were deep and irregular along their axes, and had end moraines at their seaward terminations.

The complex depositional setting on a glaciated shelf is exemplified by the shallow stratigraphy atop Georges Bank (Knott and Hoskins, 1968; Uchupi, 1970; Oldale et al, 1974; Lewis and Sylwester, 1976; Figs. 1-4). During late Tertiary or Quaternary, streams and glaciers eroded the coastal plain strata, leaving the central part of Georges Bank and removing peripheral sediments (Fig. 1). The resulting lowland east of the bank was filled partially with massive, prograded, delta-like beds that thicken to the southeast. To the west, lowland filling was more episodic. There, the sedimentary sequence contains at least five erosional surfaces, each onlapping the one directly beneath it (Fig. 2). Furthermore, each of the erosional surfaces is indented by well-developed channels that were cut and then filled. Along the northern edge of the bank, these deposits have also been disturbed by glacial loading.

Prior to the final Pleistocene ice advance, a transgression of the sea planed the bank, removing unknown amounts of Pleistocene and Tertiary sediments (Fig. 1). During the last sea-level regression, this planation surface was scoured by numerous stream channels, especially in the northeastern corner of the area. Subaerial exposure of the bank during this time apparently lasted long enough to produce complex cut-and-fill features within

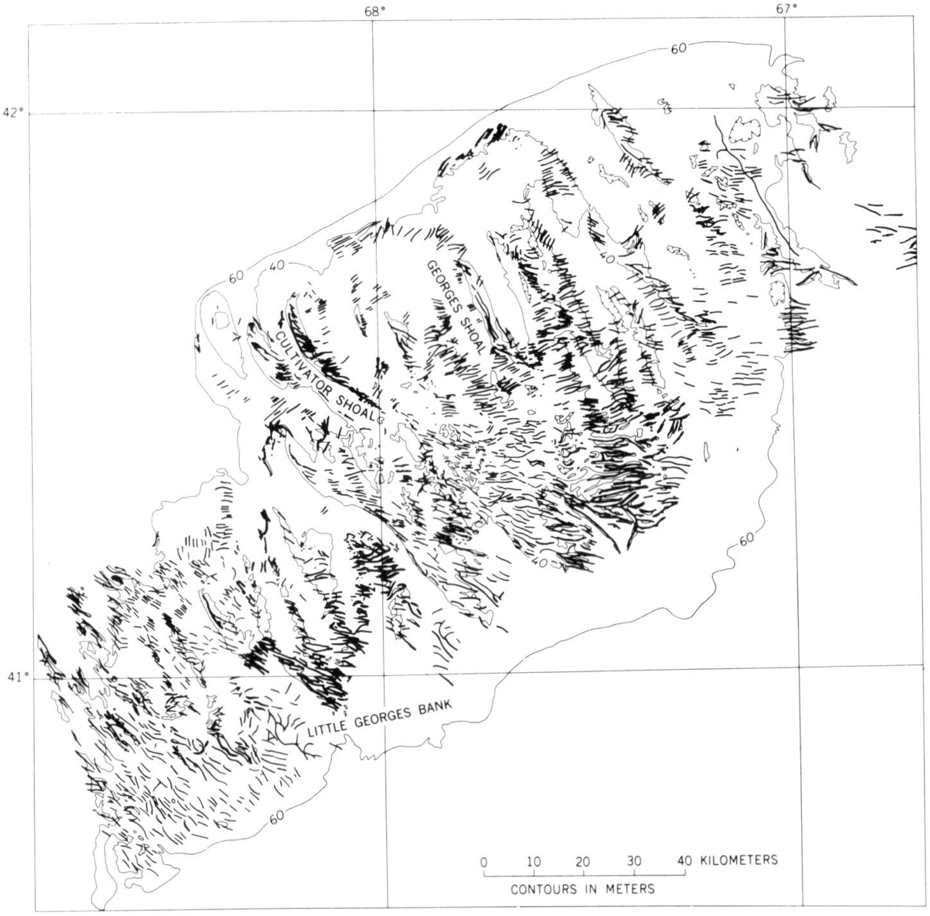

Fig. 3—Distribution of sand waves and shoals on Georges Bank. Curved lines indicate crests of sand waves. From Uchupi (1968, Fig. 4).

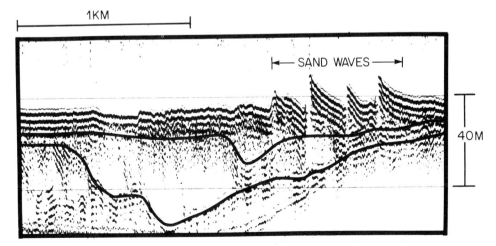

Fig. 4—High-resolution seismic-reflection profile from northeastern part of Georges Bank showing sand waves at the seafloor and cut-and-fill structures (enhanced by heavy black lines) below the bottom. Large sand shoals are not shown. From Folger et al (1978, Fig. 5).

some sediment directly above the planation surface.

Structures within the uppermost sediments atop Georges Bank are a result of reworking during the last sea-level transgression and modern sediment movement (Jordan, 1962; Uchupi, 1968; Emery and Uchupi, 1972; Hathaway et al, 1976; Folger et al, 1978; Bothner et al, 1979; Wood et al, 1979). Sediments overlying the northern half of the bank have been molded into a series of northwest-trending sand shoals (Fig. 3). These shoals are about 10 km (6 n. mi) apart, as much as 30 m (100 ft) high and 75 km (45 n. mi) long, and separated by flat floored troughs.

Superimposed on the shoals and troughs is a series of linear sand waves (Figs. 3, 4). These waves are 10 to 20 m high, 100 to 700 m apart, and 200 m to 10 km long. Their shapes vary from straight to sigmoidal to crescentic, and they may be symmetrical or asymmetrical. Characteristics of the sand waves differ from place to place in accordance with tidal currents, water depth, and probably the grain size of the sand.

Shelves with elongate sand ridges

Many shelves contain subparallel sand ridges with associated bedforms ranging from small ripples or megaripples to sand waves. These characteristic constructional bedforms are composed of Pleistocene sediments being reworked and modified within the modern environment.

The Middle Atlantic Bight is an example of a shelf covered with sand ridges. Across this shelf, the surficial sand sheet has been molded into a series of paired linear ridges and depres-

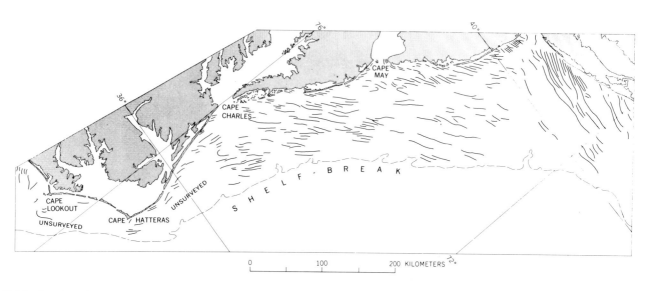

Fig. 5—Distribution of sand ridges (crestal lines) on the Middle Atlantic shelf. From Uchupi (1968, Fig. 14).

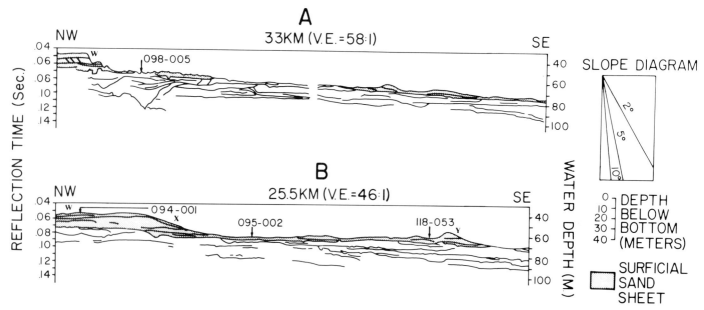

Fig. 6—Line drawings of parallel high-resolution seismic-reflection profiles across the northern part of the Middle Atlantic outer shelf. Profiles illustrate: (1) change in thickness of the surficial sand sheet; (2) morphology of ubiquitous sand ridges; (3) sedimentary structures (cross-bedding and channels) within both the surficial sand sheet and the underlying muddy unit; and (4) the change of the internal structure along a segment of the relict Fortune shore (scarp left side of profiles). Profiles have different vertical exaggerations (V. E.). Slope diagram represents vertical exaggeration of 50:1. Sound velocity in sediments is assumed to be that of water (1,463 m/sec). Profiles modified from Knebel and Spiker (1977, Fig. 3).

sions. The ridges are 2 to 18 km wide, 2 to 40 km long, and 2 to 10 m high (Veatch and Smith, 1939; Uchupi, 1968; Emery and Uchupi, 1972; Swift et al, 1972; Figs. 5, 6). The origin of these bedforms has been attributed to barrier beach-lagoon complex formation during Pleistocene lower sea levels (Veatch and Smith, 1939; Sanders, 1962; Shepard, 1963; McClennen, 1973) and to modern storm-generated waves and currents (Moody, 1964; Duane et al, 1972; Swift et al, 1972; Swift, 1976).

Perhaps the greatest problem in interpretation of ridges on the Middle Atlantic shelf has been their diverse origins even within a given shelf area. Inner shelf shoals, for example, may be related genetically not only to the lower shoreface, but to inlets, capes, and estuary mouths as well (Duane et al, 1972). Similarly, along the middle shelf, Knebel et al (1976) and Knebel and Spiker (1977) found that the near-surface sand unit accumulated under a variety of environments ranging from lagoonal to inner shelf to middle shelf and that the ridges may be either relict or modern depositional forms.

The stratigraphy beneath one inner shelf ridge is presented in Figure 7. This ridge presumably formed at the foot of the shoreface after the passage of a retreating barrier during the Holocene transgression (Stahl et al, 1974). The basal unit in the stratigraphic sequence is a lagoonal mud (layer H1) that accumulated in a coast-parallel tidal channel carved into late Pleistocene and Miocene sediments. The lagoonal mud, in turn, is overlain by a discontinuous back-barrier silty sand (layer H2) derived from lagoon washover, eolian transport, and other subaerial processes. The uppermost shelf sand sheet (layer H3) was deposited and molded into ridges after the transgressing shoreline had destroyed the barrier and perhaps part of the lagoonal sequence.

Shelves off large deltas

Sediment contributed by large rivers produces characteristic features on the shelf. Some deltas, for example, cause an appreciable seaward bulge in the shelf break, while others cause the entire shelf to build outward as deposition progresses seaward. In general, delta formation results in a large body of sediment on the shelf. This body, in turn, fosters a variety of localized environments and structures. The shelf off the Mississippi Delta illustrates these variations; details on the setting, sediments, and structures are discussed elsewhere in this volume.

Shelves with coral reefs

Coral reef development on the continental shelf generally produces small positive relief features such as the buried one in Figure 8. This kind of development occurs in tropical waters and in local patches in subtropical areas. Thus, shelves with coral reefs are quite limited. Because the emphasis of this book is on nonbiogenic and noncalcareous sedimentary structures, these shelves will be discussed no further.

Shelves bordered by rocky banks and islands

Rocky elevations are found along the outer edges of many shelves. These elevations may be either islands or rocky banks and are found on narrow or wide shelves. In many instances, basins and channels inside the bathymetric highs are partly or completely filled with sediments.

The shelf off southern California is an example of a narrow shelf bordered at the seaward side by a series of basins

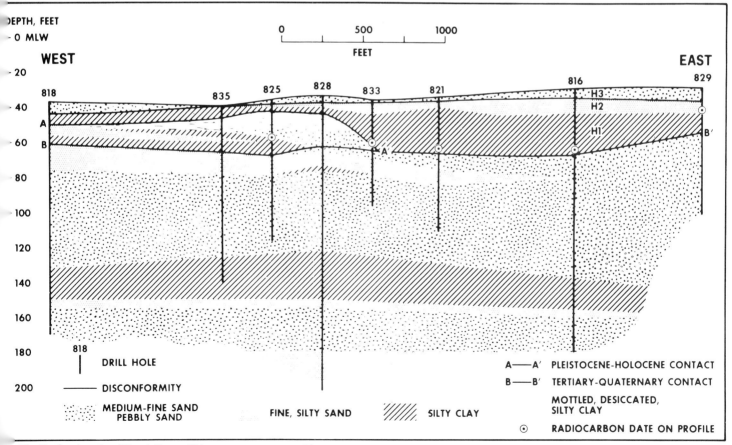

Fig. 7—Stratigraphy of Beach Haven Ridge on the inner shelf off central New Jersey coast as interpreted from drill holes, from petrographic, paleontologic, and radiometric examination of cores, and from seismic-reflection profiles. See text for interpretation of strata and inferred history. From Stahl et al (1974, Fig. 3).

and ridges (continental borderland). Some of the ridges are relatively flat, such as Cortez Bank, which contains a rock reef that is close to the surface. Other ridges have several high islands, notably Santa Catalina and San Clemente Islands (Emery, 1960; Shepard, 1977). Sediment data from the narrow shelf are limited, and an adequate discussion on the southern California borderland falls outside the scope of this chapter.

Shelves related to plate tectonics

Shelf width is generally the most prominent characteristic that can be correlated with plate tectonics. Narrow shelves are usually found on active margins, such as along the Pacific coast, whereas broader shelves are more typical for passive margins, such as the Gulf of Mexico and the Atlantic. In terms of depositional structures, however, shelves on active versus passive margins are not unique and may fall into any of the above categories.

The shelf of Kodiak Island (Alaska) can be used to exemplify this type of continental shelf. It has an average width of 70 km (40 mi) and consists of banks with transverse troughs in between (compare Fig. 9). The banks contain pebbly, clayey material of glacial origin, and are not covered by modern sediments because little recent source material is available and currents appear to prevent deposition. Troughs are

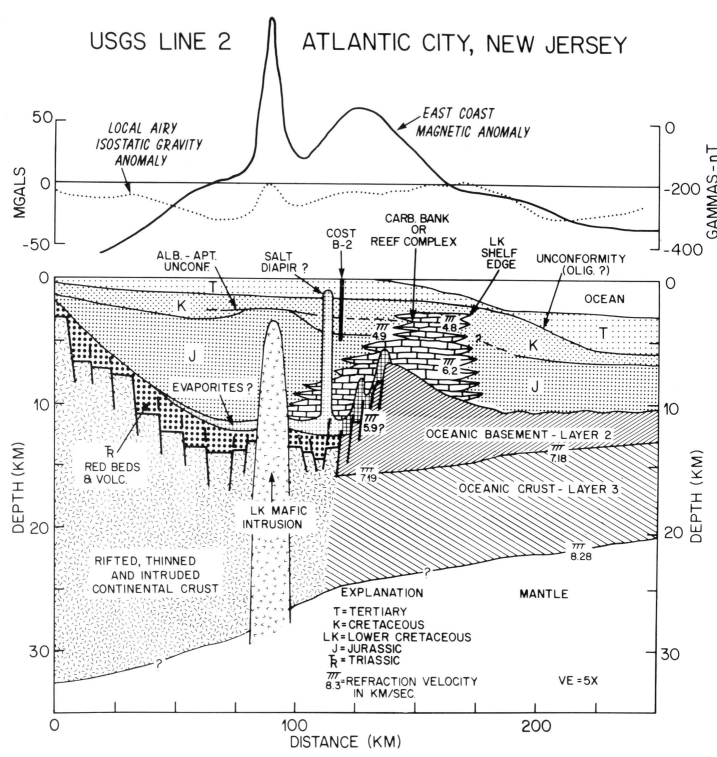

Fig. 8—Interpretative cross section along a common-depth-point (CDP) seismic reflection line obtained across the northern part of the Baltimore Canyon Trough. From Grow et al (1979, Fig. 12).

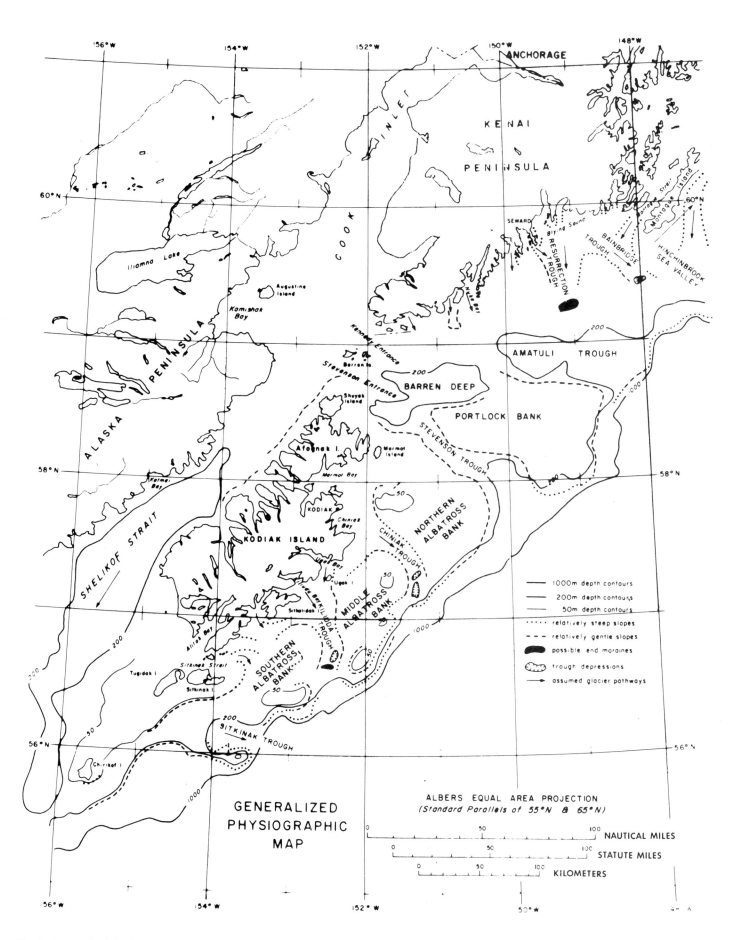

Fig. 9—Generalized physiographic map of Kodiak Shelf, Alaska. From Hampton et al (1979, Figs. 1 and 15).

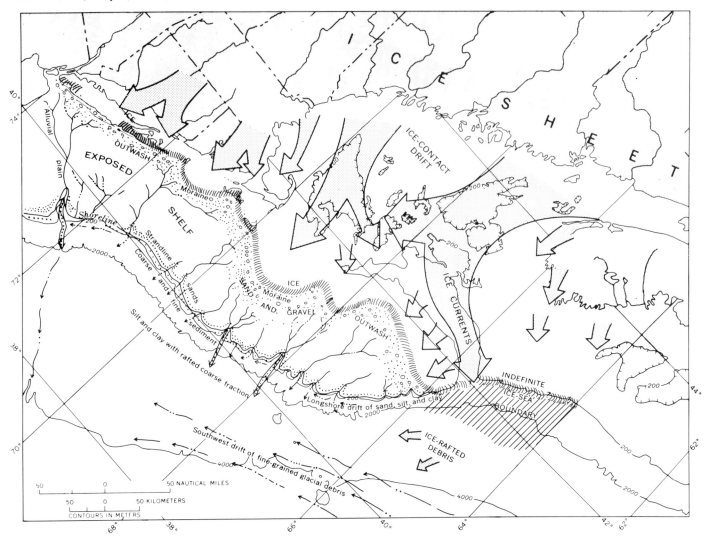

Fig. 10—Schematic map of the northeastern U.S. continental margin showing the texture and genesis of the sediments atop Georges Bank during the Pleistocene. From Schlee (1973, Fig. 39A).

partly filled with volcanic ash from the 1912 eruption of Mount Katmai. The ash may occur in a thin layer by itself, although mixing with clayey sediment due to bioturbation is common. Structural arches trend along much of the outer shelf, forming broad, physiographically high areas (shelf-break arches) both on the banks and across the troughs (Hampton and Bouma, 1977; Hampton et al, 1979).

Microfacies Within the Quaternary Shelf Environment

The types of sediments and sedimentary structures within any microfacies of the shelf environment vary from place to place. Most modern sediments brought to the shelf are silt and clay, although sand may be dominant where finer particles have been winnowed by waves and currents. In muddy sediments, laminations and rippled structures are common when coarse silt and sand stringers are present. Also, depending on the setting (such as off

deltas), one may observe a variety of slump structures. Within sandy sediments, microstructures include laminations, current ripples, and wave ripples, but larger structures may be associated with bars, sand sheets, and channels. All shelf sediments may contain a variety of biogenic structures, and primary depositional features may be destroyed by burrowing organisms as well as by migration of gas in the bubble phase.

Relict sediments on the continental shelf pose special problems for interpreting microfacies. Most of the coarse-grained sediments on the outer shelf were deposited during times of lower sea level and may represent a variety of environments such as river channels, offshore bars, or estuaries. In many places these sediments have been reworked, transported, and modified during the ensuing transgression of the sea. As a result, the internal sedimentary structures may be complicated. For example, reworked sands

from lower Cook Inlet (Bouma et al, 1977, 1979) show bioturbation and only a faint bedding. Because small ripples indicate the mechanism of sand transport over the different types of larger bedforms in that area, one occasionally finds small foresets, but these alone are not diagnostic for any one environment.

In addition to current- and wave-induced ripples, other depositional features, such as lag deposits, distinct textural and mineralogic bands, and lineations (such as sand ribbons and erosional furrows) suggest storm generated events.

Tectonic Relations

Tectonic processes have a pronounced influence on the development of sedimentary structures on the shelf. The overall gradient of the shelf, for example, may be based on its tectonic history. Similarly, the presence of tectonic or reefal highs near the shelf break determines how much of a shelf

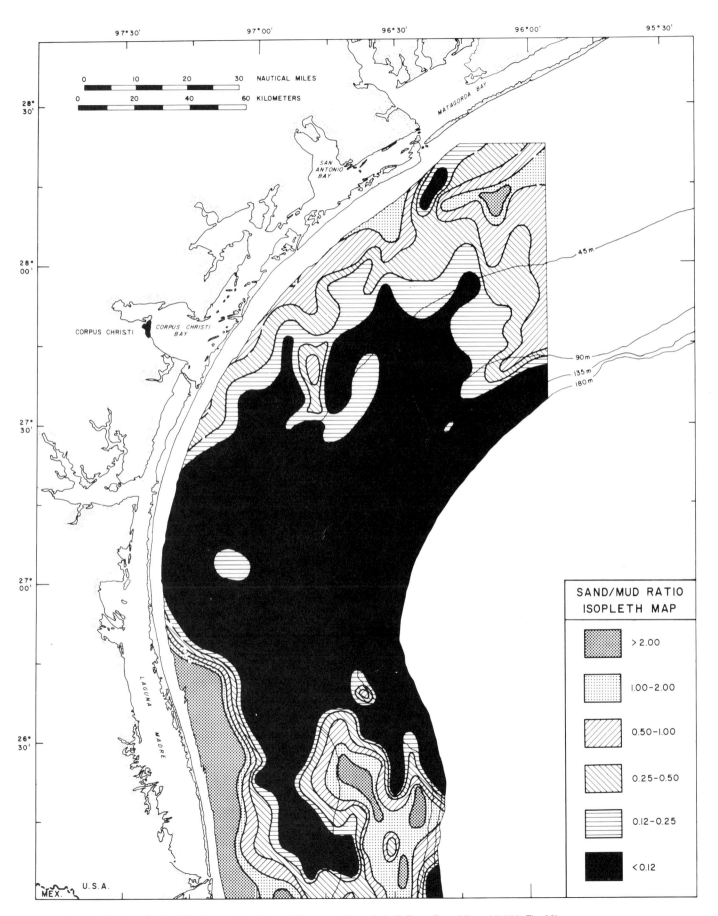

Fig. 11—Sand/mud ratios of surficial bottom sediment, south Texas continental shelf. From Berryhill et al (1976, Fig. 32).

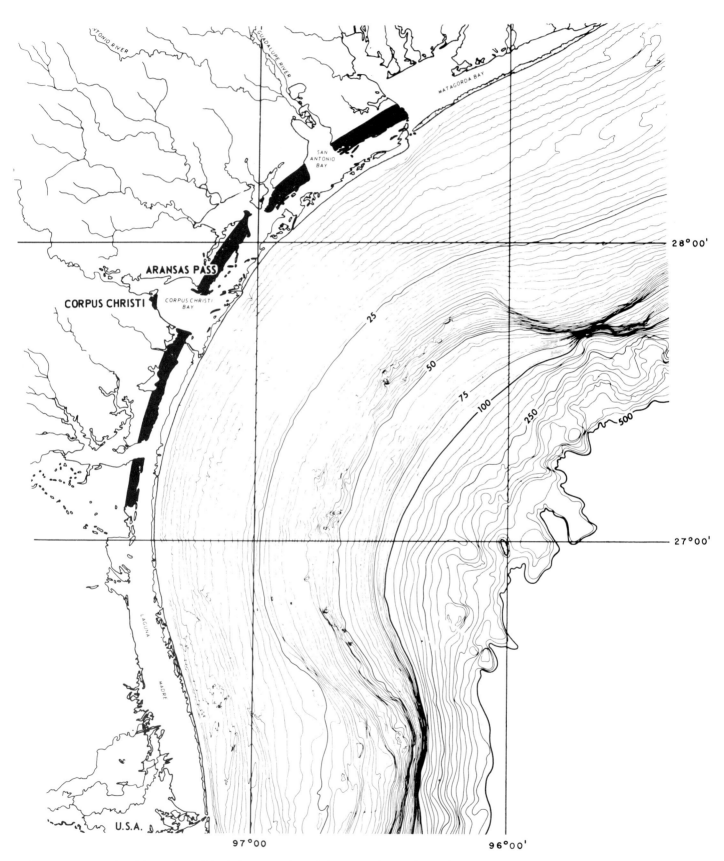

Fig. 12—Topography of the continental shelf off south Texas. Depth contours are in fathoms. The junction of the continental shelf and the continental slope approximates the 100 fathoms contour.

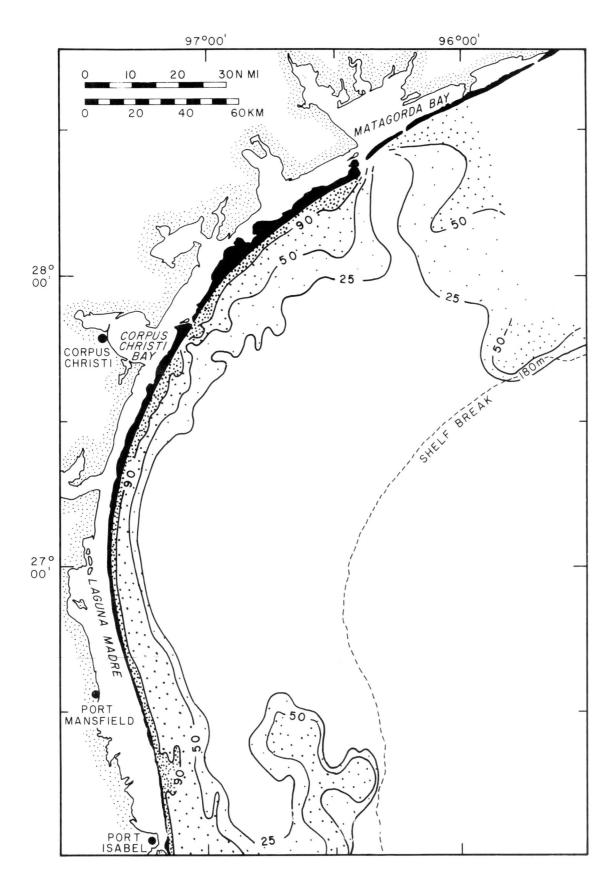

Fig. 13—Amount of the sand-sized fraction ($>63\mu$) in surficial bottom sediment, south Texas continental shelf. The percent zonation is shown by the marked contours. The barrier sand islands that mark the landward boundary of the shelf are black. Modified from Berryhill et al (1976, Fig. 29).

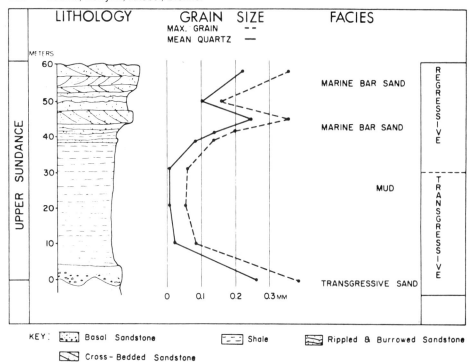

KEY: [Basal Sandstone] [Shale] [Rippled & Burrowed Sandstone] [Cross-Bedded Sandstone]

Fig. 14—Vertical sequence of sedimentary structures and grain sizes preserved in Oxfordian marine bars. Example is from the east flank of the Big Horn Mountains, Wyoming. From Brenner and Davies (1973a).

becomes exposed during a low stand of sea level. Finally, irregularities on the shelf greatly affect modern sediment transport by currents and waves. Emery (1977) used geophysical data to divide the shelf structurally into five groups. As a consequence, his categories differ from Shepard's (1977) and relate less to the topic of this publication.

SEDIMENTARY CHARACTERISTICS WITH EXAMPLES FROM MODERN AND ANCIENT SHELVES

Compositional and Textural Patterns

Several factors control or influence compositional and textural patterns on the continental shelf. Four of these factors are: reworking of relict sediments; seasonal fluctuations in sediment transport; seafloor topography; and regional current patterns. The effect of each of these factors can be illustrated for specific segments of the shelf.

Sediments atop Georges Bank primarily reflect relict processes (Schlee and Pratt, 1970; Emery and Uchupi, 1972; Schlee, 1973). The smooth southern part of the bank is covered by dominantly fine to very fine, well-sorted sand, whereas the irregular northern area is mantled by coarse to medium, poorly sorted sand containing patches of gravel. These textural differences are largely the result of depositional patterns during one of the Pleistocene glaciations. Gravelly sand over the northern flank of the bank apparently marks an ice limit, whereas finer grained sand farther south represents outwash deposits (Fig. 10). Surface sediment across the entire area was reworked during the last transgression.

On the Middle Atlantic shelf, sedimentary patterns are influenced greatly by the temporal variability of bottom current flow. Superimposed on a mean southerly bottom flow in this area (Bumpus, 1973) are weak tidal currents, low-period fluctuating currents associated with meteorological forcing, and other currents associated with forcing from the deep ocean (Scott and Csanady, 1976). In particular, fall and winter storms as well as hurricanes generate large surface waves and wind-driven currents that can mask the rather weak tidal currents and move sand that covers this part of the shelf (Butman et al, 1977). During the spring and summer, on the other hand, current speeds generally are not much greater than background levels caused by tidal flow, and the dominant sedimentary process is bioturbation (Wood and Folger, 1979).

The shelf off Kodiak Island, Alaska, shows a strong relation to the seafloor topography (Hampton et al, 1979). There, large shallow banks contain Pleistocene material, commonly coarse-grained sediment, while troughs and depressions are partly filled with unconsolidated fine-grained deposits (Fig. 10). These fine deposits consist of slightly muddy sand with large amounts of 1912 Mount Katmai ash that have been swept from the banks by currents. No recent sediment source is available to cover the banks locally.

The distribution of mud off the south Texas coast clearly reflects the present regional circulation system in that part of the Gulf of Mexico (Fig. 11). Berryhill et al (1976) shows that textural parameters exhibit consistent regional increases in sediment size shoreward, northward, and southward from the central sector. These trends are due largely to sorting and deposition of sediments in response to the convergence and attendant changes in strength of regional currents. The following description of this area helps to explain the setting and compositional patterns.

The surface of the continental shelf off south Texas is relatively smooth and has an average seaward gradient of about 2 m per km (12 ft per mi). The average width from shoreline to the point of increased gradient, marking the top of the continental slope (along the 100-fathoms isobath), is 80 km (44 mi). Topographic features are minor and consist of a series of small carbonate mounds of late Pleistocene age rising from 8 to 20 m (25 to 60 ft) above the sea floor along an arcuate line in water depths of a little less than 50 fathoms. Otherwise, the normal gradient of the sea floor is disrupted only by a sharp increase along the outer perimeters of the ancestral Rio Grande and Colorado/Brazos deltas, which built outward to the upper edge of the continental slope during the last low stand of sea level. The topography and bathymetry on the shelf off south Texas are shown in Figure 12.

Both geologically and topographically, the shelf off Texas is a seaward extension of a broad, flat, and structurally simple coastal plain province that fringes the eastern United States from Long Island to Mexico. The landward boundary of the continental shelf of Texas is a barrier coastline made up of a series of bays and lagoons separated from the Gulf of Mexico by a nearly continuous line of narrow islands of low relief. These islands are broken only by several natural inlets connecting the bays and the ocean plus a few artificial ship channels. Islands are largely unconsolidated sand; where measured by drilling, the thickness of sand is 18 m (54 ft) on northern Padre Island about 30 km (16 mi) south of Corpus Christi Bay, but 2 m (6 ft) or less toward the southern end of the island near the mouth of the Rio Grande. An older series of barrier sand islands, representing a previous higher stand of sea level, lies along the landward side of the present lagoonal system parallel with the present series (Price, 1958; Lankford and Rogers, 1969). The position of the older barrier islands is shown in Figure 12. Drill data indicate that still older sandy shoreline deposits interpreted as reworked barrier island sand lie at a sub-seafloor depth of 2 to 16 m (6 to 48 ft), a few kilometers east of the present shoreline.

In the most recently deposited sediments, sand is predominant on the inner shelf and grain size of surficial bottom sediments decreases in a seaward direction in what might be called the expected response to energy available for transport at increasing water depth and increasing distance from shore. Exceptions to this general pattern of grain-size zonation are in the southern and northeastern parts of the area where sand, coarser and better sorted than that nearshore, lies on the outer part of the shelf in water depths ranging from 45 to 100 m (135 to 300 ft). An abundance of shallow-water mollusks plus scattered lithoclasts of conglomerate and other types or rock confirm that the sand on the outer shelf is partly reworked relict deposits that accumulated on the outer shelf near the shoreline during the Wisconsin glaciation when sea level was 150 m (450 ft) lower than today. The gradual decrease in grain size seaward over the shelf off Texas was first noted by Stetson (1953, p. 18) and the relict sediments by Curray (1960, p. 244). The distribution of grain size for surficial bottom sediment in terms of sand-sized fraction ($>63 \mu$), and concentration of sand on the inner shelf, are shown by Figure 13. The map is based on 284 grab samples; the textural analysis for each sample was based on a composite of the upper 10 cm of sediment (Berryhill et al, 1976, p. 52-97).

Regional stratigraphic and sedimentologic analyses provide criteria for distinguishing shelf paleoenvironments from those related to paleoshorelines. In general, textures and structures displayed in ancient shelf-deposited rocks are similar to their coastal counterparts. However, it is difficult to obtain three-dimensional exposures of modern shelf environments that would yield data required to ascer-

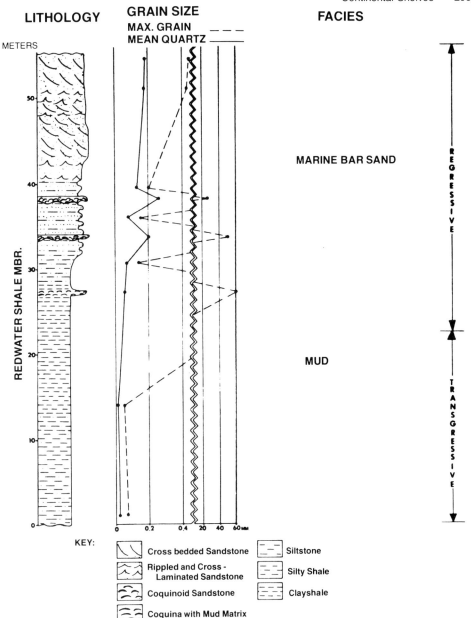

Fig. 15—Oxfordian marine bar vertical sequence interrupted by coquinoid sandstone beds. From Brenner and Davies (1974).

tain lateral and vertical textural sequences. Therefore, direct analogies are not available for many aspects of ancient shelf sediment characteristics. The Mesozoic epicontinental seaway shelf studies provide data useful in developing idealized sequences of textures and structures, thereby considering variations between shelf settings.

Vertical sequences

In marine bar-sand facies of the Oxfordian (Upper Jurassic) of the Sundance Formation in Wyoming, a very characteristic coarsening-upward regressive sequence of textures (Fig. 14) was recorded (Brenner and Davies, 1974). The lithologic sequences generally consist of bioturbated mudstone and silty sandstone at the base. Clay and silt-sized grains rapidly decrease

upward as ripple cross-laminated fine-grained sandstone becomes predominant (e.g. Fig. 62). The sequence is usually capped by large-scale festoon cross-bedded, medium-grained sandstone (e.g. Fig. 63). Multiple-bar sequences, often the rule rather than the exception, complicate vertical sequences. In this Jurassic example, coquinoid sandstone beds commonly interrupt vertical sequences (Fig. 15). These bioclastic-rich units fill channels cut into the bars (e.g. Fig. 64) and form thin sheetlike units with erosional bases (e.g. Fig. 65). Channel fillings may lack significant amounts of bioclastic material, and instead be filled with well-sorted (relative to rest of bar) sandstone and mudrock interbeds (e.g. Fig. 66).

Cretaceous examples from the shelf

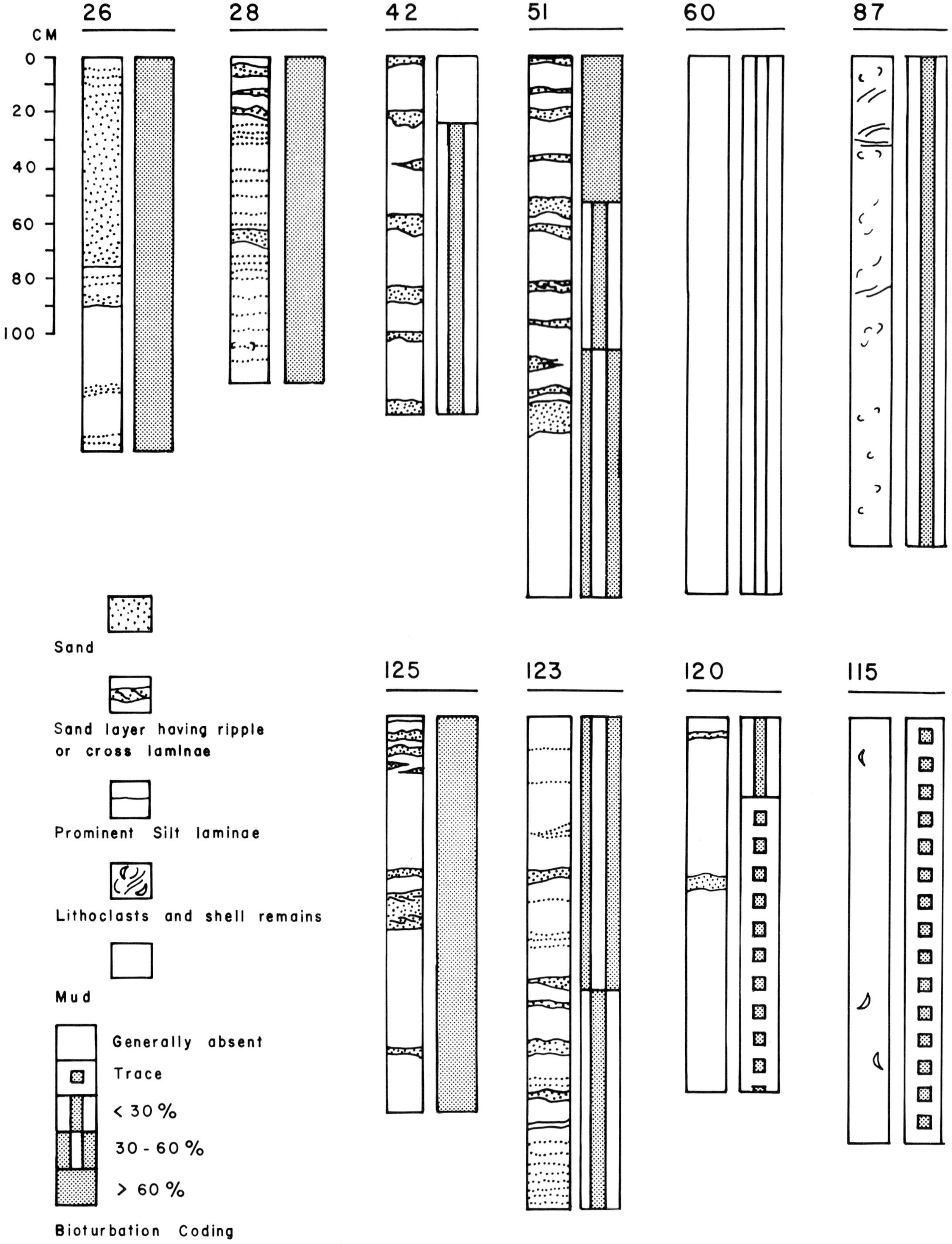

Fig. 16—Logs of gravity cores arranged in two transects across the northern part of the shelf off south Texas to show the nature of sand distribution and the extent of bioturbation. The locations of the cores are shown on Figure 18.

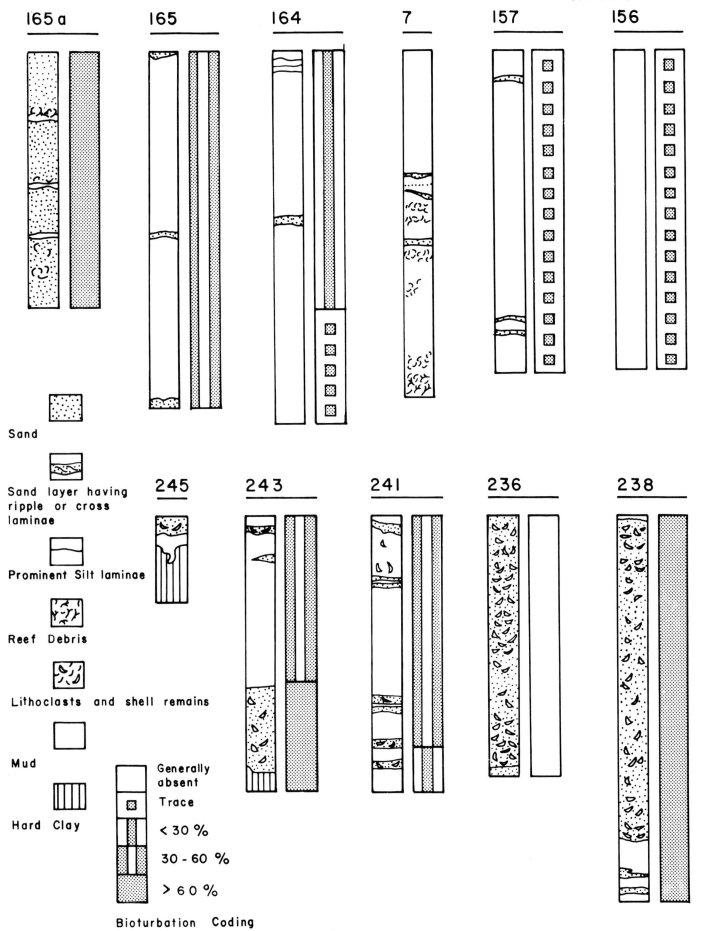

Sand

Sand layer having ripple or cross laminae

Prominent Silt laminae

Reef Debris

Lithoclasts and shell remains

Mud

Hard Clay

Generally absent

Trace

< 30 %

30 - 60 %

> 60 %

Bioturbation Coding

Fig. 17—Logs of gravity cores arranged in two transects across the southern part of the shelf off south Texas to show the nature of sand distribution and the extent of bioturbation. The locations of the cores are shown on Figure 18. Symbols are the same as for Figure 16.

of the Campanian epicontinental sea display vertical sequences similar to those described for the Jurassic example. In the Sussex Sandstone Member, mudrocks are more abundant, representing pauses in sand deposition. This increased muddiness provided widespread niches for benthic sediment "eaters," burrowers and grazers. As a result, trace fossils are well preserved in some units (e.g. Fig. 67).

In addition, cut-and-fill, channel-like features are locally floored with a shale conglomerate lag (e.g. Fig. 68). Undisturbed upper parts of bar sequences display a variety of ripples (e.g. Fig. 69)

and cross-beds (e.g. Fig. 70). Vertical sequences of textures and structures in the Shannon Sandstone Member are similar to those of the Sundance and Sussex except that they lack observable cut-and-fill channel-like features.

Ancient shelf sandstone deposits are characterized by the lack of lateral continuity of individual lithologic units. This is not surprising when considering that rapid fluctuating (e.g. tidal currents), sometimes local (e.g. storms) high-energy conditions account for most of the sandy features on open shelves.

Sedimentary Structures

Modern shelf environments

The diversity in dynamic processes, the availability and proximity of source material, and the local influence of sea-level fluctuations make it impossible to present a set of diagnostic sedimentary structures and textures for shelf deposits. Consequently, we have selected examples from different shelf areas of the United States to demonstrate this variety in shelf characteristics. In addition, two examples are given of ancient deposits that are interpreted as shelf deposits.

As primary sedimentary structures in shelf deposits are commonly destroyed by burrowing organisms, the only structures observable in shelf sections may be those created by bioturbation. For that reason, a number of drawings made from cores, in addition to photographs and radiographs are presented. One animal that may be diagnostic for many open-shelf sands is the heart urchin. Its presence has been reported from shelf sands in both the Gulf of Mexico and the Atlantic.

One structure observable in shelf sediments that may be diagnostic to a certain extent, is the storm layer. Such a layer can be very thin, consisting of a concentration of coarser grains or shell fragments, or rather thick, but still with a high concentration of particles that are normally distributed randomly throughout surrounding deposits. Such storm deposits are lags formed when storm waves resuspend surficial sediments. In ancient deposits the storm layers often are the base for the formation of concretions.

A type of sedimentary structure that at the present is not known in sufficient detail to be labeled environmentally diagnostic is the group of slump structures found in fine-grained sediment on nearly flat areas off large rivers such as the Mississippi in the northern central Gulf of Mexico and the Copper in the eastern Gulf of Alaska (see chapter on deltaic deposits).

South Texas Continental Shelf— Gravity cores show that the volume of sand carried beyond the inner shelf has varied through time, as indicated by its vertical distribution. Cores contain thin discrete sand layers indicating episodes of increased sand deposition separated by mud containing 30% or less sand-sized particles. The discrete sand beds are thicker and more numerous on the inner shelf and become less numerous and thinner seaward. A series of gravity core logs arranged in four

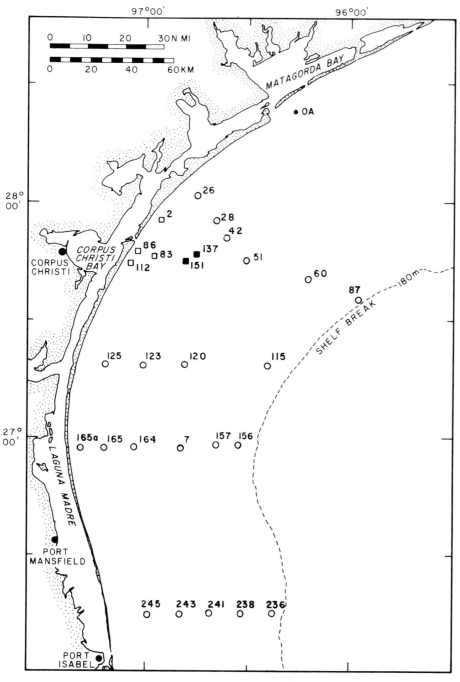

Fig. 18—Location of core stations. Core OA is detailed in Figure 20; X-radiographs of cores from stations 86, 2, 83, 112, 437 and 151 are shown in Figures 21-26, respectively.

traverses across the shelf shows the nature of the sand layers (Figs. 16, 17). Locations of the bottom stations cored are shown by the hollow circles on Figure 18. The area in which discrete sand layers occur off south Texas is shown in Figure 19.

An important aspect of sedimentation on the shelf off south Texas is bioturbation, the modification and rearranging of depositional texture by infaunal activity. Burrowing and churning of sediment appears to be a continuous process, and the grain-size fabric over many parts of the shelf alters to some degree soon after deposition. The extent of bioturbation in cores shown in Figures 16 and 17, as indicated by X-radiographs, is shown beside the lithologic log. The infaunal population is relatively abundant off Texas and bioturbation is an important aspect of the sediments.

Sand layers are typically fine to very fine sand, and little grading of grain size is evident within individual layers. Sand composition, grain size, and sorting characteristics common to most sand layers are shown in Figure 20. Common to many sand layers on the inner half of the shelf are small shell fragments that are generally but not always concentrated in the lower part of the layer. Heart urchin remains, usually whole, crushed specimens, are common in sand layers over most of the shelf.

Depositional structures within sand layers are not indigenous to the shelf, either in a geographic or a sedimentological sense. Those in typically thin layers beyond the inner shelf are apparent only in X-radiographs and are, in most cases, incompletely developed. How well these microscopic lineations would be preserved after deep burial and compaction is questionable. Two features are apparent in the many sand layers that do have internal microscopic laminae: the bases of many are in sharp contact with the underlying mud; and the internal structures indicate flow conditions rather than settling from suspension. These two major features indicate that sand is flushed outward across the shelf beyond the normal inner shelf zone of wave and littoral transport as specific events when energy for transporting sediments in deeper water is temporarily increased, and that the sand is carried mainly near the bottom or at least in the lower part of the water column.

Scour is suggested in some places. On the inner shelf, depositional structures are better developed, with cross-

lamination common and in many places resembling what has been called hummocky cross-stratification (Harms, 1975, p. 88). Other types of cross-laminations appear to be varieties of current ripples representing a range of flow velocities and directions. Obviously, depositional structures described for the inner shelf apply as well to shoreface deposits, and indeed are transitional across the inner shelf to shoreline. Examples of depositional structures within sand layers are shown by prints of X-radiographs in Figures 21 through 26.

From time to time, sea-level fluctuations during the Pleistocene have caused profound changes in the sedimentary regime on the continental shelf off Texas. During low stands of sea level, the shelf was exposed as an extension of the coastal plain. As sea level fell, deltas were built progressively seaward and formed a well-defined sequence of regressive deposits in which significant amounts of sand eventually were spread in broad fans across the shelf. As sea level rose progressively, these deposits were in turn covered by finer grained sedi-

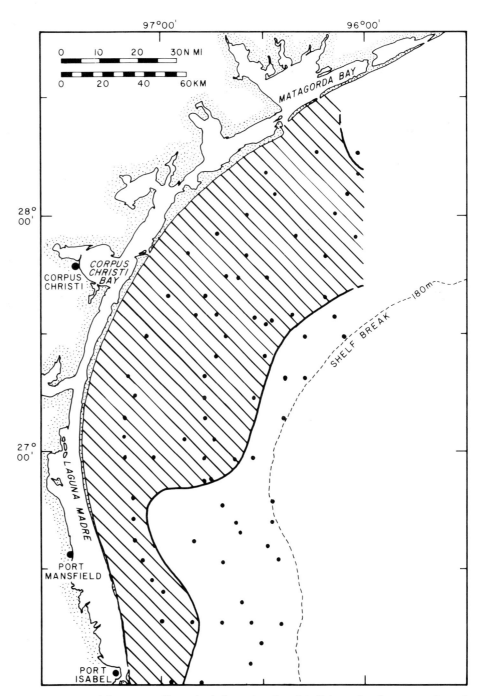

Fig. 19—Extent of discrete sand layers in shallow subsurface (late Holocene) sediments, continental shelf off south Texas. The diagonal lines indicate the area in which discrete sand layers were cored and the dots indicate the bottom station locations. From Berryhill et al. (1976, Fig. 62).

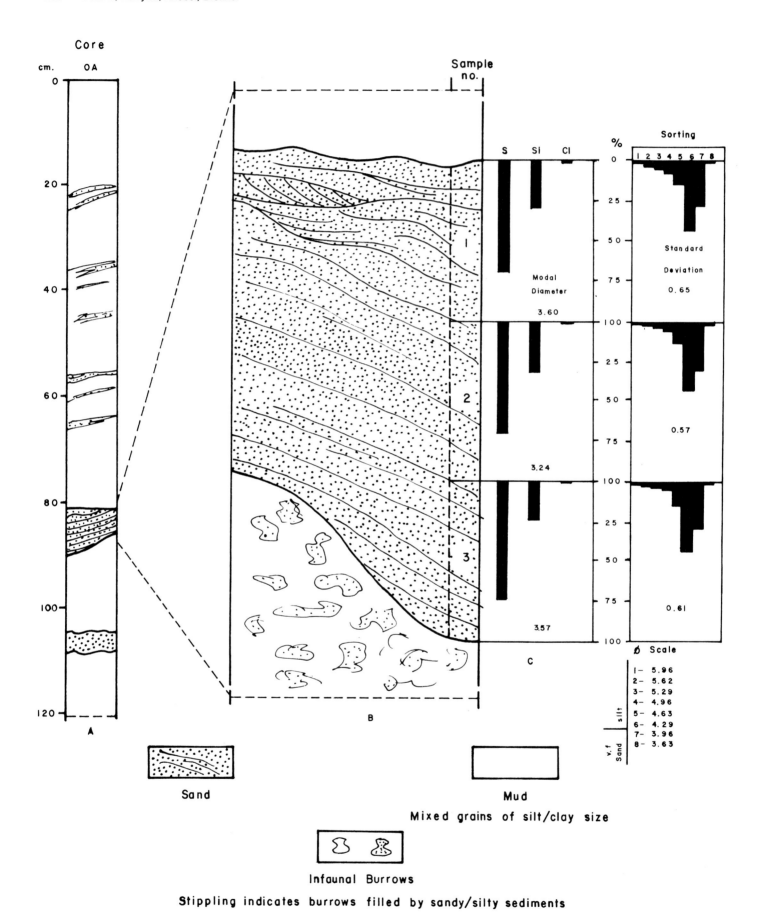

Fig. 20—Depositional structures and textural characteristics of a sand layer based on three incremental analyses from top to bottom. See Figure 18 for location. From Berryhill et al (1977, Fig. 73).

ments. Alternating stratigraphic couples of regressive and overlying transgressive sequences separated by unconformities are features of the Neogene deposits on the Texas shelf. Examples of the most recent regressive/transgressive cycle are shown in the two traverses in Figure 27. The nature of the paleogeographic surface marked by the unconformity between regressive and transgressive sequences is shown by Figure 28. The markedly different facies represented by each of the two sequences are apparent when Figure 28 is compared with Figure 13.

The following summary statements can be made regarding the environmental significance of sand on the continental shelf off south Texas:

(1) The depositional structure within sand layers is not unique to the shelf environment. Structures in the thicker sand beds of the inner shelf are better developed because of more vigorous water movement there, but, in type, they resemble (if not locally duplicate) those considered characteristic of the shoreface and surf zone.

(2) Sand alone is not an indicator of shelf deposition and must be examined in context with associated sedimentary aspects, such as: (a) interspersion as thin layers in finer grained sediments that contain open marine faunal remains; (b) regional trends in both the number and thickness of the sand layers; (c) nature of lateral facies; (d) stratigraphic juxtaposition of environmentally dissimilar deposits resulting from significant changes in sea level that cause sequences of fluvial and deltaic sands to be sandwiched between sequences of transgressive open marine muds; and (e) heart urchins, and other diagnostic animals which live only in an open marine environment where they burrow in sand, and are commonly preserved as complete, though crushed specimens.

Atlantic Continental Shelf—Core samples from the surficial sand sheet on the Atlantic Continental Shelf of the United States reveal a variety of sedimentary structures. On Georges Bank, textural variation is characteristic of the region and includes differences in grain size and sorting of the predominant sand and gravel as well as interrelations of interbedded mud layers within the sand (Bothner et al, 1979). Other features of the sand are color mottling, indistinct layering, burrow

Fig. 21—Print of X-radiograph made from a small box core (86) showing three zones of depositional structures within a single layer of sand: (1) hummocky cross bedding (note numerous well oriented small shells); (2) combined-flow ripples that contain several lenses of mud (white); and (3), irregular layering of mud and sandy mud. The thin irregular vertical lineaments through the core are worm burrows. Location shown on Figure 18.

filling, isolated clay balls, and large pieces of gravel (Bothner, 1979, personal commun.; Figs. 29, 30).

On the Middle Atlantic shelf, sedimentary structures are similar to those on Georges Bank (Stubblefield et al, 1975; Knebel and Spiker, 1977). The dominantly fine-to-coarse, well-sorted, shell sand contains scattered subrounded gravel (≤80 mm), clay stringers (1-2 cm), clay balls (≤50 mm), and large wood fragments. Distinct textural and mineralogic bands are common (Figs. 31, 32).

Off South Carolina and Georgia, the most common types of sedimentary structures are interbedded sand and mud horizons, remnant stratification, shell layers, and shell fragment layers (Pilkey et al, 1979). Individual burrows and cross-stratification are sometimes identifiable. Some burrows are the

spreite-type, and these were probably formed by heart urchins living on the open shelf (Figs. 33, 34).

Sedimentary structures on the U.S. east coast shelf reflect primarily the reworked nature of the surficial sand sheet. Sediments brought to the shelf during the late Pleistocene or Holocene are being mixed and modified by modern processes. For example, distinct textural and mineralogic bands as well as interbedded sand and mud zones and large wood fragments probably reflect episodic depositional events, such as storms. These structures were buried rapidly, as otherwise they would have been destroyed by currents, waves, and burrowing organisms. On the other hand, certain sedimentary structures have been partially obliterated either because they were not preserved by burial or because suf-

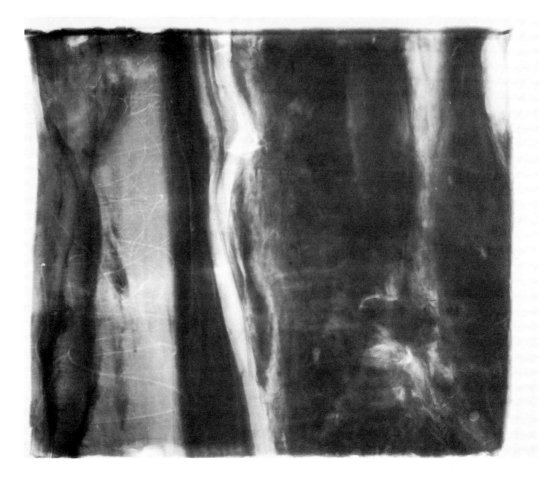

Fig. 23—Print of X-radiograph made from a small box core (83) showing a wavy lamina of mud in the upper part of the thick sand layer (dark). Wavy laminae of this type indicating unidirectional flow are common in many of the sand layers farther out on the shelf. Above the sand is a layer of mud containing some sand that reveals either combined flow ripples or crude wave ripples. Location shown on Figure 18.

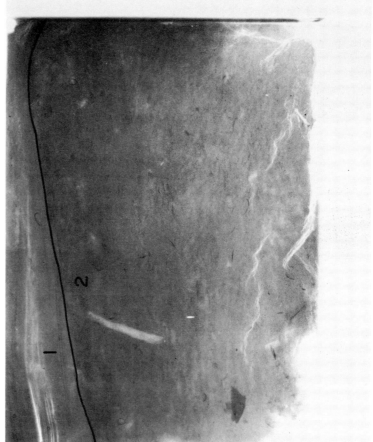

Fig. 22—Print of X-radiograph made from a small box core (2) showing a depositional contact within a single sand layer. Above the slanted contact (1), the sand is very fine grained and contains very thin clay laminae; below the contact (2), the sand is coarser and contains randomly oriented small shells. Location shown on Figure 18.

Fig. 26—Print of X-radiograph of upper part of a gravity core (151) showing very fine inclined mud laminae within a thin sand layer and the indistinct undulating base of the sand. Undulation appears to have been caused by bioturbation. The sand probably correlates with that in core no. 137 (see Fig. 25), which is about 3 km away. Location shown on Figure 18.

Fig. 25—Print of X-radiograph of upper part of a gravity core (137) showing very fine inclined mud laminae within a thin sand layer and the indistinct undulating base of the sand. Undulation probably was caused by bioturbation. The sand probably correlates with that in core no. 151 (see Fig. 26), which is about 3 km away. Location shown on Figure 18.

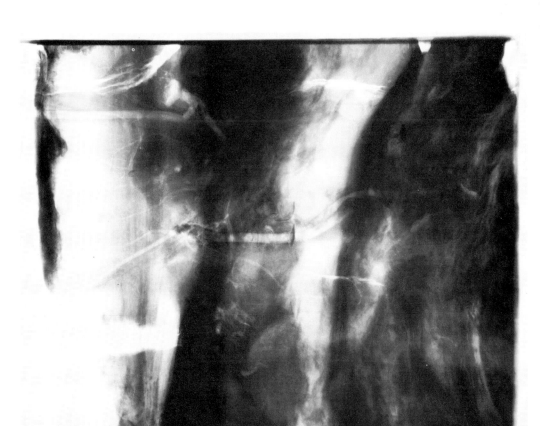

Fig. 24—Print of X-radiograph of a small box core (112) showing interlayered sand (dark) and mud that has been intermixed by bioturbation.

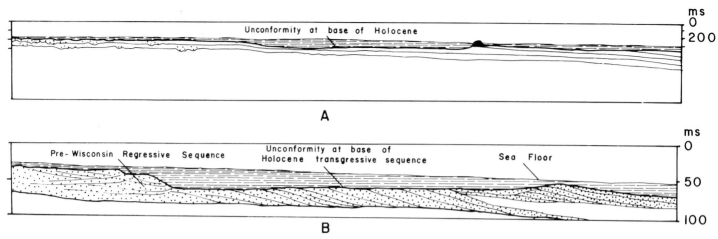

A

B

Fig. 27—Small-scale generalized cross sections constructed from high-resolution seismic-reflection profiles showing the stratigraphic relation of the transgressive Holocene sediments above the unconformity of Wisconsin age to the underlying older regressive sediments; **A,** central sector of the shelf where the regressive sequence is thin and restricted to the landward third of the shelf where buried fluvial channels are extensive; and **B,** southern part of the shelf where the deltaic regressive sediments are thick and extend as a progradational set to the outer edge of the shelf. From Berryhill et al (1976).

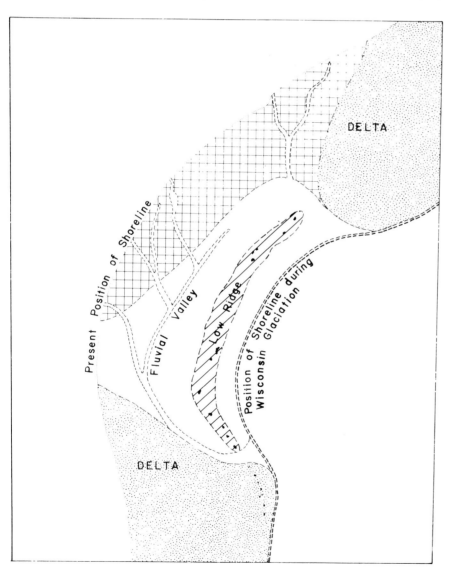

Fig. 28—Paleogeography and sedimentary facies relations on the shelf during the Wisconsin glaciation. Stippling indicates the extent of sand rich ancestral deltas built across the shelf by the Rio Grande to the south and the Colorado and Brazos Rivers; the grid pattern indicates thin alluvial fan, flood plain, and stream channel deposits, the short dashed lines indicate the location of ancient stream channels, the diagonal lines indicate a topographic ridge of low relief that formed temporary sites for the growth of carbonate reefs (black dots) as sea level began to rise in the early Holocene. Contrast this figure with Figure 13 for facies differences between the regressive and transgressive sequences.

ficient time elapsed after deposition to allow destruction by bioturbation and physical mixing. Color mottling, remnant stratification, and burrow filling are examples of this phenomenon. Finally, clay balls and gravel within the sand sheet may result from sediment erosion and winnowing. The clay balls probably stem from reworking of older, underlying muddy zones, while the scattered gravel may reflect relict fluvial, glacial, or ice-rafted sediments stranded in a matrix of mobile sand.

Lower Cook Inlet sand waves—Lower Cook Inlet, Alaska, is a large estuarine indentation of the shelf. Its surficial sediments are basically pebbly mudstone of glacial origin overlain by a thin sand blanket in its central part in water depths ranging from 40 to 70 m (120 to 210 ft). A ramp starting at a depth of about 70 m (210 ft) and sloping south brings the seafloor down to 130 to 145 m (400 to 450 ft). Part of this ramp is also covered by sand, and all of the deeper water area contains a sandy mud.

The sand is remolded into a variety of sizes and types of bedforms such as various small ripples, megaripples, dunes, sand waves, sand ridges, sand ribbons, and sand patches (Fig. 35). Maximum heights of the large bedforms are about 12 to 15 m (36 to 45 ft), with wave lengths peaking at 700 to 950 m (2,100 to 2,900 ft; Bouma et al, 1977, 1978).

Tidal currents average 4.5 to 5 knots (225 to 250 cm/sec) and appear to have little influence on the bottom. Sand transport was observed only during the last 1-2 hours of both ebb and flood tides during spring tide. Based on repeated side-scan lines and GEOPROBE

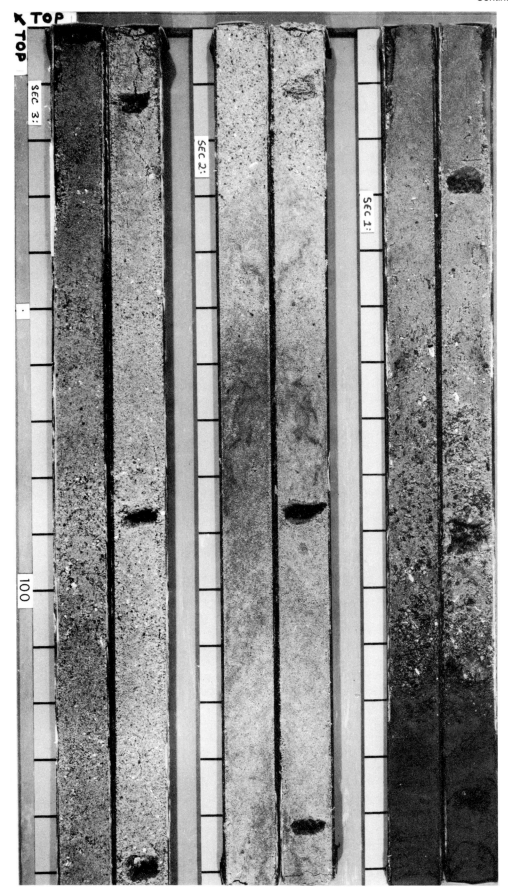

Fig. 29—Photograph of split core collected from the southwestern part of Georges Bank. Depth in core in centimeters is noted along the left half of the core. Dark patches in right half of core are voids where subsamples were removed. Note: (1) change from poorly sorted, coarse and very coarse sand to better sorted, medium sand at about 180 cm down the core (section 2); (2) indistinct layering and color mottling within the core between 180 and 230 cm (section 2); (3) preponderance of gravel near the base of the sand unit (core depth = 350 to 415 cm, section 1); and (4) silty-clay unit at the base of core (core depth = 415 to 450 cm, secton 1). Photograph courtesy of M. H. Bothner.

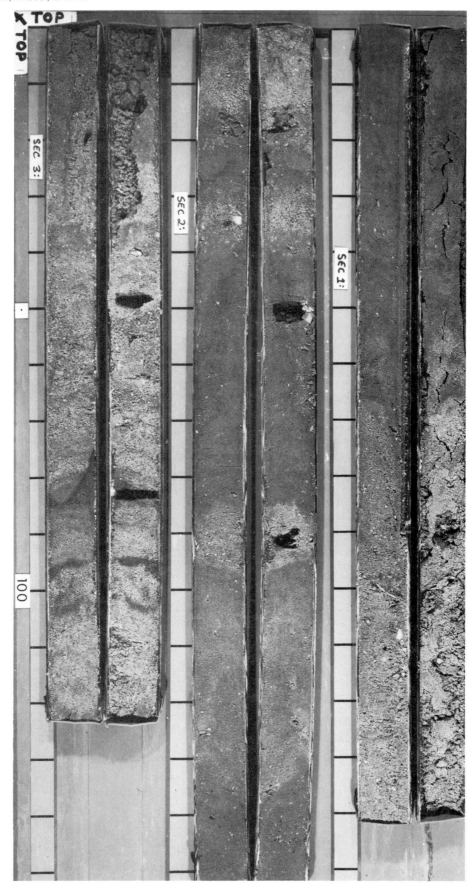

Fig. 30—Photograph of split core collected from south-central part of Georges Bank. Depth in core in centimeters is noted along the left half of the core. Dark patches in right half of core are voids where subsamples were removed. Note: (1) changes in color, texture, and sorting throughout the core; (2) distinct silty stringers (section 3) that separate a section of coarse sand above from medium sand below; and (3) indistinct layering and color mottling at 220 to 330 cm (bottom section 2 and top of section 3). Photograph courtesy of M. H. Bothner.

Fig. 31—Photograph of split core collected on the Middle Atlantic outer shelf off New Jersey. Depth in core in centimeters is noted along the left half of the core. Dark patches in right half of core are voids where subsamples were removed. Note: (1) abrupt change in lithology between surficial sand unit (0-362 cm) and underlying muddy unit (362-427 cm); (2) very coarse sand horizon containing gravel and abundant shells near top of core (10-35 cm); and (3) heavy-mineral banding near bottom of sand unit (300-320 cm).

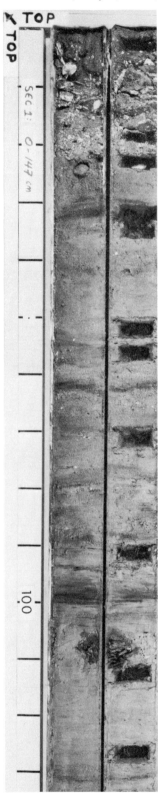

Fig. 32—Photograph of split core collected on the Middle Atlantic outer shelf off New Jersey. Depth in core in centimeters is noted along the left half of the core. Dark patches in right half of core are voids where subsamples were removed. Note: (1) coarse sand with shells and gravel at top of core (0-22 cm); (2) isolated piece of subrounded gravel at 22-23 cm in a matrix of medium to coarse sand; (3) banding (both distinct and indistinct) throughout middle part of core (30-105 cm) due to alternate zones of medium sand and sandy clay; and (4) large wood fragment just below base of layered section (106-110 cm).

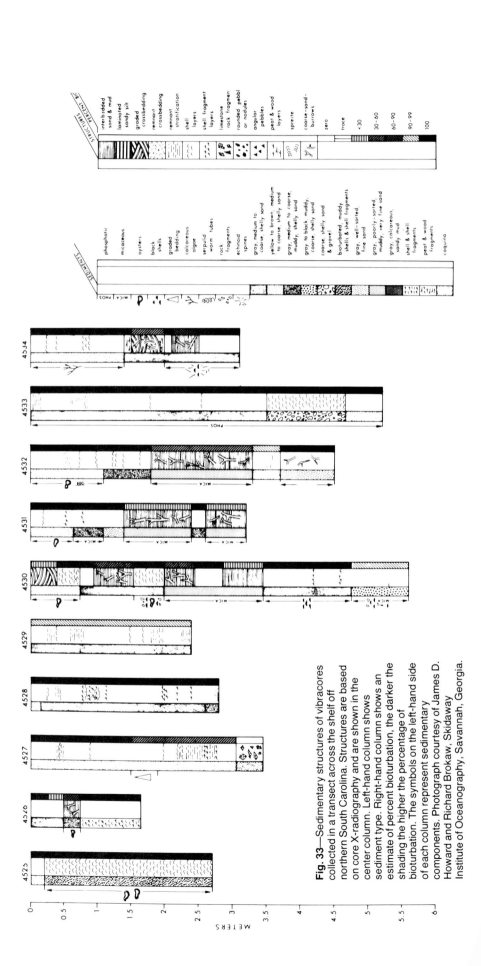

Fig. 33—Sedimentary structures of vibracores collected in a transect across the shelf off northern South Carolina. Structures are based on core X-radiography and are shown in the center column. Left-hand column shows sediment type. Right-hand column shows an estimate of percent bioturbation, the darker the shading the higher the percentage of bioturbation. The symbols on the left-hand side of each column represent sedimentary components. Photograph courtesy of James D. Howard and Richard Brokaw, Skidaway Institute of Oceanography, Savannah, Georgia.

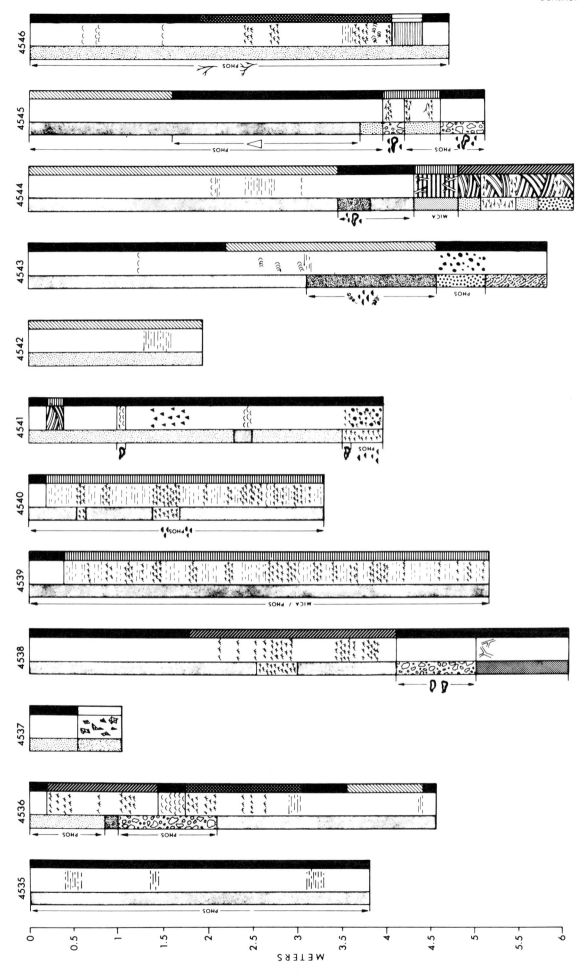

Fig. 34—Sedimentary structures of vibracores collected in a transect across the shelf of northern Georgia. Structures are based on core X-radiography and are shown in the center column. Left-hand column shows sediment type. Right-hand column shows an estimate of percent bioturbation. The symbols on the left-hand side of each column represent sedimentary components. For explanation, see key presented in Figure 33. Photograph courtesy of James D. Howard and Richard Brokaw, Skidaway Institute of Oceanography, Savannah, Georgia.

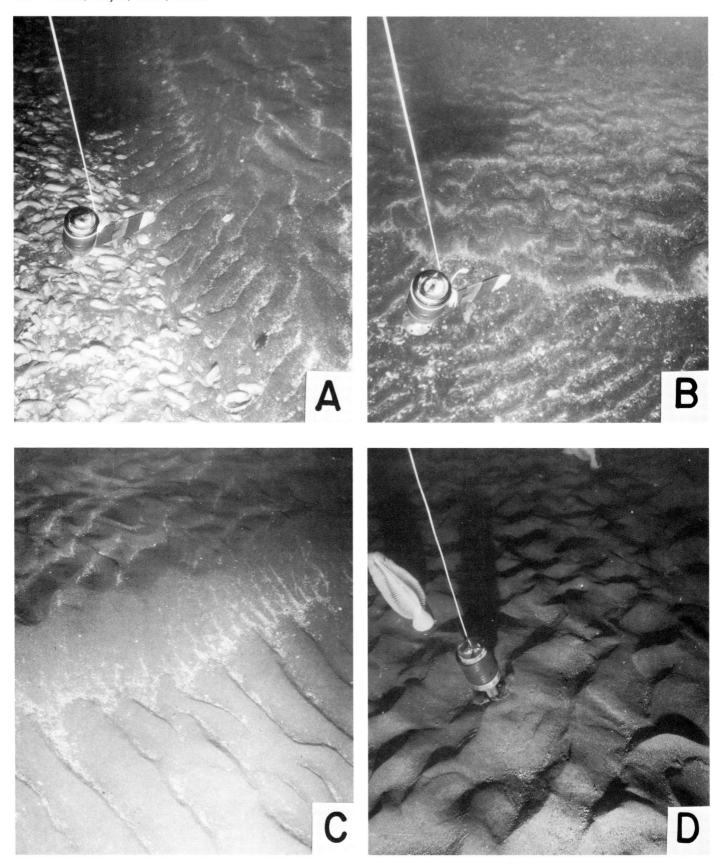

Fig. 35—Bottom photographs: **A.** Sand wave front, high slip face with transverse ripples due to cross flow in trough. Note semiorientation of shells in lag deposit in sand wave trough. **B.** Sand wave front with low slip face. Ripples on sand wave are lunate, in trough straight crested. **C.** Small sand wave that is superimposed on 8 m high sand wave. Note change in ripple pattern and lack of coarse lag. **D.** Sinuous to straight lunate ripples with slightly coarser material in troughs. Note the sea pens. **E.** Slightly sinuous ripples with interference pattern. **F.** Sinuous ripple pattern common on mid stoss sides of medium sand waves.

Except C and F, the pictures are from bedforms found in areas with insufficient sand cover over a hard bottom to form medium and large sand waves. Vertical bands on compass vane are 5 cm wide. Lower Cook Inlet, Alaska, water depth ranges from 40 to 75 m.

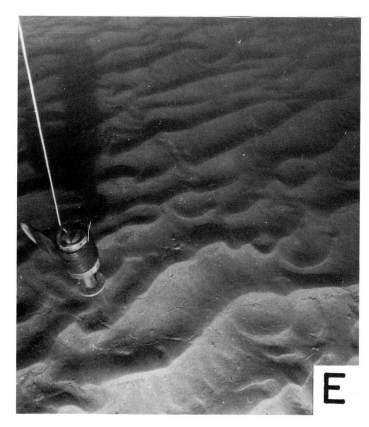
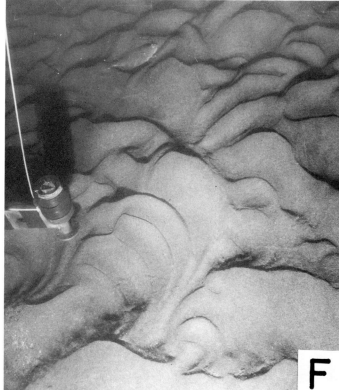

observations, storms appear to have significantly more effect (Bouma et al, 1979).

The large bedforms can easily be detected on bathymetric profiles and side-scan records (Figs. 36, 37). Ship motion normally prevents accurate detection on bathymetric records of small features, but they are visible on side scan. However, sonographs may give rise to a wrong interpretation of the height of a feature and its type of sediment (Figs. 38-41). In Figures 38, 39, and 40, the height of bedforms, based on the width of the light-colored shallow zone or the dark direct reflectance, can measure 50 to 75 cm (2 to 3 ft). However, bottom television and photographs show in many areas that the dark reflective zone is not a lee or stoss side of a bedform but is a wide gravel zone in the trough of megaripples and small sand waves that typically are less than 15 cm (0.5 ft) high (Fig. 42). Such low heights are difficult to measure accurately from sonographs even if the bedform consists totally of sand-sized materials such as those in Figures 42 and 43. These bedforms became exposed along the shoreline at very low water. Their overall morphologic characteristics differ little from those observed using bottom television.

Internal structures of these bedforms are known from only a few areas. Short cores from Cook Inlet indicate intense reworking by burrowing organisms, although some bedding can be detected locally (Fig. 44). Houbolt (1968) showed some cross-bedding, clay laminae, and coarser skeletal fragments in 75 cm long cores from sand ridges from the North Sea.

Comparing observations made on sand waves by marine geologists in the modern environment with ancient deposits is difficult because of the lack of similarity in observations. Studies on modern forms are strongly morphologic and process oriented, while those in the fossil record are primarily based on sedimentary structures and textures.

Ancient continental shelf environments

Kodiak Miocene shelf deposits—A transgressive sequence, starting with large-scale foreset beds of estuarine or deltaic origin and terminating with very muddy shelf deposits, can be seen along the coast at Narrow Cape on Kodiak Island, about 45 km south-southeast of the city of Kodiak. The series belongs to the Miocene Narrow Cape Formation (Nilsen and Moore, 1979). On the basis of the molluscan faunas, Allison (1976) suggested deposition in subtropical to warm-temperate water. Locally, burrows are oriented normal to the bedding, and some articulated clams are found in life position.

The muddy shelf deposits are characterized by parallel zones of hard concretions or shells (Fig. 45). All silicified concretions have formed around a shell or a group of shells. Their sizes vary from more or less round 3 to 4 cm balls to smooth elliptical fist and head size, to large irregular blocks consisting of smaller blocks cemented together. The sediment in a concretion zone typically is slightly more silty than the deposits below and above. Shells occur either in thick layers having an irregular scour base and a rather smooth top (Fig. 46) or in single discrete bands. Thick concentrations typically are sandy and show no preferred orientation of shells. Thin shell layers, on the other hand, may be slightly more silty than surrounding sediment while most of the shells do not touch each other and are positioned in a convex upward position (Fig. 47).

Narrow Cape sediments are thought to have been deposited in a quiet shelf environment, comparable to the north-central Gulf of Mexico, with a nearby mud source. The transgressive series indicates that normal hydrodynamic conditions were very calm and that little sand was transported far from land. Occasional storms induced bottom currents that removed part of the surficial sediments and formed a lag deposit consisting of some coarser grained material and shells. A thick layer of shells in a sandy matrix can result from a severe storm, such as a hurricane.

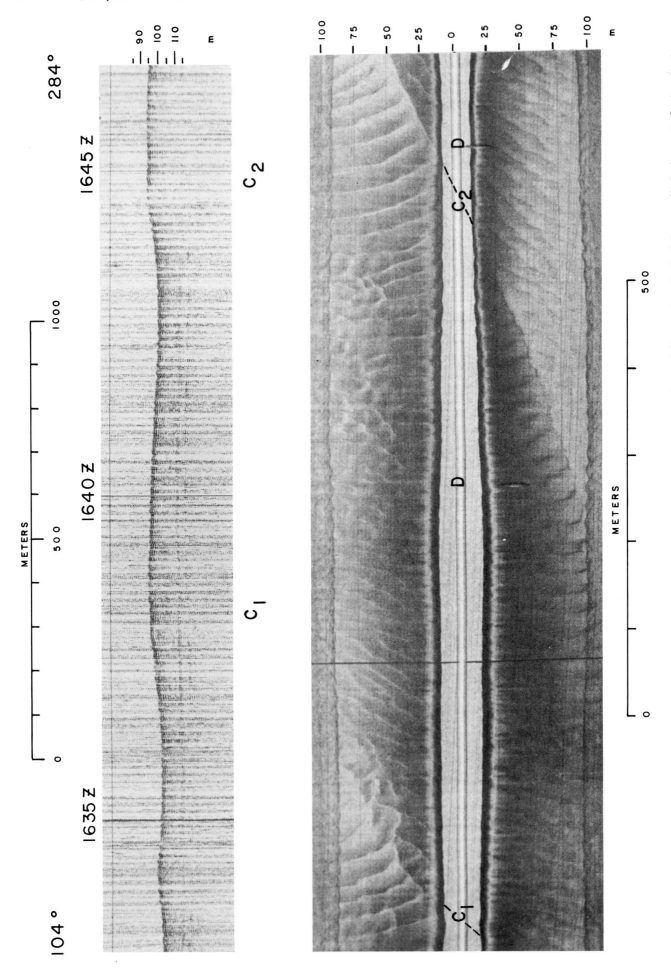

Fig. 36—High-resolution seismic-reflection profile and side-scan sonar record of sand ridges (crests parallel with main current direction) with sand waves superposed. Note the turning of sand waves crests near the sand ridge crests under the influence of helical currents. Scales of both records are different. Lower Cook Inlet, Alaska. From Bouma et al (1977, Fig. 3).

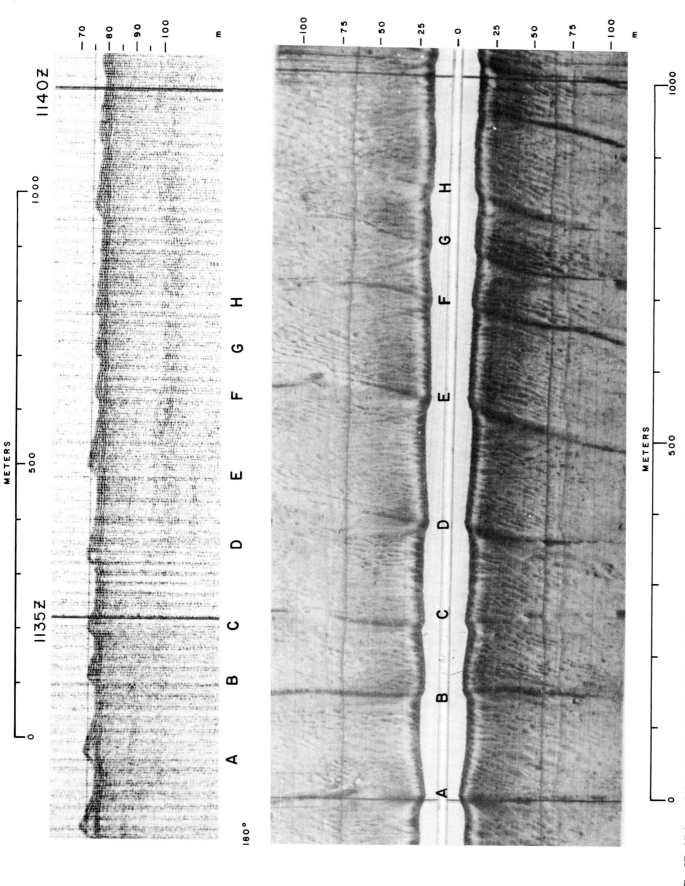

Fig. 37—High-resolution seismic-reflection profile and side-scan sonar record of two sizes of sand waves superposed on one another. Note the straight crests. Scales of both records are different. Letters indicate corresponding crests between seismic record and sonograph. Lower Cook Inlet, Alaska. From Bouma et al (1977, Fig. 4).

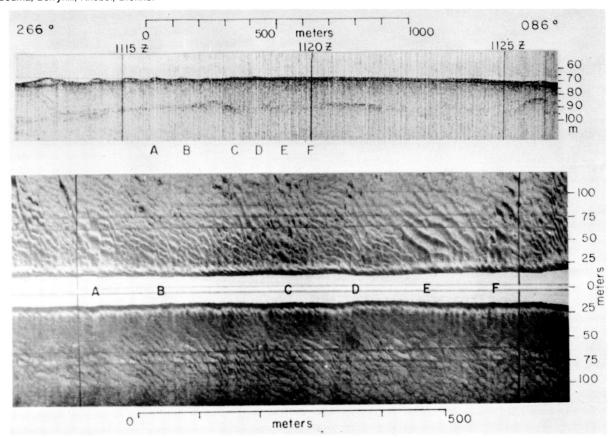

Fig. 38—High-resolution seismic-reflection profile and side-scan sonar record of a rather smooth area with small sand waves changing to a field with larger sand waves. Letters on both records indicate points of correspondence. Scales of both records are different. Lower Cook Inlet, Alaska. From Bouma et al (1977, Fig. 6).

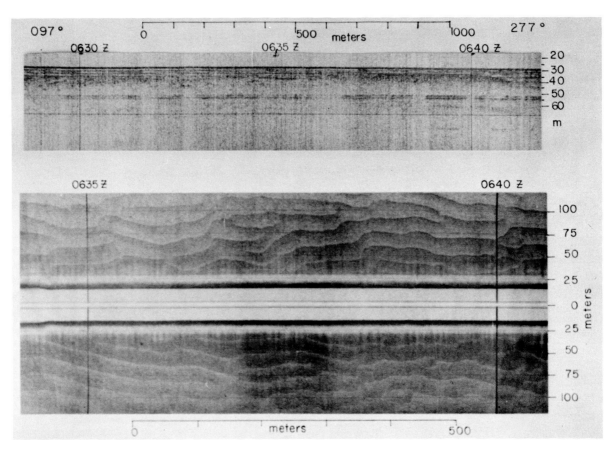

Fig. 39—High-resolution seismic-reflection profiles and side-scan sonar record of rather straight-crested sand waves. Wave heights are too low to be visible on bathymetric record. Scales of both records are different. Lower Cook Inlet, Alaska. From Bouma et al (1977, Fig. 5).

90°

270°

-100
-75
-50
-25
-0
-25
-50
-75
-100
m

500

METERS

0

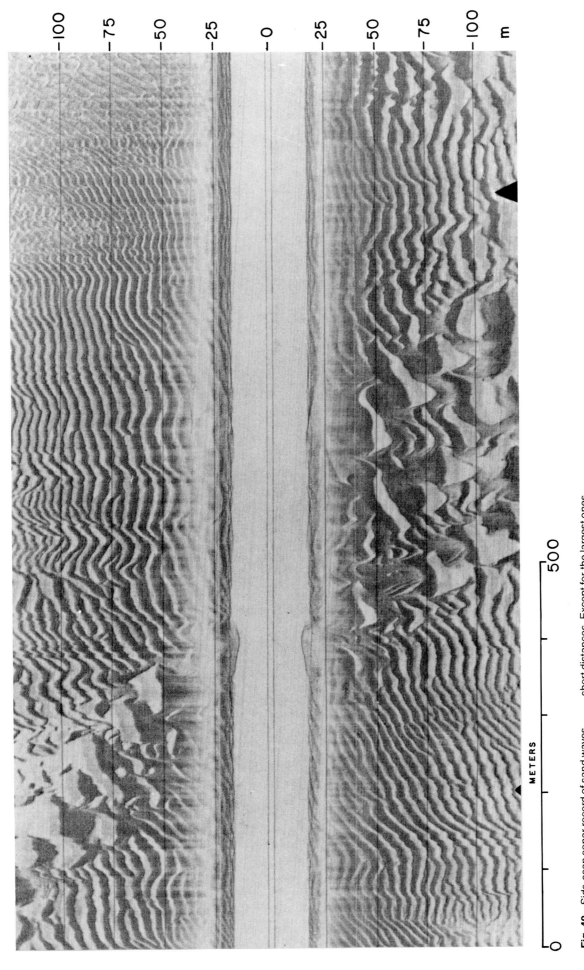

Fig. 40—Side-scan sonar record of sand waves in lower Cook Inlet, Alaska, portraying the rapid lateral changes in size and orientation over short distances. Except for the largest ones, these bedforms are less than 60 cm (2 ft) high and do not show up on bathymetric records.

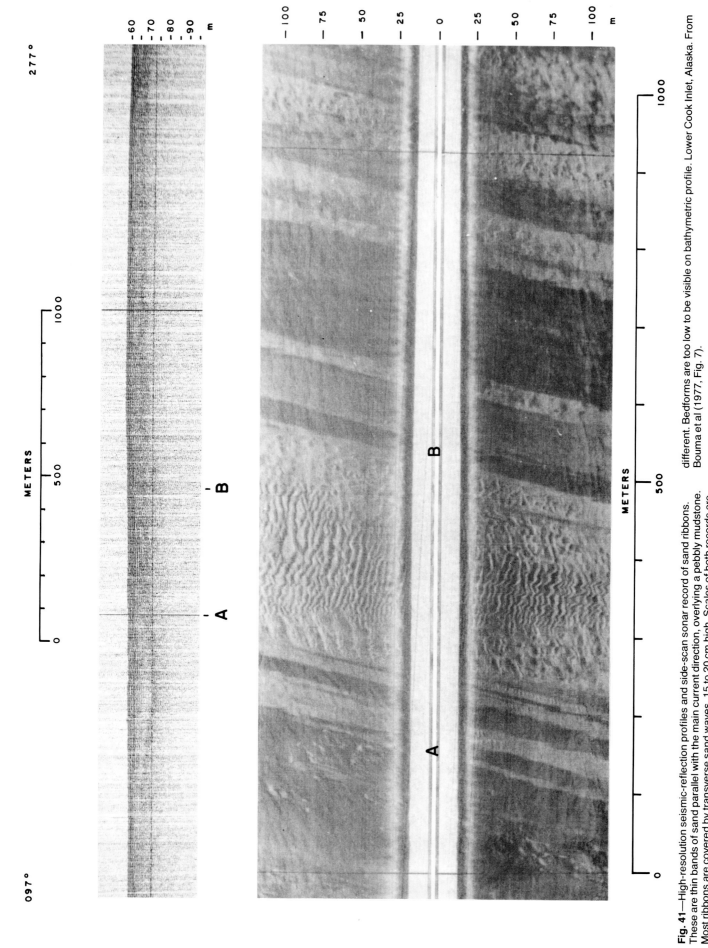

Fig. 41—High-resolution seismic-reflection profiles and side-scan sonar record of sand ribbons. These are thin bands of sand parallel with the main current direction, overlying a pebbly mudstone. Most ribbons are covered by transverse sand waves, 15 to 20 cm high. Scales of both records are different. Bedforms are too low to be visible on bathymetric profile. Lower Cook Inlet, Alaska. From Bouma et al (1977, Fig. 7).

Fig. 42—Overview of exposed small sand waves showing the sinuous crests and superposed ripples. Base of Homer Spit, Alaska, low water.

Fig. 43—Detailed view of the small sand waves shown in Figure 42. Note the sinuous character of the sand waves crests and the change from straight-crested ripples to lunate ripples from low to high on the stoss side of the sand waves. Scour pits are not typical but do occur. Low water, base of Homer Spit, Alaska.

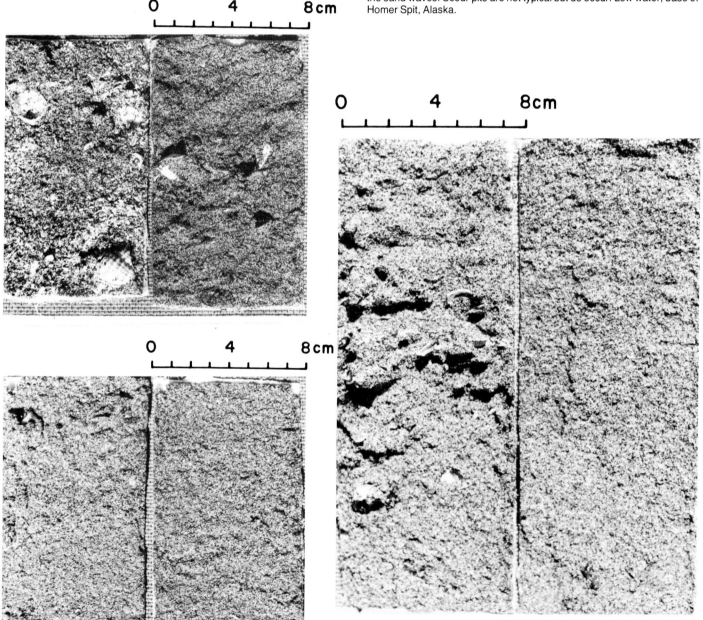

Fig. 44—Lacquer peels of six short cores collected from large sand waves in lower Cook Inlet, Alaska. Note the occasional faint bedding and abundance of shell fragments. More complete shells occur in samples collected from the trough of the large sand waves. Width of peels is 7.5 cm (3 in.).

Fig. 45—Miocene shelf deposits, Narrow Cape, Kodiak, Alaska. Note the concretion horizons parallel with the bedding. These concretions are formed around shells which in turn were concentrated in layers by storm action.

Fig. 46—Concentration of shells due to storm action. Lower boundary is irregular, upper boundary is smooth. Miocene shelf deposits, Narrow Cape, Kodiak, Alaska.

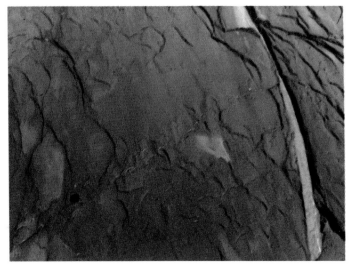

Fig. 47—Thin concentration of shells due to storm action. Note the concave upward position of the shells. Miocene shelf deposits, Narrow Cape, Kodiak, Alaska.

Fig. 48—Oligocene shelf deposits from Sespe Creek, Ventura County, California, interpreted as marine bedforms. Strong jointing obscures most sedimentary structures. Some general bedding and large-scale foresets can be distinguished. Foreset beds average 1.20 m (4 ft) in height.

Fig. 49—Same as Figure 48, in addition to some horizontal bedding planes, some foreset beds can be observed. Dip directions occur to opposing sides suggesting tidal effects.

Fig. 50—Large-scale foreset beds accentuated by concretions. These sets are at least 4 m (12 ft) high. Same general location as Figure 48.

Fig. 51—Detail of the Oligocene bedforms. Bedding planes run diagonally across the photograph from lower left to upper right. Jointing obscures the sedimentary structures. See upper left hand corner for distinct foresets. Same general location as Figure 48.

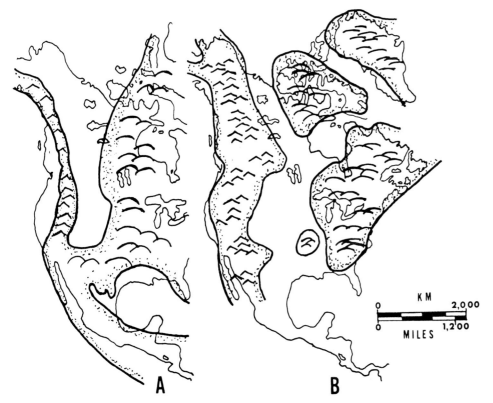

Fig. 52—Examples of Mesozoic seaway configurations in North America. **A,** Late Jurassic Oxfordian sea, open at one end. Modified from Brenner and Davies (1974). **B,** Late Cretaceous Campanian seaway, open at both ends. Modified from Brenner (1978).

straight-crested bedforms near Piedra Blanca indicate transport from the northwest, while larger lunate structures at Kimball Canyon show a paleocurrent direction from the northeast. Reid's (1978) interpretation of these deposits as submarine dunes with megaripples (Figs. 48-51) seems appropriate. The abundance of filled joints makes it difficult to see sedimentary structures unless examination takes place from close up (Fig 51). Many layers are heavily bioturbated and have a structureless appearance.

Mesozoic Epicontinental Seaways of Western North America

From Late Triassic to Late Cretaceous, the western part of North America was partially covered by an epicontinental seaway. At times the seaway was open only to the north (as in Late Jurassic Oxfordian sea, Fig. 52A), while at other times it was open both to the north and south (as in the Late Cretaceous Campanian sea, Fig. 52B). Marine offshore sandstone bodies have been described from both single and doubly opened ancient seaways.

Late Jurassic sand ridges—The upper part of the Sundance Formation in Wyoming and parts of the correlative Swift Formation in Montana contain bar-shaped marine sandstone bodies that appear to have been deposited in a shelf-like setting (Brenner and Davies, 1974). The sequences of sedimentary structures and upward coarsening textures indicate local regressive conditions, but differ from barrier island sequences because they lack evidence of emergence (Fig. 53).

The model proposed for these bars consists of a region of shallow marine sand bodies, forming a ridge and swale sea-bottom morphology (Fig. 54). Crest orientations and sediment transport directions appear to be anisotropic. Bars are frequently breached by high energy storm-generated currents, with resulting channels filled with coarse-grained coquinoid sand after storms (Brenner and Davies, 1973a). Repeated high energy events caused a coalescence of channel sands into sheets that are laterally discontinuous (Fig. 55).

Brenner and Davies (1974) called upon storm wave-generated currents, tidal currents and prevailing regional flow to account for the construction and maintenance of these Oxfordian ridge and swale morphologies. In a more detailed analysis of sedimentary structures and transport directions, Davies and Brenner (1973) proposed

Ventura County, California, Oligocene shelf deposits—The upper member of the Vaqueros Formation (upper Oligocene) along the upper Sespe Creek, Ventura County, California, represents a transgressive sequence of environments from a shallow continental shelf to outer shelf (Reid, 1978). Conspicuous in the field of mainly shaly deposits are light-colored sandstones containing cross-beds (Figs. 48 and 49). The planar and trough sets at Piedra Blanca are 1 to 3 m (3 to 10 ft) thick, and trough-shaped sets at Kimball Canyon, about 8 km (5 mi) farther east, are up to 10 m (30 ft) thick. The total member is up to 70 m (210 ft) thick. Paleocurrent directions of the smaller,

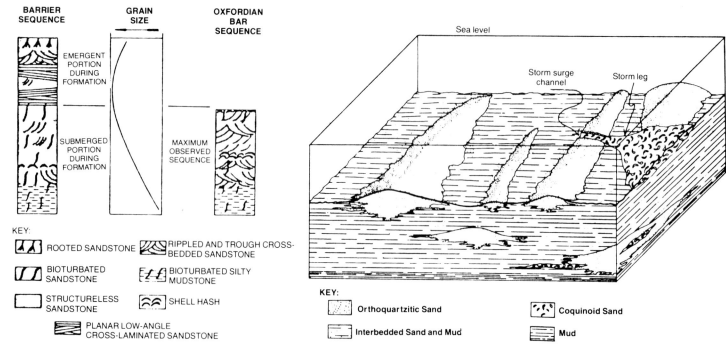

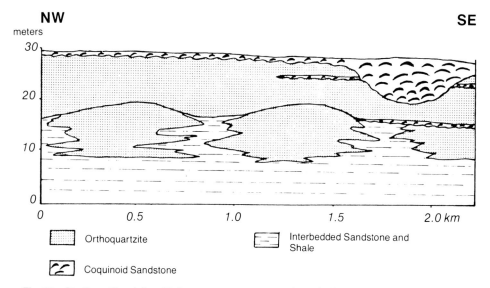

Fig. 53—Comparison between vertical sequences of sedimentary structures and grain sizes observed in a complete barrier sequence and those observed in Oxfordian offshore bars. From Brenner and Davies (1974).

Fig. 54—Block diagram illustrating the conceptual Oxfordian sea bottom in the marine bar-sand facies. From Brenner and Davies (1974).

Fig. 55—Stratigraphic relationship between storm-generated coquinoid sandstones and marine bar lithologies.

that the bars are complex sand ridges formed in a storm-dominated shelf-like setting along the western margin of the Oxfordian sea. They concluded that most sediment movement was accomplished by storm-generated currents, with fair-weather processes playing only minor roles. This work led to a comparison between this part of the seaway and parts of the modern continental shelf of the eastern United States (Brenner and Davies, 1973b). A dominant storm track in the Oxfordian seaway from northeast to southwest

would have the fetch required to generate waves large enough to disturb the sea bottom a few tens of meters below sea level. This may be somewhat analogous to storm tracks that affect modern continental shelves.

Late Cretaceous, Powder River basin—During the Campanian, part of the Powder River basin now lying in Wyoming was part of a broad, muddy shelf (Asquith, 1974) along the western margin of the Late Cretaceous epicontinental seaway (Fig. 52B). Campanian rock types in this part of the seaway

consist predominantly of mudrock and relatively thin sandstone-rich units punctuating the otherwise monotonous sequence. Two of these units, the Shannon and Sussex Sandstone Members of the Steele Shale, have been the subject of detailed outcrop and subsurface analyses (Spearing, 1976; Brenner, 1978). Biostratigraphic work by Gill and Cobban (1966) and physical stratigraphic work done by Asquith (1970, 1974) demonstrate that both the Shannon and Sussex were deposited in offshore positions, as much as 100 km east of the synchronous shoreline deposits of the Mesaverde Formation. Both sandstone units intertongue, landward as well as seaward, with marine mudrocks (Spearing, 1976; Brenner, 1978).

The model proposed by Spearing (1976) for the Shannon Sandstone Member consists of south-moving, lenticular sand bodies, driven by currents generated by storms and tides or oceanic circulation, over a progradational mud-rich shelf. Two distinct bedforms were recognized: a thin ripple-bedded series of transverse sand ribbons; and large-scale, cross-stratified longitudinal sand ridges (Fig. 56). Spearing (1976) envisioned these bedforms as migrating south-southwest as discrete sand patches in mid-shelf positions (Fig. 57). In the western part of the Powder River basin, where Spearing concentrated his efforts, two distinct

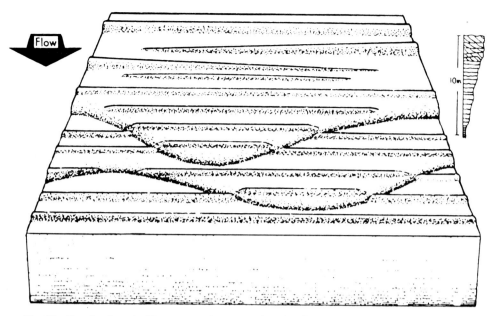

Fig. 56—Construction of a Shannon sand sequence by migrating sand patches consisting of small-scale (amplitudes from 1 to 10 cm) ripples and larger-scale (amplitudes up to 1 m) sand waves. From Spearing (1976).

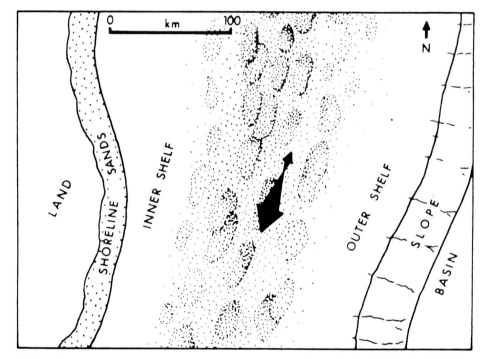

Fig. 57—Hypothetical paleogeographic reconstruction showing Shannon sand patches migrating south-southwest in mid-shelf position. From Spearing (1976).

vertical sequences also record the shift from dominance by major storm-generated currents, which winnowed the sands and formed discrete bodies, to dominance by longitudinal drift as the ridges aggraded up above fair-weather wave base.

The model proposed by Brenner (1978) for the Sussex Sandstone Member is similar in many respects to that proposed by Spearing (1976) for the underlying Shannon. The primary difference is in orientation of the sand bodies. Channel-like sequences observed on the Sussex outcrop and log patterns that suggest channels which correspond to transverse features, lead to a slightly different scenario. Sussux sandstone bodies form long (up to 60 km) linear northwest-southeast-oriented trends with lobate edges. These linear trends represent sand-ridge complexes, with cross-cutting channels that terminate into tidal deltas or washover fans (Fig. 59). Brenner's model calls on pulses of sedimentation creating a sandy mudsheet, which prograded southward and eastward from a siliciclastic source. Periodic storms winnowed out mud and piled sand-sized sediment into bars, while unusually intense storms breached pre-existing bars forming channels and channel-mouth fans. Fair-weather processes such as tides and regional drift apparently played minor roles in the creation of Sussex sand ridges, but may have played more significant roles in the southward movement of fine-grained sediment and the final shaping of ridge complexes (Brenner, 1978). It is conceivable that the geometric differences between the Shannon and Sussex sand bodies may stem from the Shannon being the product of inner shelf sedimentation, while the Sussex represents an outer shelf (D. R. Spearing, personal commun. 1978).

Both Shannon and Sussex sand bodies appear to have been formed during stillstands, which were punctuated by short-lived sediment input from a major point source to the northwest. This point source was probably a deltaic system located in present-day central Montana, perhaps related to the complex referred to by Shelton (1965) as the source of the Eagle Sandstone. Apparently a strong regional drift moved sediments southward from this point source during the Campanian.

Late Cretaceous sand bars, Denver Basin—The Hygiene Sandstone Member of the Pierre Shale contains sandstone bodies interpreted to have

coarsening-upward sequences can be recognized. Seismic data suggest that east of this study area, a third sandy interval developed at a slightly higher stratigraphic position, as the lowest sandstone "shaled-out" laterally (Brenner, 1978). These data further suggest that an offlap relationship exists between individual sandstone sequences (Fig. 58). A possible explanation of these characteristics of the Shannon, in terms of Spearing's model is that the period of shelf progradation and following sediment reworking by marine

currents was repeated at least twice. Thus, there may have been three cycles of deposition and reworking, each cycle displaced in a more seaward position as the depositional shelf continued to prograde (Fig. 58).

The multiple morphologies displayed in the Shannon suggest that at least two different flow stresses dominated. The coarsening-upward textures and upward progression of bedforms from ripples to dunes were due to progressive shoaling of the sand bodies (Spearing, 1976). Perhaps these formed as shelf sand bars, below fair-weather wave-base level in the Late Cretaceous Campanian and early Maestrichtian seaway (Porter, 1976). The vertical sequence of upward textural coarsening and burrowed to rippled to large-scale cross-bedded stratification types show striking similarities to the slightly older Shannon and Sussex Sandstone Members of the Powder River basin. At the time these sands were being deposited in northeastern Colorado, the western shoreline of the epicontinental seaway was located 100 km or more to the west, extending from eastern Utah to central Wyoming. Hygiene siliciclastics

were supplied from a large deltaic depocenter in south-central Wyoming, and fine-grained sediments were swept southeastward across the muddy shelf by "shelf currents" (Fig. 60), while storm-intensified shelf currents moved sand-sized material (Porter, 1976).

Porter's (1976) model for shelf sand bar deposition calls on storm-generated currents to winnow and redeposit sand in locally developed bars. Material scoured from interbar areas and thrown into suspension during storm activity, settled out to form a rippled, muddy layer over part of the shelf in the wake of each storm. Repeated storm events were responsible for

most of the bar movement and associated deposition. As in the Upper Jurassic Oxfordian sand bars, little if any deposition is attributed to fair-weather processes. A form of river-mouth bypass, associated with storm-driven currents, moved sediment from the deltaic depocenter into the inner shelf regime (Brenner, 1980).

Late Cretaceous sand bars, east-central Utah—The Mancos Shale in the northern part of Castle Valley, Utah, contains a sandy interval known as the Ferron Sandstone Member. Cotter (1975) divided this member into four genetic units based on environmental interpretations made from lithologic and paleontologic analyses. The paleoenvironments ranged from a nearshore lower shoreface, seaward to an offshore sand-bar complex on a shallow marine shelf. This offshore sand-bar complex, known as the Woodside unit, consists of three thin coarsening-upward sequences of sandstone separated by bioturbated dark siltstone. Cotter traced these units northwestward toward the Late Cretaceous shoreline and found that they grade laterally into marine siltstone, demonstrating their separation from lower shoreface sandstone units. Although the sandstone units are thinner, their vertical sequence of textures and sedimentary structures are similar to those found in the Sundance, Shannon, Sussex, and Hygiene sandstones (Fig. 61).

Cotter (1975) postulated that Ferron sediments were derived from the Vernal delta located to the north and west, and were transported by longshore drift in a general southwesterly direction. During storms, sediment from the nearshore regime was thrown into suspension and moved seaward where it

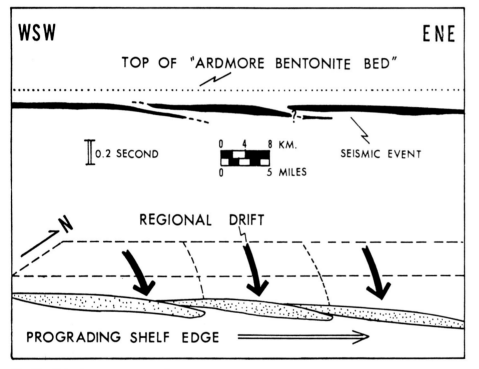

Fig. 58—Seismic events representing the tops of three sandstone units within the Shannon interval, interpreted in terms of three cycles of deposition and sediment reworking. Modified from Brenner (1978).

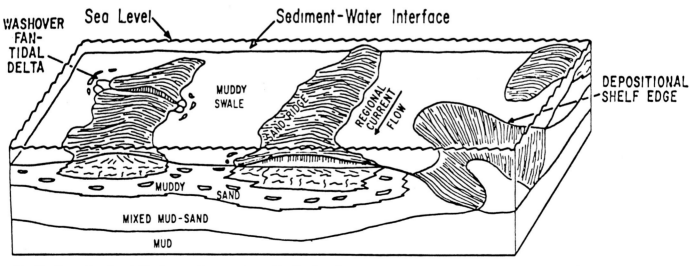

Fig. 59—Depositional model for Sussex ridge complexes. Modified from Brenner (1978).

settled out as storms abated. During fair-weather periods, biogenic activities homogenized thin storm-related strata and reworked upper parts of thicker units.

The Turonian epicontinental seaway in Utah was apparently characterized by a depositional shelf much like the Campanian shelf in Wyoming. Storm-generated currents created and maintained sand ridges similar to those of storm-dominated continental shelves of today. The complete facies pattern of the Ferron Sandstone Member shows offshore bars of the Woodside unit to be part of a nearshore-inner shelf facies tract, which as Cotter (1975) pointed out, has many similarities with the offshore zone near Sapelo Island, Georgia.

Comparisons between ancient epicontinental shelf deposits—Ripple cross-laminated, fine-grained sandstone (Fig. 62) of the Sundance Formation in Wyoming commonly has large-scale festoon cross-beds (Fig. 63) on top. Coquinoid sandstone beds commonly interrupt vertical sequences (Fig. 64), as biogenic materials fill channels cut into the bars (Fig. 65). Some channel fills consist of well-

sorted sandstone and mudrock interbeds (Fig. 66). Trace fossils may be well preserved (Fig. 67). Some of the channel features may have a shale conglomerate lag (Fig. 68). Bar sequences themselves show a variety of ripples (Fig. 69) and cross-beds (Fig. 70).

Vertical sequences of features are repeated, with modifications, from place to place. Cut-and-fill features, such as the coquinoid beds in the Sundance Formation (Fig. 64) and cross-bedded units in the Sussex Sandstone Member (Fig. 68), are the least continuous. Coquinoid sandstone beds in the Sundance Formation have cross-sectional widths from a few meters (Fig. 64A) to a few kilometers. Tracing individual co-

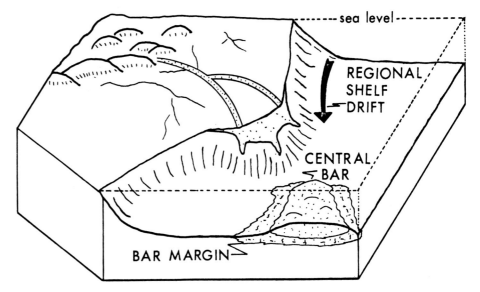

Fig. 60—Depositional model for the Hygiene bar complexes. The single bar complex is representative of many such complexes. Modified from Porter (1976).

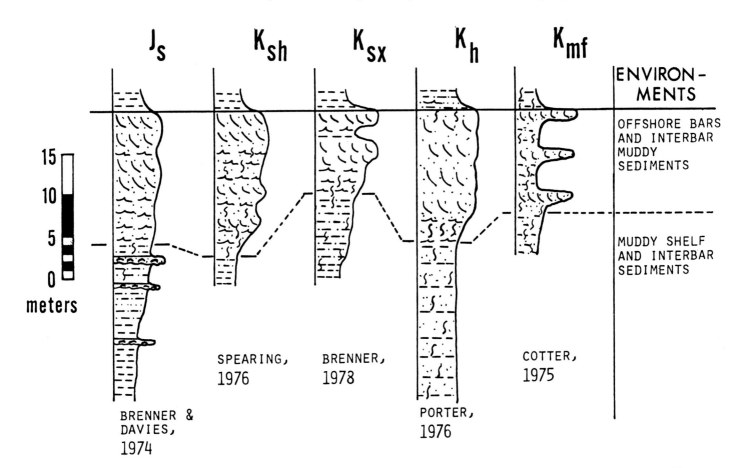

Fig. 61—Comparison of thickness and lithologic sequences for sandstones in the Sundance Formation (Js), Shannon (Ksh), and Sussex

(Ksx) Sandstone Members of the Steele Shale, Hygiene Sandstone Member of the Pierre Shale

(Kh), and Ferron Sandstone Member of the Mancos Shale (Kmf) sandstones.

Fig. 62—Ripple cross-laminated, fine-grained sandstone within an Oxfordian marine bar vertical sequence.

Fig. 63—Festoon cross-bedded sandstone from the upper portion of an Oxfordian marine bar vertical sequence.

Fig. 64—Bioclastic-rich sandstone filling channels cut into Oxfordian marine bars: **A.** Small-scale channels filled wtih coquinoid sandstone, west flank of Big Horn Mountain, Wyoming. **B.** Large-scale multiple channel-fills from the east flank of the Wind River Range, Wyoming. **C.** Close-up of basal contact of lower channel shown in B. **D.** Close-up of base of coquinoid-filled channel cutting ripple-laminated sandstone, Wyoming. (A and B from Brenner and Davies, 1973a; C from Brenner and Davies, 1974).

Fig. 65—Sheet-like coquinoid sandstone shell lag from west flank of Big Horn Mountains, Wyoming. From Brenner and Davies (1973a).

Fig. 66—Channel cut into ripple-laminated sandstone and filled with interstratified sandstone and sandy mudstone, Pryor Mountains, Montana.

Fig. 67—Bioturbation in the Sussex Sandstone Member of the Steele Shale, exposed on the southeastern flank of Salt Creek anticline, Wyoming. **A.** Calcite-cemented burrows in sandstone which is interbedded with shale. **B.** Bioturbated fine-grained sandstone and mudstone. From Brenner (1979).

Fig. 68—Shale-chip conglomerate at the base of channel sequence cutting into marine bar sandstone of the Sussex Sandstone Member of the Steele Shale. Eastern flank of Salt Creek anticline, Wyoming. From Brenner (1978).

Fig. 69—Ripple bedforms from the lower part of a Sussex marine bar sequence, Sussex Member of the Steele Shale. Eastern flank of Salt Creek anticline, Wyoming. From Brenner (1978).

quinoid beds laterally on the exceptionally continuous outcrops along the flanks of the Big Horn and Wind River mountain ranges of Wyoming has shown that they terminate either sharply, along the sides of channels (Fig. 64A), or grade gradually into coarser sandstone units of marine bars (Fig. 71). The bar-forms themselves are elongate lenses, with lengths in excess of 5 km and widths ranging from 200 m to at least 2,000 m in sections normal to sand-bar trends (Brenner and Davies, 1974).

Muddy parts of ancient shelves are generally poorly exposed in outcrops and rarely studied in the subsurface. As a result, little is known about sequences of textures and structures. The few outcrops that have been studied show features similar to those found in nearshore deposits. Finely interstratified siltstone and claystone, massive mudstone, thin interbeds of fine-grained sandstone and mudstone or shale (Fig. 72), and flaser-bedded sandstone and shale (Fig. 73) appear to be the most common features. These characteristics of muddy shelf areas from studies of Jurassic and Cretaceous rocks reflect the generally low energy conditions that prevailed, punctuated periodically by higher energy conditions (storms and perhaps spring tides) that washed silt and sand in from nearby sand bar areas. In southeastern Wyoming, Specht and Brenner (1979) have described bioclastic limestone beds (Fig. 74) which form lensoid blankets within Oxfordian shelf mudstones. These carbonate units are relatively thin (0.02-1.0 m) and discontinuous. They appear to represent the effects of sediment flushing and winnowing that took place as large storm-generated waves passed over the shelf.

Sediment Texture

Sediment types vary considerably from one shelf to another (compare Figs. 3, 4, 5, 7, 9, and 11). The study published by Berryhill et al (1976), clearly shows local variations in textural ratios and parameters within the overall rather straight forward distribution pattern of sediment types. Interaction between available sediment and dynamic conditions determines if conditions are in equilibrium. Many relict sands certainly are not currently in dynamic equilibrium with their microenvironments, and as a consequence can be moved locally. This often results in high degrees of sorting. Excellent sorting can cause an increase in heavy mineral and metal concentrations. Many of these sand deposits are thin, but their degrees of sorting and eventual coverage by fine-grained deposits can provide good stratigraphic traps for hydro-

Fig. 70—Cross-bedded sandstone in the upper part of a marine bar sequence, Sussex Member of the Steele Shale. Eastern flank of Salt Creek anticline, Wyoming. From Brenner (1978).

Fig. 71—Lateral termination of a coquinoid sandstone bed within an Oxfordian marine bar sequence. Pryor Mountains, Montana.

Fig. 72—Interstratified silty sandstone and shale from the Oxfordian portion of the Sundance Formation.

Fig. 73—Flaser-bedded, ripple-laminated sandstone and shale from the Oxfordian marine-bar facies, Montana. From Brenner and Davies (1974).

carbon accumulation (Figs. 2, 4, 6, 7, 13, 16, 17, and 27).

When the overall type of sediment is classified as a clayey silt, as in the case of the South Texas OCS area, standard deviation or sorting can be poor, although values may range from 1.51 ϕ to 3.88 ϕ (Fig. 11, Berryhill et al, 1976, Fig. 32). Figures 11 and 13 show coarse sediments are present along the shore and locally offshore, but that sorting may not follow the same trend. Well-sorted sediment can be sand, silt or clay, which cannot be seen in a plot of sorting values.

Thus, well-sorted sandy shelf deposits are found on the outer shelf and locally on other sections as river and bar deposits. All these coarse-grained shelf sands, except for coastal bars, appear to result from a regression.

Well-log Responses

By relating well cores and sequences observed on outcrops to well-log curve characteristics, Brenner (1978) calibrated well-log responses to lithologic features in the Cretaceous Sussex Sandstone Member. Five recurring characteristic log responses were recognized using gamma-ray or spontaneous potential (SP) logs in conjunction with resistivity logs. Units with high resistivities, low gamma counts, or significant SP curve development, were labeled "clean" sandstone, and represent cross-stratified, well-sorted sandstone units; ripple-laminated, shale-free, well-sorted sandstone units; and perhaps conglomerate sandstone units. Intervals characterized by irregular (wavy) but low count gamma-ray, poorly developed SP and intermediate resistivity curves, were labeled "shaly" sandstone, and represent: ripple-laminated sandstone units with shaly partings; and bioturbated shaly-sandstone units. Intervals with similar irregular (wavy) curves, but with higher gamma-counts, little if any SP curve development, and lower resistivities, were labeled "mixed" sandstone and mudrock, and represent: bioturbated sandy or silty mudstone; and shale units with thin lenses of fine-grained sandstone or siltstone. Units with high gamma counts, no SP development, and low resistivities represent shale and clayey mudstone. Extremely high gamma counts and very low (the lowest recorded) resistivities represent bentonite strata. These contrasting responses for various rock types allowed coarsening-upward bar sequences and fining-upward channel and lag deposits to be recorded on well logs (Fig. 75). Similar, upward-coarsening sequences were delineated in the Shannon Sandstone Member studied by Spearing (1976) using similar techniques.

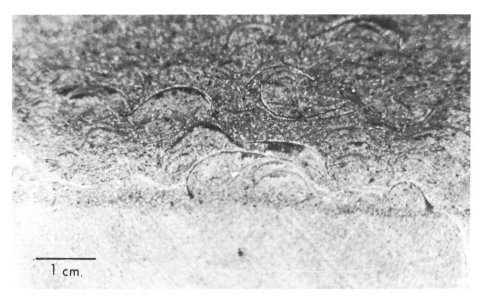

Fig. 74—Basal portion of a graded bioclastic limestone within the Oxfordian muddy shelf facies of southeastern Wyoming.

ECONOMIC CONSIDERATIONS

Petroleum Potential

It is difficult to access the economic meaning of ancient shelf deposits, because many may not have been identified as shelf deposits. However, the potential for economic oil and gas accumulations in sandstone facies of ancient shelf deposits is high. This setting, whether it be epicontinental or pericontinental, provides the four main ingredients for petroleum accumulations: (1) potential reservoirs; (2) potential hydrocarbon source rocks; (3) potential trapping situations; and (4) the time and depth of burial required to generate petroleum. Reservoir potential is high because of marine sorting mechanisms that produce relatively clean, well-sorted sands making up the upper parts of marine bars, and basal parts of channel-fills. Surrounding marine muds, representing relatively low rates of sediment accumulation under

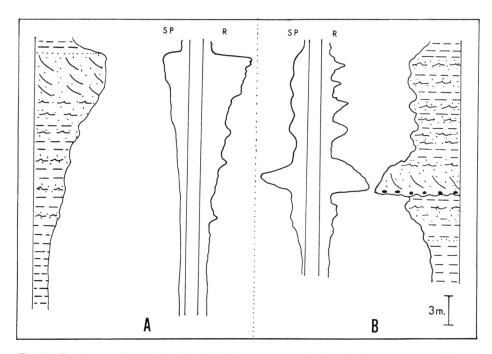

Fig. 75—Spontaneous Potential and Resistivity Log responses for **A,** marine bar sequence and **B,** channel-fill sequence of the Sussex Sandstone Member of the Steele Shale, Powder River basin, Wyoming. From Brenner (1979).

low energy conditions, may contain significant quantities of organic carbon compounds. These same muds become relatively impermeable mudrocks after compaction, forming seals for stratigraphic traps. Intrabasinal tectonism may enhance trap potential by creating structures which increase reservoir hydrocarbon capacities as well as create additional trapping situations and migration pathways. Burial, which comes with time and continued deposition, may be enhanced by tectonism, and provides the heat required to release petroleum from organic-rich source rocks.

The Cretaceous Sussex and Shannon Sandstone Members of the Powder River basin, Wyoming, were both studied, in part, because of their petroleum deposits and potential. Cumulative production from these two sandstones from eight Powder River basin fields (House Creek, Heldt Draw, Holler Draw, Jepson Draw, Triangle U, Flying E, West House Creek, and East Heldt Draw) totals 10,360,817 bbl of oil plus 4,786,346 MMcf of gas, as of March 1, 1976 (Crews et al, 1976). After considering that these numbers, small in comparison to other types of deposits in the area, represent two relatively thin sandy intervals deposited during a part of one Cretaceous age in a small part of one epicontinental seaway shelf, the overall potential of similar deposits is large.

Relation to Mineral Deposits

Because relict offshore coarse-grained shelf deposits are seldom in equilibrium with dynamic conditions, reworking is common. Heavy minerals and metals, if present, can become concentrated as lag deposits.

Placer deposits of gold are found in some of the drowned beaches off Nome, Alaska. Tin comes from alluvial deposits in Indonesia and Thailand that were transported to shelf areas during the Pleistocene. The same fluvial origin is true for diamonds off South Africa.

Phosphate nodules are found off southern California in the San Diego area, and iron is mined in small quantities near Conception Bay, Northeast Newfoundland.

REFERENCES CITED

Allison, R. C., 1976, Late Oligocene through Pleistocene molluscan faunas in the Gulf of Alaska region (abs.): First Intern. Cong. Pacific Neogene Stratigraphy, Tokyo, p. 10-13.

Asquith, D. O., 1970, Depositional topography and major marine environments, Late Cretaceous, Wyoming: AAPG Bull., v. 54, p. 1184-1224.

——— 1974, Sedimentary models, cycles, and deltas, Upper Cretaceous, Wyoming: AAPG Bull., v. 58, p. 2274-2283.

Berryhill, H. L., Jr., et al, 1976, Environmental studies, South Texas Outer Continental Shelf, 1975: Geology: National Technical Information Service, Springfield, Va., pub. No. PB 251-341, 353 p.

——— 1977, Environmental studies, South Texas Outer Continental Shelf, 1976: Geology: National Technical Information Service, Springfield, Va., pub. No. PB 289-144, 306 p.

Bothner, M., et al, 1979, Texture, clay mineralogy, trace metals, and age of cored sediments from the North Atlantic Outer Continental Shelf, in Final Report to the Bureau of Land Management: Woods Hole, Massachusetts, U.S. Geol. Survey, p. 6-1 to 6-15 (unpub.).

Bouma, A. H., M. A. Hampton, and R. C. Orlando, 1977, Sand waves and other bedforms in lower Cook Inlet, Alaska: Marine Geotechnology, v. 2, p. 291-308.

——— et al, 1978, Movement of sand waves in lower Cook Inlet, Alaska: Preprints Offshore Technology Conf., Paper 3311, p. 2271-2284.

——— 1979, Bedform characteristics and sand transport in a region of large sand waves, lower Cook Inlet, Alaska: Preprints Offshore Technology Conf., Paper 3485, p. 1083-1094.

Brenner, R. L., 1978, Sussex sandstone of Wyoming: an example of Cretaceous offshore sedimentation: AAPG Bull., v. 62, p. 181-200.

——— 1979, A sedimentologic analysis of the Sussex Sandstone, Powder River Basin, Wyoming: Earth Science Bull., v. 12, p. 36-47.

——— 1980, Construction of process-response models for ancient epicontinental seaway depositional systems using partial analogs: AAPG Bull., v. 64, p. 1223-1244.

——— and D. K. Davies, 1973a, Storm generated coquinoid sandstone: Genesis of high energy marine sediments from the Upper Jurassic of Wyoming and Montana: Geol. Soc. America Bull., v. 84, no. 5, p. 1685-1698.

——— 1973b, An ancient interior seaway and modern continental shelves: Analogies of sedimentary motif and dynamics (abs.): Geol. Soc. America Abs. with Programs for 1973 North Central Section, Columbia, Mo., p. 302.

——— 1974, Oxfordian sedimentation in Western Interior United States: AAPG Bull., v. 58, p. 444-467.

Bumpus, D. F., 1973, A description of the circulation of the continental shelf of the east coast of the United States: Progress in Oceanography, v. 6, p. 111-156.

Butman, B., M. Noble, and D. W. Folger, 1977, Long term observations of bottom current and bottom sediment movement on the Middle Atlantic continental shelf, in Middle Atlantic Outer Continental Shelf Environmental Studies, Geologic Studies, vol. III: Gloucester Point, Va., Virginia Institute Marine Sci., p. 2-1 to 2-61.

Cotter, E., 1975, Late Cretaceous sedimentation in a low-energy coastal zone: the Ferron Sandstone of Utah: Jour. Sed. Petrology, v. 45, p. 669-685.

Crews, G. C., J. A. Barlow, Jr., and J. D. Haun, 1976, Upper Cretaceous Gammon, Shannon, and Sussex sandstones, Central Powder River Basin, Wyoming: Wyoming Geol. Assoc. 28th Annual Field Conference, p. 9-19.

Curray, J. R., 1960, Sediments and history of Holocene transgression, continental shelf, Northwest Gulf of Mexico, in Recent Sediments, Northwest Gulf of Mexico: AAPG Special Pub., p. 221-266.

Davies, D. K., and R. L. Brenner, 1973, Role of storms in development of ancient marine ridge and swale system (abs.), AAPG Bull., v. 57, p. 775.

Dott, R. H., Jr., and M. A. Roshardt, 1972, Analysis of cross-stratification orientation in the St. Peter Sandstone in southwestern Wisconsin: Geol. Soc. America Bull., v. 83, p. 2589-2596.

Duane, D. B., et al, 1972, Linear shoals on the Atlantic inner continental shelf, Florida to Long Island, in D. J. P. Swift, D. B. Duane, and O. H. Pilkey, eds., Shelf Sediment Transport: Process and Pattern: Stroudsburg, Pa., Dowden, Hutchinson, and Ross, Inc., p. 447-498.

Emery, K. O., 1960, The sea off southern California, a modern habitat of petroleum: New York, John Wiley & Sons, Inc., 366 p.

——— 1977, Structure and stratigraphy of divergent continental margin, in E. McFarlan, Jr., C. L. Drake, and L. S. Pittman, eds., Geology of Continental Margins: AAPG Cont. Ed. Course Note Ser. No. 5, p. B1-B20.

——— and E. Uchupi, 1972, Western North Atlantic Ocean: Topography, rocks, structure, water life, and sediments: AAPG Mem. 17, 532 p.

Exum, F. A., 1973, Lithologic gradients in marine bar, Cadeville Sand, Calhoun Field, Louisiana: AAPG Bull., v. 57, p. 301-310.

Folger, D. W., et al, 1978, Environmental hazards on the Atlantic outer continental shelf of the United States: Preprints Offshore Technology Conf., Paper no. 3313, p. 2293-2298.

Gill, J. R., and W. A. Cobban, 1966, Regional unconformity in Late Cretaceous, Wyoming: U.S. Geol. Survey Prof. Paper 550-B, p. 20-27.

Grow, J. A., R. E. Mattick, and J. S. Schlee, 1979, Multichannel seismic depth sections and interval velocities over outer continental slope between Cape Hatteras and Cape Cod, *in* J. R. Watkins, L. Montedert, and P. W. Dickerson, eds., Geological and Geophysical Investigations of Continental Margins: AAPG Mem. 29, p. 65-83.

Hampton, M. A., and A. H. Bouma, 1977, Slope instability near the shelf break, western Gulf of Alaska: Marine Geotechnology, v. 2, p. 309-331.

———— and R. von Huene, 1979, Geo-environmental assessment of the Kodiak Shelf, western Gulf of Alaska: Preprints Offshore Technology Conf. Paper 3399, p. 365-376.

Harms, J. C., 1975, Stratification and sequence in prograding shoreline deposits, *in* J. C. Harms, J. B. Southard, D. R. Spearing, and R. G. Walker, compilers, Depositional Environments as Interpreted from Primary Sedimentary Structures and Stratification Sequences: SEPM Short Course No. 2, 161 p.

Hathaway, J. C., et al, 1976, Preliminary summary of the 1976 Atlantic margin coring project of the U.S. Geological Survey: U.S. Geol. Survey open-file rept. 76-844, 217 p.

Hobday, D. K. and H. G. Reading, 1972, Fair weather versus storm processes in shallow marine sand bar sequences in the Late Precambrian of Finmark, north Norway: Jour. Sed. Petrology, v. 42, no. 2, p. 318-324.

Houbolt, J. J. H. C., 1968, Recent sediments in the southern bight of the North Sea: Geologie en Mijnbouw, v. 47, p. 245-273.

Johnson, H. D., 1977, Shallow marine sand bar sequences: an example from the Late Precambrian of north Norway: Sedimentology, v. 24, p. 245-270.

Jordan, G. F., 1972, Large submarine sand waves: Science, v. 136, p. 839-848.

Klein, G. D., and T. A. Ryer, 1978, Tidal circulation patterns in Precambrian, Paleozoic, and Cretaceous epeiric and mioclinal shelf areas: Geol. Soc. America Bull., v. 89, p. 1050-1058.

Knebel, H. J., et al, 1976, Maps and graphic data related to geologic hazards in the Baltimore Canyon Trough area: U.S. Geol. Survey Misc. Field Studies Map MF-828, 3 sheets.

———— and E. Spiker, 1977, Thickness and age of surficial sand sheet, Baltimore Canyon Trough area: AAPG Bull., v. 61, p. 861-871.

Knott, S. T., and H. Hoskins, 1968, Evidence of Pleistocene events in the structure of the continental shelf off the northeastern United States: Marine Geology, v. 6, p. 5-43.

Lankford, R. R., and J. J. W. Rogers, 1969, Summary, *in* R. R. Lankford and J. J. W. Rogers, eds., Holocene Geology of the Galveston Bay area: Houston Geol. Soc., 141 p.

Lewis, R. S., and R. E. Sylwester, 1976, Shallow sedimentary framework of Georges Bank: U.S. Geol. Survey open-file rept. 76-874, 12 p.

McClennen, C. E., 1973, Nature and origin of the New Jersey continental shelf topographic ridges and depressions: PhD thesis, Rhode Island Univ. 94 p.

Moody, D. W., 1964, Coastal morphology and processes in relation to the development of submarine sand ridges off Bethany Beach, Delaware: PhD thesis, Johns Hopkins Univ., 167 p.

Narayan, J., 1971, Sedimentary structures in the Lower Greensand of the Weald and Bas-Boulonnais, France: Sedimentary Geology, v. 6, p. 73-109.

Nilsen, T. H. and G. W. Moore, 1979, Reconnaissance study of Upper Cretaceous to Miocene stratigraphic units and sedimentary facies, Kodiak and adjacent islands, Alaska, with a section on sedimentary petrography by G. R. Winkler: U.S. Geol. Survey Prof. Paper 1093, 34 p.

Nio, Swie-Djin, 1976, Marine transgression as a factor in the formation of sandwave complexes: Geologie en Mijnbouw, v. 55, p. 18-40.

Oldale, R. N., et al, 1974, Geophysical observations on northern part of Georges Bank and adjacent basins of Gulf of Maine: AAPG Bull., v. 58, p. 2411-2427.

Pilkey, O. H., et al, 1979, History of the Georgia embayment sediment cover: Geol. Soc. America Abs. with Programs, v. 11, No. 4, p. 208.

Porter, K. W., 1976, Marine shelf model, Hygiene member of Pierre Shale, Upper Cretaceous, Denver Basin, Colorado, *in* R. C. Epis and R. J. Weimer, eds., Studies in Colorado Field Geology: Prof. Contrib. of Colo. Sch. Mines, No. 8, p. 251-263.

Price, W. A., 1958, Sedimentology and Quaternary geomorphology of south Texas: Trans. Gulf Coast Assn. Geol. Soc., v. 8, p. 41-75.

Pryor, Q. A., and E. J. Amarel, 1971, Large-scale cross-stratification in the St. Peter Sandstone: Geol. Soc. America Bull., v. 82, p. 229-244.

Reid, S. A., 1978, Mid-Tertiary depositional environments and paleogeography along upper Sespe Creek, Ventura County, California, *in* A. E. Fritsche, ed., Depositional Environments of Tertiary Rocks along Sespe Creek, Ventura County, California: SEPM Pacific Section, Pacific Coast Paleogeography Field Guide 3, p. 27-41.

Sanders, J. E., 1962, North-south trending submarine ridge composed of coarse sand off False Cape Virginia (abs.): AAPG Bull., v. 46, p. 278.

Schlee, J. S., 1973, Atlantic continental shelf and slope of the United States—sediment texture of the northeastern part: U.S. Geol. Survey Prof. Paper 529-L, 64 p.

———— and R. M. Pratt, 1970, Atlantic continental shelf and slope of the United States—gravels of the northeastern part: U.S. Geol. Survey Prof. Paper 529-H, 39 p.

Scott, J. T., and G. T. Csanady, 1976, Nearshore currents off Long Island: Jour. Geophys. Research, v. 81, p. 5401-5409.

Shelton, J. W., 1965, Trend and genesis of lowermost sandstone unit of Eagle Sandstone at Billings, Montana: AAPG Bull., v. 49, p. 1385-1397.

Shepard, F. P., 1963, Submarine geology: New York, Harper and Row, 557 p.

———— 1977, Geological oceanography: evolution of coasts, continental margins, and the deep-sea floor: New York, Crane, Russak and Co., Inc., 214 p.

Spearing, D. R., 1976, Upper Cretaceous Shannon Sandstone: an offshore shallow-marine sand body: Wyoming Geol. Assoc. 28th Annual Guidebook, p. 65-72.

Specht, R. M., and R. L. Brenner, 1979, Genesis of bioclastic limestone lithosomes in Upper Jurassic epicontinental mudstones, southeastern Wyoming: Jour. Sed. Petrology, v. 49, p. 1307-1322.

Stahl, L., J. Koczan, and D. J. P. Swift, 1974, Anatomy of a shoreface-connected sand ridge on the New Jersey shelf: implications for the genesis of the shelf surficial sand sheet: Geology, v. 2, p. 117-120.

Stetson, H. C., 1953, The continental terrace of the western Gulf of Mexico: its surface sediments, origin and development, *in* H. C. Stetson and P. D. Trask, The Sediments of the Western Gulf of Mexico: Papers in Physical Oceanography and Meterology, Mass. Inst. of Tech. and Woods Hole Ocean. Inst., v. XII, No. 4, 120 p.

Stubblefield, W. L., J. W. Lavelle, and D. J. P. Swift, 1975, Sediment response to the present hydraulic regime on the central New Jersey shelf: Jour. Sed. Petrology, v. 45, p. 337-358.

Swett, K., G. D. Klein, and D. E. Smit, 1971, A Cambrian tidal sand body—the Eriboll Sandstone of northwest Scotland: an ancient—recent analog: Jour. Geology, v. 79, p. 400-415.

Swift, D. J. P., 1976, Continental shelf sedimentation, *in* D. J. Stanley and D. J. P. Swift, eds., Marine Sediment Transport and Environmental Management: New York, John Wiley and Sons, p. 311-350.

———— et al, 1972, Holocene evolution of the shelf surface, central and southern Atlantic shelf of North America, *in* D. J. P. Swift, D. B. Duane, and O. H. Pilkey, eds., Shelf Sediment Transport: Process and Pattern: Stroudsburg, Pa., Dowden, Hutchinson, and Ross., Inc., p. 499-574.

Uchupi, E., 1968, Atlantic continental shelf and slope of the United States—physiography: U.S. Geol. Survey Prof. Paper 529-C, 30 p.

———— 1970, Atlantic continental shelf and slope of the United States—shallow structure: U.S. Geol. Survey Prof. Paper 529-I, 44 p.

Veatch, A. D., and P. A. Smith, 1939, Atlantic submarine valleys of the United States and the Congo Submarine Valley: Geol. Soc. America Spec. Paper 7, 101 p.

Wood, S. A., et al, 1979, Submersible observations on Georges Bank, *in* Final Report to the Bureau of Land Management: Woods Hole, Ma., U.S. Geol. Survey, p. 5-1 to 5-10 (unpub.).

———— and D. W. Folger, 1979, Submersible observations of the bottom in lease areas in the Baltimore Canyon Trough, *in* Middle Atlantic Outer Continental Shelf Environmental Studies, v. III, Geologic Studies: Gloucester Point, Va., Virginia Inst. Marine Sci., p. IV-1 to IV-13.

Characteristics of Sediments on Modern and Ancient Continental Slopes

Harry E. Cook
Michael E. Field
James V. Gardner
U.S. Geological Survey
Menlo Park, CA

INTRODUCTION

The need to expand the search for energy resources in deeper marine environments has intensified the importance of better understanding the nature and origin of continental slope settings and of acquiring a working knowledge of their characteristics. It has been known for a number of years that coarse-grained mass-flow deposits beyond the shelf break can form major petroleum reservoirs (Barbat, 1958), and it is likely that these deep-water environments will continue to be future exploration targets (Hedberg, 1970; Curran et al, 1971; Gardett, 1971; Nagel and Parker, 1971; Yarborough, 1971; Cooke et al, 1972; Schlanger and Combs, 1975; Enos, 1977; Walker, 1978; Wilde et al, 1978). Recently, however, with the concept of plate tectonics, seismic stratigraphy, and advances in seismic-reflection technology, there has emerged a more sophisticated approach to understanding the developments of continental margins. This understanding has placed more emphasis on the geological history and petroleum potential of continental slopes (Burk and Drake, 1974; Weeks, 1974; Bouma et al, 1976; Thompson, 1976; Bloomer, 1977; Schlee et al, 1977; Mattick et al, 1978).

This paper presents a practical guide for recognizing continental slope sequences. Criteria are presented that we believe are common, or could be common, to all slopes regardless of whether they are adjacent to active, passive or buttressed continental margins.

PHYSIOGRAPHY AND TYPES OF CONTINENTAL SLOPES

World-Wide Distribution

Heezen et al (1959) defined continental slopes as those parts of the continental margin that have gradients of greater than 1:40 (1.5°). Continental slopes occur throughout the world as the steepest part of the descent from the continental shelf to the deep-sea floor. They cover less than 7% of the submerged surface of the earth, but their total length exceeds 110,000 km (Stanley, 1969), occupying a total area of over 60 million sq km. Continental slopes are relatively steep, typically 3 to 6°; canyons and escarpments commonly have slopes in exceess of 15°. Slopes typically begin at the shelf break at about 100 to 200 m depth where the gradient abruptly steepens

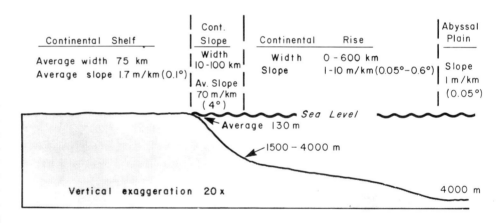

Fig. 1—The principal elements of a continental margin. (Modified after Drake and Burk, 1974.)

The authors tried to assemble a collection of illustrations documenting the stratigraphic and geographic diversity of slope sequences. Illustrations were solicited from numerous people. We are greatly appreciative of those who responded, because without their help a highly illutrated paper such as this would not have been possible. These people include: R. M. Carter; C. Cornford; R. H. Dott, Jr., R. W. Embley; G. deVries Klein; H. J. Knebel; F. Krause; F. Ricci Lucchi; and F. Picha. The donor of each illustration is acknowledged in the figure captions. A number of illustrations by the senior author are from the Eocene basin plain—submarine fan-slope system in the south-central Pyrenees of Spain. Special thanks go to Emiliano Mutti and C. Hans Nelson for sharing their ideas and preparing detailed road logs of key localities in this area.

Table 1, which provides a synthesis of this paper on continental slopes, is largely derived from Nardin et al (1979, Tables 1, 2, 3), with suggested modifications by M. A. Hampton. We have profited from discussions and reviews by P. R. Carlson, M. A. Hampton, D. Reed, P. A. Scholle, and D. R. Spearing. The writers appreciate very much the clerical help of Terry Coit, the drafting services of Lee Bailey, Jeanne Blank, and Phyllis Swenson, and the compilation of data and references by Michael White.

Table 1. Major Types of Submarine Mass Transport on Slopes and Suggested Criteria for their Recognition.[1]

TYPES OF MASS TRANSPORT		INTERNAL MECHANICAL BEHAVIOR	TRANSPORT MECHANISM AND DOMINANT SEDIMENT SUPPORT
ROCKFALL			FREEFALL AND ROLLING SINGLE BLOCKS ALONG STEEP SLOPES.
SLIDE	TRANSLATIONAL (GLIDE)	ELASTIC	SHEAR FAILURE ALONG DISCRETE SHEAR PLANES SUBPARALLEL TO UNDERLYING BEDS. SLIDE MAY BEHAVE ELASTICALLY AT TOP; PLASTICALLY AT BASE AND THIN LATERAL MARGINS.
	ROTATIONAL (SLUMP)		SHEAR FAILURE ALONG DISCRETE CONCAVE-UP SHEAR PLANES ACCOMPANIED BY ROTATION OF SLIDE. MAY MOVE ELASTICALLY OR ELASTICALLY AND PLASTICALLY.
SEDIMENT GRAVITY FLOW	DEBRIS FLOW OR MUD FLOW	PLASTIC	SHEAR DISTRIBUTED THROUGHOUT THE SEDIMENT MASS. CLASTS SUPPORTED ABOVE BASE OF BED BY COHESIVE STRENGTH OF MUD MATRIX AND CLAST BUOYANCY. CAN BE INITIATED AND MOVE LONG DISTANCES ALONG VERY LOW ANGLE SLOPES.
	GRAIN FLOW		COHESIONLESS SEDIMENT SUPPORTED BY DISPERSIVE PRESSURE. USUALLY REQUIRES STEEP SLOPES FOR INITIATION AND SUSTAINED DOWNSLOPE MOVEMENT.
	LIQUEFIED FLOW	FLUID	COHESIONLESS SEDIMENT SUPPORTED BY UPWARD DISPLACEMENT OF FLUID (DILATANCE) AS LOOSELY PACKED STRUCTURE COLLAPSES; SETTLES INTO A TIGHTLY PACKED TEXTURE. REQUIRES SLOPES $>3°$
	FLUIDIZED FLOW		COHESIONLESS SEDIMENT SUPPORTED BY UPWARD MOTION OF ESCAPING PORE FLUID. THIN (<10 CM) AND SHORT-LIVED.
	TURBIDITY CURRENT FLOW		CLASTS SUPPORTED BY FLUID TURBULENCE. CAN MOVE LONG DISTANCES ALONG LOW ANGLE SLOPES.

ACOUSTIC RECORD CHARACTERISTICS	SEDIMENTARY STRUCTURES AND BED GEOMETRY	REFERENCE TO FIGURES
Strong hummocky bottom return, hyperboale and side echoes common. Weak, chaotic internal return; structureless.	Grain supported framework, variable matrix, disorganized. May be elongate parallel to slope and narrow perpendicular to slope.	
Internal reflectors continuous and often undeformed; abrupt terminations. Strata of glide blocks may be unconformable or subparallel to underlying sediment.	Bedding may be undeformed and parallel to underlying beds or deformed especially at base and margins where debris flow conglomerate can be generated. Hummocky, slightly convex-up top, base subparallel to underlying beds; 10's to 1000's of meters wide and long.	27-31
Internal reflectors continuous and undeformed for short distances with deformation at toe and along base. Concave-up failure plane at head and subparallel to adjacent bedding at toe. Surface usually hummocky.	Bedding may be undeformed. Upper and lower contacts often deformed. Internal bedding at angular discordance to enclosing strata. Size variable.	15-26; 32-43; 56
Sea floor reflectors may be hyperbolic, irregular, or smooth. Commonly acoustically transparent with few or no internal reflectors. Mounded or lens shaped with blunt termination at head. May be chaotic internally.	Clasts matrix supported; clasts may exhibit random fabric throughout the bed or oriented subparallel, especially at base and top of flow units; inverse grading possible. Clast size and matrix content variable. Occur as sheet to channel-shaped bodies cm's to several 10's of meters thick and 100's to 1000's (?) of meters long; widths variable.	45-55; 58, 59, 62
Individual flow deposits very thin; may not be resolvable with present seismic-reflection techniques. Repeated flows may produce a sequence of thin, even, reflectors.	Massive; clast A-axis parallel to flow and imbricate upstream, inverse grading may occur near base.	68-70
	Dewatering structures, sandstone dikes, flame and load structures, convolute bedding, homogenized sediment.	72-75
Thin, even, continuous, acoustically highly reflective units; onlaps slope or raised topography. Discontinuous, migrating and climbing in channel sequences.	Bouma sequences. Mm's to several 10's of cm thick. 10's to 1000's of meters in length; widths variable.	57-61; 63-68; 76

[1]Modified from Nardin et al. 1979

Table 2. Seismic Characteristics and Environmental Facies Interpretation of Clastic Facies Units (modified after Sangree and Widmier, 1977).

Depositional Framework Interpretation	Seismic Facies Unit	Environmental Facies Interpretation	External Form of Facies Unit	Reflection Geometry at Boundaries
Shelf-margin and prograded slope	Sigmoid-progradational	Clay muds deposited by low energy turbidity currents and by hemipelagic deposition from low-velocity water currents. Shelfal undaform portions may involve wave and even fluvial transport processes.	Elongate lens to subtle fan	Concordant at the top with downlap at the base
Shelf-margin and prograded slope	Oblique-progradational	Sediment complex usually deposited in shelf margin deltaic environment: includes delta plain, delta front and prodelta processes. May also be formed in deep water associated with strong bottom currents.	Fan	Concordant at top if undaform cycles present
	Subzones undaform			
	Upper clinoform			Toplap truncation at the top
	Middle and lower clinoform			
	Fondoform			Downlap at the base
Basin-slope and basin-floor	Sheet-drape	Deep marine hemipelagic clays and oozes	Sheet-drape	Concordant at top and concordant or very slight onlap at base
Basin-slope and basin-floor	Slope-front fill	Deep-water sediment complex commonly related to submarine fans	Large fan	Concordant at top; onlaps updip and downlaps downdip
Basin-slope and basin-floor	Onlapping fill	Relatively low velocity turbidity current deposits	Basin trough channel and slope front fill	Onlap at the base and usually concordant at the top
Basin-slope and basin-floor	Mounded (fan complex)	Deep-water sediment complex commonly located at mouth of submarine canyon, composed of turbidites, mass-movement and hemipelagic deposits associated with major subaerial drainage systems.	Fan	Onlap of overlying units. Downlap at base.
Basin-slope and basin-floor	Mounded (contourite)	Deep-water sediment complex formed by deposition from deep-marine currents. Possibly composed primarily of fine-grained clastics.	Elongate mound	Truncated and concordant at top, downlap at base.
Basin-slope and basin-floor	Mounded onlapping fill	Relatively high velocity turbidity current deposits	Mounded; basin trough channel and slope front fill	Onlap at the base and concordant or erosional truncation at top
Basin-slope and basin-floor	Chaotic fill	Gravity mass transport and high energy turbidity current sediments

Lithology is function of upslope sediment source | Basin trough channel and slope front fill

Degree of mounding variable. Wavy sub-parallel chaotic pattern tends to be associated with smoother and lower mounds than contorted and discordant chaotic patterns. | Unit onlaps at base but individual onlap terminations are rare because of reflection pattern. Where preserved, reflection segments at the top may show concordance or erosional truncation. Mass transport gouge is common at base of contorted discordant pattern. |

Table 2. Continued.

Principal Internal Configuration	Lateral Relations	Other Seismic Facies Parameters		
		Amplitude	Continuity	Frequency (Cycle Breadth)
Sigmoid along depositional dip and parallel to sub-parallel along depositional strike	May grade laterally or vertically to oblique-progradational facies. Commonly onlapped by onlap-fill facies. Undaform part merges with shelf facies and fondoform part may grade to sheet-drape facies.	Generally moderate to high, relatively uniform	Normally continuous	Varies parallel to dip with broadest cycles associated with thicker beds of the middle clinoforming zone. Cycle breadth is uniform on sections parallel to depositional strike.
Parallel	Commonly merges downdip with deep basin turbidites, mass transport and hemipelagic facies. Frequently onlapped by onlapping-fill facies. May grade laterally or vertically to sigmoid-progradational facies. Undaform portion merges with parallel layered shelf facies.	Moderate to high	Generally continuous	Relatively uniform
Oblique along depositional dip and parallel or gently oblique to sigmoid parallel to depositional strike		Moderate to high	Generally moderately continuous	Fairly uniform
		Variable generally lower than other subzones	Discontinuous to moderate increases toward lower clinoform zone	Cycle breadth decreases rapidly downdip as beds thin
		Generally moderate to low	Continuous	Relatively narrow cycle breadth decreases basinward
Parallel	Commonly interbedded with turbidite sands and silts, and grades to gently divergent fondoform sediments of prograding complexes	Commonly relatively low to moderate	Continuous	Normally uniformly narrow
Parallel to sub-parallel	Thins and grades into basin floor facies, commonly pinches out updip	Variable	Variable	Variable
	Commonly grades to mounded onlap or chaotic fill facies. Alternation with other fill facies is common.	Variable	Commonly continuous	Cycle breadth increases into fill center trends to be relatively narrow
Extremely varied. Complex mounded	Located near submarine canyon. Commonly grades basinward to sheet drape facies.	Variable but tends to be low. Frequently decreases rapidly with increasing depth, suggesting high energy absorption.	Tends to be discontinuous	Highly variable
Asymmetric mounds	Thins and grades into basin floor facies	Variable	Variable	Variable
Irregularly mounded to parallel	Commonly grades to onlap fill or chaotic fill facies. Alternation with other fill facies is common.	Variable decreases as continuity decreases	Discontinuous to moderately continuous generally less than nonmounded onlap fill facies	Cycle breadth increases in fill center
Chaotic and contorted	May grade slopeward to lower and middle clinoform subzones of oblique progradational facies. Also may lie downdip of prominent detachment scars. Alternation with other fill facies is common.	Ranges from low to high in contorted discordant patterns and is generally low in wavy sub-parallel chaotic fill patterns	Very discontinuous short segments may occur in the contorted discordant patterns	Variable in contorted discordant facies reflecting internal heterogeneity more uniform in wavy sub-parallel chaotic fill pattern

DIVERGENT AND PARALLEL LAYERED

ONLAPPING FILL

OBLIQUE PROGRADATIONAL

SIGMOID PROGRADATIONAL

SHEET DRAPE

MOUNDED CHAOTIC

Fig. 2—Examples of typical seismic facies (scales vary; vertical exaggeration approximately 2.5:1). From Sangree et al (1976).

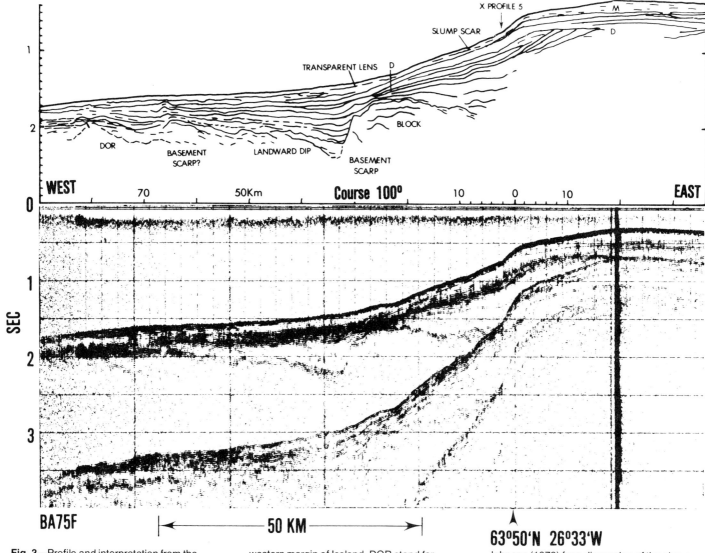

Fig. 3—Profile and interpretation from the continental shelf to the continental rise off the western margin of Iceland. DOR stand for deepest observed reflector. See Egloff and Johnson (1979) for a discussion of the above interpretation.

Table 3. Lithological Assemblage and Criteria for Distinguishing Submarine Outer Continental Margin Environments (from Stanley and Unrug, 1972).

MARINE OUTER MARGIN ENVIRONMENTS	DEPOSITIONAL THICKNESS	LATERAL STRATA PERSISTENCY	WEDGE-SHAPED DEPOSITS	CHANNEL DEPOSITS AND FLUXOTURBIDITES	SLUMP DEPOSITS	LARGE GRAVITY-SLIDE DEPOSITS	PROXIMAL TURBIDITES	DISTAL TURBIDITES	MUD BALLS	CONSTANT CURRENT DIRECTION INDICATORS	SCOUR MARKS DOMINANT	TOOL MARKS DOMINANT	FINELY LAMINATED HEMIPELAGIC SEDIMENTS	MAJOR FAUNAL BREAK	SKELETAL ABUNDANCE	REWORKED FLORA AND FAUNA	BIOTURBATE STRUCTURES	OUTCROPS OF OLDER ROCK
OUTER SHELF															15			
SHELF-BREAK																		18
UPPER & MID-SLOPE										10				14		16	17	
LOWER & BASE OF SLOPE	1		3	4	5	6	7	8	9		11	12	13					
BASIN FLOOR		2																

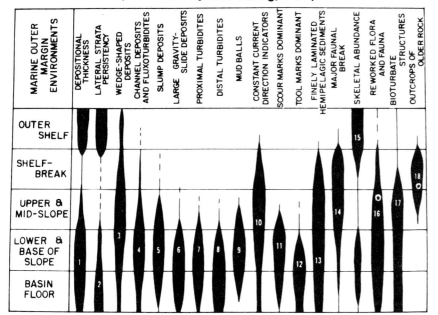

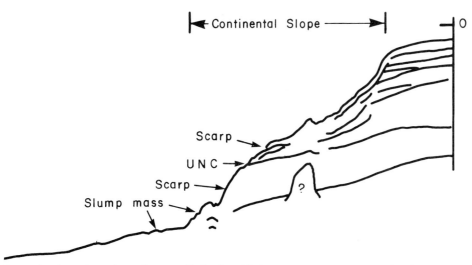

Fig. 4—Continental margin southeastern Bering Sea. UNC denotes a prominent unconformity.

(Fig. 1). They extend down to 1,500 to 4,000 m, and those adjacent to deep-sea trenches can descend to depths of 10,000 m. Widths of slopes range from 10 to almost 100 km. The base of the slope, where it grades into a continental rise or deep-sea fan, is marked by a recognizable, but less abrupt change in declivity.

Feature on Slopes

The most common features on continental slopes are networks of steep canyons and smaller gullies (Shepard, 1973). Canyons commonly traverse the entire slope perpendicular or at a high angle to the strike of the slope. Canyons and gullies are sometimes connected in tributary systems, in contrast to distributary systems that are diagnostic of deep-sea fans. Slope gullies and delta-front troughs are small, low-relief features that commonly do not extend to the base of the slope (Shepard, 1963). For additional descriptions of canyons and related features, see Emery (1960), Buffington and Moore (1963), Shepard and Dill (1966), and Kelling and Stanley (1976).

Other major features on continental slopes include basins, diapirs and uplifted fault blocks. Slope basins up to tens of square kilometers in area are formed by tectonic warping, damming of canyons, and local uplift or diapirism around a stable area (Bouma and Garrison, 1979). These small basins may vary significantly in mode of deposition and lithologic properties from the normal slope sequence. Mass transport of material is also a major process on continental slopes. Submarine slides have been reported from slopes having a broad range in depositional history, tectonic activity and gradient (see Lewis, 1971; Hampton and Bouma, 1977; Coleman and Prior, 1978). Intact slides range in size from tens of meters to tens of kilometers on a side (Carlson and Molnia, 1978; Field and Clark, 1979). Slides are found at all positions on the slope, whereas sediment gravity-flow deposits are most commonly found at the base of the slope.

Continental slopes possess essentially the same range of physiographic characteristics regardless of whether they occur on active, passive, or buttressed continental margins. Slopes do, however, exhibit differences in structural character or associated structural features on different types of margins. Active margins, those where subduction or transform faulting is occurring, typically have deep-sea terraces on the

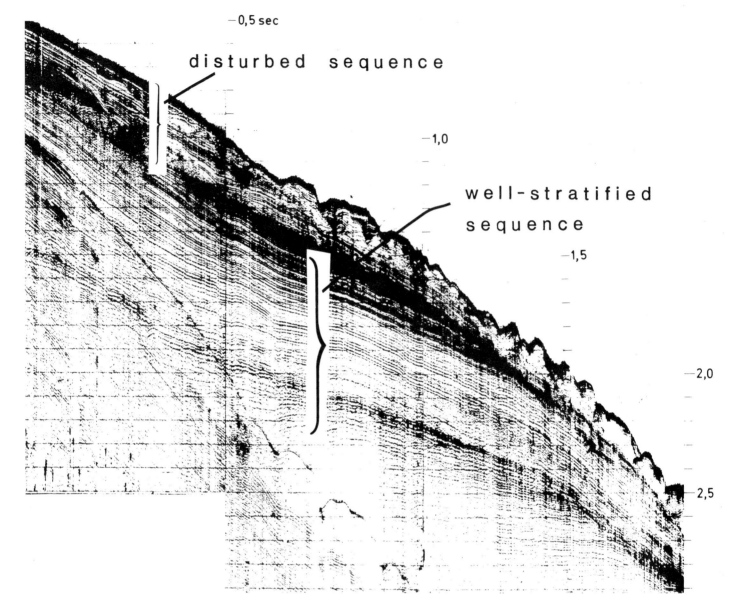

Fig. 5—Continental slope off Cape Bojador, west Africa. Note the acoustically well-stratified section below the uppermost more disturbed, slumped, and dissected surface section. After Seibold and Hinz (1974).

continental slope (Uyeda, 1974). These deep-sea terraces are not typically found on passive margins. Buttressed margins, such as the Gulf of Mexico and the southern California Borderland, commonly have horsts, diapiric intrusions or organic reefs that cause sediment to pond behind them. A continental slope with these features can be a complicated, broken-up physiographic province.

Passive margins typically have little large-scale surface structure, but rather show truncation of subsurface beds, sediment drape on the surface, and gravity-induced mass transport features on almost all scales.

MASS TRANSPORT PROCESSES ON SLOPES

Mass transport is used here for the *en masse* downslope movement of material containing various water amounts, for which gravity provides the driving force (Dott, 1963; Cook et al, 1972). A selected list of recent papers treating various aspects of mass transport includes: Dott (1963), Cook et al (1972), Hampton (1972, 1975), Middleton and Hampton (1973, 1976), Carter (1975), Wilson (1975), Lowe (1976a, b), Enos (1977), Varnes (1978), Cook (1979a), and Nardin et al (1979). The following discussion draws on these references. Table 1 summarizes the characteristics of the main types of mass transport found on slopes.

Mass transport can be divided into three types—rockfalls, slides and sediment gravity flows. Slides and sediment gravity flows can be further subdivided on the basis of their internal mechanical behavior and dominant sediment support mechanisms.

Rockfalls—also referred to as talus accumulations, are only common in marine environments at the base of steep slopes such as carbonate reefs, canyon walls, or fault scarps. Deposits of this type accumulate by the rolling or freefall of individual clasts.

Slides—can be divided into translational (glide) and rotational (slump) types (Varnes, 1978). The shear plane of a translational slide is predominantly along planar or gently undulatory surfaces parallel with underlying beds. Slumps or rotational slides exhibit concave-upward shear planes and usually a backward rotation of the slumped body. Slides can exhibit varied amounts of internal deformation. Some slides show purely elastic behavior, the

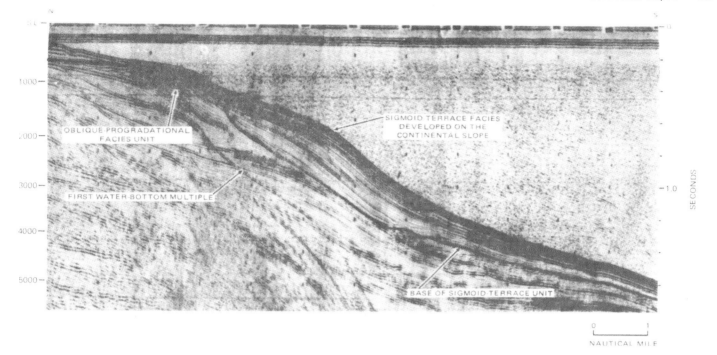

Fig. 6—Sigmoid-terrace acoustic facies on the continental slope of the Gulf of Mexico. A broad festoon of sigmoid reflections has its maximum thickness in about 550 m of water in this section, which was shot normal to the shelf edge. Note the thinning and onlap of reflectors on the front of an oblique-progradational facies. The sigmoid-terrace facies unit also offlaps downdip and thins to about 120 m at the right-hand edge of the figure. Similar sections can be seen on parallel lines from many kilometers along this part of the slope. From Sangree et al (1976).

original bedding virtually undisturbed except at the basal shear plane. Other slides behave in both an elastic and plastic manner, with semiconsolidated sediment deformed into overfolds. Some slides become so internally deformed that they are remolded into debris flows (Cook, 1977, 1979a, b).

Sediment gravity flows — (Middleton and Hampton, 1973, 1976) are divided into five types in Table 1. Their dominant internal mechanical behavior is plastic in the case of debris flows (the moisture of sediment and water has a finite strength). Liquefied flows, fluidized flows, and turbidity flows are considered to behave mainly as fluids (the sediment-water mixture has no internal strength). Grain flows behave either as a plastic or fluid. Hampton (1972, 1975), Middleton and Hampton (1976) and Lowe (1976a, b) give detailed treatments of sediment gravity flow processes.

It is important to stress that the classification shown in Table 1 represents end-member concepts. Several processes can operate simultaneously during mass movement. The name given to a deposit simply reflects the interpreted dominant transport mechanism. In addition, terminology described above has developed mainly from studies of ancient sediment. Some differences exist between mass transport observed in rocks and that on modern slopes where ephemeral or intermediate types of movement are acoustically recorded.

Mass transport processes and classification schemes are areas of active research and rapidly changing information. As a result, some parts of this paper and concepts included in Table 1 may become dated and require prudent application.

CRITERIA FOR RECOGNIZING SLOPES

As Dott and Bird (1979) aptly state, in the modern ocean "there is an enormous advantage of knowing what the environment is, but it is not always obvious what features of the modern case are preservable and which are most diagnostic of that particular environment." The ancient record "provides a much better view of features too large to be displayed in box cores but too small to be resolved at all (or only as lines and shadows) on conventional subbottom profiles. It is self-evident, therefore, that the best sedimentary models are constructed both from outstanding modern and ancient examples."

Slope sediment is sufficiently different from shallow marine shelf sediment, that confusion between the two should not pose a serious problem. This is not the case, however, with slope deposits and inner (upper) submarine-fan or continent-rise deposits, as these environments have two important things in common. First, all are sites of extensive fine-grained pelagic

and hemipelagic sedimentation and second, sediment gravity-flow deposits form a large proportion of fan, rise and slope sequences.

One of the most distinct differences between fan and rise and continental slope sediment, however, is reflected in the instability of slopes. Disruption and reorientation of sediment into slides and chaotically deformed masses may be very common on slopes. In contrast, the frequency and scale of these features on submarine fans and rises are usually relatively minor.

Large-scale sediment slides measuring several square kilometers in area have been described from many modern slope environments (Heezen and Drake, 1964; Moore et al, 1970; Lewis, 1971; McGregor and Bennett, 1977; Moore, 1977). Partly due to the resolving power of conventional seismic systems, little information has been reported on small-scale mass movement on slopes (Coleman and Prior, 1978; Field and Clarke, 1979; Knebel, 1979). Evidence from the rock record (elsewhere in this chapter; Piper et al, 1976; Cook, 1979a, b) shows that the scale of deformation in slope sediment ranges from centimeters to kilometers, and as analytical techniques improve, many of these smaller mass movements on modern slopes will probably be identified.

Sediment cores collected on modern slopes in known areas of sliding commonly show no indication of mass

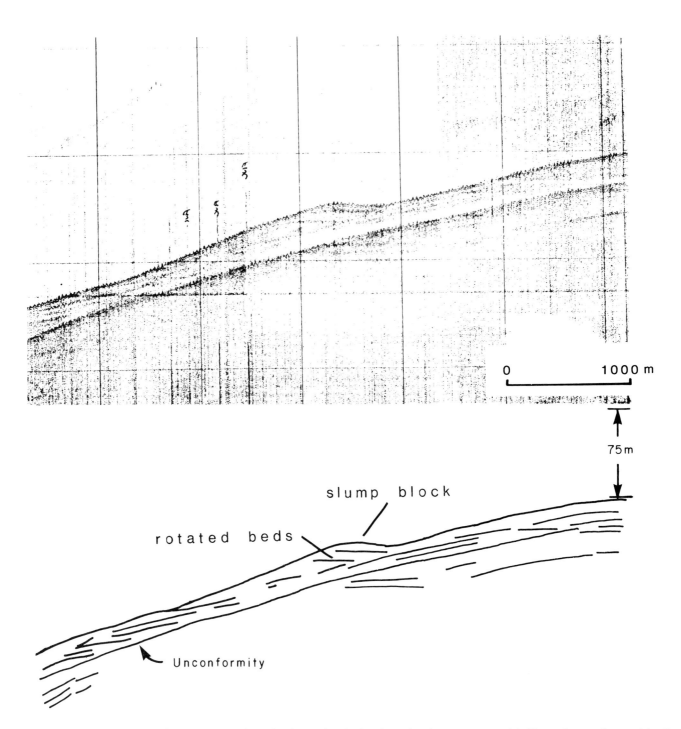

Fig. 7—Photograph and interpreted line drawing of a 1.0 kHz record of part of the continental slope of northern California. Note the regional unconformity, intraformational truncation of acoustic surfaces, downslope divergence of reflectors, and the draped material of the surface section overlying the unconformity.

transport, usually because the slide plane was not penetrated. However, bottom photographs and evidence from the rock record suggest that small-scale deformation from several centimeters to tens of meters in extent does occur and should be evident in sediment cores.

Another important difference between slopes and fans relates to the texture and geometry of sediment grav-ity-flow deposits. Slope sequences commonly include massive-bedded conglomerate and breccia containing clasts more than ten meters wide. These deposits, commonly interpreted to have been transported by debris flows, are either sheet-like or confined to relatively narrow elongate channels with little evidence of significant mean-dering. In contrast, sediment gravity-flow deposits in ancient inner fans are usually sandstone, exhibit Bouma structures, occur in well-organized thinning and fining upward sequences, and form in meandering channels. Channels can meander more easily on a fan than on a slope, and thus fan-channel deposits tend to be as much as several kilometers wide. A detailed dis-cussion of submarine fans is included in this volume (Howell and Normark). Differences between ancient slopes

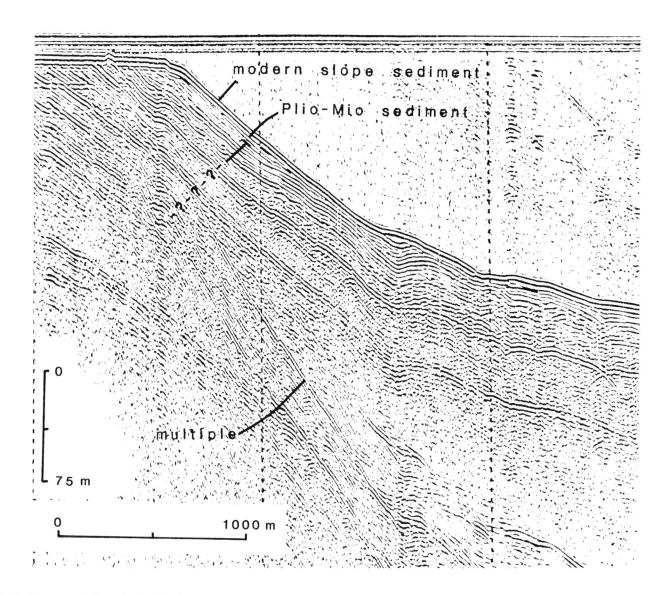

Fig. 8—Deep-penetration seismic-reflection profile (160 kJ) from the southern California continental borderland. Subsurface strata parallel the present slope. Differentiation of modern sediments on dipslope requires detailed core information.

and fans are also discussed by Mutti and Ricci Lucchi (1972) and Walker and Mutti (1973).

Another distinction between slope, fan and rise sediments is the usually greater thickness of coeval sediment on fans and rises versus slopes. Modern slopes commonly have a lower net sediment accumulation than adjacent continental shelves and rises (Emery et al, 1970). An ancient example is the early Paleozoic continental margin of Nevada; here Late Cambrian shelf sedimentary deposits are 540 m thick, slope sedimentary deposits are 130 m thick, and coeval submarine-fan deposits reach thicknesses of 1,000 to 2,000 m (Cook and Taylor, 1977; Rowell et al, 1979).

MAJOR SEDIMENTARY UNITS ON SLOPES

In the following discussion and illustrations, slope sequences are divided into three major sedimentary units: (1) undisturbed, *in situ*, pelagic and hemipelagic sediment; (2) slides and other chaotically deformed sediment; and (3) sediment gravity-flow deposits. Proportions of these three units on slopes varies, but because slopes are especially susceptible to mass failure, a high percentage of the section may contain allochthonous material. For example, up to 40 to 50% of some ancient slope sequences consist of mass transport deposits (Cook and Taylor, 1977).

Mutti and Ricci Lucchi (1972) classified submarine fan and slope sediment in the northern Apennines into seven lithofacies, A through G. Two of their facies are particularly useful for recognition of slope facies—undisturbed, *in-situ*, slope sediment (Facies G) and slides, and other chaotically deformed sediment (Facies F).

Undisturbed Sediment

Modern (Figs. 2-9; Tables 2, 3)— The major source of information regarding undisturbed modern continental slope sediment comes from seismic-reflection data and cores taken by the Deep Sea Drilling Project (DSDP). As has been shown by DSDP sampling, seismic data are often misinterpreted with respect to lithofacies. Seismic data can, however, be used to define the gross geometries of continental slope deposits.

Attempts to interpret specific litho-

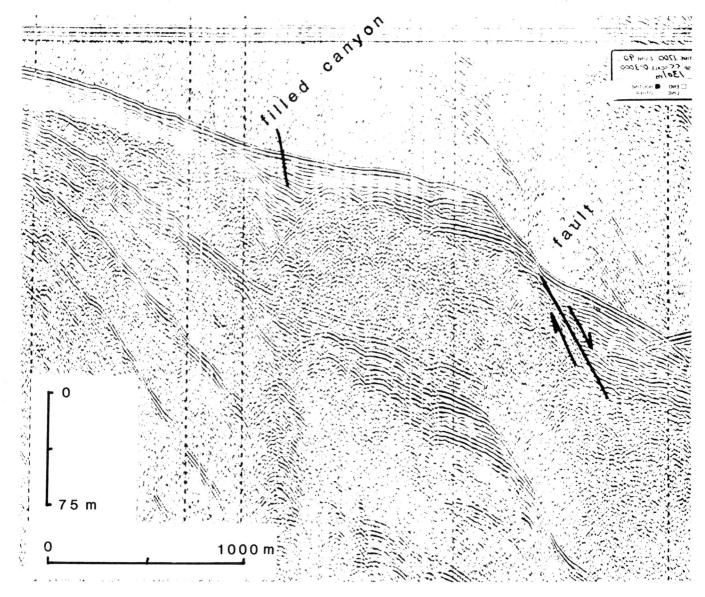

Fig. 9—Deep-penetration seismic-reflection record (160 kJ) from the southern California continental borderland. Faulted basement rocks and filled canyon underlie modern surface sediments.

Fig. 10—Fine-grained, thin-bedded argillaceous marl. Resistant beds are silty to fine-grained sandy marls. Hecho Group, Eocene, south-central Pyrenees, Spain. Photo by H. E. Cook.

Fig. 12—Fine-grained, laminated mudstone and siltstone with contorted zones. Elkton Siltstone Member, Tyee Formation, Eocene, Oregon. Photo by R. H. Dott, Jr.

Fig. 11—Fine-grained, laminated mudstone and siltstone. Elkton Siltstone Member, Tyee Formation, Eocene, Oregon. Photo by R. H. Dott, Jr.

Fig. 13—Fine-grained, laminated lime mudstone and wackestone. Lower part of Hales Limestone, Upper Cambrian, Nevada. Photo by H. E. Cook.

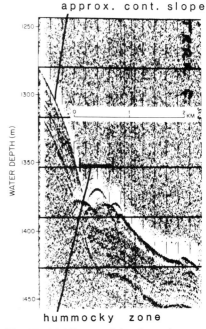

Fig. 14—Laminated lime mudstone and wackestone; pervasive sponge spicules (small light-colored blobs); larger light-colored spherules are authigenic pyrite. Bed is 4 cm thick. Lower part of Hales Limestone, Upper Cambrian, Nevada. Modified from Cook and Taylor (1977, Fig. 26).

Fig. 15—3.5-kHz record showing a slump deposit on the slope off eastern United States. Note hummocky upper surface and lack of bedding in upper part of slide. This feature may represent one of a series of slides. From Knebel and Carson (1979).

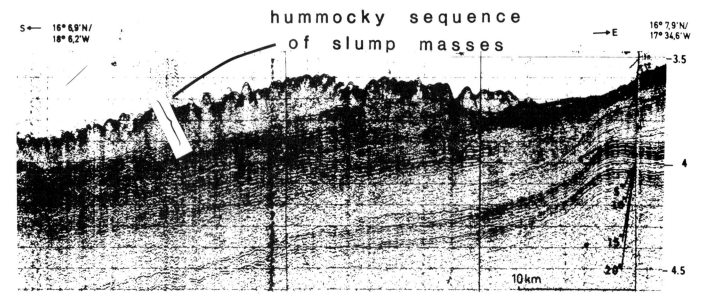

Fig. 16—Seismic-reflection profile on the continental slope off west Africa. Note the hummocky upper sequence of slump masses. From Seibold and Hinz (1974).

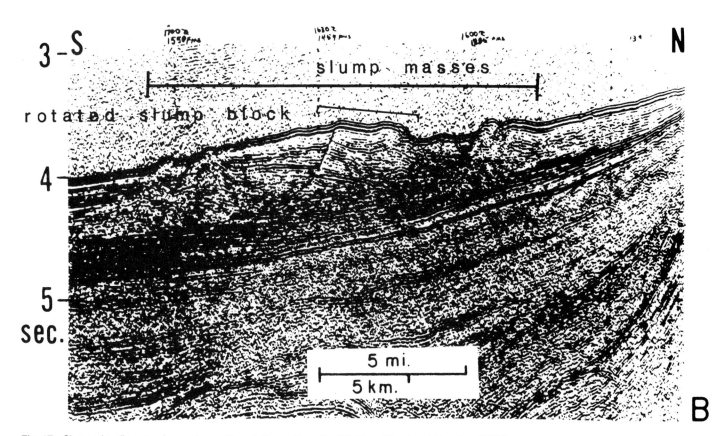

Fig. 17—Slumped sediment on lower slopes off south Texas. Kane line D1. From Stuart and Caughey (1977).

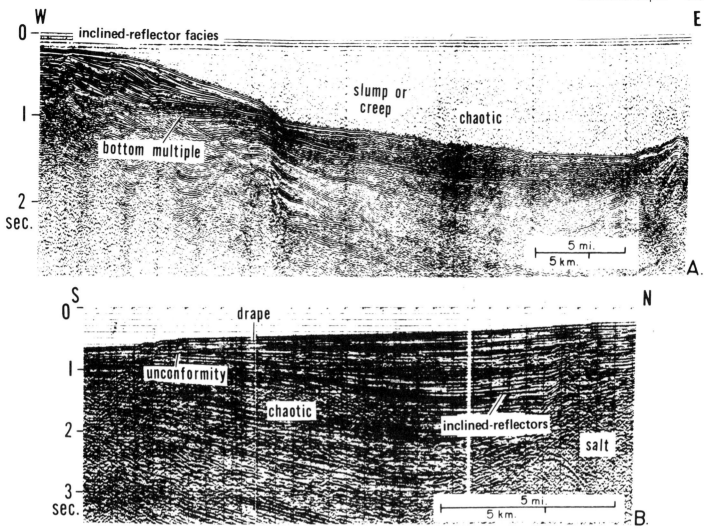

Fig. 18—Inclined-reflector seismic facies. (A) Western Gulf of Mexico, Kane line B7, (B) upper slope basin off central Louisiana. Fault may have influenced generation of inclined reflectors. Twenty-four-fold profile, from Stuart and Caughey (1977).

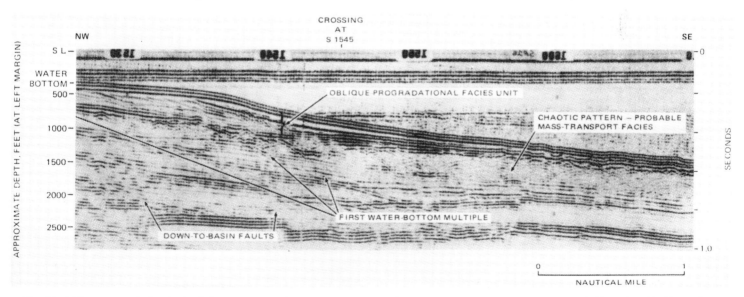

Fig. 19—Oblique-progradational seismic facies in near-surface beds form the shelf-slope margin, southeast of Matagorda Bay, Texas. An oblique-progradational facies unit lies on an older, chaotic mass-transport facies unit in the upper 300 m of sediments. Although the seismic-signal triplet at the water bottom obscures the upper 45 m of section, the unit is at least 120 m thick near the shelf edge. The middle-zone reflections dip 3° to 6° on the shelf and flatten to less than 1° where they join the outer zone of reflections. The entire unit thins downdip and ends in a depositional pinchout about 1.5 km downdip from the right margin of the figure. Note that the oblique-prograding facies infills the hummocky upper surface of the mass-transport unit. Extensional faulting at the shelf margin is contemporaneous with deposition of the oblique-progradational unit, as shown by an increase in thickness on the downthrown side of the fault. (Vertical exaggeration approximately 2.5:1.) From Sangree et al (1976).

facies from seismic-reflection data collected in the Gulf of Mexico (see Sangree and Widmier, 1977) have produced tables of observed geometries and lithofacies (Table 2). Certainly lithologic and stratigraphic correlations should be attempted with seismic-reflection data from continental slopes; however, caution should be used because the work of Sangree and Widmier (1977) may only be pertinent to their region.

Seismic-reflection data from continental slopes show a variety of geometric relations (Fig. 2). Typical passive continental slopes are shown in Figures 3 and 4. These profiles show prominent shelf-slope breaks and relatively steep continental slopes and upper continental rises. Reflectors show intraformational truncation surfaces within the slope sediment; subsequent sediment drapes over the truncations. Divergence of reflectors in a downslope direction is also apparent in these profiles. Figures 5 and 6 show other seismic-reflection geometries that are commonly found in modern slope sequences. The data cited previously are all low-resolution data (airgun or sparker data with wavelengths of about 15 m or greater). High-resolution data (3.5 kHz to 800 Hz systems with wavelengths of 0.5 to 2 m) show a much more detailed picture but with many elements found in low-resolution data. For instance, Figure 7 shows a 1.0 kHz record of the upper continental slope off northern California. This profile shows intraformational truncation surfaces, downslope divergence of reflectors, and a section of draped (or semi-draped) material over an intraformational truncation surface. Unfortunately, most piston cores would not penetrate much of the section shown in Figure 7. Thus, our knowledge of lithofacies types on continental slopes is highly biased toward surficial sediment.

The main acoustic characteristics of modern continental slopes can be summarized as follows. Reflectors are parallel to subparallel in a single "sedimentation regime." They overlie older strata conformably (Fig. 8) or unconformably overlie subjacent strata on faulted or eroded slopes (Fig. 9). Individual reflector groups are "thin bedded" to "thick bedded." Inter-reflector units vary slightly in thickness both along slope and downslope. These units often thicken (diverge) downslope (Field and Clarke, 1979) and are sigmoidal in configuration (Fig. 2).

The idea that continental slopes are plastered with hemipelagic sediment (an admixture of terrigenous and pelagic components) is well instilled in marine geology literature (Stanley, 1969 for instance). However, the few cores recovered from continental slopes by DSDP contain an almost pure pelagic facies (Hollister, et al, 1972; Lancelot, et al, 1977). Minor amounts of disseminated silt occur, but not in amounts justifying the name "hemipelagic facies." One summary of data compiled before the DSDP drilling is shown in Table 3. This table gives a compilation of data from piston and gravity cores representing only the uppermost 10 to 15 m of the slope section (Kelling and Stanley, 1976). DSDP has drilled only a few continental slopes, but those cores represent much more of the individual section than do the piston and gravity cores.

Three DSDP sites are on passive margins (Sites 108 and 369) or once-active margins that are now inactive (Site 468). Site 108 was drilled off the eastern United States and a pelagic section was recovered that contains no terrigenous components (Hollister, Ewing, et al, 1972). However, only two cores were recovered. Photographs of the cores show no contacts, sediment deformation, or any other sign of *in situ* disturbance.

DSDP site 369 was drilled to almost 600 m depth into the lower continental slope off Spanish Sahara, northwest Africa (Lancelot, Seibold, et al, 1977). Again, the sediment is mainly pelagic with only minor amounts of terrigenous material. This section does show, in places, slump features including in-

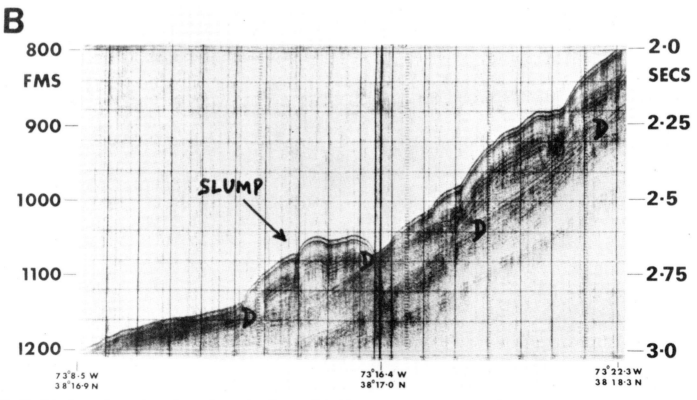

Fig. 20—Seismic record across lower slope and upper rise off eastern United States illustrating slump blocks. From Kelling and Stanley (1976).

clined and contorted bedding, plastic flow, microfractures, and thick sequences of mixed, chaotic lithologies. This sediment has calculated accumulation rates (uncorrected for compaction) of between 4 and 30 m/million years. These rates of accumulation compare to open-ocean pelagic accumulation and not with the higher rates which should occur from an added influx of terrigenous material.

The pelagic nature of continental slope sediment drilled by DSDP is discussed by Gardner et al (1977). They suggested three alternative explanations for the absence of a terrigenous component: (1) The coarse material is captured by and confined to large submarine-canyon systems thereby bypassing the slope; (2) little coarse debris is shed off the continent and shelf;

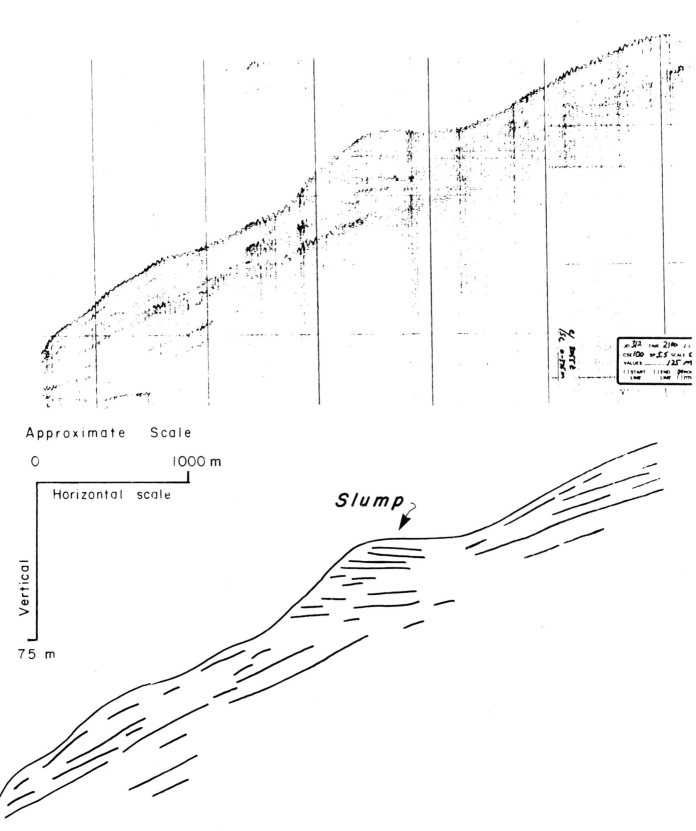

Fig. 21—High-resolution seismic record and interpreted line drawing showing slump block on the northern California upper slope at an approximate water depth of 250 m.

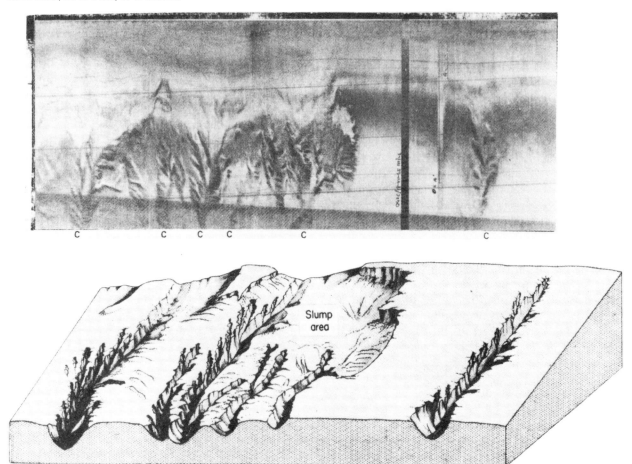

Fig. 22—Sonograph (at top) and block diagram (below) of an area near the top of the continental slope at the western edge of the Celtic Sea (northeastern Atlantic), showing canyon head axes (c on top photograph), subsidiary gullies, and large slumped areas. From Belderson and Stride (1969) and reprinted in Kelling and Stanley (1976).

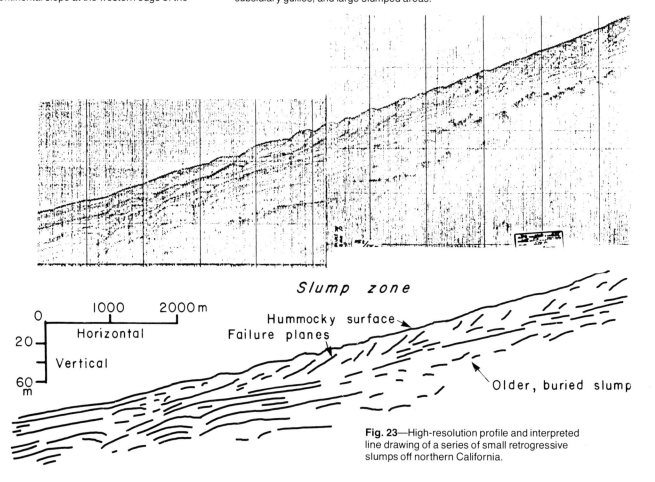

Fig. 23—High-resolution profile and interpreted line drawing of a series of small retrogressive slumps off northern California.

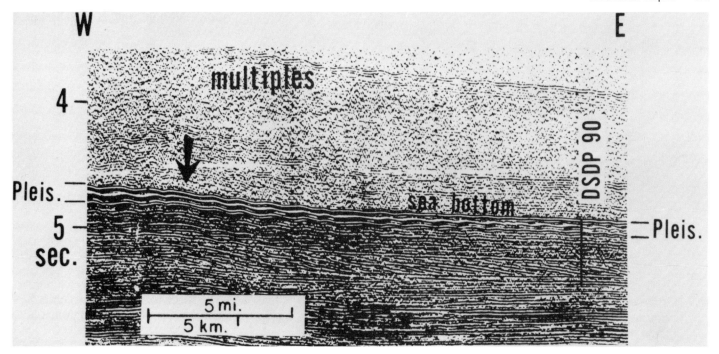

Fig. 24—Pseudomegaripple pattern (arrow) seen in a seismic-reflection profile caused by closely spaced slump faults. These features suggest slump fault rather than sedimentary megaripple origin. From Stuart and Caughey (1977).

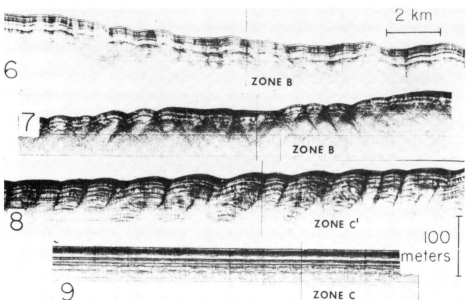

Fig. 25—3.5-kHz profiles off northeastern South America showing possible fluid-flow bedforms. From Embley and Langseth (1977).

or (3) coarse terrigenous sediment may be trapped in estuaries and on continental shelves adjacent to the slope. At this time, no one has attempted to synthesize a sediment-budget model for a continental margin, so details of slope sedimentation are unknown.

A 415-m section was cored at DSDP site 468 on the continental slope off southern California. A summary of Site 468 (Haq, Yeats, et al, 1979) reports the section containing pelagic ooze and glauconitic silty sand. Brecciated zones were also recovered, presumably recording the time when active subduction occurred along this margin.

The only other slopes drilled by DSDP are those in active continental margins, that is, trench slopes. These sites include Site 434 on the lower slope and Site 435 on the upper slope of the Japan Trench (Langseth, Okada et al, 1978) and Sites 438, 439 and 440 on the deep-sea terrace of the Japan Trench slope (von Huene, Nasu, et al, 1978). These sites all contain material and contacts implying considerable downslope movement (i.e. hackly fractures with slickensides on the fracture faces, isolated erratic subangular to rounded pebbles and cobbles, and massive sand bodies). In all cases, however, sediment is dominantly pelagic, not terrigenous. Terrigenous components are either distributed in a subtle way throughout the sediment or occur discretely within a pelagic sequence.

Ancient (Figs. 10-14)—Throughout the geologic column undisturbed slope sediment has numerous features in common. Beds exhibit contacts ranging from planar, nearly parallel and continuous for tens of meters (Figs. 10, 11), to more wavy and discontinuous. Undisturbed sediment is further characterized by its thin bedding to millimeter-thick laminae (Figs. 12-14). Typical rocks are fine grained and include lime mudstone, argillaceous mudstone, marlstone, and siltstone. Slope rocks often have large contents of organic matter. Picha (1979) reports organic carbon contents up to 9% for Tertiary canyon fill mudstone in the Carpathians of Czechoslovakia. Pyrite (Fig. 14) and concretions of siderite and ankerite are locally common (Cook and Taylor, 1977; Picha, 1979).

The Eocene Hecho Group in the south-central Pyrenees of Spain is an excellent sequence for study of a variety of facies and facies associations (Mutti, 1977, Fig. 1). Virtually continuous exposures document a lateral succession from shallow marine to slope to basin plain settings. In the Hecho Group slope, facies can be compared with adjacent submarine fan, basin plain, and shallow marine shelf sedimentary rocks.

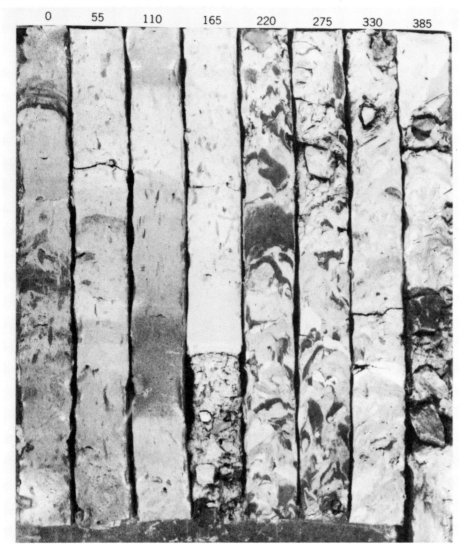

Fig. 26—Slump(?) structures in multicolored Tertiary clay. Core sections from 3,615 m, Campeche Escarpment, Gulf of Mexico (core sections in centimeters). From Heezen and Hollister (1971).

Slides, Slumps, and Soft-Sediment Deformation

Modern (Figs. 15-26)—Most evidence of mass movement and, in particular, slumps on modern slopes has come from surface-towed seismic-reflection profiling. Data from side-scan sonar, bottom photographs, sediment cores, deep-towed acoustic systems, and observations from manned submersibles have added perspective. Slumps (rotational slides) have been identified more frequently than translational slides, perhaps because the rotational nature of movement makes them more easily recognized. Although slumps commonly are not internally deformed, some permanent deformation takes place along the basal shear plane and the leading edge or toe of the slump. Many features in modern marine environments have been observed and classified as slumps that show no rotation but do show apparent internal deformation (Fig. 15). The emplacement mechanism of these types of features is unclear, but they may result from several failures that do not conform to standard definitions of slides or flows. Such features have been termed debris slides by Varnes (1978) and Hampton and Bouma (1977). Figure 15, from Knebel and Carson (1979), shows a feature exhibiting a tranparent, hummocky, upper surface and rotated, unbroken, internal structure, indicating a complex history of slumping followed by subsequent sliding.

Fig. 27—Seaward-prograding slope sequence. Translational slide, No. 1, is 10 m thick and 400 m wide. No. 2 is a smaller rotational slide (slump). Downslope transport direction was southwest, obliquely out of the photo to the left. Lower part of Hales Limestone, Upper Cambrian, Nevada. From Cook (1979a, Fig. 2).

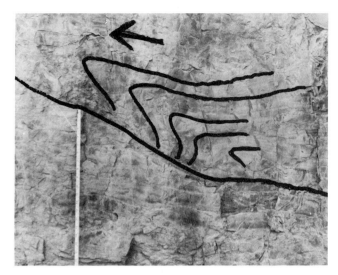

Fig. 28—Contact between basal shear plane of translational slide in Figure 27 and underlying hemipelagic limestone. Arrow points in direction slide moved. Tape is 50 cm long. From Cook (1979, Fig. 4).

Fig. 30—Base of a 3.5-m-thick translational slide showing near-horizontal bedding remolding into tabular clasts. Tape is 20 cm long. Lower part of Hales Limestone, Upper Cambrian, Nevada. Modified from Cook (1979a, Fig. 14).

Fig. 29—Interior part of translational slide in Figure 27 showing large open overfolds developed in semiconsolidated hemipelagic limestone. From Cook (1979a, Fig. 7).

Fig. 31—Translational slide 50 cm thick. Tape is 15 cm long. Lower part of Hales Limestone, Upper Cambrian, Nevada. Modified from Cook (1979a, Fig. 24).

Large-scale slumping on the continental slope off northern Senegal, Africa, and off south Texas, as shown in Figures 16 and 17, shows characteristic patterns of hummocky surfaces and rotated bedding planes. The rotation and disruption of beds commonly results in chaotic, completely disrupted patterns on seismic records (Figs. 18, 19). Note in Figures 18 and 19 that even where no surface expression is evident and slumps are buried at depth, chaotic facies can be used to recognize the slumped zone.

Slump blocks off the eastern United States (Fig. 20) and northern California (Fig. 21) show characteristic patterns of surface relief, rotated bedding, and no apparent internal deformation. Zones of sliding are commonly associated with canyons, gullies, scarps and other such features on the slope, often complicating and obscuring the geom-

Fig. 32—Rotational slide (slump), 10 m thick, shown in Fig. 27. Black lines show concave-up basal shear plane and the upper contact with an overlying translational slide. From Cook (1979a, Fig. 18).

Fig. 33—Soft-sediment deformation (slump?) in slope facies. Waitemata Beds, lower Miocene, Whangaparoa Point, North Island, New Zealand. Photo by G. deVries Klein.

Fig. 34—Soft-sediment deformation in a slump(?). Marnoso-arenacea Formation, Miocene, Apennines, Italy. From Ricci Lucchi (1975, Fig. 54).

Fig. 35—Soft-sediment deformation slump(?) in limestone. DSDP Site 318, slope zone on south side of Tuamotu Ridge, lower Oligocene. Scale in centimeters. Photo by G. deVries Klein.

Fig. 36—Sequence of interbedded thin sandstone and mudstone exhibiting soft-sediment deformation. Submarine canyon fill, Paleogene, Czechoslovakia. Bar scale is 2 cm. Modified from Picha (1979, Fig. 11).

Fig. 39—Argillaceous marl and fine-grained sandstone. Sandstone bed (center) shows incipient stages of deformation during sliding(?). Exposed section is about 5 m thick. Hecho Group, Eocene, south-central Pyrenees, Spain. Photo by H. E. Cook.

Fig. 37—Small-scale sliding(?) in laminated siltstone. Elkton Siltstone Member, Tyee Formation, Eocene, Oregon. Photo by R. H. Dott, Jr.

Fig. 40—Load-casted basal contact of slump(?). Slope facies, Embetsu Formation, Pliocene, Utakosui, Hokkaido, Japan. Scale is 2 m long. Photo by G. deVries Klein.

Fig. 38—Argillaceous marl that underwent sliding(?) and disruption into ball and pillow features. Exposed section is about 10 m thick. Hecho Group, Eocene, south-central Pyrenees, Spain. Photo by H. E. Cook.

etry and nature of the mass movement (Fig. 22).

Figure 23, a high-resolution seismic record from the northern California continental margin, shows details of a large-scale retrogressive slump progressing in an upslope direction (retrograde) at an unknown rate. The zone is composed of a series of small slump blocks clearly separated by shear planes. Rhythmic hummocks are characteristic of these types of slumps, but some workers speculate that the rhythmic topography is a type of bed form resulting from fluid movement. Figure 24 shows a similar type of slump zone as viewed on a deep-penetration seismic-reflection record. Similar surface features recorded on high-resolution records from the South American continental margin (Fig. 25) are referred to as waves and believed by Embley and Langseth (1977) to result from fluid

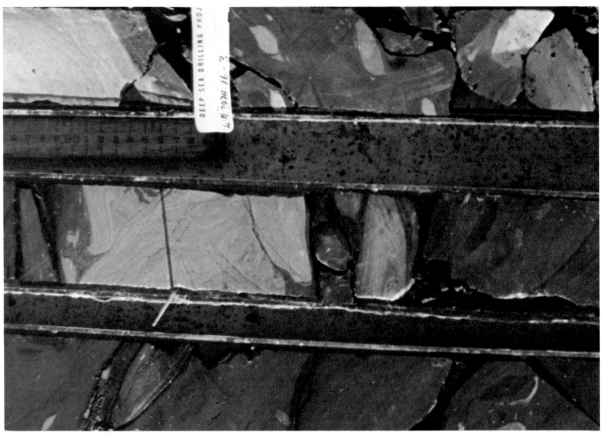

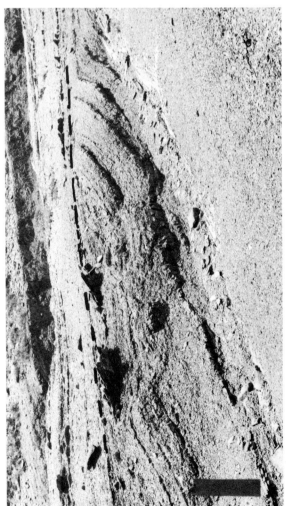

Fig. 43—Interbedded continental rise, hemipelagic sediment (light colored) and deformed slope mudstone (dark colored). The deformed mudstone is interpreted to have moved as a slide from the outer shelf or upper slopes to its present position on the rise. An outer-shelf or upper-slope origin for the deformed mudstone is based on its microfauna assemblage. This mudstone originated within an oxygen minimum layer (see Fig. 44). Hemipelagic sediment contains 0.3% organic carbon; mudstone contains 1.7% organic carbon. DSDP Site 397A, core 16, upper continental rise, eastern Atlantic, south of the Canary Islands, Miocene. Photo by C. Cornford.

Fig. 41—Marlstone (gray) and interbedded sandstone turbidites (brown) showing intraformational truncation surface (between units 1 and 2), in situ slope sediments (unit 2), slide(?) with internal folding of beds (unit 3), in situ slope sediment (unit 4). Verghereto Formation, Miocene, northern Apennines, Italy. Pick for scale between untis 1 and 2. Photo by F. Ricci Lucchi.

Fig. 42—Intraformational truncation surface (dashed line) in argillaceous marlstone. Hecho Group, Eocene, south-central Pyrenees, Spain. Bar scale is 1 m. Photo by H. E. Cook.

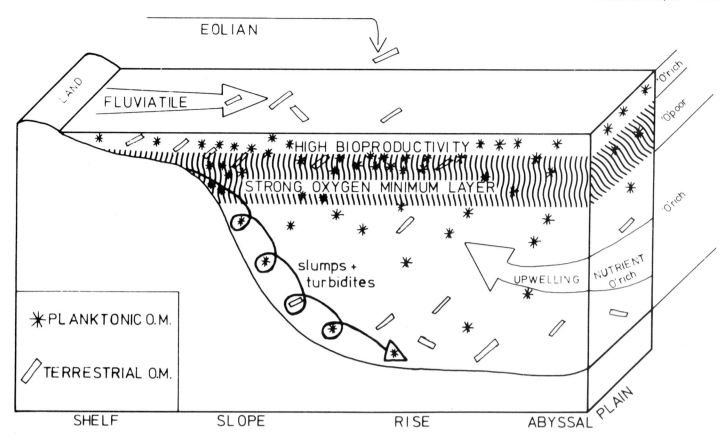

Fig. 44—Factors affecting organic accumulation on a continental margin. Model across a passive continental margin based on DSDP Site 397 off Cape Bojador, North Africa, south of the Canary Islands. Plant derived terrestrial organic matter (rectangles) and marine planktonic organic matter (stars). Normally most organic matter is degraded as it settles through oxidizing waters or during benthic reworking. But in areas of upwelling the development of a water layer of reduced oxygen concentration at the base of the zone of high bioproductivity allows the accumulation of organic-rich sediment under reducing conditions. Such sediment is potential hydrocarbon source material. This organic-rich sediments can be transported downslope by slides and gravity flows into the continental rise. On the continental rise these source beds and coexisting coarser clastics can form an ideal hydrocarbon source-reservoir rock combination. Drawing and interpretation by C. Cornford.

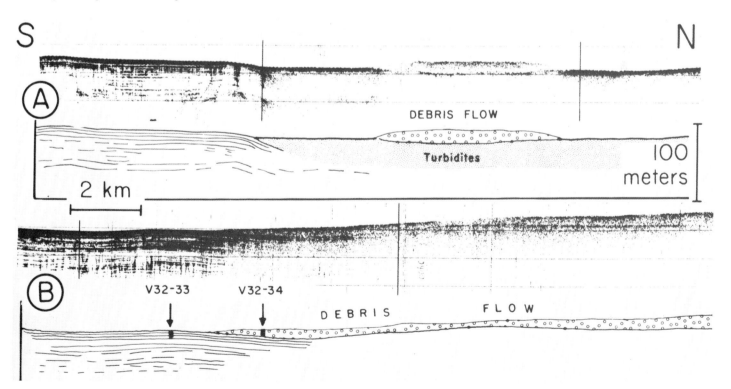

Fig. 45—3.5-kHz profile and line-drawing interpretation showing transition from stratified hemipelagic sequence (left) to ponded, highly reflective turbidites to lens-shaped debris flow. Note onlapping of debris flow over hemipelagic sequence. Reflector at bottom of debris flow is top of hemipelagic sequence. V32-33 and V32-34 indicates relative position of piston cores in Figure 46. From Embley (1976).

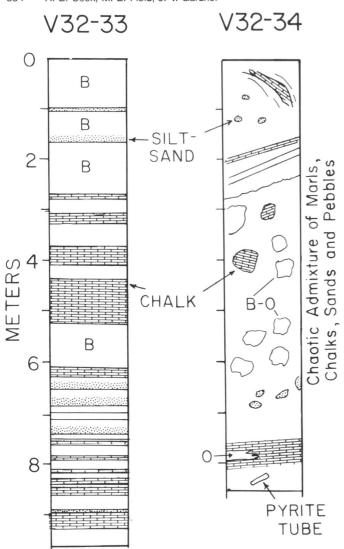

Fig. 48—Debris-flow deposit. Channel axis of 12-m-thick, 400-m-wide limestone-conglomeratic debris-flow deposit. Black solid line is contact between debris-flow deposit and in situ hemipelagic sediment. Top of channel not shown. Dashed line encloses 3 × 15 m clast of hemipelagic limestone surrounded by randomly oriented tabular clasts. Clasts and matrix are all slope-derived material. This deposit was derived from a slide that originated on the slope. This debris flow deposit occurs within the slope sequence shown in Figure 27. Lower part of Hales Limestone, Upper Cambrian, Nevada. From Cook (1979a, Fig. 21).

Fig. 46—Lithologic logs of cores V32-33 and V32-34 whose positions are shown in Figure 45. Core V32-33 was taken in stratified hemipelagic section and V32-34 in debris flow 4 km away. Core V32-33 consists entirely of parallel-bedded marl, chalk, and thin silt-sand layers of Pleistocene age. A 3-cm pyrite concretion, believed to have come from continental rise off Spanish Sahara, was found at 830 cm depth in V32-34. B = brown marl; O = olive-gray clay and marl. From Embley (1976).

Fig. 47—Box core of bioturbated debris flow conglomerate. Base of continental slope, Baltimore Canyon, eastern United States. Lamont-Doherty Cruise, *Robert Conrad* 19-49. Photo by G. deVries Klein.

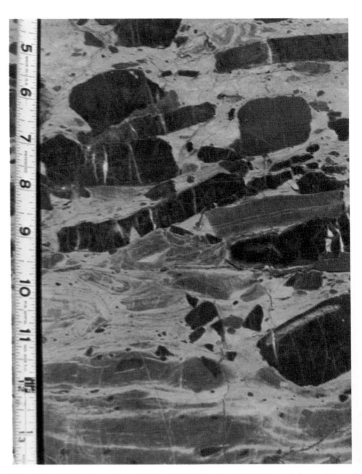

Fig. 49—Debris-flow deposit, 2 m thick, lateral part of channel deposit shown in Figure 48. In thin lateral margins of channel deposits, tabular clasts are oriented subparallel and matrix supported. Scale in inches. Photo by H. E. Cook.

Fig. 50—Debris-flow deposit. Limestone debris bed expands and narrows at irregular intervals. 1 and 2 show the "tail" and "wave" of the debris-flow deposit respectively. Similar features have been observed in subaerial deris flow and in flume experiments with subaqueous flows. 3 shows erosional base. Thickness of deposit at "wave" is about 12 m. Sekwi Formation, Lower Cambrian, Northwest Territory, Canada. From Krause and Oldershaw (1979, Fig. 3a).

Fig. 51—Debris-flow deposit. Shale-chert conglomerate in channel. Elkton Siltstone Member, Tyee Formation, Eocene, Oregon. Photo by R. H. Dott, Jr.

Fig. 52—Debris-flow deposit. Volcaniclastic debris-flow conglomerate. DSDP Site 290, slope facies, West Philippine Basin near Central Basin fault, lower Oligocene. Scale in centimeters. Photo by G. deVries Klein.

Fig. 53—Debris-flow deposit. Volcaniclastic debris-flow conglomerate. DSDP Site 313, small-slope basin, Mid-Pacific Mountains, Upper Cretaceous. Scale in centimeters. Photo by G. deVries Klein.

Fig. 54—Debris-flow deposit. Sandstone bed with larger clasts of Paleozoic rocks derived from slope canyon walls. Large clast at top of core is projecting above the top of the bed. Supporting sand matrix shows both debris-flow and turbidity-current deposits. Submarine canyon-fill, Paleogene, Czechoslovakia. From Picha (1979, fig. 10a).

Fig. 55—Debris-flow deposit. Shallow-water limestone clasts admixed with deep-water clasts and dark argillaceous lime mud. Deposit forms a petroleum reservoir on the eastern margin of the Midland Basin, west Texas, Permian (Wolfcampian). Core is about 7.5 cm wide. Photo by H. E. Cook.

flow. Features produced by differing mechanisms, such as fluid flows and slides, can be quite similar in geometry. This subtlety makes it difficult to determine which process causes a given feature.

Cores collected from the Campeche Escarpment, Gulf of Mexico, clearly show the effects of mass transport (Fig. 26). Intense deformation in this core suggests that original beds plastically deformed and began to break up into individual clasts. A seismic-reflection profile across this area would probably show a chaotic facies, as in Figures 18 and 19.

Ancient (Figs. 27-44)—This paper uses the terminology of slides and slumps defined by Varnes (1978). Much of the literature on mass tranport of ancient submarine sediment does not distinguish between deformed strata, that have moved along discrete shear planes (slides, Table 1, Figs. 27, 32), and deformed strata, with no obvious basal shear plane (Figs. 34-39). Literature on

Fig. 56—Turbidites and soft-sediment deformation. (1) Outer shelf calcarenite and marl; (2) calcareous turbidites; (3) soft-sediment folds (slide?). Dashed line is the contact between outer shelf sediment and retrogradational slope sediment. Cerrogna Formation, Miocene, central Apennines, Italy. Modified from Ricci Lucchi (1975, Fig. 46).

Fig. 57—Turbidite. Channelized deposit in slope facies showing normal grading of cobble-sized clasts. Hecho Group, Eocene, south-central Pyrenees, Spain, pick for scale. Photo by H. E. Cook.

Fig. 58—Conglomerate. Debris-flow deposit (1) containing boulder 1.2 m across overlain by normally graded conglomerate turbidite (2). Same locality as Figure 57. Photo by H. E. Cook.

Fig. 59—Conglomerate. Massively bedded 50-cm-thick conglomerate overlain by 30-cm-thick sandstone turbidite. Discordant attitudes between sediment gravity-flow deposits are common. Same locality as Figure 57, black vertical bar scale 25 cm. Photo by H. E. Cook.

Fig. 60—Turbidites. Upper-slope channel deposits with stratigraphic top is to the right. Massive bed at base. Facies A (Mutti and Ricci Lucchi) overlain by facies B and C. Murihiku Supergroup, Jurassic, Sandy Bay, South Island, New Zealand. Bar scale 1 m. Photo by R. M. Carter (see Carter et al, 1978).

Fig. 61—Turbidites. Upper-slope channel deposits with stratigraphic top is to the right. Facies B and C (Mutti and Ricci Lucchi) overlain by facies D. This represents the stratigraphic top of Figure 60. Same locality as Figure 60. Black line is about 5 m long. Photo by R. M. Carter (see Carter et al, 1978).

submarine slides often fails to differentiate between translational slides and rotational slides (slumps; Table 1). Authors commonly use the term "slump" for any type of feature exhibiting soft-sediment deformation without clear basal features. Therefore, some "slumps" illustrated in this section may be translational slides rather than rotational slides (slumps), or some "slumps" may be only deformed strata with no sharp upper and lower boundaries. If the genesis of the sediment mass is questionable, its interpreted origin is followed by a question mark in the figure caption.

Features described as slides and slumps range in thickness from a few centimeters (Figs. 35-37) to tens of meters or more (Fig. 34). Maximum three-dimensional geometries of ancient submarine slides are usually inaccurately

Fig. 62—Slope channel deposit. Thick-bedded to massive channel deposit cut into thin-bedded shelf sediment. Solid step-like line is base of channel. Channel is filled with chaotically bedded sandy breccia-conglomerate beds 1-3 m thick. Dotted line encloses large rectangular block. Height of cliff is about 80 m. McPhee Cove Conglomerate, Glenomaru Group, Jurassic, South Island, New Zealand. Modified from Carter et al (1978, Fig. 23-26).

known because of limited exposures.

The degree of internal deformation in slides ranges from only slight (Fig. 32), to moderate (Figs. 28, 29), to complete disruption of bedding (Fig. 31). In the last case, shear strength of the sediment was exceeded and the mass began movement as a highly viscous debris flow (Cook, 1977, 1979a, b). All gradations indicating intensity of internal deformation probably appear in ancient slides. The notion, so often stated in the literature, that modern submarine slides are "essentially undeformed" does not appear true for ancient submarine slides. We believe that some modern slides may have more internal deformation than is reported. Their "undeformed nature" may in some cases reflect the problem of limited resolution of conventional seismic-reflection systems.

Features showing intraformational truncation surfaces are included in this section (Figs. 41, 42). These erosional surface features are interpreted to represent slide scars.

Under certain circumstances slides may play an important role in transporting organic-rich sediment (potential hydrocarbon source rocks) onto the continental rise (Figs. 43, 44).

Sediment Gravity Flow Deposits

Modern (Figs. 45-47)—Few observations of mass flows on modern slopes are available, partly because of the need for corroborating evidence from sediment cores. Information provided from seismic-reflection profiling tells little about matrix and framework, but does show geometry and internal stratifications. Embley (1976) obtained both seismic records and sediment cores from the toe of debris-flow deposits on the lower continental rise which allowed cross-correlation and provided supportive evidence on the nature of the movement (Figs. 45, 46). Box cores collected at the base of the continental slope in Baltimore Canyon also have textures interpreted as debris-flow deposits (Fig. 47). Cores from the Campeche Escarpment show features that may be transitional between slides and debris-flow deposits (Fig. 26).

Fig. 63—Conglomerate turbidite. Normally graded turbidite with only Bouma division A. Slope facies of the Hecho Group, Eocene, south-central Pyrenees, Spain. Photo by H. E. Cook.

Fig. 64—Conglomerate turbidite. Normally graded limestone turbidite, 50 cm thick, contains both shelf- and slope-derived clasts. Tabular clasts with subparallel orientation. Only Bouma division A is developed. Lower part of Hales Limestone, Upper Cambrian, Nevada. Modified from Cook and Taylor (1977, Fig. 36) and Cook (1979, Fig. 30). Located in slope sequence as shown in Figure 27.

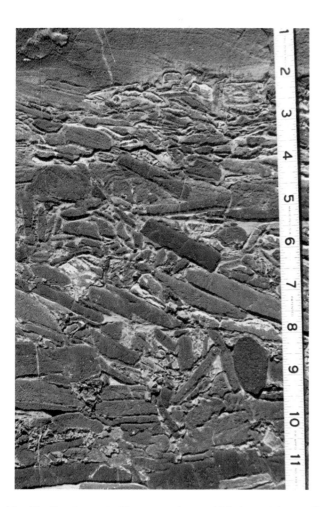

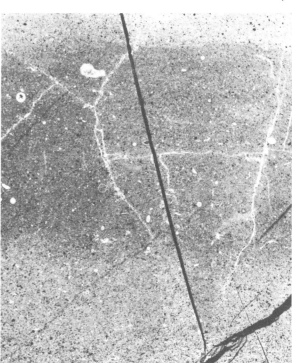

Fig. 66—Mudstone turbidites. Three normally graded fine-grained turbidites. Each bed is about 1 cm thick. Middle bed contains abundant calcispheres. Slope facies, Jurassic, South Island, New Zealand. Photo by R. M. Carter.

Fig. 65—Conglomerate. Upper part of 1.5-m-thick channel deposit. Clasts are a mixture of slope- and shelf-derived limestone. Clasts are normally graded, imbricated in an upslope direction at the top of the channel and oriented subparallel below the top of the channel. A rippled Bouma division C caps the bed. Located in slope sequence shown in Figure 27. Lower part of Hales Limestone, Upper Cambrian, Nevada. Modified form Cook and Taylor (1977, Fig. 38) and Cook (1979, Fig. 29).

Fig. 67—Sandstone turbidite. Normally graded sandstone turbidite. Located at the outer margin of a channel deposit in an overbank position. Elkton Siltstone Member, Tyee Formation, Eocene, Oregon. Photo by R. H. Dott, Jr.

Fig. 68—Grain-flow deposit or turbidite. Upper part of thick, porous, rather massive bed; overlain by mudstone with lamina and lenses of fine-grained sandstone. Bar scale is 2 cm. Submarine canyon-fill, Paleogene, Czechoslovakia. From Picha (1979, Fig. 9d).

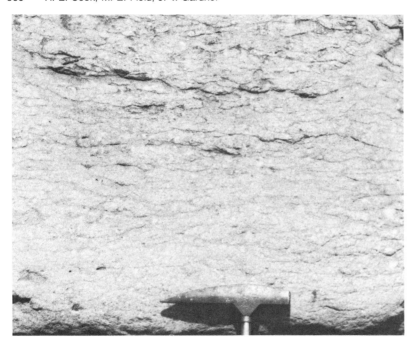

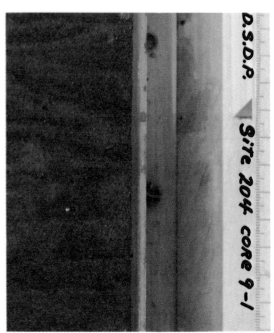

Fig. 69—Dish structures in sandstone; possible grain-flow deposit. Same locality as Figure 57, slope facies of Hecho Group, Eocene, south-central Pyrenees, Spain. Photo by H. E. Cook.

Fig. 70—Dish structures in sandstone; possible grain-flow deposit. DSDP Site 204, upper trench slope, Pacific plate, east side of Tonga-Kermadec Trench, scale in centimeters. Photo by F. deVries Klein.

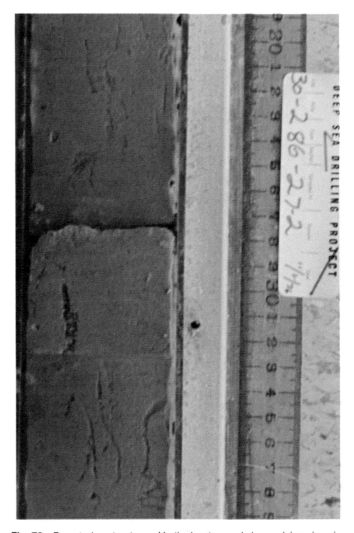

Fig. 71—Pull-apart fractures in volcanic sandstone suggest that sediment has been subjected to mass movement. DSDP Site 189, slope zone, north flank of Aleutian Ridge, Bering Sea, upper Miocene. Scale in centimeters. Photo by G. deVries Klein.

Fig. 72—Dewatering structures. Vertical water- and air-expulsion pipes in clay, suggest liquefied or fluidized flow. DSDP Site 286, slope zone of New Hebrides marginal basin, middle Eocene, scale in centimeters. Photo by G. deVries Klein.

Ancient (Figs. 48-77)—Sediment gravity flows are commonly divided into five end-member types (Table 1). Of these five types, debris flow (Figs. 48-55, 58) and turbidity-current flows (Figs. 57, 58, 60, 61, 63, 64, 66-68) are the best documented and appear to be the dominant processes for transporting large volumes of sediment-fluid mixtures downslope. It is generally well accepted what constitutes field evidence for debris flow (Cook et al, 1972; Hampton, 1975; Walker, 1975; Middleton and Hampton, 1976; Cook, 1979a) and turbidity-current flow (Middleton and Hampton, 1976). However, this is not true for grain flow.

The theoretical concept of grain flow is clear (Bagnold, 1954, 1956; Lowe, 1976a), but types of field criteria suitable for grain flow production are still debatable (Middleton and Hampton, 1976). Reverse grading and dish structures (Stauffer, 1967) are currently two of the most commonly cited structures attributed to a grain-flow process (Figs. 69, 70). Liquefied flow and fluidized flow have been rigorously evaluated by Lowe (1976b). Features such as fluid escape pipes, convolute bedding, and load structures are considered valid evidence of previous sediment liquefication or fluidization (Figs. 72-75). These sedimentary structures

commonly form after deposition. However, fluidized flow and liquefied flow are unlikely major 'long-distance' transporting agents (Lowe, 1976).

It should be emphasized that sediment gravity flow deposits often show features indicating sediment transport and deposition by more than one process (Fig. 65). During mass transport of sediment, one process will dominate at any one point in space or time; however, a sequence of mass-transport processes may operate before actual sediment deposition. The rock record records final transportational and depositional event(s).

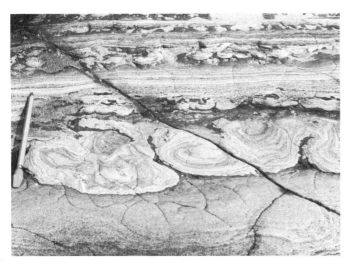

Fig. 73—Load deformation in well-bedded sandstone-siltstone. Overbank(?) deposits. Usually considered evidence that sediment was in a liquefied or fluidized state. Elkton Siltstone Member, Tyee Formation, Eocene, Oregon. Photo by R. H. Dott, Jr. (see Dott and Bird, 1979, Fig. 7).

Fig. 74—Load structures at base of a channel deposit. Same locality and interpretation as Figure 73, pick in lower right for scale. Photo by R. H. Dott, Jr.

Fig. 75—Mudstone dike. Liquefied or fluidized flow of mud downward forming a dike. Marnoso-arenacea Formation, Miocene, Apennines, Italy. Modified form Ricci Lucchi (1975, Fig. 55a).

REFERENCES CITED

Bagnold, R. A., 1954, Experiments in the gravity-free dispersion of large spheres in a Newtonian fluid under shear: Royal Soc. London Proc. Ser. A, v. 225, p. 49-63.

——— 1956, The flow of cohesionless grains in fluid: Royal Soc. London Phil. Trans. Ser. A., v. 249, p. 235-297.

——— 1966, An approach to sediment transport problem from general physics: U.S. Geol. Survey Prof. Paper 422-I, 37 p.

Barbat, W. F., 1958, The Los Angeles Basin area, California, *in* Habitat of Oil: AAPG, p. 62-77.

Belderson, R. H., and A. H. Stride, 1969, The shape of submarine canyon heads revealed by Asdic: Deep-Sea Research, v. 16, p. 103-104.

Bloomer, R. R., 1977, Depositional environments of a reservoir sandstone in west-central Texas: AAPG Bull., v. 61, p. 344-359.

Bouma, A. H., et al, 1976, Gyre Basin, an intraslope basin in northwest Gulf of Mexico, *in* Beyond the shelf break: AAPG Marine Geol. Comm. Short Course, v. 2, p. E-1 to E-28.

——— and L. E. Garrison, 1979, Intraslope basins, Gulf of Mexico (abs.): Geol. Soc. Amer. Abstracts with Programs, v. 11, p. 392.

Buffington, E. C., and D. G. Moore, 1963, Geophysical evidence on the origin of gullied submarine slopes, San Clemente, California: Jour. Geology, v. 71, p. 356-370.

Burk, C. A., and C. L. Drake, 1974, Geologic significance of continental margins, *in* C. A. Burk and C. L. Drake, eds., The geology of continental margins: New York, Springer-Verlag, p. 3-10.

Carlson, P. R., and B. F. Molnia, 1978, Submarine faults and slides on the continental shelf, northern Gulf of Alaska: Marine Geotechnology, v. 2, p. 275-290.

Carter, R. M., 1975, A discussion and classification of subaqueous mass-transport with particular application to grain-flow, slurry-flow, and fluxoturbidites: Earth Science Reviews, v. 11, p. 145-177.

——— et al, 1978, Sedimentation patterns in an ancient arc-trench-ocean basin complex: Carbniferous to Jurassic Rangitata Orogen, New Zealand, *in* D. J. Stanely and G. Kelling, eds., Sedimentation in submarine canyons, fans, and trenches: Stroudsburg, Penn., Dowden, Hutchinson and Ross Pub., p. 340-361.

Coleman, J. M., and D. B. Prior, 1978, Contemporary gravity tectonics—an everyday catastrophe? (abs.): AAPG Ann. Mtg., v. 62, p. 505.

Cook, H. E., 1979a, Ancient continental slope sequences and their value in understanding modern slope development, *in* O. H. Pilkey and L. S. Doyle, eds., Geology of continental slopes: SEPM Spec. pub. 27, p. 287-305.

——— 1979b, Small-scale slides on intercanyon continental slope areas, Paleozoic, Nevada (abs.): Geol. Soc. of Am. Ann. Mtg., v. 11, p. 405.

——— et al, 1972, Allochthonous carbonate debris flows at Devonian bank ("reef") margins, Alberta, Canada: Bull. Canadian Petroleum Geol., v. 20, p. 439-497.

——— and M. E. Taylor, 1977, Comparison of continental slope and shelf environments in the Upper Cambrian and Lowest Ordovician of Nevada: SEPM Spec. Pub. 25, p. 51-82.

Curran, J. F., K. B. Hall, and R. F. Herron, 1971, Geology, oil fields, and future petroleum potential of Santa Barbara Channel area, California, *in* Future petroleum provinces of the United States—their geology and potential: AAPG Mem. 15, p. 192-211.

Dott, R. H., Jr., 1963, Dynamics of subaqueous gravity depositional processes: AAPG Bull., v. 47, p. 104-128.

——— and K. J. Bird, 1979, Sand transport through channels across an Eocene shelf and slope in southwestern Oregon: SEPM Spec. Pub. 27, 49 p.

Egloff, J., and G. L. Johnson, 1979, Erosional

Fig. 76—Contourite(?) bed. Possible contourite (C) overlain by several normally graded mudstone beds. *Chondrites* mottled mudstone directly over contourite. Upper trench slope facies of Murihiku Supergroup, Jurassic, South Island, New Zealand. Contourite (C) is about 1 cm thick. Photo by R. M. Carter.

Fig. 77—Contourites(?). Composed of well-sorted sand-sized (0.25 mm) shallow-water derived limestone grains (Nuia, a possible alga). Ripple forms have wave lengths of about 9 cm and 0.5 to 1.0 cm amplitudes. Both base and top have sharp contact with enclosing hemipelagic mudstone. Upper part of Hales Limestone, Lower Ordovician, Nevada, scale in centimeters. Photo by H. E. Cook.

and depositional structures of the southwest Iceland insular margin: thirteen geophysical profiles, *in* J. S. Watkins, L. Montadert, and P. W. Dickerson eds., Geological and geophysical investigations of continental margins: AAPG Mem. 29, p. 43-63.

Embley, R. W., 1976, New evidence for occurrence of debris flow deposits in the deep sea: Geology, v. 4, p. 371-374.

———— and M. G. Langseth, 1977, Sedimentation processes on the continental rise off northeastern South America: Marine Geology, v. 25, p. 279-297.

Emery, K. O., 1960, The sea off southern California: New York, John Wiley and Sons, 366 p.

———— et al, 1970, Continental rise off eastern North America: AAPG Bull., v. 54, p. 44-108.

Enos, P., 1977, Tamabra Limestone of the Poza Rica Trend, Cretaceous, Mexico, *in* H. E. Cook and P. Enos, eds., Deep-water carbonate environments: SEPM Spec. Pub. 25, p. 273-314.

Field, M. E. and S. H. Clarke, Jr., 1979, Small-scale slumps and their significance for basin-slope processes southern California borderland, *in* O. H. Pilkey and L. S. Doyle, eds, The geology of continental slopes: SEPM Spec. Pub. 27.

Gardett, P. H., 1971, Petroleum potential of Los Angeles, California *in* Future petroleum provinces of the United States—their geology and potential: AAPG Mem. 15, p. 298-308.

Gardner, J. V., W. E. Dean, and L. Jansa, 1977, Sediments recovered from the northwest African continental margin, Leg 41, Deep Sea Drilling Project, *in* Initial Reports of the Deep Sea Drilling Project: U.S. Govt. Printing Office, Washington, D. C., v. 41, p. 1121-1134.

Hampton, M. A., 1972, The role of subaqueous debris flow in generating turbidity currents: Jour. Sed. Petrology, v. 42, p. 775-793.

———— 1975, Competence of fine-grained debris flows: Jour. Sed. Petrology, v. 45, p. 834-844.

———— and A. H. Bouma, 1977, Slope instability near the shelf break, western Gulf of Alaska: Marine Geotechnology, v. 2, p. 309-332.

Haq, B., et al, 1979, Eastern Pacific boundary currents: Geotimes, v. 24, p. 30-31.

Hedberg, H. D., 1970, Continental margins from the view point of the petroleum geologist: AAPG Bull., v. 54, p. 3-43.

Heezen, B. C., and C. L. Drake, 1964, Grand Banks slump: AAPG Bull., v. 48, p. 221-225.

———— and C. D. Hollister, 1971, The face of the deep: New York, Oxford Univ. Press, 659 p.

———— M. Tharp, and M. Ewing, 1959, The floors of the oceans, *in* The North Atlantic: Geol. Soc. Amer. Spec. Paper 65, 122 p.

Hollister, C. D., et al, 1972, Initial reports of the Deep Sea Drilling Project: U.S. Govt. Printing Office, Washington, D.C., v. 11, 1077 p.

Kelling, G., and D. J. Stanley, 1976, Sedimentation in canyon, slope, and base-of-slope environments, *in* D. J. Stanley and D. J. P. Swift, eds., Marine sediment transport and environmental management: New York, John Wiley and Sons, p. 379-435.

Knebel, H. J., and B. Carson, 1974, Small-scale slump deposits, middle Atlantic continental slope off eastern United States: Marine Geology, v. 29, p. 221-236.

Krause, F. F., and A. E. Oldershaw, 1979, Submarine carbonate breccia beds—a depositional model for two-layer, sediment gravity flows from the Sekwi Formation (Lowre Cambrian), Mackenzie Mountains, Northwest Territories, Canada: Canadian Jour. Earth Sci., v. 16, p. 189-199.

Lancelot, Y. E., et al, 1977, Initial reports of the Deep Sea Drilling Project: U.S. Govt. Printing Office, Washington, D.C., v. 41, 1259 p.

Langseth, M. L., et al, 1978, Transects begun near the Japan Trench: Geotimes, v. 23, p. 22-26.

Lewis, K. B., 1971, Slumping on a continental slope inclined 1-4°: Sedimentology, v. 16, p. 97-110.

Lowe, D. R., 1976a, Subaqueous liquefied and fluidized sediment flows and their deposits: Sedimentology, v. 23, p. 285-308.

———— 1976b, Grain flow and grain flow deposits: Jour. Sed. Petrology, v. 46, p. 188-190.

Mattick, R. E., et al, 1978, Petroleum potential of U.S. Atlantic slope, rise, and abyssal plain: AAPG Bull. v. 62, p. 592-608.

McGregor, B. A., and R. H. Bennett, 1977 Continental slope sediment instability northeast of Wilmington Canyon: AAPG Bull., v. 61, p. 918-928.

Middleton, G. V., and M. A. Hampton, 1973, Mechanics of flow and deposition, *in* G. V. Middleton and A. H. Bouma, eds., Turbidites and deep water sedimentation: SEPM Pacific Sect. Short Course, Anaheim, California, p. 1-38.

———— 1976, Subaqueous sediment transport and deposition by sediment gravity flows, *in* D. J. Stanley and D. J. P. Swift, eds., Marine sediment transport and environmental management: New York, John Wiley and Sons, p. 197-217.

Moore, D. G., 1977, Submarine slides, *in* B. Voight, ed., Rock slides and avalanches: Developments in Geotechnical Engineering, v. 1, p. 563-604.

Moore, T. C., et al, 1970, Large submarine slide off northeastern continental margin of Brazil: AAPG Bull., v. 54, p. 125-128.

Mutti, E., 1977, Distinctive thin-bedded turbidite facies and related depositional environments in the Eocene Hecho Group (south-central Pyrenees, Spain): Sedimentology, v. 24, p. 107-131.

———— and F. Ricci Lucchi, 1972, Le torbiditi dell' Appennino settentrionale: introduzione all' analisi di facies: Memorie della Societa Geologica Italiana, Rome, v. 11, p. 161-200.

Nagel, H. E., and E. S. Parker, 1971, Future oil and gas potential of onshore Ventura Basin, California, *in* Future petroleum provinces of the United States—their ge-

ology and potential: AAPG Mem. 15, p. 254-297.

Nardin, T. R., et al, 1979, A review of mass movement processes, sediment and acoustic characteristics, and contrasts in slope and base-of-slope systems versus canyon-fan-basin floor systems: SEPM Spec. Pub. 27, p. 61-73.

Picha, F., 1979, Ancient submarine canyons of Tethyan continental margins, Czechoslovakia: AAPG Bull., v. 63, p. 67-86.

Piper, D. J. W., W. R. Normark, and J. C. Ingle, Jr., 1976, The Rio Dell Formation: a Plio-Pleistocene basin slope deposit in northern California: Sedimentology, v. 23, p. 309-328.

Ricci Lucchi, F., 1975, Miocene paleogeography and basin analysis in the Periadriatic Apennines, *reprinted from* C. Squyres, ed., Geology of Italy: Petroleum Exploration Soc. of Libya, Tripoli, 111 p.

Rowell, A. J., M. N. Rees, and C. A. Suczek, 1979, Margin of the North American continent in Nevada during late Cambrian time: Am. Jour. Sci., v. 279, p. 1-18.

Sangree, J. B., et al, 1976, Recognition of continental slope seismic facies offshore Texas-Louisiana, *in* Beyond the shelf break: AAPG Short Course, v. 2, p. F-1 to F-54.

———— and J. M. Widmier, 1977, Seismic stratigraphy and global changes in sea level, part 9: Seismic interpretation of clastic-depositional facies, *in* C. E. Payton, ed., Seismic stratigraphy—applications to hydrocarbon exploration: AAPG Mem. 26, p. 165-184.

Schlanger, S. O., and J. Combs, 1975, Hydrocarbon potential of marginal basins bounded by an island arc: Geology, v. 3, p. 397-400.

Schlee, J., et al, 1977, Petroleum geology on the U.S. Atlantic Gulf of Mexico margins, *in* Exploration and economics of the petroleum industry: Southwestern Legal Found. Proc., v. 15, p. 47-93.

Seibold, E., and K. Hinz, 1974, Continental slope construction and destruction, west Africa, *in* C. A. Burk and C. L. Drake, eds., The geology of continental margins: New York, Springer-Verlag, 1009 p.

Shepard, F. P., 1963, Submarine geology, second edition: New York, Harper and Row Pub., 557 p.

———— 1973, Submarine geology, third edition: New York, Harper and Row Pub., 517 p.

———— and R. F. Dill, 1966, Submarine canyons and other sea valleys: Chicago, Ill., Rand-McNally, 381 p.

Stanley, D. J., 1969, Lecture 8: Sedimentation in slope and base-of-slope environments, *in* D. J. Stanley ed., Concepts of continental margin sedimentation: Am. Geol. Institute Short Course Lecture Notes, p. 8-1 to 8-25.

———— and Unrug, R., 1972, Submarine channel deposits, fluxoturbidites and other indicators of slope and base-of-slope environments in modern and ancient marine basins, *in* J. K. Rigby and W. K. Hamblin, eds., Recognition of ancient sedimentary environments: SEPM Spec.

Pub. 16, p. 287-340.

Stauffer, P. H., 1967, Grain flow deposits and their implications, Santa Ynez Mountains, California: Jour. Sed. Petrology, v. 37, p. 487-508.

Stuart, C. J., and C. A. Caughey, 1977, Seismic facies and sedimentology of terrigenous Pleistocene deposits in northwest and central Gulf of Mexico, *in* C. E. Payton, ed., Seismic stratigraphy—applications to hydrocarbon exploration: AAPG Mem. 26, Tulsa, Oklahoma, p. 249-275.

Thompson, T. L., 1976, Plate tectonics in oil and gas exploration of continental margins: AAPG Bull., v. 60, p. 1463-1501.

Uyeda, S., 1974, Northwest Pacific trench margins, *in* C. A. Burk and C. L. Drake, eds., The geology of continental margins: New York, Springer-Verlag, p. 473-491.

Varnes, D. J., 1958, Landslide types and processes, *in* E. G. Eckel, ed., Landslides and engineering practice: Highway Research Board, Natl. Acad. Sci., Spec. Rept. 29, p. 20-47.

———— 1978, Slope movement types and processes, *in* R. L. Schuster and R. J. Krizek, eds., Landslides: analysis and control: Transportation Research Board, Natl. Acad. Sci., Spec. Rept. 176, p. 11-33.

von Huene, R., et al, 1978, Japan Trench transected on Leg 57: Geotimes, v. 23, p. 16-21.

Walker, R. G., 1975, Generalized facies models for resedimented conglomerates of turbidite association: Geol. Soc. Amer. Bull., v. 86, p. 737-748.

———— 1978, Deep-water sandstone facies and ancient submarine fans: Models for exploration for stratigraphic traps: AAPG Bull., v. 62 p. 932-966.

———— and E. Mutti, 1973, Turbidite facies and facies associations *in* G. V. Middleton and A. H. Bouma, eds., Turbidites and deep water sedimentation: SEPM Pacific Sect. Short Course, Anaheim, California, p. 119-158.

Weeks, L. G., 1974, Petroleum resources potential of continental margins, *in* C. A. Burk and C. L. Drake, eds., The geology of continental margins: New York, Springer-Verlag, p. 953-964.

Wilde, P., W. R. Normark, and T. E. Chase, 1978, Channel sands and petroleum potential of Monterey deep-sea fan, California: AAPG Bull., v. 62, p. 967-983.

Wilson, J. L., 1975, Carbonate facies in geological history: Berlin, Springer-Verlag, 471 p.

Yarborough, H., 1971, Sedimentary environments and the occurrence of major hydrocarbon accumulations (abs.): Gulf Coast Assoc. Geol. Socs. Trans., v. 21, p. 82.

Sedimentology of Submarine Fans

D. G. Howell
W. R. Normark
U. S. Geological Survey
Menlo Park, California

INTRODUCTION

Subaqueous fan deposition occurs in a variety of settings from the deep sea to freshwater lakes. The mode of fan deposition includes a whole family of mass flow processes; thus the so-called "turbidite deep-sea fan" is a misnomer. Submarine fan is a more appropriate term that needs modification for the special case of a nonmarine fan.

Data derived from modern oceanographic surveys and studies of ancient submarine fans are used to construct generalized physiographic models composed of these parts: slope/feeder channel/canyon, inner (upper) fan, middle fan, outer (lower) fan, and basin plain; subenvironments of deposition include channels, channel thalwegs, levees, and interchannel and fringe areas.

Ancient fan deposits can be classified into seven broad lithofacies on the basis of grain size, bed fabric and thickness, and associated sedimentary structures. The environments represented by an ancient fan can be reconstructed from the association of particular lithofacies and the lateral and vertical character of bedding cycles.

These sedimentary structures, lithofacies, and bedding cycles are described and illustrated in this study with comparisons of subenvironments defined for modern an ancient fans.

During the last several decades, the recognition and classification of submarine fans has enabled geologists to better understand certain aspects of sedimentologic and tectonic dynamics, particularly along continental margins. In reconstructing an ancient fan, the spatial relations of lithofacies, rather than formations, need to be mapped. Excellent papers summarize widely applicable lithofacies-classification schemes (Mutti and Ricci Lucchi, 1972; Walker, 1978). Such lithofacies criteria have been used successfully in redescribing well-known submarine fan sequences (e.g., Ricci Lucchi, 1975a, b; Ingersoll, 1978). However, these lithofacies are not restricted to submarine fan depositional settings.

Photographic illustrations are not specifically referenced in this chapter. The order of photos follows the general outline of the text, and the caption for each illustration is self-explanatory.

Historical Review

Recognition of turbidity currents dates back at least to Florel (1885), who made reference to undercurrents flowing into Lake Geneva. Daly (1936) suggested that turbidity currents flowing off continental shelves could be erosive agents forming submarine canyons, while Stetson and Smith (1938) felt that turbidity currents were responsible for carrying fine sediments into the open ocean. Bramlette and Bradley (1942) attributed coarse, graded sand beds from deep sea cores to turbidity currents. Bell (1942) reported velocities for density currents in Lake Mead and discussed experimental work on the properties of turbidity currents. Kuenen (1937) experimentally tested the theory of turbidity currents; his later research focused on identification of hydraulic and transport processes responsible for graded bedding and movement of terrigenous sediments through canyons to ocean basins (e.g., Kuenen and Menard, 1952). Turbidity current origins for ancient marine sediments were proposed for beds in the Apennines (Migiorini, 1944), the Alps (Crowell, 1955), and southern California (Natland and Kuenen, 1951). In a classic paper (Kuenen and Migliorini, 1950), Kuenen also argued that the application of turbidity-current concepts explained to a large degree most of the then enigmatic features of flysch.

The papers of Daly and Kuenen, in particular, stimulated the interest of geologists, who during the next two decades launched many field and laboratory studies focusing on sediments of inferred density-current origin. Wood and Smith's (1957) study of the Aberystwyth Grits marked the start of a long line of authors who successfully applied Kuenen's theories to particular flysch sequences. Bagnold (1954) indicated that under certain conditions, mass flow (grain suspensions) may be sustained by an upward dispersive pressure rather than turbulence, so "grain flows" became a companion to "turbidites." Dott (1963), taking the concept further, proposed that a whole spectrum of mass flow processes occurred, with important associations in-

In selecting the illustrations for this chapter, we have tried to present a collection that expresses the geographic and stratigraphic diversity of submarine fans. To achieve this, slides were requested from a number of collegues. The donor of each illustration not taken by the senior author is acknowledged, and this effectively references the information contained in the respective caption. We would like to thank the following for sharing their submarine-fan photos with us: A. H. Bouma, Joanne Bourgeois, R. M. Carter, R. H. Dott, Jr., R. M. Egbert, M. D. Hicks, R. V. Ingersoll, J. P. Kern, G. de Vries Klein, M. G. Laird, M. H. Link, T. C. Mackinnon, Hugh McLean, C. H. Nelson, T. H. Nilsen, F. Ricci Lucchi, P. A. Scholle, Finn Surlyk, G. van der Lingen, and G. R. Winkler. In preparing this paper we have benefited by discussion and reviews by R. M. Carter, A. H. Bouma, T.H. Nilsen, and I.G. Speden.

cluding slumps, grain flows, and turbidity currents. Sanders (1965) correctly stressed that several such flow processes may be operating during the evolution of a single flow; the deposits of which were too often, then and now, casually referred to as "turbidites."

From field observations Bouma (1962) systematized many of the sedimentary structures known to be associated with turbidity currents *(sensu stricto)*. The full or partial Bouma T_{a-e} sequence is now a universally used term. Dzulynski and Walton (1965), among others, described and illustrated a host of sedimentary structures useful in identifying sedimentary rocks of density-current origin and allowing paleo-flow directions to be inferred. A further advance in understanding at this time resulted from the synthesis of field-based observation and contemporary flume studies, allowing interpretation of the Bouma sequence in terms of the hydrodynamic behavior of a waning current (Walker, 1965; Walton, 1967). Building on such understanding and using other sedimentologic criteria, Walker (1967) proposed a model to determine the relative proximal and distal aspect of turbidites. His was one of the first attempts to systematize sedimentary structures and lithofacies data in order to reconstruct the geometry of a flysch basin. Sullwold (1960) was one of the first workers to recognize the submarine fan geometry of ancient sediments using lithologic and paleocurrent criteria. Jacka et al (1968) developed an "idealized plan view of a deep-sea fan," based primarily on induction from sedimentalogic relations, proving remarkably similar to the geomorphic pattern of modern fans surveyed for the first time in the late 1960s (e.g., Shepard et al, 1969; Normark, 1970; Haner, 1971; Nelson and Kulm, 1973).

Ingersoll (1978) pointed out that the classifications of submarine fans developed by marine geologists are based on surface morphologies and lateral variations in surface sediments (e.g., Normark, 1974). In contrast, land-based geologists utilize the Waltherian principle of vertical sequence analysis combined with the lateral distribution of lithofacies (e.g., Mutti and Ricci Lucchi, 1972; Ricci Lucchi, 1975). From these geologic data one can infer the sedimentologic regimes and constructional dynamics of a submarine fan. Minor nomenclatural and conceptual problems exist between the marine and land-base fan models (see Fig. 1), but a working knowledge of both is essential. One looks down on known modern submarine fans and sees a time-correlative surface and surficial sediments, whereas the other infers ancient fans from within, recording changes through time with no direct knowledge of surface morphology.

Fan Occurrence

The stratigraphic association of fans is varied. Fan strata may overlie oceanic crust, other marine strata, cratonic crystalline rock, or even be interstratified with nonmarine beds. Fans constructed of resedimented massflow material are known from lacustrine, continental slope and rise, and abyssal plain settings; thus the bathymetrically nonrestrictive term "submarine fan," rather than the often used "deep-sea" fan, is preferred for marine occurrences, and "sublacustrine fan" for the nonmarine occurrences. In this study we are especially concerned with submarine fans.

Characteristics of any particular submarine fan result from the complex interplay of numerous variables, including at least: basin configuration (e. g., confined or unconfined); basin-axis gradient; tectonic setting (both local and global); source sediment type, size, and flux; climate; and sea-level stability. Although every fan is therefore unique, underlying models nonetheless emerge. The generalized fan model (Fig. 2a-d), though still very similar to the earlier models of Jacka et al (1968) and Normark (1970), incorporates the data available from a wide variety of recent studies of known modern and inferred ancient submarine fans. Schematic representations for several modern fans (Fig. 2e-h) show a wide range of geomorphic expressions. The Navy fan example (Fig. 2e) is the smallest and most thoroughly studied, showing many of the features of the "suprafan" model (Fig. 2d). The narrow and irregular basin shape restricts lateral migration in the mid-fan region of Navy fan, and the lower fan settings are found in two separate arms of the basin (Normark et al, 1979). The larger modern fans (Fig. 2f-h) have little in common and show much less agreement with any of the models due to prominent valley and channel systems. This lack of agreement emphasizes that the models in Figure 2, which are based on smaller fans built in basins, may represent only one part of a larger spectrum of fan deposition.

Whereas each submarine fan has unique features, there are apparently no sedimentary structures or lithofacies relations that are restricted to the submarine fan setting. For instance, "nearly all of the structures that have been found in alluvial channels are present in deep-sea channels of the

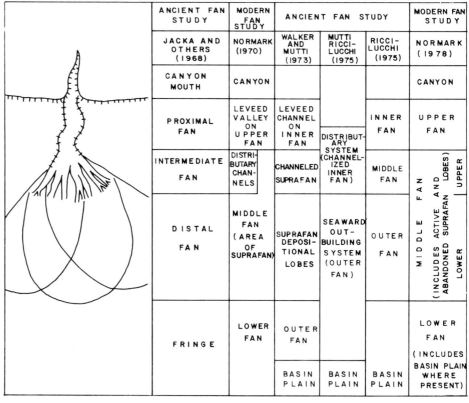

ANCIENT FAN STUDY	MODERN FAN STUDY	ANCIENT FAN STUDY			MODERN FAN STUDY
JACKA AND OTHERS (1968)	NORMARK (1970)	WALKER AND MUTTI (1973)	MUTTI RICCI-LUCCHI (1975)	RICCI-LUCCHI (1975)	NORMARK (1978)
CANYON MOUTH	CANYON				CANYON
PROXIMAL FAN	LEVEED VALLEY ON UPPER FAN	LEVEED CHANNEL ON INNER FAN	DISTRIBUT-ARY SYSTEM (CHANNEL-IZED INNER FAN)	INNER FAN	UPPER FAN
INTERMEDIATE FAN	DISTRI-BUTARY CHAN-NELS	CHANNELED SUPRAFAN		MIDDLE FAN	MIDDLE FAN (INCLUDES ACTIVE AND ABANDONED SUPRAFAN LOBES) UPPER / LOWER
DISTAL FAN	MIDDLE FAN (AREA OF SUPRAFAN)	SUPRAFAN DEPOSI-TIONAL LOBES	SEAWARD OUT-BUILDING SYSTEM (OUTER FAN)	OUTER FAN	
FRINGE	LOWER FAN	OUTER FAN			LOWER FAN (INCLUDES BASIN PLAIN WHERE PRESENT)
		BASIN PLAIN	BASIN PLAIN	BASIN PLAIN	

Fig. 1—Comparison of published submarine fan classifications modified from Ingersoll (1978).

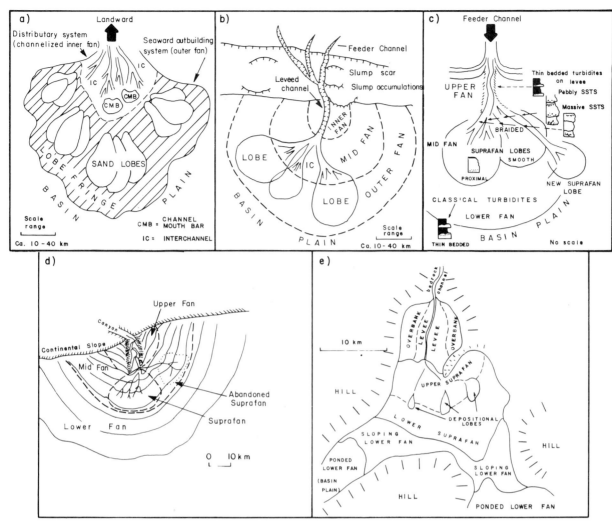

MODERN DEEP-SEA FANS

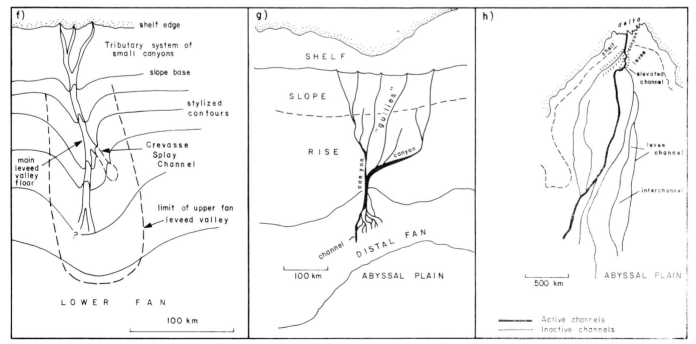

Fig. 2—Submarine fan models and growth patterns of selected modern fans: **A, B,** and **C** inferred from the rock record and represent restricted borderland-like basins (**A** from Mutti [1977], **B** from Ricci Lucchi [1975], and **C** from Walker [1978]). **D** is model for small modern submarine fans (from Normark, 1978). **E, F, G,** and **H** are simplified sketches of modern fans. **E** is Navy fan in the California borderland (from Normark et al, 1979). **F** is patterned after the Noyo lobe of Delgada fan (from Normark and Hess, 1980). **G** is Hatteras fan from Cleary and Conolly (1974), and **H** the Bengal fan from Curray and Moore (1974).

Delaware Basin Group. Indeed, from a single exposure it might be impossible to distinguish a deep-sea channel from an alluvial one" (Jacka et al, 1968, p. 4).

Despite the variables and lack of any unique diagnostic features, most ancient fan deposits can nonetheless be readily recognized because they possess a large number of characteristic sedimentary structures and lithofacies. By carefully recording both the stratigraphic and geographic locations of such features, the physiography of an ancient fan can be approximated. Thus, the purpose of this study is to present selected illustrations of the sedimentary structures and rock types belonging to submarine fan deposits, and then relate these features to specific settings within the idealized fan model. Examples are chosen from a variety of inferred ancient submarine fans composed of different rock types.

Studies of modern sedimentation and facies analysis of ancient submarine-fan systems indicate that two broadly contrasting basin settings can be defined: submarine fans generally less than 100 km across occupying restricted basins floored by continental crust (e.g., borderland fans of southern California); and submarine fans greater than 100 km across in deep water on oceanic crust (e.g., Astoria, Monterey, Bengal fans, and the Amazon cone; Nelson and Kulm, 1973). In smaller restricted basins, sediment is commonly coarse grained, well sorted, and matrix depleted; it is derived from littoral drift cells and passes into the basin via submarine canyons. Larger submarine fans of open-ocean basins comprise finer grained, matrix-rich turbidites and a much larger hemipelagic mud component. These systems generally feed through major river systems, and a higher matrix content in the massflow deposits results in a seaward displacement of the coarse-grained facies (Damuth and Kumar, 1975).

Ancient analogs of restricted basin fans are well preserved and are known petroleum reservoirs (e.g., southern California and the North Sea) whereas ancient examples of deep-sea fans built on oceanic crust are highly dismembered and probably poor petroleum reservoirs due to the effects of tectonic accretion onto continental margins. For this reason, the focus of this paper is submarine fan systems of small restricted basins, though the lithofacies discussed may apply to some larger, open-ocean systems (e.g., Bouma and Nilsen, 1978).

TERMINOLOGY

Sedimentation Processes

Within any submarine fan, sediment dispersal is governed by a variety of mechanisms including rolling, suspension, and saltation of traction flow and intergranular mass flow (Middleton and Hampton, 1976). For most fans, terrigenous sediment flux is high and sedimentation is dominated by mass flow processes. The resulting deposit is considered a "resedimented bed," because the material has had a previous history of deposition, commonly in fluvial, coastal, or shelf settings, before being remobilized into a mass flow. Possibly six major types of processes are applicable to resedimented, mass-transported sediments; they are slump creep, slurry flow, grain flow, fluxoturbidity current, turbidity current, and fluidized flow (see Carter, 1975). At the moment of deposition a given bedform reflects the specific flow process then operational (Fig. 3). However, most redeposited sediments are probably involved in two or more of these mass-transport mechanisms during their downslope journey to the fan depositional site (Fig. 4).

The volumes of mass flow deposits range from a few tens of cubic meters to at least as large as 11 cu km, which is the largest flow unit presently on record (Elmore et al, 1979). This enormous deposit, the Black Shell turbidite, has been traced from the Hatteras fan out across a large part of the Hatteras abyssal plain and measures at least 100 by 500 km. As is probably typical of most large mass-flow deposits, the Black Shell turbidite was initiated as a massive shelf-edge slump, and resulting deposits become progressively more organized in a downcurrent direction, as reflected by better sorting, more continuous vertical grading, better developed Bouma sequences, and a lower mud content (Elmore et al, 1979). Menard (1964) calculated a maximum velocity of 19.1 m/sec for a turbidity current that followed the huge Grand Banks slump east of Newfoundland, and velocities ranging from 3 to 6 m/sec have been estimated for density currents depositing material in the Cascadia Channel of the northeast Pacific (Griggs and Kulm, 1970). A turbidite moving 5 m/sec requires 5.5 hrs to travel 100 km.

The grain-size competence of density currents depends on the density and height of the current, the density of the clasts, the gradient, and the bottom stress exerted by the flow, which is in part a function of the current velocity (Komar, 1970). Inferring reasonable current densities and dimensions, density currents with velocities approaching 20 m/sec are capable of carrying clasts up to large cobble sizes.

Reworking of sediment by bottom or contour currents that form traction-carpet deposits occurs in some fans, but is thought to play a minor role. Sediment rain of fine-grained material is ubiquitous, though the volume of this material, which accumulates in different fan settings, is highly variable. As a result of evolution of calcareous plankton in the Cretaceous, it is often possible to distinguish between pure pelagic debris and the finer grained terrigenous material of a mass flow that settles out in the waning stages of deposition of a resedimented bed. This distinction can be conveyed by adding to the Bouma sequence the T_{ep} interval for pure pelagic material, in contrast to turbidite mud T_{et}.

Primary fan physiographic divisions

Primary surface regions of a submarine fan as used in the lithofacies review given later are depicted on Figure 2a-b. Many fans are fed by canyons, slope-feeder channels, or deltaic sources that tend to localize sediment supply. Seaward from these source areas most fans can be subdivided into an inner, middle (mid), and outer fan. Figures 1 and 2 both indicate problems in comparing physiographic settings between modern and ancient fans. These difficulties reflect basic differences in types of data used to contruct fan models. Morphologic characteristics are among the most easily recognized features of modern fans, and distinctions of upper and mid fan divisions are commonly (but not always) tied to key morphologic characters. On ancient fans, the divisions of inner, middle, and outer fans must be based on facies interpretations because the morphology of the fan at any given time cannot be determined. Thus, it is not surprising that the mid fan division for one worker is termed lower fan by another.

Distal and Proximal

The terms "distal" and "proximal," correctly used in reference to the relative separation between a sediment and its source, are often used as general lithofacies descriptors. Specifically, "distal" is used for thin-bedded,

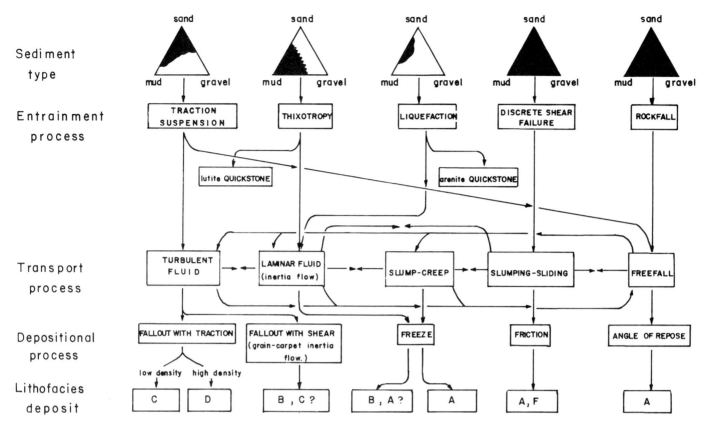

Fig. 3—Flow-diagram for the common types of sediment mass-transport processes and their products. Identical sedimentary products may result from a variety of different paths across this diagram, as shown by the many different flow-paths indicated by arrows, modified from Carter (1975).

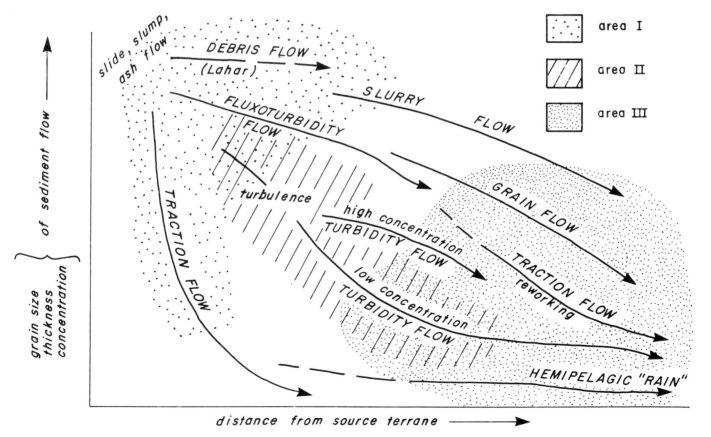

Fig. 4—Inferred dynamic relations of flow processes that operated during the deposition of the different middle Miocene lithofacies, southern California borderland: Santa Cruz Island (areas I and II) and Santa Rosa Island (area III), California. Modified from Howell and McLean (1976).

muddy turbidites and "proximal" for thicker bedded, sandy turbidites. However, such usage is either ambiguous or incorrect and should be discouraged. For example, so-called "distal turbidites" are now known to occur throughout a fan, particularly in overbank settings, and "proximal turbidites" are commonly well-developed in a lower fan association. Thus any index of degree of distal or proximal position, if based primarily on bed thickness and grain size, must be applied with caution (Bouma and Nilsen, 1978; Ingersoll, 1978).

Progradation and Retrogradation

Few submarine fans occur as static sedimentary regimes; more commonly an active fan migrates through space and time. The sense of migration is readily determined in an ancient sequence through vertical (stratigraphic) sequence analysis. This technique applies Walther's law of succession of facies, which states simply, the vertical stacking of lithofacies at any given location reflects the migrating lateral positions of these facies over a period of time (see discussion on p. 187-188 in Blatt et al, 1973).

In a progradational sequence, ascending lithofacies relfect successively more proximal locations on the fan (e.g., see Fig. 5). Circumstances that may cause a progradation include: an increase in the sediment flux (due to either climatic or tectonic processes); a lateral or forward movement of the fan environment (caused perhaps by change in location of the feeding channel); a rising of the continental freeboard (isostatic or tectonic); or eustatic sea-level lowering (due principally to glaciation or diminishing rates of mid-ocean ridge spreading). Thus a progradational event is not necessarily related to a regressing shoreline.

In a retrogradational sequence, ascending lithofacies indicate a progressive change to more distal depositional settings on the fan (e.g., see Fig. 6). This may be caused by the same factors of a progradation, but working in reverse. Commonly in a vertical sequence of an ancient submarine fan, progradational and retrogradational cycles will be superposed in a pattern that reflects the tectonic and sedimentary history of the basin.

Morphology and distribution of young sedimentary deposits on modern fans show that detrital deposition generally is not uniformly distributed. Few turbidity currents are capable of spreading across the entire basin or

fan. Active deposition is confined to a given sector, and fan shape is gradually formed as the active locus of deposition shifts. Within each active depositional sector, turbidites show a general decrease in both bed thickness and grain size away from the source (Fig. 7). Each progradational or retrogradational cycle may be composed of a series of partially overlapping depositional packets showing a systematic variation of sedimentological characters within each packet.

Geomorphic Terms and Ancient Rocks

Certain geomorphic features easily identified on modern submarine fans cannot be easily recognized in ancient turbidite deposits. Thus, use of geomorphic terminology in discussing facies associations for ancient rocks may be misleading. For example, the term levee refers to a depositional ridge that bounds a channel. In most

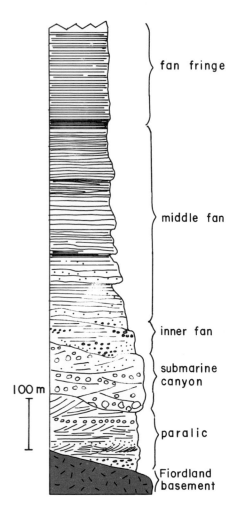

Fig. 6—An example of a retrogradational sequence, Balleny Group, Eocene and Oligocene, Fiordland, New Zealand. Modified from Carter and Lindqvist (1977).

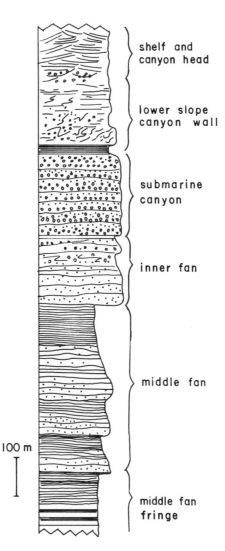

Fig. 5—An example of a progradational sequence, Pigeon Point Formation, Upper Cretaceous, California.

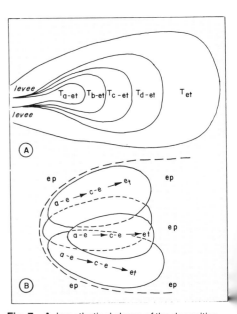

Fig. 7—**A,** hypothetical shape of the deposition cone of a single turbidite beyond the confines of a channel, no scale. **B,** hypothetical filling of a basin by turbidites showing that each depositional cone does not fill the entire basin and that succeeding turbidites seldom lay completely on top of each other, no scale factors. Figure modified from Bouma (1972).

turbidite exposures, it is impossible to recognize any original depositional relief and local slopes that might define a natural levee. Differential compaction can result in coarser channel-fill sediments appearing as positive relief features in ancient sequences. Because overbank deposition can occur over wide areas without building any perceptible ridge along a fan channel, the general term overbank deposit is preferred to levee deposit. Channel deposits are more confidently identified because of common local erosional relief along margins and floor of the channels.

The term suprafan refers to an area of characteristic morphology on some modern submarine fans. A suprafan generally exhibits an overall convex-upward bulge on a radial profile of the fan and has a rough surface of numerous channels, channel remnants, isolated depressions, and large depositional bedforms. The suprafan is also the area of active sand deposition on modern fans. These characters may be difficult to use in recognizing suprafan deposits in ancient turbidites, especially because suprafans extend over large areas and suprafan channels may be confused with upper fan valleys. As a result, some ambiguity exists in equating the suprafan concept with fan classifications based on ancient lithofacies relations (Fig. 1).

MODERN FAN MORPHOLOGIES AND DIMENSIONS

The primary factors controlling morphology and dimensions of submarine fans are size and shape of the basin of deposition and total volume, rate, and grain-size distribution of sediment supplied to the basin. Recent work shows that the character of the sediment supplied is probably the most important factor controlling development of geomorphic features on upper and middle fan areas; most critical is the proportion of coarse-grained sediment (primarily sand) and ability of the turbidity currents to transport sandy material efficiently (Normark, 1978). Thus several models for fan deposition must be considered, and some of the seman-

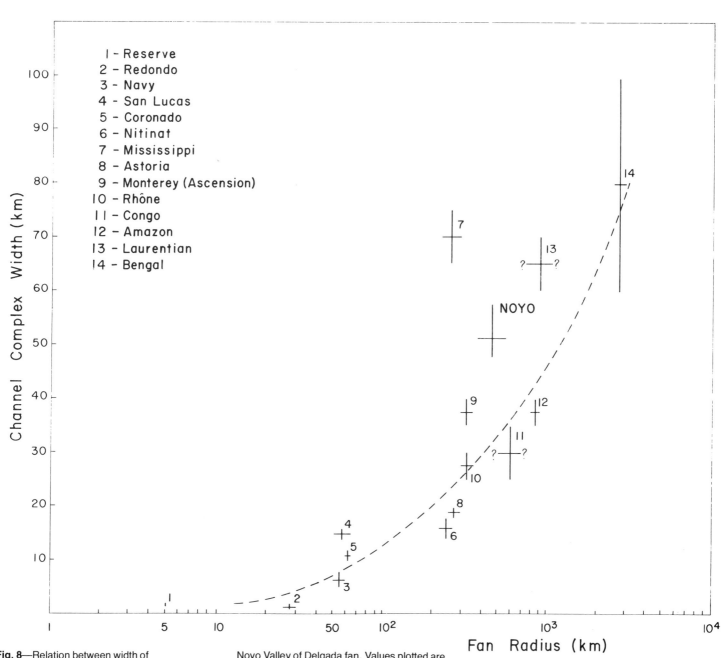

Fig. 8—Relation between width of fan-valley/levee complex and fan radius modified from Normark (1978) with addition of Noyo Valley of Delgada fan. Values plotted are for fan-valley crossing on uppermost fan.

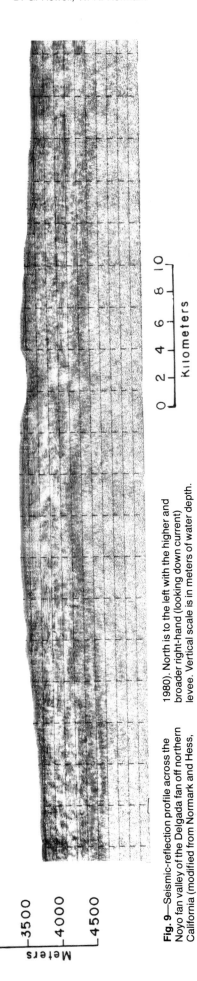

Fig. 9—Seismic-reflection profile across the Noyo fan valley of the Delgada fan off northern California (modified from Normark and Hess, 1980). North is to the left with the higher and broader right-hand (looking down current) levee. Vertical scale is in meters of water depth.

tic and conceptual problems depicted in Figure 1 may result from real differences in fan morphology (Fig. 2d-g). Because many workers have attempted to fit all observations into one model, additional research is needed to resolve the primary differences in types of submarine fans. Some of these differences are noted in the discussions of geomorphic features and subdivision of a fan surface into upper, middle, and lower fan with associated basin plain and abyssal plain regions.

Fan Dimensions and Submarine Canyons

Provided the basin of deposition is large enough, the size of a submarine fan is proportional to the amount of sediment supplied. Thus, the largest known present-day fans form offshore from major sediment-carrying rivers that spill into expansive continental-margin settings. The Bengal fan is fed by the Ganges and Brahmaputra Rivers and is the largest in the modern oceans (more than 2,500 km in radius). Other large fans (>500 km radius) form in deep water below deltas of the Indus and Amazon Rivers. Smaller submarine fans are found in enclosed basins along continental margins as in the California Continental Borderland; sediment sources tend to be local and major sediment influxes seasonal, with fans commonly 20 to 60 km in radial length. Comparison with the even smaller sublacustrine fans (radial length less than 10 km) suggests that fan size itself is not important in determining geomorphic characters of a submarine fan (Normark, 1978).

Fans receiving sediment from deltaic areas may not have a well-defined canyon acting as a point source for sediment. The Laurentian fan has numerous small slope gullies feeding downslope into three large fan valleys several kilometers wide and as deep as 800 m (Stow, 1977). A small man-made sublacustrine fan also has major leveed valleys with no feeding canyon (Normark, 1974).

Dimensions of submarine canyons rarely show proportionality to the associated fan. Canyon size is controlled locally by the nature of the bedrock underlying the shelf and slope areas; relatively large canyons could develop in prograding slope areas where deposition of silt and mud sediment (facies G, Mutti and Ricci Lucchi, 1972) is rapid. Canyons cut in plutonic or metamorphic basement may enlarge slowly. Thus, canyon size and its local geologic setting are not necessarily reliable indi-

cators of either fan size or geomorphic features developed on the fan.

Upper Fan and Fan Valley

The upper fan division on a modern fan is most easily recognized by the presence of a prominent leveed valley (Fig. 1). Where a single submarine canyon feeds sediment to the fan, a single leveed valley develops and the down-fan termination of the valley marks the transition to the mid fan (Normark, 1978). The size of the valley complex, which includes open channel and associated levee sediments, exceeds 70 km in width on large fans. Size of the valley complexes is larger on the larger fans (Fig. 8) and seismic-reflection profiles show that most sediments of the valley complex are overbank or "levee" deposits (Fig. 9). Overbank and hemipelagic sediments dominate on the upper fan because the valley floor forms a small percentage of the fan surface.

Leveed valleys form elevated pathways for sediment transported across the head of the fan, and thus indicate aggradation of both levees and channel floor. Valley floors are commonly aggraded a hundred meters or more above the adjacent fan surface. The gradient of the levee flank away from the channel axis is 2 or 3 times greater than the average channel thalweg gradient. Large-scale bedforms with wavelengths of a kilometer or more commonly develop on the outer slopes of the levees on large fans, e.g. Monterey and Amazon fans (Normark et al, 1980; Damuth and Kumar, 1975; Figs. 9, 10). Detailed studies show large scour features around levee terminations (Normark et al, 1979). Although surficial relief on levees varies, the overall internal structure shows nearly parallel bedding that thins from the crest outward. Wedging of reflectors within a levee section is gradual and occurs over distances of several kilometers to tens of kilometers; thus, it would require excellent exposures and an absence of differential compaction to recognize this geometry in ancient fan sequences.

Local relief on valley floors seen on deep-tow, narrow-beam sounding profiles indicates active channeling. Smaller channels cutting valley floors range from several meters in depth and a few tens of meters in width for smaller fans to thalweg channels 20 m deep and hundreds of meters wide on larger fans. Local steps and isolated depressions are also common on valley floors. Sand- and gravel-size sediments are commonly cored from upper fan-valley floors, and during aggradation of the valley, thick coarse-grained beds are deposited. The most common sedimentary facies ascribed to the upper fan are coarse-grained channel-fill sequences (Nelson and Kulm, 1973; Walker and Mutti, 1973; Walker, 1978), although overbank silts, muds and hemipelagic sediments are more abundant volumetrically.

Available sedimentologic data from a wide size range of submarine fans show that those fans receiving high proportions of sand or coarser sediment tend to have shorter and less elevated leveed valleys. This may account for the common occurrence in ancient sequences of medium- to coarse-grained mass flow deposits cropping out in bundles of thick-bedded flow units interbedded with thin-bedded turbidites or hemipelagic mud. For without the presence of levees, laterally shifting turbidites will not be expressed as thinning-upward sequences. The best developed leveed valley systems are found on modern fans fed from deltaic sources where much of the coarser sediment remains in the deltaic environment (Indus, Ganges, or Mississippi fans) or from shelf areas where much of the coarse material is trapped in estuaries (the Laurentian fan) or simply left behind in areas of wide shelves with no canyons.

Middle Fan

The effects of rate and size distribution of sediment supplied to a submarine fan appear to produce the greatest variability in geomorphic features of the middle fan region. In many modern fans along the western margin of North America, the middle fan area is the site of rapid sand deposition. At the end of the leveed valley of the upper fan, where a channelized sand body appears as an irregular bulge on a radial fan profile (Fig. 11), this depositional bulge is termed a suprafan (Fig. 1). Mutti and Ricci Lucchi (1972) described ancient sandy, unchannelized depositional lobes in an outer (lower) fan region, and fan facies associations used in this study follow the general Mutti-Ricci Lucchi model. Some ancient fans, however, are composed wholly of overlapping channelized and unchannelized sandstone bodies that depart from this model, reflecting instead a suprafan-like depositional setting (e.g., early Tertiary Cantua submarine fan of California, Nilsen, 1979; and Late Triassic Torlesse submarine fan(s) of New Zealand, Howell, 1979).

Initially, it was thought that much of the local relief seen on suprafans reflected numerous well-defined distributary channels. All detailed studies of suprafans, however, show a wide variety of channels, channel remnants, and other mesotopographic features that are not resolvable by conventional survey techniques. Thus, little is known about geomorphic features that range in size from a few meters to several hundred meters in width. Only high-resolution, deeply towed geophysical instruments (Spiess and Tyce, 1973) and multibeam-sonar-sounding systems can determine the structure and relief of modern midfan areas. Available deep-tow surveys of several modern fans show that these distributary channels are generally without levees and are several hundred meters wide, 1 to 10 m deep, and only several kilometers long. These distributary channels are more sinuous than the larger, leveed valleys on the upper fan. The interchannel areas may have rough topography including depositional bedforms and giant scours (Normark et al, 1979). In distal areas of active sand deposition, surface relief is generally too small to resolve even with deep-tow techniques (features of less than two meters vertical relief and of lengths less than 20 m). Depositional lobes build up at the end of the distributary channels on the upper suprafan, and except for channels around margins of lobes there is little relief on this area of the suprafan (Fig. 12). The lower suprafan is also a site of sand deposition, and sediments here might resemble channel-free lower fan sand lobes of the Mutti-Ricci Lucchi model.

Lower Fan to Basin Plain

In the Mutti-Ricci Lucchi model, sand lobes characterize the lower (outer) fan (Fig. 2). The best described example of sand-lobe geometry is the Laga Formation in east-central Italy. These sandstone bodies range from 3 to 15 m in thickness and are more than 10 km across (Mutti et al, 1978). Similar features are not well known from modern fans except perhaps for the Astoria fan (Fig. 13; Nelson, 1976), primarily because no coring device has been available to penetrate sand beds of these thicknesses near the surface of modern fans. In modern lower fan settings, nonchannelized muddy turbidites predominate. Morphologically, it is difficult or impossible to distinguish lower fan from basin-plain or abyssal-plain regimes, because slope of the fan surface, shape of fan, and sedimentary characters change subtly across this environment. Deposition in small basins, however, tends to enhance development of basin-plain facies at the expense of the morphologic lower fan.

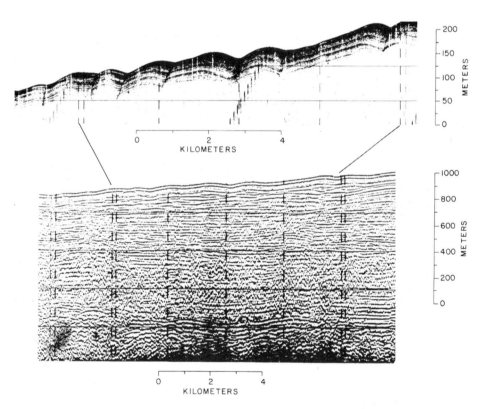

Fig. 10—High-resolution reflection profile (upper) and single-channel seismic-reflection profile (lower) of sediment waves on right-hand levee of valley on upper Monterey fan (modified from Normark and others, in press).

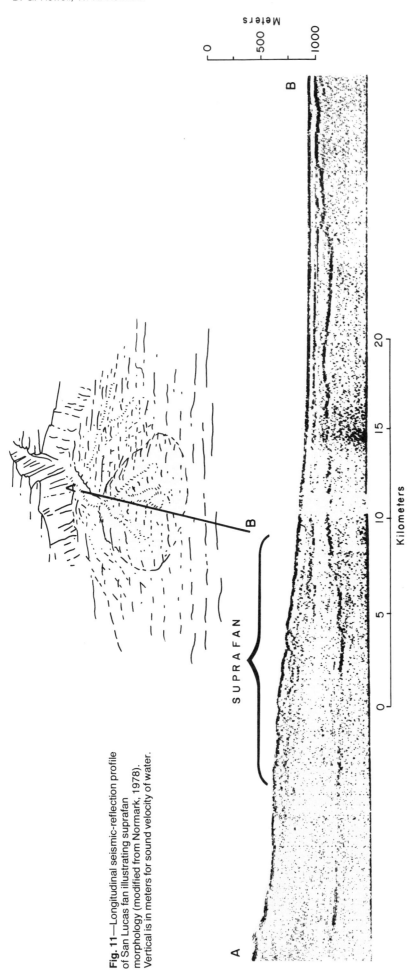

Fig. 11—Longitudinal seismic-reflection profile of San Lucas fan illustrating suprafan morphology (modified from Normark, 1978). Vertical is in meters for sound velocity of water.

For many modern fans, the terms lower fan or basin plain, and possibly abyssal plain (assuming the appropriate bathymetry) may be applied to the same depositional regime.

SUBMARINE FAN LITHOFACIES

A host of published and unpublished schemes are used for shorthand descriptions and facies designations of submarine fan deposits. Most incorporate characteristics of grain size, sand-to-shale ratio, bed thickness, bedding continuity, and associated sedimentary structure. An ideal classification should be objective and all embracing, but the test of any taxonomy is its utility for inferring genetic relations. In this study we follow the classification of lithofacies first proposed by Mutti and Ricci Lucchi (1972), because most lithologies of a submarine fan are included, and one can generally infer the depositional setting of ancient submarine fans from the stratigraphic or map pattern of the Mutti-Ricci Lucchi lithofacies classification.

Mutti and Ricci Lucchi (1972), Walker and Mutti (1973), and Ricci Lucchi (1975a, b) classify fan sediments into seven lithofacies (A through G). Ingersoll (1978) and Bouma and Nilsen (1978) adopted these facies designations for description of Upper Cretaceous strata of California and Kodiak Island, Alaska, respectively. The following summary of characteristics, and many of the accompanying illustrations of these facies are taken directly from the above-mentioned authors. For a more elaborate description of each facies, the reader should refer directly to the above papers. It must be emphasized, however, that rock units conforming to all or part of this lithofacies scheme do not necessarily require a submarine-fan reconstruction, i.e. flysch or "turbidites" are not by themselves *prima facie* evidence for a fan morphology.

Facies A

Facies A consists of conglomerate and coarse-grained to pebbly sandstone. Bed thicknesses are generally greater than one meter, but lateral variations in thickness are frequent. Scour and channeling is typical; however, in distal parts of a fan, facies A conglomerate beds often have planar bases. Most facies A outcrops display a succession of composite beds. Individual flow units are identified by distinct grain-size variations or alignment of plant debris or intraformational mud

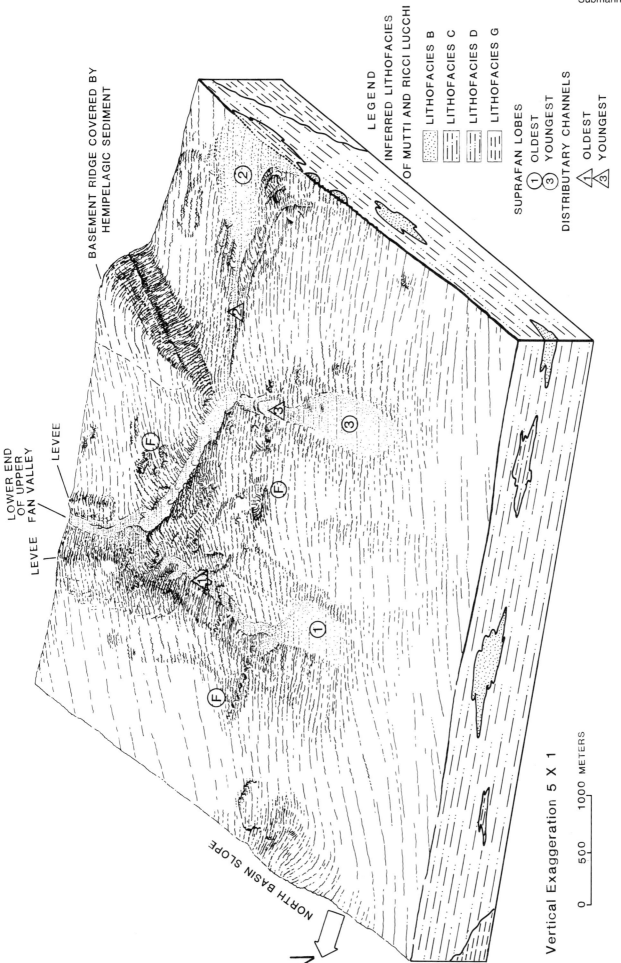

LEGEND

INFERRED LITHOFACIES
OF MUTTI AND RICCI LUCCHI

LITHOFACIES B

LITHOFACIES C

LITHOFACIES D

LITHOFACIES G

SUPRAFAN LOBES

① OLDEST   ③ YOUNGEST

DISTRIBUTARY CHANNELS

△₁ OLDEST   △₃ YOUNGEST

Ⓕ FLUTE-SHAPED DEPRESSIONS

Vertical Exaggeration 5 X 1

0   500   1000 METERS

BASEMENT RIDGE COVERED BY
HEMIPELAGIC SEDIMENT

LOWER END
OF UPPER
FAN VALLEY

LEVEE

LEVEE

NORTH BASIN SLOPE

N

Fig. 12—Oblique map of the middle fan region of Navy fan (modified from Normark et al, 1979; drawn by T. R. Alpha). Mutti-Ricci Lucchi facies are depicted for the suprafan area with distributary channels leading to small, smooth depositional lobes. Only one distributary is active at any time and circled numbers show sequence of lobe development (1 is oldest, 3 is youngest).

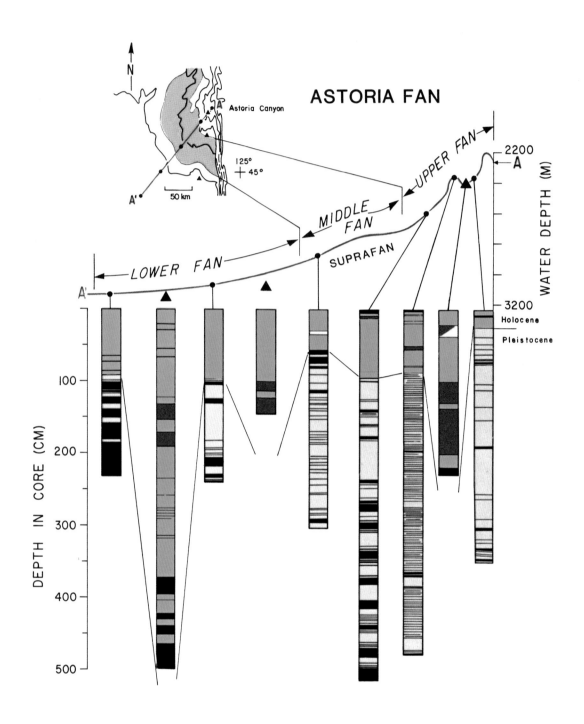

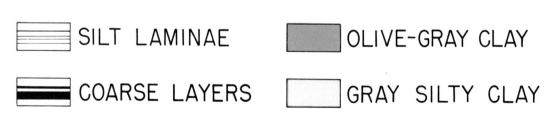

Fig. 13—Lithology and stratigraphy from cores on Astoria fan (modified from Nelson, 1976). The middle fan region along this profile shows most of the morphologic characters used to define a suprafan and was an active area for sand deposition in the Pleistocene.

clasts. A given bed may be nongraded or show normal, reverse, coarse-tail, or distribution grading (cf. Middleton, 1967). Conglomerate beds with cobble-size material may be normally or inverse to normally graded. Facies A strata are most commonly associated with rocks of facies B and E.

Special attention should be given to the fabric of conglomerate beds of Facies A (Walker, 1975; Kelling and Holyroyd, 1978; and Howell and Link, 1979). Clasts may be either framework or matrix supported, and a succession of flow units may include both massive and stratified units. "Disorganized" and "organized" are sometimes useful bed descriptor terms. The former applies to massive beds with clasts that are randomly oriented and matrix supported, while the latter denotes stratified beds, possibly graded, where clasts are either matrix or framework supported and may be imbricated.

For beds within facies A, the Bouma sequence is generally not applicable though for some coarse sandstone beds, T_a or T_{ae} is appropriate. Facies A beds principally result from slurry or debris-flow processes (disorganized beds) or from turbulent and traction fallout or grain-flow processes (organized beds) (Middleton and Hampton, 1973).

Pebbly mudstones of Crowell (1957) are generally not considered part of facies A. This is especially true where slump folding is clearly evident (see facies F).

Facies B

Facies B is generally composed of coarse- to medium-grained sandstone in thick, massive and often composite bed sequences. Some scour and channel features occur, but lateral bed continuity is greater than in facies A. A typical bed includes granules or mud chips along a basal scour surface, with faint parallel laminations, dish structures, and elutriation scars in the remainder of the bed (Fig. 14). Individual flow units are often difficult to determine due to uniformity in grain size in an outcrop. Fresh surfaces are generally required to identify zones of amalgamation for a composite sequence of flow units.

The Bouma sequence is not applicable to beds of this facies. Facies B strata are commonly interbedded with E and sometimes with A, C, and D facies strata. When A and B, or B and C beds occur together, all gradations of characteristics may also occur, rendering it difficult or arbitrary to subdivide the unit. Facies B generally occurs in a channelized setting, particularly but not exclusively on the inner and middle fan. Facies B beds reflect hydraulic processes of grain flow, and when they are transitional to facies C in character, fluxoturbidity currents are suggested (Carter, 1975; Middleton and Hampton, 1976).

Facies C

Facies C comprises coarse- to fine-grained sandstone commonly interbedded with thin layers of mudstone. The sandstone beds are the classical turbidites of Bouma (1962). The sixfold subdivisions of an ideal turbidite is not always fully developed. Bouma's original fivefold division (T_a to T_e; Fig. 15) has been modified to distinguish between turbidite and pelagic mud intervals in the T_e division; these are referred to as T_{et} and T_{ep} divisions respectively. The T_{et} may be graded, though recognition of this requires special circumstances such as differential surficial drying of water saturated beds or a T_{et} of terrigenous mud overlain by a T_{ep} of calcareous ooze. Interval T_d is commonly absent; or it may be difficult to see or distinguish between the T_d and T_c or T_{et} and T_{ep} intervals. For description of a given bed the notation Bouma T_{a-e} is useful (T_{a-e}, T_{ace}, T_{bc}, etc.).

Sandstone beds of facies C are generally 0.25 to 2.5 m thick, though thinner beds with a complete Bouma sequence are common. Facies C beds are of uniform thickness for great lateral distances. Mud chips or pebbles may lie along the basal surface, normal grading is common, and sole markings are well developed.

Facies C is generally associated with the upper parts of channel-fill sequences and with such nonchannelized settings as middle fan fringe, outer fan or even basin plain. These sandstone beds are primarily deposited by turbidity currents (Kuenen and Migliorini, 1950; Middleton, 1967).

Facies D

Facies D consists of thin interbeds of sandstone and mudstone with sandstone beds being tabular and persisting laterally for great distances. Each sandstone bed is typically graded and displays the upper part of the Bouma sequence T_{cde} or T_{ce}. Bed thickness is generally 0.05 to 0.25 m, and sole marks are commonly well developed. Thick facies D strata are transitional with thin facies C beds, and the two are often interbedded. Facies D strata are traditional "distal turbidites" though such thin-bedded turbidites actually occur in nearly all parts of a submarine fan as well as on the basin plain.

Facies D sandstone beds in general represent deposition by low-density turbulent flows; often the T_{et} interval of a turbidite will be much finer grained and thicker than the underlying T_c or T_d interval of the same bed.

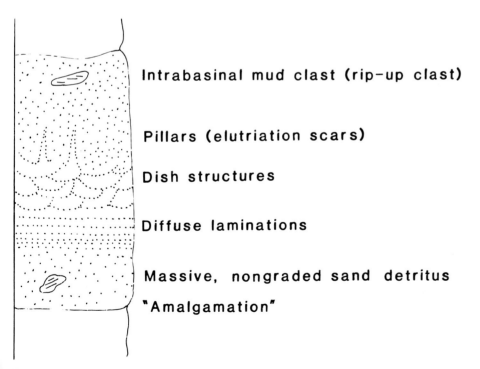

Intrabasinal mud clast (rip-up clast)

Pillars (elutriation scars)

Dish structures

Diffuse laminations

Massive, nongraded sand detritus

"Amalgamation"

Fig. 14—Idealized sequence of structures within a facies B sandstone, grainflow bed or fluxoturbidite, modified from Stauffer (1967).

Facies E

Facies E consists of thin interbedded sandstone and mudstone with a variety of internal-bedding characteristics including flaser bedding, massive sand, graded sand, and climbing ripples. It can be differentiated from D beds by the following characteristics: (1) the sandstone is coarser grained than in facies D beds of equivalent thickness, (2) E has a higher overall sandstone-to-shale ratio, (3) E has thinner but more numerous sandstone beds, and (4) E commonly contains wavy and discontinuous sandstone bedding. The Bouma sequence may not always apply to facies E, however a T_{ce} cycle, with a pronounced grain-size discontinuity, is a likely representation.

Facies E beds are characteristically associated with channelized environments of a submarine fan. The facies B/E couplet is typical of inner to middle fan associations, and facies E along with D, G and F compose most overbank or levee deposits. Facies E beds represent high-concentration gravity and traction flow processes near channel margins.

Facies F

Facies F consists of remobilized deposits exhibiting mass slumping and localized resedimentation processes. Facies F beds reflect a stage of gravity emplacement characterized by only a limited amount of downslope movement within or surrounding a submarine fan. Typical examples of facies F are localized zone of slump folds, sequences of pebbly mudstone where the matrix shows flow and deformation features rather than stratification, and zones of isolated and enclosed slump blocks. Laterally extensive exposures of disorganized debris flows or olistostromes do not fit these characteristics and should instead be classified as facies A.

Facies F forms by sediment failure, gravity slumping and sliding, and is typically found near the lower slope or along channel margins of inner and middle fan environments.

Facies G

Facies G material comprises pelagic and hemipelagic detritus that tends to blanket all areas of a submarine fan.

Bedding, where discernible, is generally thin and parallel. Facies G is best developed in slope and interchannel settings and less commonly as fill in abandoned channels. Facies G may be interbedded with D and E facies units. Dispersal mechanisms for this fine-grained material are poorly understood and may involve all or a combination of: settling of true pelagic material; deposition from nepheloid layers; or possible deposition from contourites.

FAN FACIES ASSOCIATIONS

This abbreviated summary of fan facies associations is taken principally from Mutti and Ricci Lucchi (1972, 1975) with supporting data from Walker and Mutti (1973), Ricci Lucchi (1975a, b), Bouma and Nilsen (1978), Ingersoll (1978), Carter et al (1978), and Howell and Vedder (1978). Regardless of the size of a submarine fan, many modern and ancient fans can be subdivided physiographically into inner, middle, and outer fan portions, with a basin plain fringing the fan and a slope with feeder channels or submarine canyon setting shoreward and abutting the fan apex. The fan environments reviewed here, however, do not necessarily correspond to the configuration of all modern fans surveyed to date.

Slope, Feeder Channels and Submarine Canyons

Slope deposits are principally mudstone (facies G) with rare or widely spaced interbedded sandstone beds. Because of the gradient, inherent weakness of the sediment, and (possibly) as a result of pressure fluctuation from breaking of internal ocean waves, sediment failure is common. Slump scars may occur in an upper slope setting and slump accumulations in the lower slope. Where the slope is incised by a feeder channel, a lensoid body of sandstone (facies B, C and D) is encased in the mud.

In areas of submarine canyons, deposition usually occurs in the waning stages of the canyon's life, typically initiating a retrogradational event. Lithofacies A and F are common constituents, but nested channelized facies of B, C, D and E can also occur. Sudden abandonment may result in a facies G fill.

Inner Fan

The inner fan is recognized by massive deposits of lithofacies A, B and F, generally with A dominant. On modern

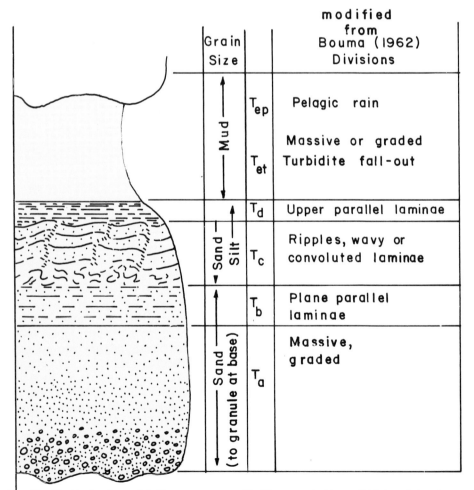

	Grain Size		modified from Bouma (1962) Divisions
	Mud	T_{ep}	Pelagic rain
		T_{et}	Massive or graded Turbidite fall-out
	Sand Silt	T_d	Upper parallel laminae
		T_c	Ripples, wavy or convoluted laminae
		T_b	Plane parallel laminae
	Sand (to granule at base)	T_a	Massive, graded

Fig. 15—The idealized complete T_{a-e} sequence, modified from van der Lingen (1969), and showing turbidite and hemipelagic subdivisions for the T_e unit.

fans, however, facies A sediment constitutes a small proportion of the morphologically defined upper fan, and fine-grained overbank sediment dominates. The walls of inner fan valleys are generally so deep that all mass flow material is limited to the confines of the valleys, and the more laterally extensive levee and interchannel areas are typified by facies G fine-grained material.

Bedding cycles are rare within the inner fan. Subordinate amounts of facies E correspond to overbank deposits of an intervalley channel thalweg. Coarse-grained channel deposits tend to be thick, with limited lateral continuity (<1 to 5 km wide). The thalweg of the main channel(s) of the inner fan may be straight, braided or sinuous, each channel pattern probably resulting in slightly different sedimentologic patterns as seen in corresponding fluvial analogs.

Middle Fan

The middle fan area is typically represented by channelized sandstone bodies of facies A, B, and less commonly C. Because the channels are not as deep as in the inner fan, levee, overbank and interchannel deposits (facies E and D) often accompany the thicker and coarser grained deposits. Failure along the flanks of channels may result in facies F, characteristically as zones of contorted facies E strata. The lateral shifting of numerous distributary channels of the middle fan results in thinning- and fining-upward sequences; individual cycles may range from only a few to more than eighty meters thick. Incompletely developed thinning-upward cycles of principally facies E, D and G represent middle fan-fringe areas.

Outer Fan

The outer fan as defined here begins beyond the limits of distributary channels. Lithofacies C and D characterize the outer fan, and this environment is recognized by constructional lobes, a progradational process resulting in coaesning- and thickening-upward sequences. The top layers of such a cycle may be lithofacies B (more rarely A) grading up to beds with midfan characteristics. Depositional lobes are generally laterally extensive on a mesoscopic scale, with deposits usually less than 30 m thick. Poorly developed thickening-upward units of principally lithofacies C, D and G are typical of an outer fan fringe.

Basin Plain

The basin plain is represented by hemipelagic and pelagic mudstone (facies G) with laterally extensive facies D beds of uniform thickness and some rare facies C turbidites. These turbidite beds do not display vertical cycles of bed thickenss or grain-size variations. The basin-plain deposits may be transitional with those of the outer-fan fringe and slope areas of deposition.

Other Facies Associations

Suprafan areas on modern fans are probably underlain by coarsening- and thickening-upward sequences in the form of a depositional lobe (Normark et al, 1979). Suprafan surfaces are characterized by channels and channel-like depressions, and suprafan sequences should be distinguished from the outer fan sand lobes of Mutti and Ricci Lucchi (1972) by association with these channel-fill sequences. Suprafan thickening-upward sequences if they exist, are indicative of *midfan* environments (Fig. 1). The sandy sediments found on modern suprafans make coring difficult and prevent acoustic penetration by high-resolution reflection systems; thus, little is known about depositional sequences underlying modern suprafans or channelized sand lobes on lower fans.

Thickening-upward cycles might form on the upper fan as a crevasse-splay deposit if the levee is breached. As erosion deepens the levee crevasse, more of the turbidity current can exit

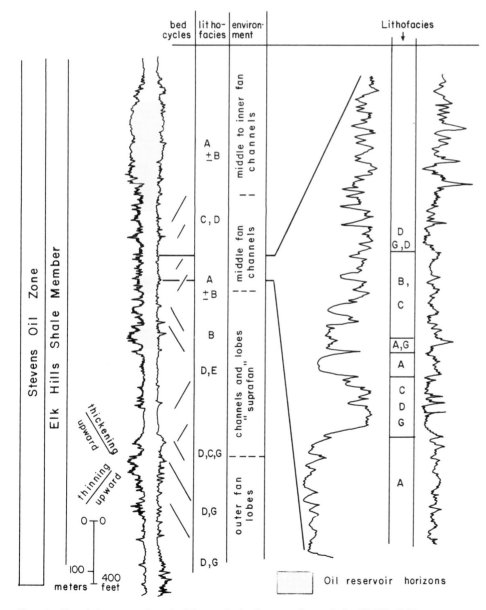

Fig. 16—Electric log curves from the Miocene in the Stevens oil zone in the Elk Hills field. Lithofacies designations made by L. F. Krystinik based on rock cores. Submarine fan environments are based on bed cycle patterns inferred from induction log SP curve and the nature of the lithology. The facies A beds are principally pebbly sandstone.

the valley and coarser material may be deposited. If the crevasse splay forms on the upper fan it is likely succeeded by progradation of a leveed valley across the splay.

Thining-upward sequences are not restricted to channel-fill units. A shifting of the primary depositional area on the fan would result in a thinning- and fining-upward succession as the area became more distal to the active depositional lobe. Even slow lateral migration of the leveed valley on the upper fan would produce a thinning-upward sequence in overbank deposits.

Recognition of most or all submarine fan lithofacies in a depositional basin does not necessarily constitute formation of a submarine fan. Whether a fan shape and dispersal system develops depends on the size and shape of the basin as well as the tectonic environment. Turbidite deposition in small slope basins or in a trench exhibit most of the lithofacies described (Moore and Karig, 1976), but continued deformation and narrow basin shape make it difficult to recognize a simple fan unit. Trenches and trench-slope basins are probably dominated by axial-flowing turbidity currents, and axial transport, in addition to multiple sediment sources along the trench slope, would tend to prevent development of a single-fan unit.

Associated Fossil Assemblages

In situ death assemblages comprising macrofossils are rarely reported from submarine fan sequences. This is generally due to two factors: (1) small fans in restricted basins have impoverished fauna, as they are prone to have poor water circulation and low-oxygen bottom water; and (2) large deep-sea fans where oxygen supply may be enhanced lie below calcium carbonate compensation depth (ca. 4,000 to 5,000 m below sea level). Body fossils commonly occur as isolated specimens suggesting downslope transportation and deposition. Larger assemblages of fossils are, however, found in thick sandstone or conglomerate beds, and may represent a mixture of fossils endemic to several shallow marine or even nonmarine environments.

Microfossil collections are preserved best in finer grained lithofacies, most particularly in the T_{ep} interval of a turbidite. These faunules of microfossils lend themselves to paleobathymetric as well as chronostratigraphic analyses, though fossils reworked from older strata in the T_{et} interval may be a problem.

Trace fossils are locally well developed in submarine fan sequences, particularly in middle- and outer-fan settings. Middle-fan environments often contain trace fossils along the soles of sandy turbidites, and form-genera are commonly "shallow water" indicators (Crimes, 1977). This apparent anomaly results from fast flowing depositional currents. The episodes of high oxygenation and slow buildup of organic detritus in the middle fan resembles high-energy neritic environments. Grazing forms on the tops of beds (trails, fecal strings and shallow burrows) typical of "deep water" increase in number in outer-fan settings (Ksiazkliewicz, 1970), suggesting this more distal area offers a relatively tranquil, organic-rich environment for sediment eaters and grazers (Crimes, 1977).

Hydrocarbon Accumulation and Submarine Fans

Subsurface exploration in basins with submarine-fan deposits utilizes electric logs (Fig. 16), seismic-reflection data (see Payton, 1977), and cored or ditch samples to reconstruct depositional patterns. The tracing of sand-filled channels and predicting areas of sand lobes are primary objectives.

In restricted basins the poor circulation and oxygen-depleted bottom water favors rich accumulations of organic debris in hemipelagic muds. These potential hydrocarbon source beds are interbedded with the coarser grained massflow deposits of fans. Source beds are most abundant in more distal parts of the fans. The thick-bedded sands and gravels of channels and depositional lobes provide ideal conditions for oil migration and storage, given a favorable clastic mineralogy (Nelson and Wright, 1979). Interchannel and lobe-fringe fine-grained deposits provide impermeable zones that can provide seals for reservoir rocks, and enhanced compaction of mud compared to sand may result in local closure surrounding post-compaction high-standing sandstone bodies. On a still larger scale, concentration of sand deposition in the inner, middle or suprafan areas results in extensive post-compaction highs that in terms of seismic stratigraphy are called "mounded onlapping-fill facies" (Sangree and Widmier, 1977). In southern California nearly all recovered oil to date has come from late Tertiary, turbidite-facies strata of fan systems in restricted-borderland basins (Weser,

stricted-borderland basins (Weser, 1977). In the North Sea, principal producing zones of the Frig and Forties oil fields are inferred Eocene suprafan depositional lobes of restricted basin fans (Stanley and Bertrand, 1979); however, the hydrocarbon source rocks are probably underlying Cretaceous and Jurassic shales (Heritier et al, 1979).

Open-ocean deep-sea fans probably have a lower potential for petroleum owing to the poorer quality of hydrocarbon source strata compared to shallower water fans. The average weight percent organic carbon from DSDP cores, Legs 1 to 23, is 0.3%; 17% of the measurements exceeded the minimum cirteria of organic richness for source beds (0.5%) suggested by Dow (1979) and only 4% of the samples indicated organic carbon concentrations above 1.0% (McIver, 1975). These low values of total organic matter reflect oxidation of organic material in water above and at the site of deposition. Areas of higher rates of sedimentation, such as along continental rises and on upper fans, show a higher concentration of organic carbon (Wilde et al, 1978), and average values commonly exceed 1% on the upper Monterey fan. Conditions that can favor development of hydrocarbon source beds in the deep ocean are high organic productivity associated with areas of deep-water upwelling, the oxygen minimum zone along the upper slope of most continental margins, and anoxic waters in intraslope- or rise-depressions resulting from salt diapirism or tectonic forces.

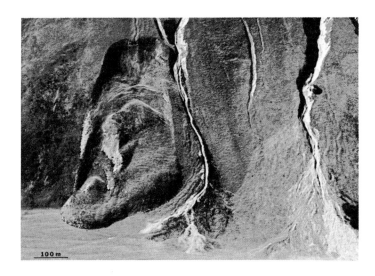

Photo 1. Oblique airphoto, ca. 1.0 km wide, showing a subaerial analog for a subsea fan in a slope setting note the feeding canyons, constructional fans, and adjacent region of slope failure, with slump scar and basal accumulation. Photo by R. M. Carter, Oreti River, New Zealand.

Photo 2. Small-scale analog of multiphase construction of mass emplaced (sand) fans, parts of which have been reworked by traction currents; red pencil for scale. Photo by R. M. Carter.

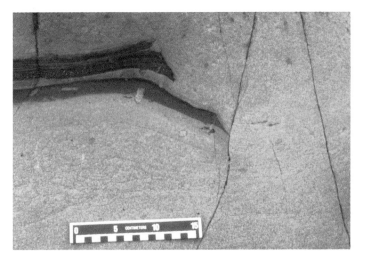

Photo 3. Bed truncation on left and sandstone bed amalgamation on right. Houstenaden Creek sequence, Upper Cretaceous, Seal Island Point, Oregon coast. Photo by Joanne Bourgeosis.

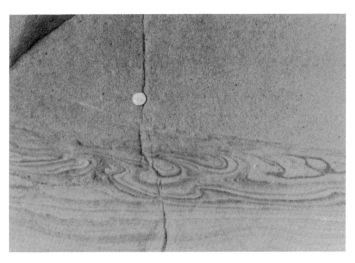

Photo 4. Amalgamated sandy turbidites; Bouma T_{b-c} overlain by the $T_{a'}$, U.S. quarter for scale. Upper Cretaceous strata, San Rafael Mountains, California.

Photo 5. Channelized amalgamation of sandstone and pebbly sandstone. Upper Cretaceous strata, San Rafael Mountains, California.

Photo 6. Top part of a turbidite (Bouma $T_{b,c,d,e}$) with an erosional (amalgamation) surface between the T_c and T_d intervals. Makara Formation, Micene, Hawkes Bay region, New Zealand. Photo by G. van der Lingen.

Photo 7.   Scour channel along base of turbidite, Bouma T_a zone of central bed is normally graded, but note inverse grading within scour depression. Balleny Group, Eocene and Oligocene, Gates Harbour, Fiordland, New Zealand.

Photo 8.   Graded turbidite, Bouma $T_{c-e'}$, color banding due to differential oxidation reflecting different porosities and permeabilities within a given depositional unit; top to the left. Pliocene Pico Formation, Ventura, California.

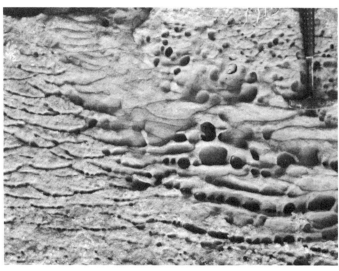

Photo 9.   Small-scale cavernous weathering with enhancement of dish structure due to leaching. Unnamed Eocene sandstone, San Nicolas Island, California.

Photo 10.   Well developed dish structures with vertical pillars (fluid escape routes). Upper Cretaceous strata, San Rafael Mountains, California.

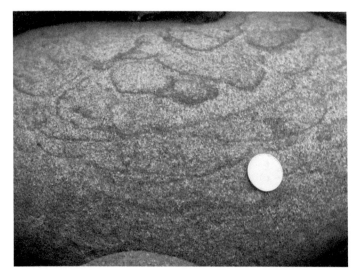

Photo 11.   Dish and pillar structures with a view subparallel with a top surface. Houstenaden Creek sequence, Upper Cretaceous, Seal Island Point, Oregon coast. Photo by Joanne Bourgeosis.

Photo 12.   Well developed laminations in a mass-emplaced, thick-bedded sandstone; local disturbances due to fluid escape or drag. Guipozcoa flysch, Eocene, Spain. Photo by F. Ricci Lucchi.

Photo 13. Climbing ripples, flow left to right. Hunters Cove Formation, Upper Cretaceous, Cape Sebastian, Oregon coast. Photo by Joanne Bourgeosis.

Photo 14. Basal part of a coarse sandy turbidite overlying a turbidite Bouma $T_{a,b,c'}$, where plane-parallel laminations pass upward to climbing ripples. Pico Formation, Pliocene, Ventura Avenue oil field area, California. Photo by P. A. Scholle.

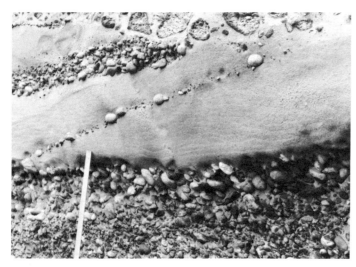

Photo 15. In ascending order: inversely graded cobble conglomerate with imbricated clasts, two composite sandstone beds with dish structures and cobble-lag deposit, overlain by disorganized conglomerate with normally graded conglomerate and sandstone bed with distribution grading. Scripps Formation, Eocene, San Diego, California. Photo by M. H. Link.

Photo 16. Truncated cross-beds below coarse grained sandstone (facies B) overlying thin-bedded sequences with small-scale cross-beds (facies E). Carmelo Formation of Bowen (1965), Paleocene, Point Lobos, California.

Photo 17. Large-scale cross-stratified cobble conglomerate, a deep-water fan-channel deposit. Lago Sofia conglomerate of the Cerro Torro Formation, Upper Cretaceous, southern Chile. Photo by R. H. Dott, Jr.

Photo 18. Large-scale flutes up to 1.5 m high (modified by loading) at base of a thick bed of conglomerate. Lago Sofia conglomerate, Upper Cretaceous, southern Chile. Photo of and from R. H. Dott, Jr.

Photo 19.   Large flute casts flow from top left towards bottom right. Sillery Group, Cambrian, Gaspe, Quebec. Photo by P. A. Scholle.

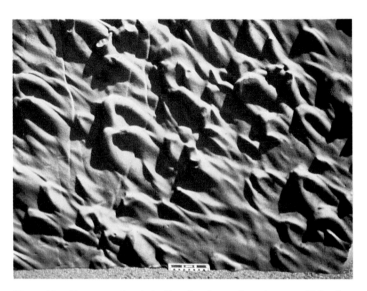

Photo 20.   Flute casts (scale is 15 cm) on base of sandstone turbidite, flow from top left towards bottom right. Hunters Cove Formation, Upper Cretaceous, Crook Point, Oregon coast. Photo by Joanne Bourgeosis.

Photo 21.   Flutes with spiral-eddy casts, flow left to right. Torlesse graywackes, Upper Triassic, Mt. Hutt Range, New Zealand.

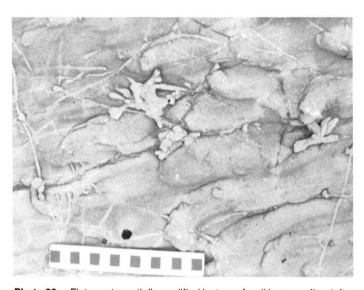

Photo 22.   Flute casts partially modified by trace-fossil burrows, flow left to right. Upper Paleocene flysch strata, Zumaya, northern Spain. Photo by P. A. Scholle.

Photo 23.   Small flattened flute casts stratigraphically just above grooves on bases of thin turbidites, facies D, flow right to left. Balleny Group, Eocene and Oligocene, Fiordland, New Zealand.

Photo 24.   Prod cast on lower bedding plane, flow towards the left. Smithwick Shale, Marble Falls, Texas. Photo by A. H. Bouma.

Photo 25. Crescent cast (ca. 5 cm) around mudstone pebble, with grooves, flow towards the right. Marnoso-arenacea Formation, Miocene, northern Apennines, Italy. Photo by F. Ricci Lucchi.

Photo 26. Prod and bounce casts of varied orientation, flow principally toward bottom and lower right. Rheinishe Schiefergebinge, Devonian, West Germany. Photo by F. Ricci Lucchi.

Photo 27. Mega grooves, see hammer on right margin for scale. Orca Group, Paleocene and Eocene(?), Naked Island, Alaska. Photo by G. R. Winkler.

Photo 28. Younger grooves deforming older grooves, diverse flow directions indicated. Orca Group, Paleocene and Eocene(?), Hinchinbrook Island, Alaska. Photo by G. R. Winkler.

Photo 29. Crossing groove casts, basal Eocene flysch strata, near Zumaya, northern Spain. Photo by P. A. Scholle.

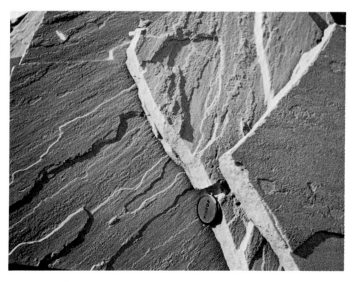

Photo 30. Parting lineation on detached blocks in a quarry. Marnoso-arenacea Formation, Miocene, Northern Apennines, Italy. Photo by F. Ricci Lucchi.

Photo 31. Delicate grooves and striations covered by a deformed (liquified?) skin. Laga Formation, Miocene, Central Apennines, Italy. Photo by F. Ricci Lucchi.

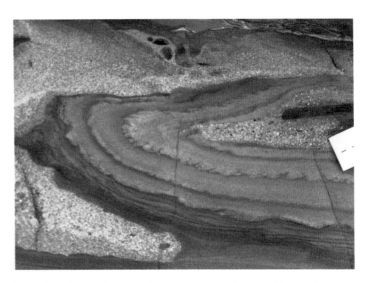

Photo 32. Rip-up flap of sediment composed of thin turbidites with climbing ripples, graded bedding, and small flame structures; scale in centimeters. Pigeon Point Formation, Upper Cretaceous, north of Pigeon Point, California.

Photo 33. Turbidite with facing flames, flow most likely oblique to perpendicular to outcrop surface. Balleny Group, Eocene and Oligocene, Puysegur Point, Fiordland, New Zealand.

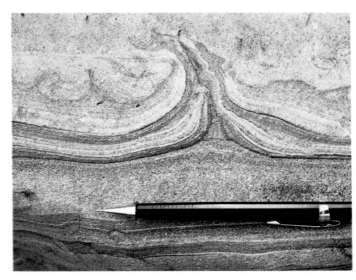

Photo 34. Small flame structures with fluid escape pipe (not a reliable flow or slope indicator). Balleny Group, Eocene and Oligocene, Big River, Fiordland, New Zealand.

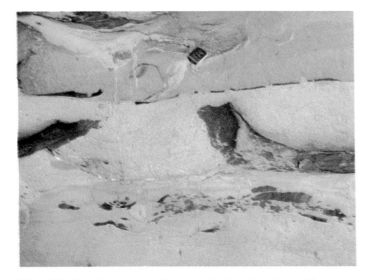

Photo 35. Sandstone, channeling, and plucking of mudstone with intrabasinal ("rip-up") clasts, match book for scale. Unnamed Upper Cretaceous sandstone, San Rafael Mountains, California.

Photo 36. A multiple flame structure of silty marl protruding into an arkosic sandstone (dotted with encrusting modern mollusca). Balleny Group, Eocene and Oligocene, Gates Harbour, Fiordland, New Zealand.

Photo 37. Cross section of a diapiric structure (convolute laminae) within the Bouma T_c division of a turbidite bed. Makara Formation, Miocene, Hawkes Bay region, New Zealand. Photo by G. van der Lingen.

Photo 38. Thin-bedded turbidites with small flame structures and post-depositional, fluid-escape scar. Balleny Group, Eocene and Oligocene, Big River, Fiordland, New Zealand.

Photo 39. Large-scale flame structures of sandstone protruding into conglomerate, scale in centimeters. Houstenaden Creek sequence, Upper Cretaceous, Cape Sebastian, Oregon. Photo by Joanne Bourgeosis.

Photo 40. Conglomerate channeling into and plucking sandstone. Carmelo Formation of Bowen (1965), Paleocene, Point Lobos, California.

Photo 41. Large-flame structure of mudstone (Bouma T_e) protruding into sandstone (Bouma T_a). Pico Formation, Pliocene, Ventura Avenue oil field, California. Photo by P. A. Scholle.

Photo 42. Dislodged ("rip-up") block of arkosic sandstone within contorted zone of marl, overlain by thin-bedded turbidite and pelagic chalk. Balleny Group, Eocene and Oligocene, Gates Harbour, Fiordland, New Zealand.

Photo 43. Armored mud-ball within diamictite, Houstenaden Creek sequence. Upper Cretaceous, near Pistol River, Oregon coast. Photo by Joanne Bourgeosis.

Photo 44. Exposure of turbidites that are the reservoir rocks in a repeated section at a lower structural horizon. Pico Formation, Pliocene, Ventura Avenue oil field, California.

Photo 45. Sandy turbidite with relatively even spacing of the Bouma T_{a-e} units. Waitemata Group, Miocene, Whangaparoa Peninsula, North Island, New Zealand.

Photo 46. Turbidite (Bouma T_{a-e}) with a thick T_e interval. Waitemata Group, Miocene, Whangaparoa Peninsula, North Island, New Zealand.

Photo 47. Bouma T_{c-e} turbidites where the T_{et} and T_{ep} are clearly differentiated. The T_{ep} interval is composed principally of bioclastic pelagic ooze whereas the T_{et} interval is composed of terrigenous clay. McIvar Formation, Miocene, Waian basin, Southland, New Zealand. Photo by R. M. Carter.

Photo 48. Bouma $T_{a,b}$ mega-turbidite. Pigeon Point Formation, Upper Cretaceous, near Pigeon Point, California.

Photo 49. Bouma T_a pebbly turbidites, graded beds. Houstenaden Creek sequence, Upper Cretaceous, near Seal Island Point, Oregon. Photo by Joanne Bourgeosis.

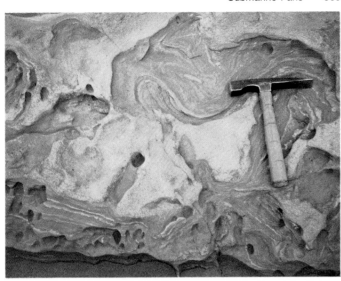

Photo 50. Liquefaction of a thick-bedded turbidite and the underlying hemipelagic bed. Balleny Group, Eocene and Oligocene, Puysegur Point, Fiordland, New Zealand.

Photo 51. A thin but complete Bouma T_{a-e} turbidite. Pico Formation, Pliocene, Venture Avenue oil field, California. Photo by P. A. Scholle.

Photo 52. Turbidite Bouma $T_{b,c,e}$ with T_c division composed of convolutions. Makara Formation, Miocene, Hawkes Bay region, New Zealand. Photo by G. van der Lingen.

Photo 53. Red radiolarian chert (Bouma T_{ep}?) with interbedded dolomitized carbonate turbidites. Mesozoic strata, Shikoku, Japan. Photo by P. A. Scholle.

Photo 54. Turbidites 20 to 30 cm thick Bouma $T_{a,b,e}$. Flame structure near the base and intrabasinal mudstone clasts at top of the T_b interval. Balleny Group, Eocene and Oligocene, Puysegur Point, Fiordland, New Zealand.

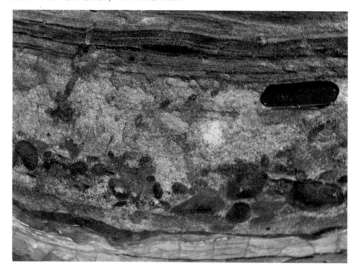

Photo 55.   A thin Bouma $T_{a,b}$ turbidite with rounded intrabasinal rip-up clasts. Pigeon Point Formation, Upper Cretaceous, Pescadero Beach, California.

Photo 56.   Three turbidites in ascending order: Bouma $T_{a,c,et,ep}$; T_{et-ep}; and $T_{c,d,et,ep}$. Note erosional contact at base of the fine-grained T_e turbidite. Balleny Group, Eocene and Oligocene, Gates Harbour, Fiordland, New Zealand.

Photo 57.   Facies A, a thick bed of graded conglomerate (Bouma T_a turbidite). Lago Sofia conglomerate of the Cerro Torro Formation, Upper Cretaceous, southern Chile. Photo by R. H. Dott, Jr.

Photo 58.   A thick interval of facies A graded-conglomerate beds (Bouma T_a turbidites) overlying coarse-grained arkosic turbidites of facies C. Upper Cretaceous strata, San Rafael Mountains, California.

Photo 59.   Facies A, organized, stratified and normally graded conglomerate beds with white stringers of lapilli tuff, midfan association of ·volcaniclastic fan. Blanca Formation, Miocene, Santa Rosa Island, California.

Photo 60.   Facies A, inverse- to normally-graded conglomerate bed overlying sandstone and intrabasinal rip-up clasts, 15-cm-long chisel for scale. Note imbrication of cobbles indicating flow directed toward the right, inner-fan channel deposit, Eocene strata, Santa Cruz Island, California.

Photo 61. Facies A, conglomerate with large-scale foreset beds overlying a sequence of thick- and thin-bedded turbidites and mudstone, middle fan region. Upper Cretaceous strata, San Miguel Island, California.

Photo 62. Facies A, disorganized conglomerate composed of volcaniclastic material, a proximal debris flow. Blanca Formation, Miocene, Santa Cruz Island, California.

Photo 63. Facies A, disorganized conglomerate, a debris flow. Eocene strata, Isle de Margarite, Venezuela. Photo by A. H. Bouma.

Photo 64. Facies A, cobble conglomerate (mass-flow, thalweg deposit) disconformably overlying a slump-block levee deposit (facies E) composed of mudstone and sandstone. Cabrillo Formation, Upper Cretaceous, Tourmaline Bay, San Diego, California.

Photo 65. Interbedded diamictite, facies A, of mud and cobbles (disorganized conglomerate, a debris flow) with massive to bedded mudstone, facies G. Pico Formation, Pliocene, Santa Paula Creek, Ventura, California.

Photo 66. Thick-bedded sandstones (facies B), thinning-fining upward into facies C turbidites. Caban Group, Silurian, Quarry near Rhyader, central Wales. Photo by R. M. Carter.

Photo 67.   Packet of amalgamated sandstone beds, (channel sands, facies B) overlying and incising silts and mudstone (overbank or levee deposits, facies E). Brushy Canyon Formation, Permian Delaware basin, Texas. Photo by H. E. Cook.

Photo 68.   Facies B, composite of fluxoturbidites, top to the right, ca. 10 m of section exposed. Franciscan assemblage, Cambria slab, Upper Cretaceous, Cambria, California.

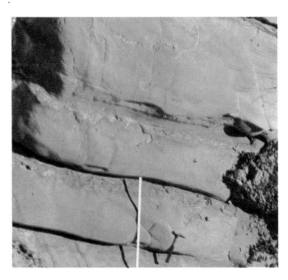

Photo 69.   Facies C, literally continuous turbidites with local slump folds. Makara Formation, Miocene, Hawkes Bay region, New Zealand. Photo by R. M. Carter.

Photo 70.   Close-up of facies C stacked turbidites, meter stick for scale. Most beds in this sequence are Bouma $T_{a,b,c}$ Eocene strata, Santa Rosa Island, California.

Photo 71.   Facies C, color banded calc flysch (thick-bedded turbidites, Bouma T_{d-ep}), hammer 80 cm long. Blackmount Formation, late Oligocene, Waiau River, New Zealand. Photo by R. M. Carter.

Photo 72.   Facies D, thin-bedded turbidites that show lateral persistence in bed thickness, interchannel environment of deposition. Juncal Formation, Eocene, Santa Ynez Mountains, California.

Photo 73. Facies D, thin- to medium-bedded turbidites, thinning- and thickening-upward cycles (lower suprafan), youngest to right, ca. 8 m of section exposed. Torlesse graywackes, Upper Triassic, Ohau Ski Basin, Southern Alps, New Zealand. Photo by M. D. Hicks.

Photo 74. Facies D, close up of thin-bedded turbidite (Bouma $T_{c,d,e}$) with channelized bases and lateral variation in bed thickness, levee or overbank environment of deposition. Waitemata Group, Miocene, Whangaparoa Peninsula, North Island, New Zealand.

Photo 75. Facies E, probable turbidites (Bouma $T_{c,e}$) represented by starved ripples and mudstone, not local channelized zone of coarse sand, overbank environment of deposition. Pigeon Point Formation, Upper Cretaceous, Pescadero Beach, California.

Photo 76. Facies E, thin-bedded wavy (cross-bedded) sandstone, volcaniclastic debris (levee or channel margin deposit). Blanca Formation, Miocene, Santa Rosa Island, California.

Photo 77. Facies E, thin, graded sandstone beds (levee or channel margin deposits). Brushy Canyon Formation, Permian, Delaware basin, Texas. Photo by R. M. Carter.

Photo 78. Sequence of facies E strata, locally with slump folds, from a channel margin region. Brushy Canyon Formation, Permian, Delaware basin, Texas (see photos 67 and 77 for associated strata). Photo by R. M. Carter.

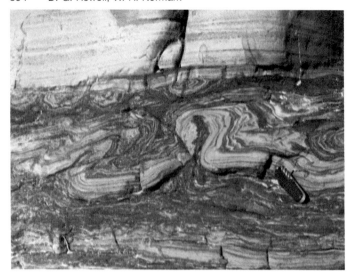

Photo 79.   Local slump (facies F) involving facies E strata, thin turbidites (Bouma $T_{c,e}$, or $T_{d,e}$, levee or channel margin deposit). Pigeon Point Formation, Upper Cretaceous, Pescadero Beach, California.

Photo 80.   Slumped overbank deposits (facies E) from a lacustrine fan sequence. Ridge Basin strata, Miocene and Pliocene, Castaic Canyon, California. Photo by P. A. Scholle.

Photo 81.   Slide and slump roll-over structures (facies F) involving turbidites in a 4-m-thick thinning-upward cycle. Topanga Formation, Miocene, Santa Monica Mountains, California.

Photo 82.   Megaslump (facies F) of thin-bedded turbidites (facies D). Hecho Group, Eocene, Southern Pyrenees. Photo by F. Ricci Lucchi.

Photo 83.   Facies F, a zone of broken and contorted bedding, associated with thinning-upward cycles of turbidites. Topanga Formation, Miocene, Santa Monica Mountains, California.

Photo 84.   Thick zone of broken and contorted bedding (facies F) that overlies a thick sequence of facies C turbidites. Waitemata Group, Miocene, Whangaparoa Peninsula, North Island, New Zealand.

Photo 85.   Inner-fan slump deposits (facies F) consisting of dislodged blocks of facies B sandstone and cobbles floating in a medium-grained sandstone matrix. Eocene strata, Santa Cruz Island, California.

Photo 86.   Slump folds of conglomerate and mudstone, facies F, (a pebbly mudstone in the making). Pigeon Point Formation, Upper Cretaceous, Pigeon Point, California.

Photo 87.   Dislodged block of sandstone in a facies C turbidite sequence. Waitemata Group, Miocene, Whangaparoa Peninsula, North Island, New Zealand.

Photo 88.   Facies F, pebbly mudstone with isolated sandstone and conglomerate blocks. Pigeon Point Formation, Upper Cretaceous, Pigeon Point, California.

Photo 89.   Facies F, large slump resulting in overturning of thin-bedded turbidites and siltstone (levee or overbank deposits) overlain by thick-bedded facies B strata, man for scale. Valle Formation, Upper Cretaceous, Vizcaino Peninsula, Baja California, Mexico.

Photo 90.   An approximately 1-m-thick zone of crumped bedding (facies F) of thin-bedded turbidites (facies D). Eocene strata, San Rafael Mountains, California.

Photo 91.   Thick zone of massive mudstone (facies G). Pico Formation, Santa Paula Creek, Ventura, California.

Photo 92.   Facies G mudstone with interbedded thin (less than 10 cm thick) turbidite sandstone beds. Upper Creataceous strata, Sacramento Valley, California. Photo by R. V. Ingersoll.

Photo 93.   In the skyline, view of El Capitan, a shelf-carbonate reef complex capping (flanking) a clastic sequence displaying slope and feeder-channel characteristics (Cherry Canyon Formation). Present relief represents the original submarine topography. Permian, Delaware basin, Texas. Photo by R. M. Carter.

Photo 94.   Massive mudstone with sandstone lenses, entire sequence cut by volcanic and sandstone dikes. Inferred pull-apart basin margin, outer shelf/upper slope sequence, basal part of a thick retrogradational unit. Topanga Formation, Miocene, Santa Monica Mountains, California.

Photo 95.   Slope association, local discordance due to slumping. Marnoso-arenacea Formation, Miocene, Northern Apennines. Photo by F. Ricci Lucchi.

Photo 96.   Lenticular beds of sandstone and conglomerate in a slope channel sequence. Kialaguik Formation, Middle Jurassic, Puale Bay, Alaska Peninsula, Alaska. Photo by R. M. Egbert.

Photo 97. Margin of Doheney Channel, a major feeder channel cut into silts and mud, inner fan setting. Miocene strata, Dana Point, California. Photo by W. R. Normark.

Photo 98. Lenses of conglomerte discordantly overlying a mud/silt overbank deposit (facies G and E) inner fan channels. Cabrillo Formation, Upper Cretaceous, Tourmaline Bay, San Diego, California. Photo by M. H. Link.

Photo 99. An inner-fan or submarine-canyon channel-fill sequence at least 300 m thick composed of interbedded organized, well-stratified conglomerate and sandstone. Pigeon Point Formation, Upper Cretaceous, Pigeon Point, California.

Photo 100. Massive organized conglomerate channel (facies A) approximately 3 m deep cutting into amalgamated (composite) sandstone fluxoturbidites (facies B), inner fan channel. Eocene strata, Santa Cruz Island, California.

Photo 101. Part of a large channel fill sequence, inner to middle fan association. Sandstone on top cutting into sandstone and mudstone turbidites. Marnoso-arenacea Formation, Miocene, Northern Apennines, Italy. Photo by F. Ricci Lucchi.

Photo 102. Two thinning- and fining-upward sequences, locally up to 80 m thick, facies A ascending to C, D, E, and G, deposition in a channelized part of a submarine fan. Torlesse graywackes, Upper Triassic, Rugged Range, Southern Alps, New Zealand.

Photo 103. Packets of thick-bedded facies B and C sandstone overlain by thin-bedded facies D and E strata, probably representing middle fan channel and overbank environments, respectively. Aberystwyth Grits, Silurian, coast section south of Aberystwyth. Photo by R. M. Carter.

Photo 104. Thinning-upward megasequence up to 50 m thick, facies C sandstone passing upward to facies D beds and capped by mudstone (facies G). Intermediate thickening upward sequence ca. 5 m thick may represent local crevasse-splay progradational pulses within the overall middle-fan channel megasequence. Topanga Formation, Miocene, Santa Monica Mountains, California.

Photo 105. Facies C, D, and E sandstone turbidites in lower part arranged in 2- to 6-m thick thinning-upward cycles. Overlying facies C and D turbidites are basal parts of a thicker thinning-upward cycle, middle fan channel and interchannel environments. Upper Cretaceous strata, Point San Pedro, California.

Photo 106. Middle fan association, interchannel mudstone (facies G) overlain by sandstone turbidites, facies D, overbank or levee deposits. Upper Cretaceous strata, Sacramento Valley, California. Photo by R. V. Ingersoll.

Photo 107. Organized and stratified conglomerate (facies A) interbedded with composite facies B sandstone overlying, with little or no scour, thin-bedded turbidites, a middle or outer fan association. Eocene strata, San Miguel Island, California.

Photo 108. Sandstone turbidites in thickening-upward cycles, typical of the depositional lobes at the base of distributary channels, outer fan environment, sequence youngest to the right. Upper Triassic, Torlesse graywackes, Ohau Ski Basin, Southern Alps, New Zealand.

Photo 109.   Thin-bedded turbidites in a poorly developed thickening upward cycle, outer-fan fringe environment. Marnoso-arenacea Formation, Miocene, northern Apennines, Italy. Photo by F. Ricci Lucchi.

Photo 110.   Laterally persistent thin-bedded turbidites (facies D) interbedded wtih hemipelagic mudstone (facies G). Basin-plain association, truck for scale. Upper Cretaceous strata, Sacramento Valley, California. Photo by R. V. Ingersoll.

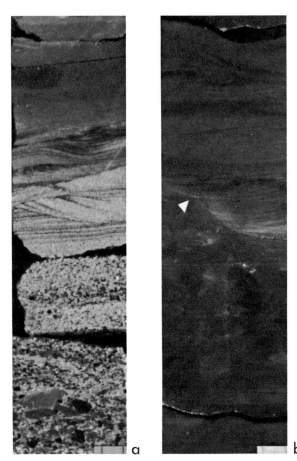

Photo 111a.   A volcaniclastic-carbonate turbidite (Bouma T$_{a-e}$), part of a middle-fan channel sequence. Lower Eocene strata, Daito Basin, Pacific Ocean, DSDP site 446, 1 cm bar scale. Photo by G. de Vries Klein.

Photo 111b.   Mud/silt turbidite (Bouma T$_c$ interval showing) cutting into pelagic mud (see arrow), middle fan environment. Middle Miocene strata, South Fiji Basin, Pacific Ocean, DSDP site 285, 1 cm bar scale. Photo by G. de Vries Klein.

Photo 111c.   Graded silt (dark) interbedded with hemipelagic clay (light), Bouma T$_{et-ep}$ turbidites, interchannel environment. Toyama submarine fan, Holocene, Sea of Japan, DSDP site 299, 1 cm bar scale. Photo by G. de Vries Klein.

Photo 111d.   Graded nannoplankton ooze overlain by graded intervals of clay, sand, and silt (medium, light, and dark gray, respectively), interchannel environment. Coral Sea Basin, early Pleistocene, Pacific Ocean, DSDP site 310, 1 cm bar scale. Photo by G. de Vries Klein.

Photo 111e.   Foram-nanno, calc-turbidite (Bouma T$_{d,et,ep}$) with slump fractures near the base, submarine fan. Lower Miocene strata, Daito Ridge, Pacific Ocean, DSDP site 445, 1 cm bar scale. Photo by G. de Vries Klein.

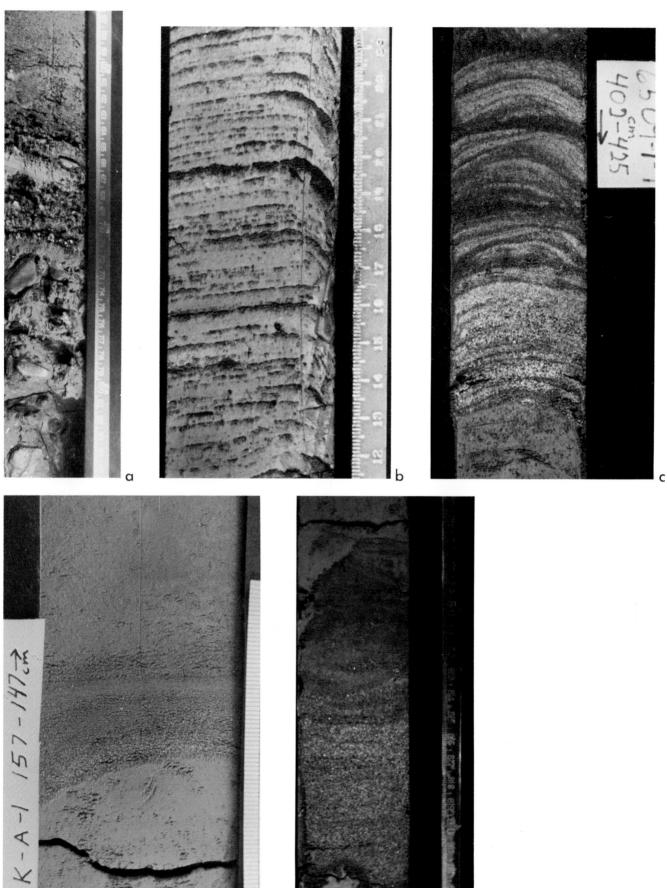

Photo 112. Pleistocene and Holocene cores from the Astoria fan, northeast Pacific Ocean: (a) disorganized and graded conglomerate beds, facies A, from an inner fan channel; (b) thin-bedded turbidites (Bouma $T_{c,e}$), facies E, inner fan levee; (c) sandstone turbidite (Bouma $T_{b,c,d,e}$), facies D from a middle fan channel; (d) fine-grained turbidite (Bouma $T_{d,e}$) overlying a hemipelagic mudstone, middle fan environment, (e) well graded turbidite (Bouma $T_{a,b,c,e}$) from an outer fan fringe environment. All scales in centimeters. Photos by C. H. Nelson.

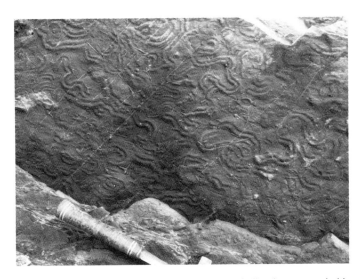

Photo 113. *Taphrhelminthopsis*, surface trail or shallow burrow, probably from echinoids, preserved on sole of sandstone turbidite. Paleocene flysch, Zumaya, northern Spain. Photo by P. A. Scholle.

Photo 114. Trails on a surface of an inner fan sandstone bed. *Scolicia* (probably from echinoids) and *Granularia (?),* burrow entrances. Scale in centimeters. Greifenstein Sandstone, Eocene, Vienna Woods, Austria. Photo by J. P. Kern.

Photo 115. Outer-fan or basin-plain setting with *Chondrites* traces. Monte Antola flysch, Upper Cretaceous, northern Apennines, Italy. Photo by P. A. Scholle.

Photo 116. *Spirophycus,* partially displaced and broken fecal string preserved on the sole of a sandstone turbidite, middle-fan channel environment, scale in centimeters. Griefenstein Sandstone, Eocene, Vienna Woods, Austria. Photo by J. P. Kern.

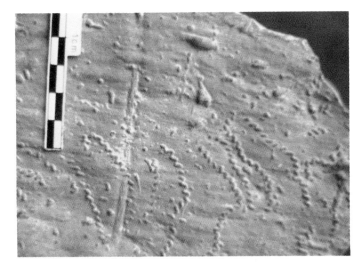

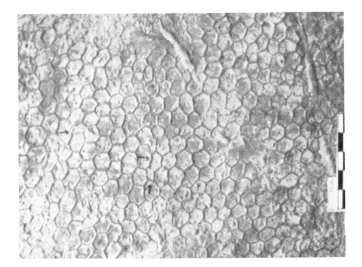

Photo 117. *Helicolithus,* shallow bedding-plane feeding burrow, excavated in mud (Bouma T$_e$) and preserved on the sole of an overlying sandstone turbidite, inner fan channel environment, scale in centimeters. Magura Sandstone, Eocene, Homrzyska, Poland. Photo by J. P. Kern.

Photo 118. *Paleodictyon,* shallow bedding-plane feeding burrow, excavated in mud and preserved on sole of overlying sandstone turbidite, scale in centimeters, middle to outer fan environment. Beloweza Formation, Eocene, Grybow, Poland. Photo by J. P. Kern.

Photo 119. *Lophoctenium,* feeding burrow excavated at interface between mud and sand layers, inner-fan channel environment. Greifenstein Sandstone, Eocene, Vienna Woods, Austria. Photo by J. P. Kern.

Photo 120. *Cosmorhaphe,* shallow bedding-plane feeding burrow excavated in mud and preserved on sole of subsequently deposited sandy turbidite, middle to outer fan environment, scale in centimeters. Beloweza Formation, Eocene, Lipnica Wielka, Poland. Photo by J. P. Kern.

Photo 121. Sand filled horizontal burrows excavated in the Bouma T_e mud layer between two thin-bedded turbidites, middle- to outer-fan environment, Pigeon Point Formation, Pescadero Beach, California.

REFERENCES CITED

Bagnold, R. A., 1954, Experiments on a gravity-free dispersion of large solid spheres in a Newtonian fluid under shear: Royal Society [London] Proc., A225, p. 49-63.

Blatt, H., G. V. Middleton, and R. Murray, 1972, Origin of sedimentary rocks: Prentice-Hall, Englewood Cliffs, New Jersey, 634 p.

Bell, H. D., 1942, Density currents as agents for transporting sediments: Jour. Geology, v. 50, p. 512-547.

Bouma, A. H., 1962, Sedimentology of some flysch deposits, a graphic approach to facies interpretation: Elsevier, Amsterdam, 168 p.

—— and T. H. Nilsen, Turbidite facies and deep-sea fans, with examples from Kodiak Island, Alaska: Offshore Technology Conference, Paper no. 3116, Houston, Texas, p. 559-567.

Bowen, O. E., 1965, Stratigraphy, structure, and oil possibilities in Monterey and Salinas quadrangles, California: AAPG Pacific Sec., Symposium of Papers, no. 40, p. 48-67.

Bramlette, M. N., and W. H. Bradley, 1942, Lithology and geologic interpretations, *in* Geology and biology of North Atlantic deep sea cores between Newfoundland and Ireland, Part 1: U.S. Geol. Survey Prof. Paper 196-A, p. 1-34.

Carter, R. M., 1975, A discussion and classification of subaqueous mass-transport with particular application to grain-flow, slurry-flow and fluxoturbidites: Earth Sci. Rev. 11, p. 145-177.

—— J. K. Lundqvist, 1977, Balleny Group. Chalky Island, southern New Zealand: an inferred Oligocene submarine canyon and fan complex: Pacific Geology, v. 12, Tokai University Press, Japan, p. 1-46.

—— M. D. Hicks, R. J. Norris, and I. M. Turnbull 1978, Sedimentation patterns in an ancient arc-trench-ocean basin complex: Carboniferous to Jurassic Rangitata Orogen, New Zealand, *in* D. J. Stanley, and G. Kelling, Sedimentation in submarine canyons, fans, and trenches: Dowden, Hutchinson, and Ross, Stroudsburg, Pennsylvania, Chapter 23, p. 340-361.

Cleary, W. J., and J. R. Conolly, 1974, Hatteras deep-sea fan: Jour. Sed. Petrology, v. 44, no. 4, p. 1140-1154.

Crimes, T. P., 1977, Trace fossils of an Eocene deep-sea sand fan, northern Spain, *in* T. P. Crimes, and J. C. Harper, eds., Trace fossils 2: Geological Journal Spec. Issue No. 9, Seel House Press, Liverpool, p. 71-90.

Crowell, J. C., 1955, Directional current structures from pre-Alpine flysch, Switzerland: Geol. Soc. America Bull., v. 66, p. 1351-1384.

—— 1957, Origin of pebbly mudstones: Geol. Soc. America Bull., v. 68, p. 993-1010.

Curray, J. R., and D. G. Moore, 1974, Bengal deep-sea fan, *in* C. A. Burke, and C. E. Drake, eds., The geology of continental margins: Springer Verlag, New York, 1009 p.

Daly, R. A., 1936, Origin of submarine canyons: Am. Jour. Sci. 5th series, v. 31, p. 410-420.

Damuth, J. E., and N. Kumar, 1975, Amazon cone—morphology, sediments, age, and growth pattern: Geol. Soc. America Bull., v. 86, p. 863-878.

Dott, R. H., 1963, Dynamics of subaqueous gravity depositional processes: AAPG Bull., v. 47, p. 104-128.

Dow, W. G., 1979, Petroleum source beds on continental slopes and rises, *in* J. S. Watkins, L. Montardert, and P. W. Dickerson, eds., Geological and geophysical investigations of continental margins: AAPG Mem. 29, p. 423-442.

Dzulynski, S., and E. K. Walton, 1965, Sedimentary features of flysch and graywackes: Elsevier, Amsterdam, 274 p.

Elmore, R. D., O. H. Pilkey, W. J. Cleary and H. A. Curran, 1979, Black Shell turbidite, Hatteras abyssal plain, western Atlantic Ocean: Geol. Soc. America Bull. Pt. I, v. 90, p. 1165-1176.

Florel, F., 1885, Les ravins sous-lacustres des fleuves glaciaries: Comptes Randus de l'Academie des Sciences, Paris, v. 101, p. 725-728.

Griggs, G. B., and L. D. Kulm, 1970, Sedimentation in Cascadia deep-sea channel: Geol. Soc. of America Bull., v. 81, p. 1361-1384.

Heritier, F. E., P. Lossel, and E. Wathne, 1979, Frigg field—large submarine fan trap in lower Eocene rocks of North Sea Viking Graben: AAPG Bull., v. 63, no. 11, p. 1999-2020.

Howell, D. G., 1979, Depositional configuration of redeposited sediments as a function of sediment accumulation rate, South Island, New Zealand: Geol. Soc. America, Abs. with Programs, v. 11, no. 7, p. 447.

———— and M. H. Link, 1979, Eocene conglomerate sedimentology and basin analysis, San Diego and the southern California borderland: Jour. Sed. Petrology, v. 49, no. 2, p. 517-540.

———— and Hugh McLean, 1976, Middle Miocene paleography, Santa Cruz and Santa Rosa Island, *in* D. G. Howell, ed., Aspects of the geologic history of California Continental Borderland: AAPG Pacific Sect. Misc. Pub. 24, p. 266-293.

Ingersoll, R. V., 1978, Submarine fan facies of the Upper Cretaceous Great Valley sequence, northern and central California: Sed. Geology, v. 21, p. 205-230.

Jacka, A. D., R. H. Beck, L. St. Germain, and S. C. Harrison, 1968, Permian deep-sea fans of the Delaware Mountain Group (Guadalupian) Delaware basin, *in* Guadalupian facies, Apache Mountain area, west Texas Permian basin section: SEPM Pub. no. 68-11, p. 49-90.

Kelling, G., and J. Holyroyd, 1978, Clast size, shape, and composition in some ancient and modern fan gravels, *in* D. J. Stanley, and G. Kelling, eds., Sedimentation in submarine canyons, fans and trenches: Dowden, Hutchinson, and Ross, Stroudsburg, Pennsylvania, p. 138-159.

Komar, P. D., 1970, The competence of turbidity current flow: Geol. Soc. America Bull. v. 81, p. 1555-1562.

Kuenen, Ph. H., 1937, Experiments in connection with Daly's hypothesis on the formation of submarine canyons: Leidsche Geol. Media, v. 8, p. 327-335.

————, and Menard, H. W., 1952, Turbidity currents, graded and non-graded deposits: Jour. Sed. Petrology, v. 22, p. 83-97.

————, and Migliorini, C. I., 1950, Turbidity currents as cause of graded bedding: Jour. Geology, v. 58, p. 91-127.

Ksiazkiewicz, M., 1970, Observations on the ichnofaunas of the Polish Carpathians, *in* T. P. Crimes, and J. C. Harper, eds., Trace fossils: Jour. Geology Special Issue No. 3, Seel House Press, Liverpool, p. 283-322.

McIver, R. D., 1975, Hydrocarbon occurrence from JOIDES Deep Sea Drilling Project: 9th World Petroleum Cong. Proc., v. 2, p. 269-280.

Menard, H. W., 1964, Marine geology of the Pacific: McGraw-Hill Publishers, New York, 271 p.

Middleton, G. V., 1967, Experiments on density and turbidity currents, III, Deposition of sediments: Can. Jour. Earth Science, v. 4, p. 475-505.

———— and Hampton, M. A., 1976, Subaqueous sediment transport and deposition by sediment gravity flows, *in* D. J. Stanley, and D. J. P. Swift, eds., Marine sediment transport and environmental management: John Wiley and Sons, Inc., p. 197-218.

Migliorini, C. I., 1944, Sul modo di formazione dei complessi tipo macigno: Soc. Geol. Italiana Bull., v. 62, p. 48-49.

Moore, G. F., and D. E. Karig, 1976, Development of sedimentary basins on the lower trench slope: Geology, v. 4, p. 693-697.

Mutti, E., 1977, Distinctive thin-bedded turbidite facies and related depositional environments in the Eocene Hecho Group (South-Central Pyrenees, Spain): Sedimentology, v. 24, p. 107-131.

———— and Ricci Lucchi, F., 1972, Le torbiditi dell 'Apennino settentrionale: introduzione all 'analisi di facies: Memoir Society Geology Italy, 11, p. 161-199. English translation in International Geology Review, 1978, v. 20, no. 2, p. 125-166.

Natland, M. L., and Ph. H. Kuenen, 1951, Sedimentary history of the Ventura Basin, California, and the action of turbidity currents: SEPM Spec. Pub. 2, p. 76-107.

Nelson, C. H., 1976, Late Pleistocene and Holocene depositional trends, processes, and history of Astoria deep-sea fan, northeast Pacific: Marine Geology v. 20, p. 129-173.

———— and L. D. Kulm, 1973, Submarine fans and deep-sea channels, *in* G. V. Middleton, and A. H. Bouma, eds., Turbidites and deep-water sedimentation: SEPM Pacific Sec., p. 39-78.

———— and A. Wright, 1979, Application of fan sedimentary facies models to petroleum geology: AAPG Fall Educ. Conf. Notes, v. II, p. 126-141.

Nilsen, T. H., 1979, Comparative sedimentology of the Lower Tertiary Butano and Cantua deep-sea fan sequences, Central California: Geol. Soc. America, Abs. with Program, v. 11, p. 120.

Normark, W. R., 1970, Growth patterns of deep-sea fans: AAPG Bull., v. 54, p. 2170-2195.

———— 1974, Submarine canyons and fan valleys: factors affecting growth patterns of deep-sea fans, *in* R. H. Dott, Jr., and R. H. Shaver, eds., Modern and ancient geosynclinal sedimentations: SEPM Spec. Pub. 19, p. 56-68.

———— 1978, Fan valleys, channels and depositional lobes on modern submarine fans: characters for recognition of sandy turbidite environments: AAPG Bull., v. 62, no. 6, p. 912-931.

———— and G. R. Hess, 1980, Quaternary growth patterns of California submarine fans, *in* M. E. Field, A. H. Bouma, and I. P. Colburn, eds., Quaternary depositional environments of the Pacific Coast: SEPM, Pacific Sec., Symposium volume, p. 201-210.

————G. R. Hess, D. A. V. Stow, and A. J. Bowen, 1980, Sediment waves on the Moneterey fan levee: a preliminary physical interpretation: Marine Geology, v. 37, p. 1-18.

———— D. J. W. Piper, and G. R. Hess, 1979, Distributary channels, sand lobes, and mesotopography of Navy submarine fan, California Borderland, with applications to ancient fan sediments: Sedimentology, v. 26, p. 749-774.

Payton, C. E., ed., 1977, Seismic stratigraphy—applications to hydrocarbon exploration: AAPG Mem. 26, 516 p.

Ricci Lucchi, F., 1975a, Depositional cycles in two turbidite formations of northern Apennines (Italy): Jour. Sed. Petrology, v. 45, p. 3-43.

———— 1975b, Miocene paleogeography and basin analysis in the Periadriatic Apennines, *in* H. S. Coy, ed., Geology of Italy: Earth Science Society of Libyan Arab Republic, Tripoli, p. 5-111.

Sanders, J. E., 1965, Primary sedimentary structures formed by turbidity currents and related resedimentation mechanisms, *in* Primary sedimentary structures and their hydrodynamic interpretation: SEPM Spec. Pub. 12, p. 192-219.

Sangree, J. B., and J. M. Widmier, 1977, Seismic interpretation of clastic depositional facies, *in* C. E. Payton, ed., Seismic stratigraphy—applications to hydrocarbon exploration: AAPG Mem. 26, p. 165-184.

Shepard, F. P., R. F. Dill, and U. von Rad, 1969, Physiography and sedimentary processes of La Jolla submarine fan and fan-valley, California: AAPG Bull., v. 53, p. 390-420.

Spiess, F. N., and R. C. Tyce, 1973, Marine physical laboratory deep-tow instrumentation systems: Scripps Institution of Oceanography Reference 73-4, 37 p.

Stanley, D. J., and J. P. Bertrand, 1979, Submarine slope, fan, and trench sedimentation—New concepts and problem solving (Penrose Conference Report): Geology, v. 7, no. 1, p. 49-52.

Stauffer, P. H., 1967, Grain-flow deposits and their implications, Santa Ynez Mountains, California: Jour. Sed. Petrology, v. 37, p. 487-508.

Stetson, H. C., and J. F. Smith, 1938, Behavior of suspension currents and mud slides on continental slopes: Am. Jour. Sci., v. 35, p. 1-13.

Stow, D. A. V., 1977, Late Quaternary stratigraphy and sedimentation on the Nova Scotian outer continental margin: Ph.D. dissert., Dalhousie University, Nova Scotia, 360 p.

Van der Lingen, G. J., 1969, The turbidite problem: New Zealand Jour. Geol. and Geophysics, v. 12, p. 7-50.

Walker, R. G., 1965, The origin and significance of the internal sedimentary structures of turbidites: Proc. Yorks Geol. Soc., v. 35, p. 1-32.

———— 1967, Turbidite sedimentary structures and their relationship to proximal and distal depositional environments: Jour. Sed. Pet., v. 37, p. 25-43.

———— 1975, Generalized facies for resedimented conglomerates of turbidite associations: AAPG Bull., v. 59, p. 737-748.

———— 1978, Deep-water sandstone facies and ancient submarine fans: models for exploration for stratigraphic traps: AAPG Bull., v. 62, no. 6, p. 932-966.

———— and E. Mutti, 1973, Turbidite facies and facies associations *in* G. V. Middleton, and A. H. Bouma, eds., Turbidites and deep-water sedimentation: SEPM Pacific Sec., p. 119-157.

Walton, E. K., 1967, The sequence of internal structures in turbidites: Scottish Jour. Geology, v. 3, p. 306-317.

Wilde, P., W. R. Normark and T. E. Chase, 1978, Channel sands and petroleum potential of Monterey deep-sea fan, California: AAPG Bull., v. 62, p. 967-983.

Explanation of Indexing

A reference is indexed according to its important, or "key," words.

Three columns are to the left of a keyword entry. The first column, a letter entry, represents the AAPG book series from which the reference originated. In this case, M stands for Memoir Series. Every five years, AAPG will merge all its indexes together, and the letter M will differentiate this reference from those of the AAPG Studies in Geology Series (S) or from the AAPG Bulletin (B).

The following number is the series number. In this case, 31 represents a reference from Memoir 31.

The last column entry is the page number in this volume where this reference will be found.

Note: This index is set up for single-line entry. Where entries exceed one line of type, the line is terminated. The reader must sometimes be able to realize keywords, although commonly taken out of context.